30th Ethylene Producers Conference 2018

Topical Conference at the 2018 AIChE Spring Meeting and 14th Global Congress on Process Safety

Orlando, Florida, USA
22 - 25 April 2018

ISBN: 978-1-5108-6437-5

Printed from e-media with permission by:

Curran Associates, Inc.
57 Morehouse Lane
Red Hook, NY 12571

Some format issues inherent in the e-media version may also appear in this print version.

Printed by Curran Associates, Inc. (2018)

For permission requests, please contact AIChE
at the address below.

AIChE
120 Wall Street, FL 23
New York, NY 10005-4020

Phone: (800) 242-4363
Fax: (203) 775-5177

www.aiche.org

Additional copies of this publication are available from:

Curran Associates, Inc.
57 Morehouse Lane
Red Hook, NY 12571 USA
Phone: 845-758-0400
Fax: 845-758-2633
Email: curran@proceedings.com
Web: www.proceedings.com

TABLE OF CONTENTS

Some abstracts in this book refer to images that were not made available by the publisher

AIChE Paper Number: 4a

PLANT-WIDE OPTIMIZATION FOR HIGH VALUE ADDED PRODUCTS – A STEAM CRACKING CASE STUDY

Kevin M. Van Geem

Laboratory for Chemical Technology

Ghent University

Guy B. Marin

Laboratory for Chemical Technology

Ghent University

Andrés E. Muñoz G.

R&D Manager

AVGI

Yu Zhang

Key Laboratory of Advanced Control and Optimization for Chemical Processes

East China University of Science and Technology

Wenli Du

Key Laboratory of Advanced Control and Optimization for Chemical Processes

East China University of Science and Technology

Feng Qian

Key Laboratory of Advanced Control and Optimization for Chemical Processes

East China University of Science and Technology

Prepared for Presentation at the 2018 Spring National Meeting

Orlando, Florida, April, 23, 2018

PLANT-WIDE OPTIMIZATION FOR HIGH VALUE ADDED PRODUCTS – A STEAM CRACKING CASE STUDY

Kevin M. Van Geem

Laboratory for Chemical Technology

Ghent University

Guy B. Marin

Laboratory for Chemical Technology

Ghent University

Andrés E. Muñoz G.

R&D Manager

AVGI

Yu Zhang

Key Laboratory of Advanced Control and Optimization for Chemical Processes

East China University of Science and Technology

Wenli Du

Key Laboratory of Advanced Control and Optimization for Chemical Processes

East China University of Science and Technology

Feng Qian

Key Laboratory of Advanced Control and Optimization for Chemical Processes

East China University of Science and Technology

Abstract:

Recent progress in accurate modeling of steam cracking plants has enabled global optimization to be performed at a plant-wide level. This paper presents a real time optimization (RTO) system developed by East China University of Science and Technology (ECUST) to maximize the net profit of the ethylene plant of Sinopec Shanghai Petrochemical Company (SSPC). The system has a three-layer structure consisting of the global RTO, the local RTO and the advanced process control (APC) layer from top to bottom, among which the APC layer has already been applied to SSPC and is currently in service.

COILSIM1D-RTO is used to simulate the steam cracking reactors. A simplified radiant box model and a convection section model are developed to account for the fuel consumption and superheated high pressure steam generation. The downstream compression and separation units are modeled in Aspen Plus to evaluate the maximum allowed process gas load and energy consumption.

As the plant-wide RTO system is currently under implementation and no real time optimization results can be obtained, a case study is performed using historical operating data from year 2017 to illustrate the benefit that this plant-wide RTO system can deliver. Optimization results show that most of the crackers should be operated under higher cracking severity to improve yields of the high value added products. By optimizing the cracking severity and dilution steam ratio, the estimated increase of net profit of the ethylene plant in Case 1 and Case 2 is about 29.50 million and 30.75 million yuan per year, respectively (i.e. approx. 4.7 million USD and 4.9 million USD, at current exchange rates).

Introduction

Real time optimization (RTO) has become common practice in industrial ethylene plants for many years. In most of these applications, optimization is done for individual steam cracking units with the aim of increasing the yield of light olefins. Such optimization can lead to inferior decisions from a plant-wide perspective. For instance, more ethylene can be produced with higher coil outlet temperature (COT) and dilution steam ratio. But this will increase the cost of the ethylene plant in terms of energy and steam consumption. Moreover, the maximum load of the cracked gas compressor is also an important issue to be considered [1]. Therefore, plant-wide real time optimization in which the cracking furnaces, the downstream units, as well as the energy system are optimized jointly is necessary to improve the performance of the whole ethylene plant.

To meet the rapidly increasing demand of ethylene and propylene worldwide, many ethylene plants have been built in recent years in Asia. Also in North America various projects for new furnaces are being evaluated and executed. Nowadays, the capacity of a typical ethylene plant is about 1 to 1.4 million tpy [2]. Multiple steam cracking furnaces run in parallel and are connected to a single separation train. Thanks to the abundance of cheap (shale) gas, ethane is the most widely used cracking feedstock in USA. This makes the operation optimization of steam cracking furnaces relatively simpler than when doing it for crackers that are fed with heavier hydrocarbons such as naphtha, which is primarily used in Europe [3]. The steam cracking landscape of China is certainly more complex, mainly because of the following aspects:

Firstly, multiple feeds ranging from ethane, propane, liquefied petroleum gas (LPG) to naphtha and hydrocracking vacuum gas oil (HVGO) are often used. In addition, these different feedstocks are sometimes mixed before entering steam cracking furnaces. Switching feeds without decoking is also a common practice.

Additionally, the furnaces are usually designed to be able to deal with a board range of feedstocks, given the need of frequent changes in hydrocarbon feedstocks. Moreover, many ethylene plants in China were built some 30 years ago and have been retrofitted for higher capacity from time to time. It is therefore not surprising to encounter sites with more than 5 different types of cracking furnaces.

Plant-wide RTO of such a plant is quite a challenging task. It requires not only accurate simulations of the cracking process of diverse feedstocks and their mixtures in different cracking furnaces, but also adequate models of the downstream separation train to calculate the loss of light olefins in the distillation columns. Moreover, it is crucial to compute the maximum load of the separation units based on the optimized operating conditions. These maximum loads need in turn to be used as constraints of the optimization problem. At the same time, the energy consumption of the cracking section and the separation section should be evaluated, to ensure an adequate trade-off.

In this paper, a plant-wide RTO strategy is introduced. The optimization is made on the basis of numerical simulations of steam cracking furnaces and separation train using well-established software tools, namely COILSIM1D and Aspen Plus.

The former, as a steam cracker simulation package, calculates the yields of the valuable light olefins, the consumption of dilution steam and fuel, and the production of high pressure steam. The latter gives the operating constraints of the downstream processing train, according to the product distribution predicted by COILSIM1D. The plant-wide RTO solution is provided by East China University of Science and Technology (ECUST), which is currently under implementation to an industrial ethylene plant of Sinopec Shanghai Petrochemical Company (SSPC) in China. The optimization system will be completely online at the end of 2018, and this case study is therefore conducted using previously acquired data of the site. Comparison between the industrial data and the simulation results are in good agreement. Based on the historical operating conditions, plant-wide optimization is performed and the estimated benefit is about 30 million yuan per year.

Configuration of the Ethylene Plant

SSPC has increased its ethylene capacity from the original design to about 750,000 tpy as a result of continuous revamping. It has 14 furnaces in total, from different licensors. They are fed with naphthas, HVGOs, as well as recycled ethane and propane. The feed and capacity of all furnaces is summarized in Table 1.

There are two sites in SSPC, namely the so-called 'old site' and a 'new site'. The old site was built in 1987 and consists of 10 cracking furnaces. The construction of the new site was completed in 2002, which brought an additional capacity of 300,000 tpy to SSPC. Both sites share the same feedstock storage tank zone but are independent in the downstream by having their own separation units. As shown in Figure 1, a typical front-end demethanizer refrigeration system is used in the old zone, while the new zone utilizes a front depropanizer arrangement.

Table 1. Overview of the optimized furnaces

Name	Feedstock type	Capacity (ton/h)	Location
BA101	NAP/HVGO	26	Old site
BA102	NAP/HVGO	26	Old site
BA103	LNAP/NAP	20	Old site
BA104	LNAP/NAP/LPG	20/20/14	Old site
BA105	NAP/HVGO	26	Old site
BA106	NAP/HVGO	26	Old site
BA107	LNAP/NAP/LPG	20/20/14	Old site
BA108	NAP/LPG	10/8	Old site
BA110	NAP/HVGO	26	Old site
BA111	LNAP/NAP	22	Old site
BA2101	NAP/LPG	48/40	New site
BA2102	NAP/LPG	48/40	New site
BA2103	LNAP/NAP	48	New site
BA2104	LNAP/NAP	48	New site

The composition of the feed is available in real time. For LPG, the volume fraction of the components from H_2 to C_5 hydrocarbons is measured by online gas chromatography. For light and mid-range naphtha, as well as HVGO, only commercial indices such as PIONA, distillation curve and density are measured offline every two days. Some key components of the cracked gas, including H_2, CH_4, C_2H_4, C_3H_6, C_2H_6, and C_3H_8 are measured by online analyzers. The time delay of the cracked gas composition measurement is about 10 minutes.

All typical operating variables of the cracking furnaces and separation train, such as feedstock flow rate, dilution steam ratio, coil inlet/outlet temperature (CIT/COT), coil outlet pressure (COP), fuel gas consumption, reflux ratio of the distillation column and other parameters are provided by a DCS system. COT control, COT balancing control and throughput control are in good service.

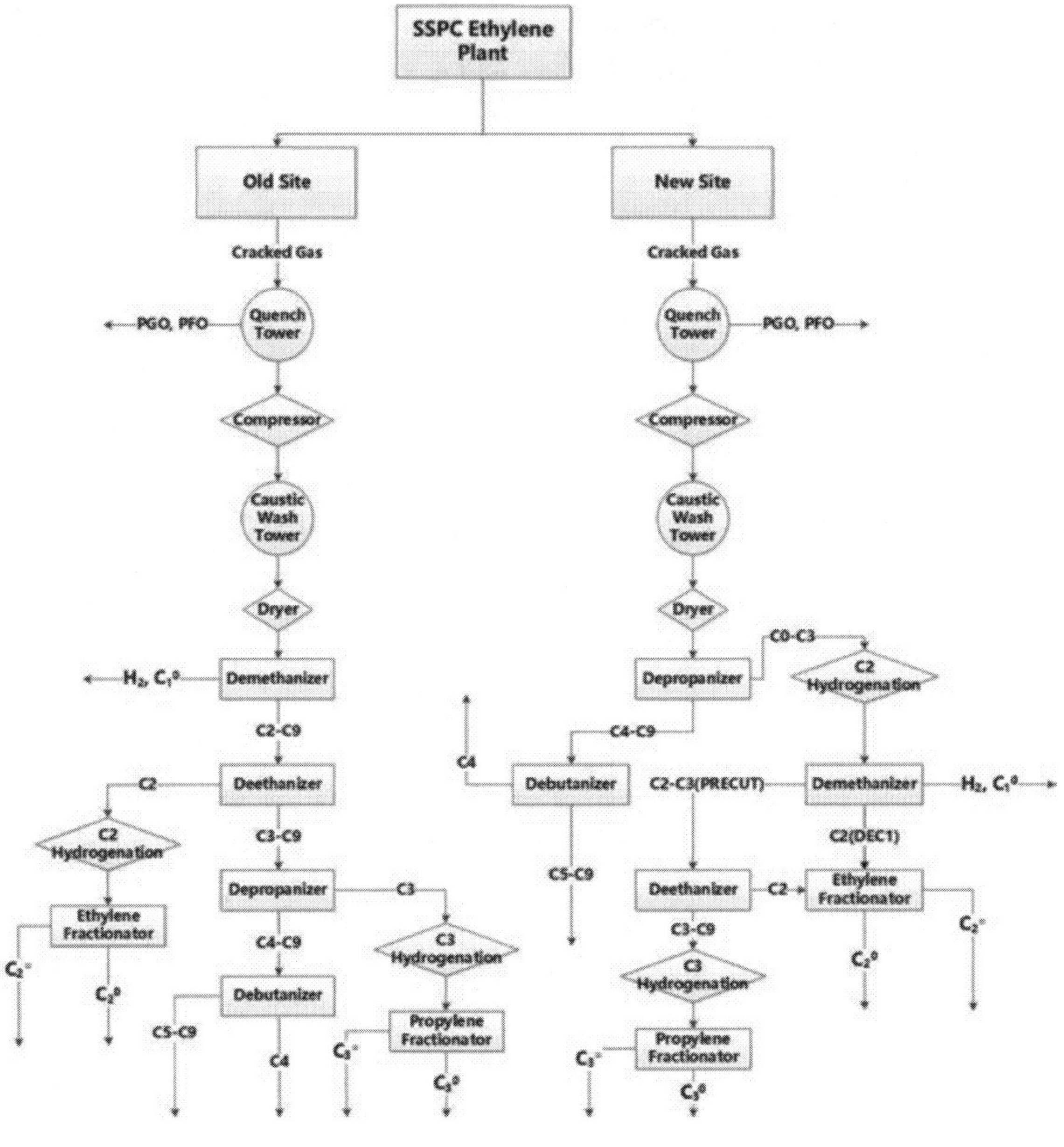

Figure 1. Schematic flow chart of the ethylene plant

Modeling Approach

Cracking Reactor Modeling with COILSIM1D

COILSIM1D is a comprehensive commercial software package for steam cracking simulations. The software is an extremely accurate tool for cracking reaction modeling, due to an extensive and sophisticated reaction network consisting of more than 700 molecular species, 40 radical species and over 300,000 elementary reactions [4]. The corresponding kinetics and thermodynamic parameters are obtained by ab-initio calculations and validated using an extensive dataset originated from pilot plant experiments and industrial furnace measurements [5-10].

The software combines the foregoing single-event kinetic network with a one-dimensional reactor model. The model can be used for any configuration of tubular reactor from laboratory scale to industrial steam crackers. In addition, reactor simulations can be performed for a wide range of operating conditions and various kinds of hydrocarbon feeds, ranging from ethane up to HVGO.

Recently, AVGI (a Ghent University spin-off) [11] has acquired all technical and commercial rights to COILSIM1D. On one hand, this helps to increase the speed of the software development in close collaboration with industry, helping also to continuously improve the user-friendliness of the GUI and to incorporate radiant box and convection section models, which enables users to carry out fully coupled simulations of the whole steam cracking furnace [12]. On the other hand, interaction on a business-to-business structure is further strengthened, allowing for customization of packages to meet the specific needs of the users.

In this work, COILSIM1D-RTO, a tailored version of COILSIM1D for the application of RTO to SSPC, is used to simulate the reactor coil and the transfer line exchanger (TLE) of the 14 cracking furnaces. The geometric parameters of the tubular reactors are customized to be in line with those of the industrial cracking furnaces of SSPC.

Feedstock Reconstruction

In order to take full advantage of the abovementioned flexible and accurate fundamental kinetic model, detailed molecular composition of the cracking feedstock is necessary. However, liquid feedstocks such as naphtha and HVGO are complex mixtures of a broad range of hydrocarbons, whose detailed composition cannot be straightforwardly determined by online measurements. For these feeds, only a limited number of the so-called commercial indices, including specific density, average molecular mass, PIONA analysis, etc., measured by means of relatively fast and low cost standardized analytical procedures are necessary. Therefore, feedstock reconstruction methods are needed to calculate detailed composition of a complex feedstock based on these commercial indices.

In this work, SIMCO, a powerful tool for rapid and accurate feedstock reconstruction that is part of the COILSIM1D pack, is adopted to compute detailed composition of liquid feedstock. Two methodologies, namely the Shannon entropy maximization (SEM) method [13-14] and an artificial neural network (ANN) [15], are used in SIMCO. Based on the available commercial indices, SIMCO decides whether the feedstock is comparable to those in the database of ANN. For similar feeds, the detailed composition of the unknown feedstock is calculated by ANN. Otherwise the SEM method will be used. SIMCO is seamlessly incorporated in COILSIM1D and the reconstructed feedstock can be directly imported into reactor simulations.

Modeling of Radiant Box and Convection Section

COILSIM1D is currently in the final stage of incorporating a furnace model, which integrates a radiant box and a convection model to allow simulations of the full steam cracker to be performed [12]. However, for the present work it is unnecessary to carry out fully coupled simulations, because all the boundary conditions required for cracking reactor simulations can be gathered from the DCS system in real time, instead of being calculated by Radiant box and/or convection section models, which would add some computational time.

The reason for introducing models of the radiant box and the convection section is to estimate the consumption of fuel gas and the production of superheated high pressure steam in the course of operating optimization. They have to be taken into account due to their effects on the net profit of the ethylene plant.

For fuel consumption calculations, a simplified radiant box model based on incident radiative heat flux (IRHF) [16] is used. This model correlates the incident radiative heat flux to the flue gas bridge wall temperature obtained from an overall zero-dimensional heat balance of the radiant box. Making use of the net heat flux and tube metal temperature (TMT) profiles of the cracking reactor computed by COILSIM1D-RTO, the incident radiative heat flux profile is back-calculated using heat transfer equations. Overall heat balance calculations are then performed to obtain the fuel flow rate under new operating conditions. The development and validation of the IRHF-based model are detailed in literature [16].

A convection section model is used to compute the production of superheated high pressure steam in the current work. It is reduced from a more sophisticated convection model [17]. Instead of using a combination of flow pattern map and gas-liquid two-phase heat transfer equations to describe the flow boiling process of liquid feedstocks in convection banks, the reduced convection section model calculates heat balance of the process gas side and the flue gas side in each bank, the thermal efficiency of which is estimated using real time data of every steam cracker in SSPC.

Modeling of the Separation Train

Downstream of the cracking furnaces, the cracked gas is quenched, compressed and dried to separate the heavy fractions and the dilution steam from the light hydrocarbons. The effluent is then directed towards a fractionation system where a sequence of distillation columns is used to fractionate and isolate the desired products. It is obvious that the hydrocarbon distribution of the cracked gas will have a significant impact on the compressor duty and the distillation columns. As mentioned before, the capacity of a single train steam cracker is usually limited by the maximum load of the compressor. Hence it is crucial to take into account the constraints from the compression and separation sections in plant-wide RTO to ensure the applicability of the optimization results. Moreover, energy consumption in terms of the heat input to the reboiler of distillation columns should also be considered.

In this work, the downstream units of the old and the new site of SSPC are modeled using Aspen Plus [18]. For the light fractions, 28 molecular species including the most important cracking products are considered in the simulation. The heavier fractions can be lumped into three main categories, i.e. pyrolysis gasoline, pyrolysis gas oil, and pyrolysis fuel oil. According to the initial and final boiling point values, n-decane, n-tetradecane, and n-heneicosane are selected to represent these heavy mixtures.

C2 and C3 hydrogenation units in the fractionation system are simulated as well. Chemical kinetics of the selective hydrogenation are adopted from a comprehensive study of the process where the parameters are obtained by fitting the simulation results to industrial data of hydrogenation reactors [19].

Finally, with the necessary information such as mass flow rate and composition of the cracked gas, Aspen simulations of the entire downstream separation system can be performed to calculate energy consumption, loss of light olefins and the maximum load of each unit.

Real Time Optimization System

The plant-wide RTO system developed by the Key Laboratory of Advanced Control and Optimization for Chemical Processes (KLACOCP) at ECUST is the result of long term research on model-based process control and optimization for industrial ethylene plants. Figure 2 shows the hierarchical structure of the system, which consists of three layers from top to bottom, i.e. global RTO, local RTO and advanced process control (APC). The APC and local RTO layers have been widely applied to many ethylene plants in China, providing the basis of the global RTO.

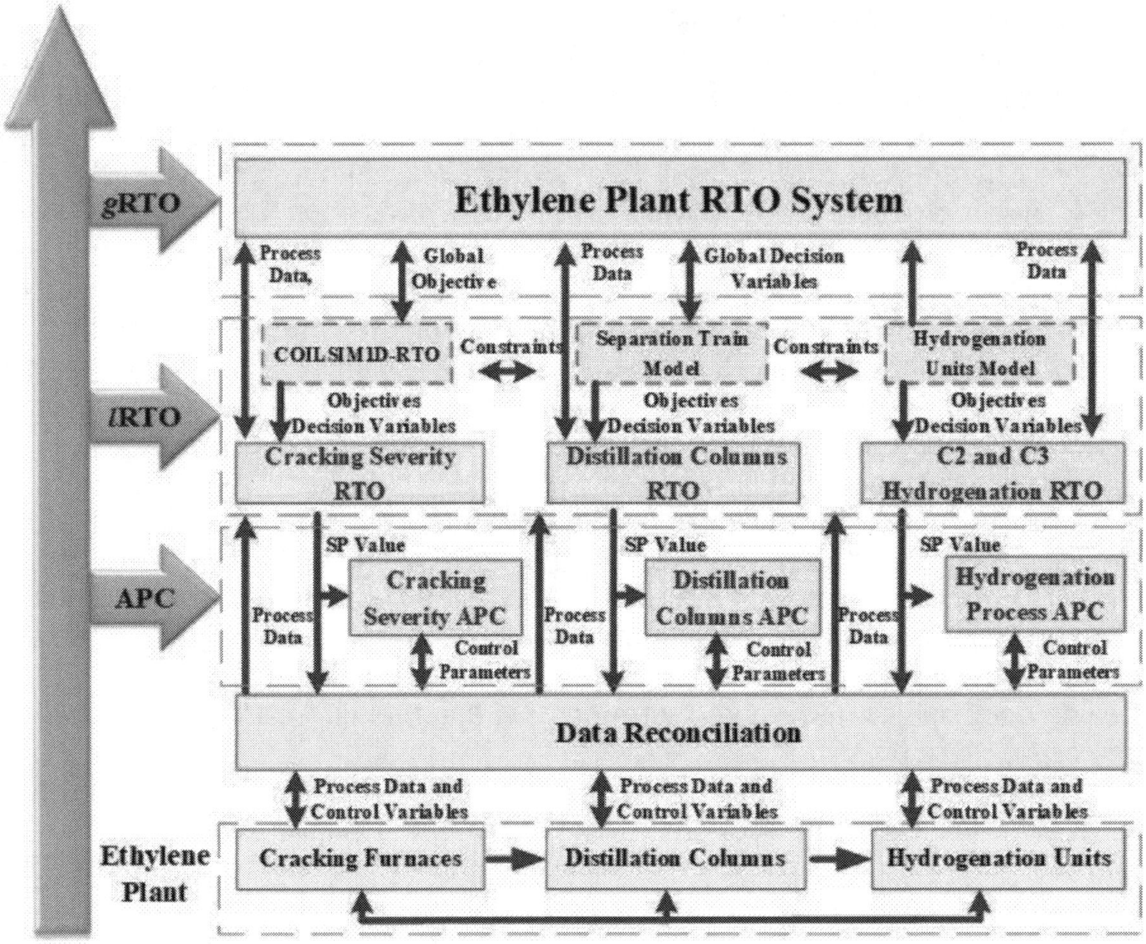

Figure 2. Hierarchical structure of the plant-wide RTO system

Global RTO Layer

The objective of the global RTO is to maximize the net profit of the ethylene plant. In this specific case, the contributions of all high value added products including ethylene, propylene, 1,3-butadiene, hydrogen, benzene as well as the superheated high pressure steam to the income of the plant are considered. The price of each product is used to calculate its benefit (in Chinese yuan). The costs include the usage of process feeds, fuel, dilution steam and the energy consumption in compression and separation sections. The main constraints of the global RTO are the maximum capacity, maximum allowed TMT and pressure drop of the cracking reactor, upper limit of the load of the compressors and distillation columns.

The steam cracking furnace is the core of an ethylene plant, as it determines the product distribution of the cracked gas, making it the most important unit to be optimized. Feedstock flow rate, dilution steam ratio and cracking severity are the main control variables. As the final objective is a combination of many factors such as feedstock type, product yield, energy consumption, no gradient information can be obtained straightforwardly. Therefore, only the input-output relationship is accounted for in the course of optimization using a pattern search algorithm. The optimization algorithm treats the models as "black boxes" which provide the net profit of the plant and constraints based on a given set of operating variables.

Ideally, the global RTO should search for the entire decision space (three dimensional in this specific case) to find the optimal operating conditions. However, for every single searching step, simulations of the reactor, as well as radiant and convection sections have to be performed for 14 crackers, followed by two separate process simulations of the entire downstream system for the old and the new sites respectively. This usually takes about 3-5 minutes, depending on the complexity of the calculation. The global RTO needs about 30-60 searching steps in the three-dimensional space to reach the optimal point, which cannot meet the requirement of obtaining optimization results in less than 2 hours, as required by the client.

To remedy this problem, the searching space is reduced by removing the dilution steam ratio and cracking severity from the global RTO layer. This implies that only the feedstock flow rate is considered for global RTO, based on the current dilution steam ratio and cracking severity for all steam crackers, while the latter two are optimized at a local RTO level. The reason is that feedstock flow rate has a significant impact on the load of the downstream units, specially the cracked gas compressor. Thus it must be taken into account in global RTO to ensure the practical applicability of the optimization results. On the other hand, nearly 96% of the dilution steam is removed from the cracked gas before entering the compressor and cracking severity mainly affects the product distribution. Moreover, changes of these two variables in furnace optimization are within a small range (dilution steam ratio is practically always fixed in operation) so that they have quite limited effects on the load of downstream units.

Therefore, the load of different feedstocks and their dispatching in different furnaces are used as the control variables in global RTO. The computed optimal set points for feedstock flow rate will be directly passed down to the control level of the cracking furnace system.

Local RTO Layer

Three local RTO modules are incorporated in this layer, as shown in Figure 2. The cracking severity optimization module aims at maximizing the profit of the cracking furnaces. The hydrogenation reactor optimization module is used to optimize the selectivity of ethylene and propylene in the C2 and C3 hydrogenation units. Minimal energy consumption is the objective of the local RTO module for the ethylene and propylene rectifying columns.

Due to the similarities between the applied optimization strategies in these local RTO modules, the cracking severity optimization module will be detailed as an example to illustrate the key features of the local RTO layer.

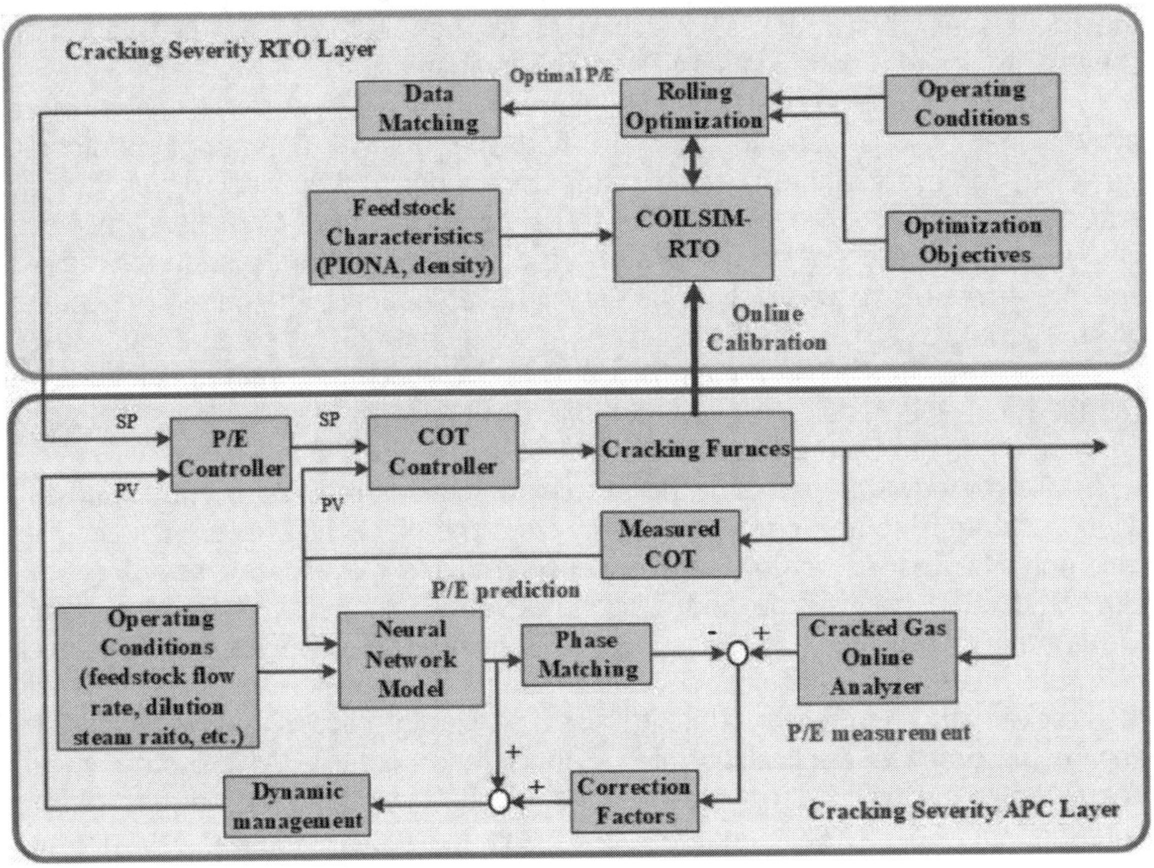

Figure 3. Schematic diagrams of the cracking severity RTO and APC layers

The upper part of Figure 3 is the schematic diagram of the local cracking severity optimization module, in which dilution steam ratios and cracking severities (P/E ratio) are treated as control variables to maximize the profit for each furnace individually. The RTO system will keep collecting operation data from the DCS system, calculating the average value of each operating variable over the last 10 minutes and using them as inputs and outputs for real time calibration of the cracking furnace models. Meanwhile, the calibrated models are used in furnace optimization to provide desirable dilution steam ratio and cracking severity. These set point values will then be passed down to the APC layer. Considering the simulation time and the stress on the DCS system derived from model calibration and data exchange, local RTO is performed every 30 minutes.

Advanced Process Control Layer

It can be seen in Figure 2 that the APC layer also consists of three modules, namely the process control for cracking severity, hydrogenation reactor and ethylene and propylene rectifying columns. The APC system makes use of the concept of model predictive control to reduce the fluctuations of the operating variables. All modules shown in Figure 2 had already been implemented in the ethylene plant DCS system of SSPC through an ongoing cooperation between ECUST and SSPC. This work highlights details of the most representative one, i.e. the cracking severity APC module.

The lower part of Figure 3 shows the structure of the cracking severity control system. It is a model-assisted cascade control system, including control loops for severity and COT. The reason for using this structure is that cracking severity, expressed as a P/E ratio, has been shown to better characterize the product distribution than COT [20]. Since other operating conditions such as feedstock properties, CIT, and feedstock flow rate can change during the run, it is necessary to keep the cracking severity unchanged instead of COT to maintain the product distribution.

However, control of cracking severity cannot fully rely on the P/E value measured by online gas chromatography because of the long time it takes, compared to COT measurements. To solve this, a neural network-based inferential model for yield prediction is implemented in the APC system to determine the real time response of the cracking severity to changes in operating conditions. The weight factors of this model are trained offline using results of COILSIM1D simulations performed under historically collected operating conditions. When online, the model also utilizes real time data for self-calibration so that the drift or mounting deficiency can be remedied. In addition to the neural network model, a comprehensive logic combining the online analyzer feedback and predictive properties is used to determine the process value for cracking severity controller.

By applying the advanced cracking severity control, the COT and P/E variation can be reduced to 1 °C and 0.02, respectively, within the set point values computed by the local optimization layer using COILSIM1D-RTO, respectively. Information about the inferential model, controller performance and benefits of the cracking severity APC system is detailed elsewhere [21].

Graphical User Interface

Figure 4 shows the graphical user interface (GUI) of the plant-wide RTO system, which is published as a dynamic webpage. OLE for Process Control (OPC) technology is used for communication between the RTO system, DCS system and the underlying process models of the steam crackers and separation section. Process data collected from the DCS system are sent to the process models for optimization. The results are displayed together with the process data in the GUI of the RTO system to compare the profit of the whole plant before and after optimization. Meanwhile, the optimal operating set values are passed down to the APC layer.

The GUI consists of four main menus. The overview menu displays the operation cost and profit at a plant-wide level. Cracking furnace menu tracks the current operating conditions and the calculated optimal operating conditions of all 14 steam crackers. The status of the compressors, distillation columns, hydrogenation reactors is shown in the separation section menu. Energy consumption and generation in terms of superheated high pressure steam, high pressure steam, fuel consumption and others are summarized in the energy consumption menu.

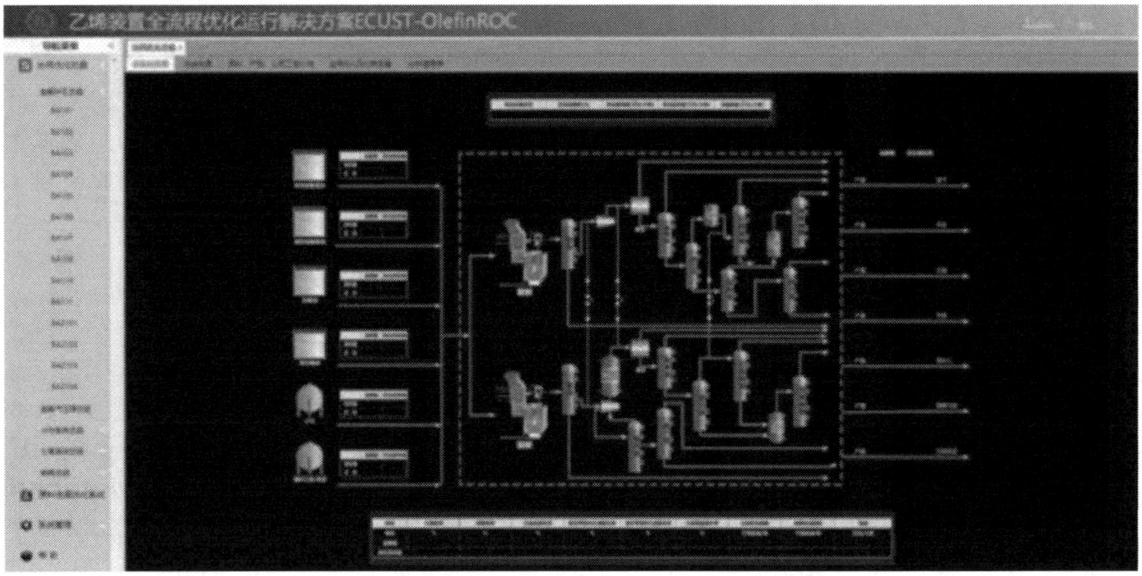

Figure 4. Graphical user interface of the plant-wide RTO system

Results and Discussion

Case Study

Since the plant-wide RTO system is still under implementation and no real time optimization results can be obtained at the current stage, historical operating data of SSPC are used to demonstrate the accuracy of the model and to estimate the potential marginal profit when applying the RTO system.

Two cases are studied in the current work using operating data of 2017. The operating conditions such as feedstock type, online measured COT and P/E are summarized in Table 3, where the P/E and COT are 24-hour average values of the corresponding day. COILSIM1D-RTO is used to perform reactor simulations of the 14 steam crackers. For each furnace, P/E value is set as target of the shooting method so as to match the cracking severity. The COT simulation results are also listed in Table 3. It can be observed that the average difference between the measured and the simulated COTs is about 8K, and the maximum difference less than 20K. Because of the uncertainties of the online measurements, the simulation results can be considered as in good agreement with the industrial data.

Table 3. Comparison of measured and simulated COT of two cases

	Case 1 (2017.3.8)				Case 2 (2017.9.16)			
Name	Feed	P/E	COT Measured (°C)	COT Simulated (°C)	Feed	P/E	COT Measured (°C)	COT Simulated (°C)
BA101	HVGO	0.597	797	797	NAP	0.660	834	822
BA102			Decoking		HVGO	0.449	818	836
BA103	LNAP	0.590	839	845	NAP	0.464	844	855
BA104	LPG	0.342	860	857	LPG	0.217	867	874
BA105	HVGO	0.517	804	816	HVGO	0.545	798	815
BA106	HVGO	0.548	798	809	HVGO	0.569	795	802
BA107	LPG	0.354	860	862	LPG	0.342	858	859
BA108	LPG	0.430	861	857	LPG	0.257	857	863
BA110	NAP	0.432	834	848			Shut down	
BA111	LNAP	0.489	837	851	LNAP	0.513	833	840
BA2101			Decoking				Decoking	
BA2102	LPG	0.256	858	843	LPG	0.347	859	866
BA2103	NAP	0.564	836	834	LNAP	0.540	838	847
BA2104	LNAP	0.552	840	844	NAP	0.532	839	847

In order to accelerate the operating optimization, the cracking reactor models are calibrated so that they can reflect the status of the crackers in real time. This is done by recording the temperature difference between the online measurement and the model prediction, and assuming that this difference remains unchanged during the time horizon of each optimization. Taking BA103 of Case 1 as an example: At the beginning of the optimization, the values of COT, P/E, as well as other operating variables are collected from the DCS system and used to perform cracking reactor simulations, obtaining a simulated COT value of 845 °C, which is 6 degrees higher than that of the measurement. The RTO system stores the temperature difference and starts calculating the optimal feedstock flow rate, dilution steam ratio, cracking severity and its corresponding COT to maximize the profit of the cracking furnaces. In addition to directly sending the three optimal values to the DCS and APC systems, the previously recorded temperature difference will be added to the optimal COT obtained from the model, yielding a reference optimal COT for the APC system to help the cracking furnace reach the desired cracking severity faster.

Table 4. P/E and COT of all furnaces before and after optimization

Name	Case 1 (2017.3.8)				Case 2 (2017.9.16)			
	Original		Optimized		Original		Optimized	
	P/E	COT (°C)	P/E	COT (°C)	P/E	COT (°C)	P/E	COT (°C)
BA101	0.597	797	0.606	792	0.660	834	0.575	861
BA102	Decoking				0.449	818	0.538	793
BA103	0.590	839	0.512	862	0.464	844	0.484	838
BA104	0.342	860	0.254	879	0.217	867	0.151	883
BA105	0.517	804	0.567	789	0.545	798	0.579	781
BA106	0.548	798	0.577	783	0.569	795	0.593	782
BA107	0.354	860	0.243	882	0.342	858	0.238	878
BA108	0.430	861	0.281	886	0.257	857	0.203	876
BA110	0.432	834	0.451	827	Shut down			
BA111	0.489	837	0.412	854	0.513	833	0.431	854
BA2101	Decoking				Decoking			
BA2102	0.256	858	0.190	887	0.347	859	0.276	879
BA2103	0.564	836	0.489	859	0.540	838	0.472	855
BA2104	0.552	840	0.444	871	0.532	839	0.513	843

In the current study, the total feedstock flow rate in the two studied cases is not changed during optimization because the cracked gas compressors are operating at a fairly high load. The original and the optimized P/E and COT values are compared in Table 4. It can be seen that the COT for most of the steam crackers is increased after optimization, indicating that the cracking severities were lower than the optimal values. It is well known that the higher cracking severity is favorable to ethylene yield but usually causes decrease of propylene yield. The cracking furnaces also need more heat input to achieve higher COT, which on one hand increases the cost due to the fuel consumption, on the other hand generates more superheated high pressure steam (SS). Therefore, the marginal profit of the steam crackers depends on the relative price of these key products and materials. As shown in Table 5, ethylene, hydrogen and 1,3-butadiene all have higher price than that of propylene. This explains why a higher COT is desirable as it increases the yield of these three more valuable products. In addition to cracking severity, the dilution steam ratio of each cracker is also optimized. The results suggest that a higher dilution steam ratio should be used, which is in line with the expectation since it not only reduces the partial pressure of hydrocarbon feedstock but also decreases the coking rate.

Table 5. Production and consumption of the key products and materials in steam cracking furnace

Name	Price (yuan/ton)	Production/Consumption rate (kg/h)			
		Case 1		Case 2	
		Original	Optimized	Original	Optimized
Hydrogen	12278	3670.4	3891.7	3565.9	3706.8
Ethylene	8630	91314.2	95316.4	92728.4	96110.3
Propylene	6310	42301.6	39225.2	43130.9	41347.7
1,3-butadiene	7713	13348.5	13694.4	13936.0	14641.3
Benzene	5827	16508.9	18165.4	15881.2	17107.2
Dilution Steam	170	220176.5	246799.9	218418.3	237820.2
Fuel	2838	40107.1	42540.5	41376.6	43065.1
SS	226	354092.3	377492.0	365174.6	381285.5

The effect of the changes in operating conditions on the compressors and separation train are also evaluated in this study. The energy consumption in the downstream process is calculated using the models built in Aspen Plus, and is then used as input to the cracked gas compressor, ethylene compressor and propylene compressor to obtain the consumption and generation of superheated high pressure steam (SS), high pressure steam (HS) and middle pressure steam (MS). Table 6 summarizes the energy consumption in the separation section in terms of steam, where positive and negative values refer to production and consumption, respectively.

Table 6. Consumption and production of steam

Name	Price (yuan/ton)	Production/Consumption rate (kg/h)			
		Case 1		Case 2	
		Original	Optimized	Original	Optimized
SS	226	-436.90	-438.54	-436.96	-438.15
HS	186	164.23	164.88	163.73	164.14
MS	170	70.40	70.57	70.79	70.95

Table 7. Net profit of the ethylene plant

	Case 1	Case 2
Net profit (yuan/h)	3687.84	3844.28
Annual benefit (million yuan/yr)	29.50	30.75

Putting together the cost and income of the steam crackers, compression section and separation section, the net profit of the ethylene plant for both studied cases can be calculated, as shown in Table 7. According to the average operating time of the crackers, which is about 8000 hours per year, the estimated benefit for the ethylene plant is 29.50 million and 30.75 million yuan per year, respectively.

Conclusions

In this work, the potential of a plant-wide RTO system is illustrated. The system makes use of numerical models to simulate the whole ethylene plant. COILSIM1D-RTO is used to perform simulations of the cracking furnaces, to obtain yields of the key products. The associated heat flux profile is used to estimate fuel consumption, using a simplified radiant box model. A simplified convection section model is used to calculate the production of superheated high pressure steam from the steam crackers. Two models for the downstream units of the old and the new site of SSPC are implemented in Aspen Plus. The maximum allowed process gas load of the compressors and the distillation columns are used as optimization constraints. The energy consumption of the separation section is also estimated in the models.

The RTO system is composed of three layers, namely the global RTO, local RTO and advance process control (APC). The global RTO, in which the load of different feeds and their distribution in different furnaces are considered as the control variables, intends to maximize the net profit of the ethylene plant with the constraints of the maximum load of the compressors and distillation columns. The local RTO for the crackers is a cracking severity optimization module, which treats the dilution steam ratios and cracking severities, expressed as propylene to ethylene ratio (P/E) as control variables to maximize the profit for each furnace individually. The APC layer is a model-assisted cascade control system including control loops for severity and COT. This layer had been already implemented into the DCS system of the ethylene plant previously in the plant by East China University of Science Technology (ECUST).

Two cases are studied using historical operating data in the year of 2017 as to calculate the potential marginal profit of applying the plant-wide RTO system. The reason is that the RTO system is still under implementation and no real time optimization results can be obtained. It is shown that, by optimizing the cracking severity and dilution steam ratio, the estimated benefit for the ethylene plant of SSPC in Case 1 and Case 2 is about 29.50 million and 30.75 million yuan per year, respectively.

References

1 . Buffenoir M. H., J. Aubry J. M. and Hurstel X., (2004), "Large ethylene plants present unique design, construction challenges", *Oil Gas Journal,* 102, pp. 60-65.

2 . Cole D., (1996), "Olefin economics impacts of the technology supplier", in: *Proceedings of the 8th Ethylene Producers' Conference*, New York, NY (United States).

3 . Djokic M. R., Van Geem K. M., Heynderickx G. J., Dekeukeleire S., Vangaever S., Battin-Leclerc F., Bellos G., Buysschaert W., Cuenot B., Faravelli T. and Henneke M., (2017), "IMPROOF: Integrated model guided process optimization of steam cracking furnaces", in: *International Conference on Sustainable Design and Manufacturing,* Springer, Cham, pp. 589-600.

4 . Van Geem K. M., Reyniers M. F. and Marin G. B., (2008), "Challenges of modeling steam cracking of heavy feedstocks", *Oil & Gas Science and Technology-REV IFP,* 63(1), pp. 79-94.

5 . Sabbe M. K., De Vleeschouwer F., Reyniers M. F., Waroquier M. and Marin G. B., (2008), "First principles based group additive values for the gas phase standard entropy and heat capacity of hydrocarbons and hydrocarbon radicals", *The Journal of Physical Chemistry A,* 112(47), pp.12235-12251.

6 . Sabbe M. K., Reyniers M. F., Van Speybroeck V., Waroquier M. and Marin G.B., (2008), "Carbon - Centered Radical Addition and β - Scission Reactions: Modeling of Activation Energies and Pre-Exponential Factors" *ChemPhysChem,* 9(1), pp.124-140.

7 . Sabbe M. K., Reyniers M. F., Waroquier M. and Marin G. B., (2010), "Hydrogen Radical Additions to Unsaturated Hydrocarbons and the Reverse β - Scission Reactions: Modeling of Activation Energies and Pre-Exponential Factors", ChemPhysChem, 11(1), pp.195-210.

8 . Sabbe M. K., Vandeputte A. G., Reyniers M. F., Waroquier M. and Marin G. B., (2010), "Modeling the influence of resonance stabilization on the kinetics of hydrogen abstractions", *Physical Chemistry Chemical Physics,* 12(6), pp.1278-1298.

9 . Saeys M., Reyniers M. F., Marin G. B., Van Speybroeck V. and Waroquier M., (2004), "Ab initio group contribution method for activation energies for radical additions", *AIChE Journal,* 50(2), pp.426-444.

10 . Saeys M., Reyniers M. F., Van Speybroeck V., Waroquier M. and Marin, G. B., (2006), "Ab initio group contribution method for activation energies of hydrogen abstraction reactions", *ChemPhysChem,* 7(1), pp.188-199.

11 . COILSIM1D. http://www.avgi.be/.

12 . Vervust A., Amghizar I., Munoz A. E., Van Geem K. M., Marin G. B., (2016), "Full furnace simulations and optimization with COILSIM1D", in: *AIChE Spring Meeting,* Houston, TX, USA.

13 . Hudebine D. and Verstraete J. J., (2011), "Reconstruction of petroleum feedstocks by entropy maximization: Application to FCC gasolines", *Oil & Gas Science and Technology–Revue d'IFP Energies nouvelles,* 66(3), pp.437-460.

14 . Van Geem K. M., Hudebine D., Reyniers M. F., Wahl F., Verstraete J. J. and Marin G. B., (2007), "Molecular reconstruction of naphtha steam cracking feedstocks based on commercial indices", *Computers & chemical engineering,* 31(9), pp.1020-1034.
15 . Pyl S. P., Van Geem K. M., Reyniers M. F. and Marin G. B., (2010), "Molecular reconstruction of complex hydrocarbon mixtures: An application of principal component analysis", *AIChE Journal,* 56(12), pp.3174-3188.
16 . Zhang Y., Reyniers P. A., Du W., Qian F., Van Geem K. M. and Marin G. B., (2017), "Incident radiative heat flux based method for the coupled run length simulation of steam cracking furnaces", *Industrial & Engineering Chemistry Research,* 56(14), pp.4156-4172.
17 . Yuan B., Zhang Y., Peng X., Du W., Hu G. and Qian F., (2017), "Simulation of the Evaporation Tube Banks in the Convection Section of a Steam Cracking Furnace Using an Evaporation Model", *Industrial & Engineering Chemistry Research,* 56(38), pp.10813-10825.
18 . Jiang A., Shao Z., Chen X., Fang X., Geng D., Zheng X. and Qian J., (2006), "Simulation and optimization of distillation column sequence in large-scale ethylene production", *Journal of Chemical industry and engineering China,* 57(9), p.2128.
19 . Du W., Bao C., Chen X., Tian L. and Jiang D., (2016), "Dynamic Optimization of the Tandem Acetylene Hydrogenation Process", *Industrial & Engineering Chemistry Research,* 55(46), pp.11983-11995.
20 . Van Geem K. M., Reyniers M. F. and Marin G. B., (2005), "Two severity indices for scale-up of steam cracking coils", *Industrial & Engineering Chemistry Research,* 44(10), pp.3402-3411.
21 . Wang H. G., Wang Z. L., Mei H., Qian F. and Tang Z. W., (2011), "Intelligent severity control stabilizes ethylene crackers", *Oil & Gas Journal,* 109(6), pp.104-109.

MINLP Optimization Approach for Managing Integrated Petrochemical Business Planning

J.A. Ortiz-Cruz*, J.Carreno-Zamora, S.A. Cardoso-Palmerin*, G. Chavez-Segura

Braskem-Idesa
Nanchital, Veracruz, Mexico

Abstract

In this presentation, we describe an optimization approach for managing polyolefin (PE) planning problem with its upstream ethylene plant as the source of its main raw materials. The main challenge it this integrated business planning is to optimize profits based on the allocation of ethylene to be used in various PE product grades. We consider the inventory of PE grades, monthly demands in the production planning and all possible product sequences that can be produced each month. Instead of solving the planning problem as two separate businesses (ethylene and PE), here we solve the entire supply chain from ethane as the ethylene main feedstock. The advantages for solving as integrated business planning, we do not need to use 'transfer price' for the ethylene as the main raw material for PE products. The profit optimization is based on true market value for the raw materials and products, with considering all production constraints.

The modeling approach used in the planning model is based on mathematical formulation of mass, energy balances, together with nonlinear cracker yield model and PE production sequences. The model describes the entire production material flow from the ethane feedstock to all by products of ethylene, utility and three PE plants. The entire ethylene product with annual capacity of roughly one million ton annually is used for PE production. The main challenge in this model approach is that the upstream unit requires nonlinear model due to component mass balances in mixing, reactor, distillation unit operations and other non-linear constraints; and the downstream requires binary variables to manage production sequence selection for feasible production operation. The resulting model formulation is in the form of large scale Mixed Integer Nonlinear Programming (MINLP) model. This model formulation is a novel approach in petrochemical industry, where the supply chain is usually represented as an LP or NLP. Here, we will discuss the practical usage of optimization of MINLP planning model for a world class ethylene producer with its polymer businesses.

Introduction

In recent years, building ethylene plant based on ethane feed is very attractive option due to the attractive ethane price as compare to liquid feedstock. If the main objective is to produce mainly ethylene product, this is a very attractive investment option as compare to more costly and complex process for handling liquid feed that can produce wider ranges of products. With increasing supplies of ethane from shale gas discovery, there are many more new ethane feed ethylene plants being built in America. The majority of the ethylene are used for producing polyethylene (PE). With more similar plants are being built in recent years, one of the key to profitability is in optimizing business operation and its supply chains.

Braskem-Idesa recently built a brand new ethylene plant with a world class ethylene capacity of more than 1 million ton annually. All of the ethylene is used internally to produce various grades of high and low density PE. Here are the challenges for making optimal decisions regarding the supply chain operation:

- Ethane supply:
  - Although there is a contract on the ethane supply, the delivery may vary based on the availability of the supplier. The key issue here is understand the impact on setting ethylene production policy that impacts the yield and the allocation of ethylene to downstream PE production plan. Furthermore, there is limited feedstock storage facility in the plant site.
- Ethylene plant operation policy:
  - With the variability of feedstock supply and prices, the plant operation policy also needs to be optimized based on the economic objective function (maximizing profit) by pushing the operation to plant constraints. The key decision here is on how to set the optimal severity policy of the ethylene plant operation. The severity will impact on the yield, recycle streams (from ethane to C5s) and utility cost. Many case study scenarios may be needed to evaluate the risk and impacts on the downstream PE productions and optimizing the ethylene production cost.
- PE product allocation and load:
  - With three production lines and variability of product grades that can be produced in each line, there are opportunities to allocate specific product and quantity to be produced in each line. The main objective is to reduce production downtime for certain production sequence of certain grades and product off grades. The optimization is more challenging when there is limited ethylene feed. Prioritizing production optimization is needed, considering product demands and inventory policy.
- Utility plant and fuel purchases:
  - The plant has its own utility plant to produce steams and electricity to supplement external supply. Ethylene production also produces fuel gas that is used as fuel in the power plant. Utility planning to ensure sufficient utility balance is important and needs to be planned and optimized together with the production planning.

At Braskem-Idesa, we have implemented advanced optimization planning application using SCMart supply chain software from Optience. The planning model covers the entire business scope, from the feed supply to product distribution.

Modeling and Optimization Approach

The workflow for typical planning optimization of this type of business is done based on sectional. There will be a separate ethylene planning optimization, then it is followed by downstream (PE) planning optimization. This type of approach is done mainly due to limited capability of planning optimization tool that can optimize the entire problem as a single problem. This approach uses a critical assumption: the ethylene price. The price is decided based on the production cost of the ethylene plant operation. Based on this production price or some transfer price, the PE production planning is done. This approach may produce local optimal for each business section. If the business belong to the same company, it would be better if we can optimize the entire supply chain problem as a single optimization without making assumption on the key product price (transfer price).

For accurate representation of the supply chain problem in the process industry, we need to be able to develop mathematical model than can represent the fundamental mass, energy balance equations and operation/business constraints. The challenge in the ethylene process modeling is the accuracy of the yield model and the chemical compositions of flows in the process. This requires a nonlinear mathematical model representation due to mixing of streams with different compositions. In addition, the reactor modeling of PE process require operation constraints that involves product sequencing and rules of identifying certain sequence will produce off-grade product quality and downtime for cleaning. To represent this type of constraints, we need to introduce binary variables in the mathematical model. If we model these business separately, the ethylene and utility plants can be formulated as nonlinear programming (NLP) model. The PE section requires mixed integer programming (MILP) model. If we want to solve these two sections (upstream and downstream) together as one model, then we need to combine the model as mixed integer nonlinear programming (MINLP) model. Optimizing MINLP model is much more difficult and challenging. Here, we have implemented simultaneous approach for the entire problem as one model for first time it has been reported in industrial planning application using SCMart software.

Here, we describe briefly the key features of SCMart system from Optience for building our planning and scheduling applications. SCMart modeling platform utilize graphical flowsheet modeling approach. The basic building blocks consist of unit operation model library that can be used to construct a supply chain problem in the process industry. The library model generates the mathematical equations for the basic mass/energy balance automatically. The model can be as linear, nonlinear or mixed integer models. It depends on the flowsheet structure that we build to represent the problem.

Here are key SCMart modeling and optimization concepts for building the planning and scheduling models:

- Material modeling:
  - It supports composition weight fraction, material properties and product based representation. This approach allows for building a complete model from upstream process where the stream are mixture of chemicals that later need to be separated to downstream (polymers) processes with solid products of different properties.
- Reactor modeling:
  - It supports nonlinear model to represent complex reactions where the yield is function of operation conditions, feedstock properties and quantities.
- Unit operation library models:
  - Each unit has prebuilt fundamental constraints for satisfying mass balance and other operation rules. Additional equations can be added to define specific constraints. Depending on the specification, the resulting mathematical model can vary from linear to nonlinear or mixed integer model.
- Flowsheet modeling approach:
  - It utilizes process flowsheet modeling concept. Each unit operation indicates an activity where some changes to the material occurs. The condition of the output stream of a unit operation is transferred to the connecting input of another unit operation. A collection of unit operations that represent a process can be grouped and identified in the model. The resulting flowsheet can be identified as collections of processes. Figure 1 shows an example of planning model where each process or section can be identified as block by block diagram.
- Model representation and optimization approach:
  - The flowsheet model is used a master representation of planning/scheduling model. This model will be used as the base for generating actual planning/scheduling problem. The application can generate multi-period case problem to be optimized.
  - Economic based objective function of the model is generated automatically based on parameters that are selected in each unit operation.
  - The equations are generated automatically during the optimization run as part of constraints in mathematical programming framework. Depending on the equation type, an appropriate optimization solver will be used to solve the resulting formulation. It can support LP, NLP, MILP and MINLP models.
  - The user can manage the degrees of freedom to be optimized in the model by specifying the variable bounds in the model.

Implementation

Implementing advanced planning system requires full understanding about the current planning workflow. From how we collect all required information for making planning decision to how we use it. Optimization planning approach requires additional information as input. One of the most important information are the limits (minimum and maximum) of critical information such as raw material supply, process capacity, inventory and product demand. This information will set the feasible region for the optimization to find the optimal solution within the limits.

Due to significant amount of data to be collected and prepared for the case, we need to pay attention on how to design data collection, validation and preparation for easy use in the planning application. In addition, we also need to design views for analyzing the results and prepare a report template in order to reduce manual work after we run the optimization case. The optimization runs very fast, in the order of less than a few minutes, most of the user time will be spend in preparing data and analyzing the results. The implementation of data integration is critical part of the new planning system.

Case Study and Results

To illustrate the benefits of our planning optimization approach, we set up a multi-period case study with 5 monthly periods. The case cover the entire supply chains at Braskem Idesa ethylene plant. Here are conditions in this case study:

- Ethane fresh feed supply to ethylene plant:
  - Assuming less than maximum capacity of the design (~90%) for the maximum supply for any period. In this scenario, we will see more interesting results as it try to optimize the severity in order to reach the constraint on the ethylene fractionator as the real constraint in the cold section.
- PE plant:
  - We assume there is a partial shutdown in the third period, so the capacity is roughly half of normal capacity.
  - There are three production lines with two lines for HDPE and one line for LDPE. The lines for HDPE can share the production of some grades.
  - There are more than 50 product grades that can be produced. The model includes the modeling of prime and off grades.
  - The production for each line consider the all possible production sequences.
- PE product demands:
  - We include list all product demands, contract and spot. In addition, we also consider the logistics cost/market area in order to reflect the real pricing and costs.

- Utility
    - The plant can sell excess electricity.

Period	Ethane Fresh Feed	Total Feed to Furnaces	Ethylene Production	Ethylene Inventory	Ethylene to PE
0				144.4%	
1	100.0%	100.0%	100.0%	175.0%	98.6%
2	95.2%	96.0%	96.0%	194.1%	98.8%
3	42.9%	43.0%	43.0%	100.0%	104.1%
4	95.2%	95.7%	95.7%	100.0%	99.3%
5	100.0%	100.1%	100.4%	100.0%	99.3%

Table 1: Case Study for Ethylene Plant

Table 1 show the results from the ethylene plant section. The results have been normalized using the following rules:

- Ethane fresh and total feed rate for period 1 is assumed as 100%
- Ethylene production for the first period is also used as the base
- Minimum Inventory is used as the base (100%)
- Ethylene to PE is scale based on ratio to ethylene production

The fresh feed on the third period is reduced due to lower capacity in the downstream PE plants. The result of solving multi-period simultaneous optimization case show the role of inventory as part of degree of freedom to be optimized. It builds up inventory to be used in later period where it can produce a higher profit due to pricing and demands.

Period	Plant 1 Utilization	Plant 2 Utilization	Plant 3 Utilization	Production 1 Sequence	Production 2 Sequence	Production 3 Sequence
1	100%	100%	90%	Sq078	Sq014	Sq010
2	100%	100%	86%	Sq088	Sq001	Sq006
3	91%	96%	100%	Sq060	Sq013	Sq012
4	96%	69%	97%	Sq061	Sq004	Sq006
5	94%	93%	81%	Sq054	Sq001	Sq015

Table 2: Case Study for PE Plant

In Table 2, we list the percentage of each plant utilization as percentage of maximum production load. Since there are so many grades, the production of each grade should be done based on possible sequence that minimize the production of 'off-grade' products and transition downtime between grades. Each sequence consist of unique production patterns that can be executed in a specific production line. The optimization will select the most profitable sequence to be implemented for each line based on product inventories and demands. This model requires significant numbers of binary variables (> 2,000) to represent the logic and production sequences. The optimal sequence produce the maximum profit by fulfilling the most profitable product demands and at the same time trying to reduce the holding inventory to the minimum. From Table 2, we can see the selection of production sequence is quite challenging without optimization. The possibility for selecting 'sub-optimal' decision due to the complexity is quite real if we rely on manual decision. The financial benefits of this approach is significant because we can easily study the impact of introducing new grades on the existing operations. The profitability of product grades vary significantly, therefore we need to continue to challenge our business operation to explore new business opportunity by producing grades that will improve our profitability.

Conclusion

Although ethane based ethylene production is simpler compare to liquid based, we have shown here that there are significant benefits can be obtained by applying optimization based planning system. This is more important for ethylene producers where ethylene is used for internal downstream plant consumption. In this scenario, the business problem can be much more complex. The complexity is mainly in allocating the ethylene to produce the most profitable downstream products as integrated business. In this case, advanced planning optimization system is indispensable part of decision support tool for helping our business competitiveness.

The results of case study shows the impact of multi-period planning. Inventory is part of critical degree of freedom that can help us to improve our operation and profitability. Therefore inventory policy in the optimization approach for multi-period planning should be considered in making a good planning decision.

CONTROL SYSTEM MIGRATION PROGRAM FOR DOW'S HYDROCARBONS FACILITIES

Eddy Yap, Dow Chemical Canada ULC, Fort Saskatchewan, AB, Canada
Marinus Van Roosmalen, Dow Benelux B.V., Terneuzen, Netherlands

Abstract

Dow's proprietary MOD™5 system has been the control platform of choice for most businesses within Dow over the past 38 years. Since its rollout in 1980, it has enabled a renaissance age for innovation, optimization and leveraging among engineers allowing for year over year improvements in asset utilization and production capability. The perfect storm comprising component obsolescence, expertise loss, and more stringent regulations has accelerated retirement of our MOD™5 platform. A company-wide Migration Program was initiated to drive this effort to convert 150+ plants from the existing MOD5™ controllers to commercial systems, primarily towards ABB 800xA.

In this paper, we will describe the migration process of MOD™5 to ABB 800xA, and the plans of converting 13 of our hydrocarbons facilities to the new platform. We will also cover challenges related to schedule, hardware design, project resources, application development and operator training.

Introduction

The MOD™5 system was designed by Dow personnel in 1978 to contain redundant input/output boards, each communicating the field signals to independent computers. The two computers communicated with each other and compared all input signals for discrepancies. Both computers calculated outputs and sent these signals to field devices based on the program. With two computers running simultaneously, it became possible to load a new program without shutting the MOD™ system (or the plant) down.

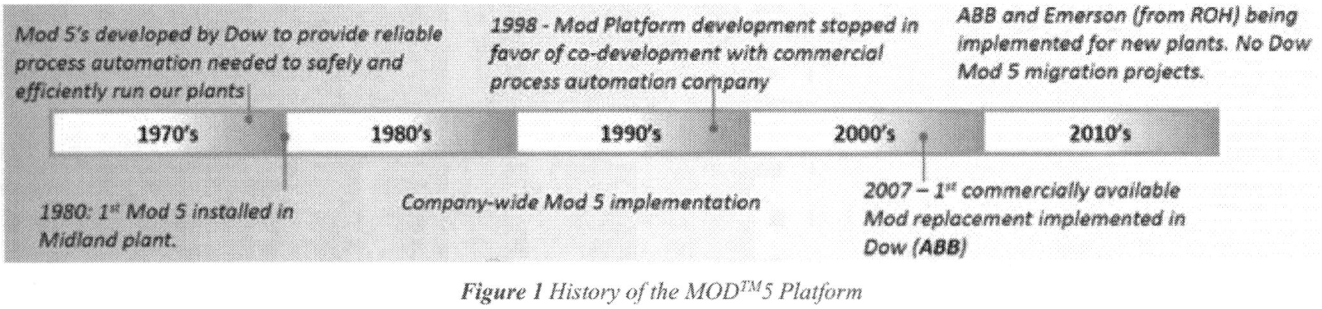

Figure 1 History of the MOD™5 Platform

In 1980, the first MOD™5 system was put into production service [1]. Over the next 25 years, Dow installed them in over 150+ plants at 48 sites around the world, comprising more than 500,000 inputs and outputs (I/O). By 1998, we knew the MOD™5 system would not be sustainable over the long-term without significant investment due to attrition of expertise, evolution of the operating software, increasingly sophisticated security measures, unavailability or uncompetitive rates for replacement of specific components. Dow decided to cease further in-house development of the system and instead opted to partner with a preferred control system vendor to co-develop or improve upon the commercially available offerings.

Despite the MOD™5 technology being at the upper end of benchmarked reliable life cycle of 20-30 years, new MOD™5 systems continued to be added until an alternative control platform was finally approved and installed in 2007. ABB 800xA became the new go-to control platform for legacy DOW businesses, while the original Rohm and Haas facilities have gone with Emerson Delta-V systems. For Dow's hydrocarbons business, we have decided to standardize on the ABB 800xA control platform with our first implementation in 2007 at the Octene plant in Tarragona, Spain. To date, we have more than 54,000 I/O already running on ABB 800xA controllers.

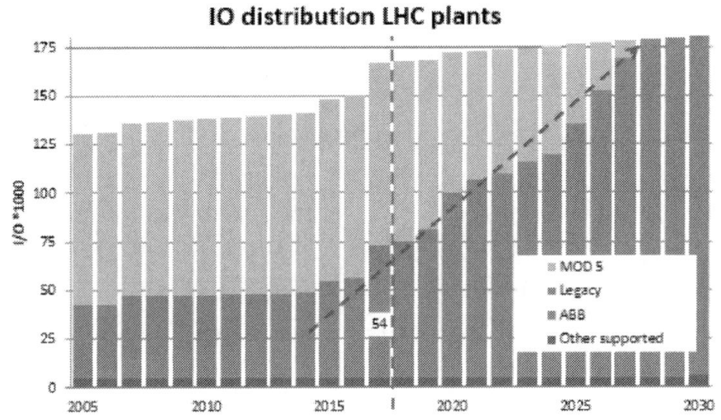

Figure 2 I/O distribution per control system for Hydrocarbons Facilities

Case for Change

The task of converting such a large installed base of MOD™5 systems to a commercial system is undoubtedly complex and expensive. To determine how long before failures of our MOD™5 systems overwhelm our ability to sustain reliable plant operation, we enlisted the help of the University of Maryland in 2012. They have had experience in assessing and quantifying failure rates of electronic systems, essentially creating reliability models to predict failure rates into the future. In 2012, they predicted that the existing inventory of MOD™5 spare parts would be depleted by 2022 if we chose not to migrate. Shortly after 2025, Dow would run out of, or be unable to repair components within 72 hours based on evaluations of our supply chain model, if no migration effort were initiated. MOD™5 components and subsequent system failures are expected to

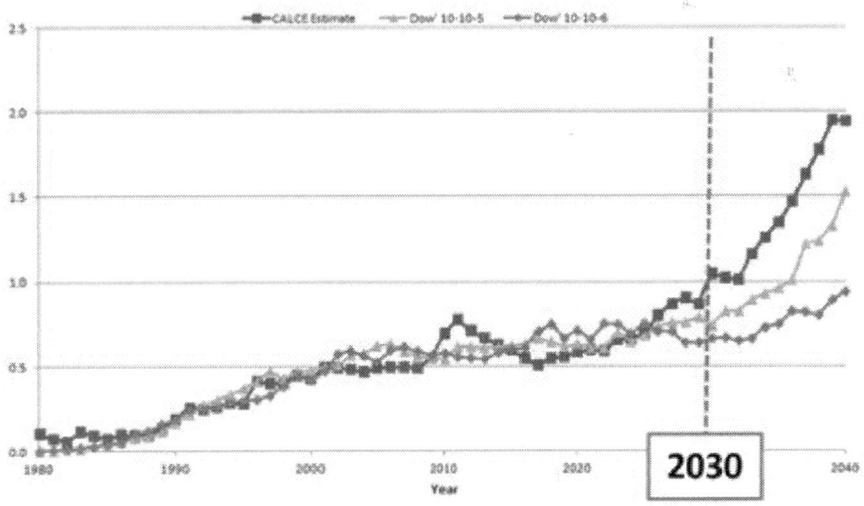

Figure 3 Predicted Failure Rate of MOD™5 Systems

increase significantly by 2030, as illustrated in Figure 3. This does not mean that all or large amounts of our MOD™5 systems will immediately fail, but that we will see more and more failures culminate to interrupt portions or whole plant operations. When the reliability model was recently updated with data up till 2017, the predicted failure rates doubled that of the original predictions made in 2012, increasing the urgency of the migration program.

Corporate Program

A corporate "Migration Program" office was created to spearhead a multi-pronged approach to extend the useful life of MOD™5 platform through 2030, and to transition out of MOD™5 in a manageable and cost-effective manner. As shown in the diagram below, the program strategy encompasses 7 different tracks: (1) Migration strategy, (2) Migration planning, (3) MOD™5 Competency, (4) VMS Environment, (5) Reliability, (6) Intellectual Property, and (7) Replacement and Installation Preparation. This paper will only address Migration strategy and Migration planning. There is some reference to Track 7 – Replacement and Installation Preparation in the Hardware migration strategy section.

Of the 20 projects that have been initiated under this program, 4 have been successfully completed and the rest are at various stages of completion. Lessons learned, upgraded tools, team integration practices, enhanced work processes, and refined implementation standards are being leveraged by the Program Implementation Team to the existing projects. This team is also responsible for developing the cost estimates for each project to obtain initial funding. They take a holistic view of all the scope and work

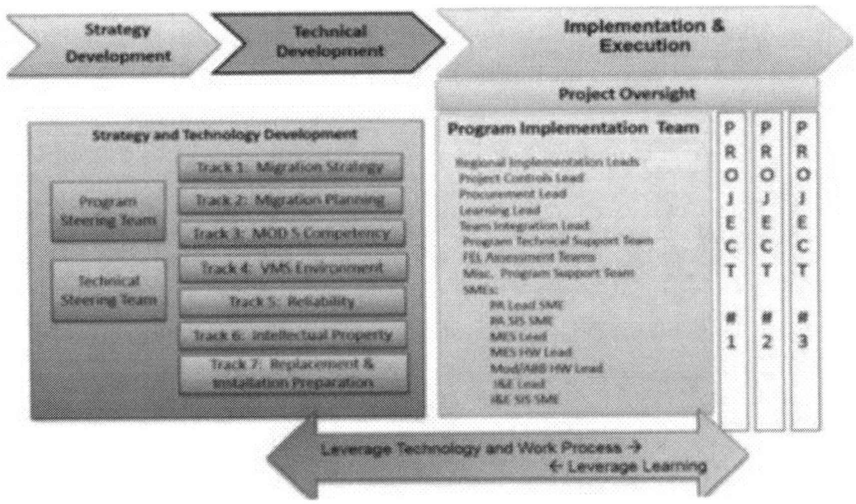

Figure 4 Migration Program Organization and Strategy

closely with each business to set the migration plans for every facility in order to properly balance resources and spending. The sheer program size, complexity of the work, and variety of stakeholders necessitates a consistent and structured approach for planning, estimating, tracking, and communicating.

Migration strategy

The ultimate goal of the migration strategy is to "provide a Solid Platform which enables a Profitable, Reliable and Safe Operation for the next 20 years". Achievement of this goal for every one of the 150+ projects along the journey relies on our ability to execute consistently, but yet be able to adapt quickly to new discoveries, and to religiously incorporate lessons learned into the overall software and hardware migration work processes.

Software migration strategy

Over the years, Dow has gradually learnt from past incidents to improve control system response, and also figured out how to best optimize operations. Culmination of these learnings to improve safety, reliability and asset utilization of the facility are directly incorporated into the control strategies that run the plants today. This compendium of knowledge presents a specific challenge that impacts all aspects of the migration projects - how do we transfer the value of these innovations to the new ABB system? Failing to do so essentially amounts to value destruction and loss of significant intellectual property invested into these facilities.

During the Front-End-Loading assessment, subject matter experts from the Migration Program sat down with plant-knowledgeable engineers for a detailed review of the existing control logic. For each unit operation in the migration scope, they determined the general state of the application such as software version, use of templates, operational issues, and I/O count in defining the complexity factor and whether the existing strategies have to be uplifted. To determine the amount of effort required, software work is categorized as follows:

(1) Like-for-like – duplicate all control functionality into the new system,
(2) Uplift – duplicate some and upgrade sections of the code where deemed problematic, or
(3) Re-write – throw out the existing code and re-design with new control strategies.

It is arguable as to which method takes the most energy. The like-for-like conversion approach preserves the existing control strategies, and is also the simplest. Plant knowledge is seldom required thereby putting less demand on experienced process control engineers. Contractors or entry-level engineers can be taught to read the MOD5 code, apply the migration tools and translate to ABB logic. Uplift requires that the process control engineer has some understanding of the process and its current operational state. Otherwise it is difficult to figure out exactly what has to be improved upon. As some parts still have to go through the like-for-like process, the uplift activity adds complexity and we estimate it requires at least 10% more time. Rewrite is quite straight-forward provided the process is well understood and there are good examples to be leveraged from. For example, cracking furnaces are somewhat similar technology, so we opted to rewrite because there is much we could leverage from the newly installed furnaces in Freeport. The step sequencing, library structure, calculations, burner management, control strategies and even the graphics could be adapted from one furnace to another. In other cases where the unit operation is unique and there are significant issues with the existing control strategies driving for a rewrite, the estimate must account for consultation with Operations and Process Engineers to ensure these issues are not repeated. This could take up to 25% more effort as compared to a like-for-like process. One of the hydrocarbons migration projects currently in flight has a scope of 7,500 I/O's that will see 90% like-for-like, 7% uplift and 3% rewrite. Another has 11,600 I/O's, of which 38% will be rewritten, 2% uplifted and 60% converted like-for like. The higher ratio of rewrite for the 2nd project is largely due to the Furnace scope, where we intend to leverage in from previous projects.

Hardware migration strategy

Speed is of the essence for the hardware migration strategy. Every day the control system is down for migration is another day of lost production. To convert 10,000 I/O every year is a challenge to timely design, build, test, ship, store, and install over 80 cabinets per year. Each cabinet replacement has to be planned and executed flawlessly before commissioning can start. Several key attributes in the current MOD™5 hardware setup have allowed for quick replacement: (1) independent marshalling cabinets, (2) consistent grouping of signals, (3) ribbon connectors between marshalling and I/O enclosures, and (4) standardized cabinet sizes. Some projects or vendors have opted to reduce the number of cabinets by terminating the field cables directly to the same cabinet of the I/O modules. Fortunately, in our standard MOD™5 hardware architecture, the field cables are terminated in independent marshalling cabinets whereby cross-wiring can be done. This allows for grouping and segregation of signals according to unit operations, instrument type such as RTDs or pulse instruments, and obviously, signal type – AI, AO, DI, DO's. This is a blessing from a control system migration perspective because the project can change out the control modules without impacting field wiring and cross-wiring. Imagine having to gut the existing cabinets and re-terminate all the wiring in a turnaround window. The ribbon connectors provided plug-and-play capability which simplified the inter-cabinet interfaces and allowed for early manufacturing of the I/O enclosures. Last but not least, Dow collaborated with ABB to design a new style of cabinets that could fit within the same footprint of the MOD™5 cabinets that will be replaced. Several prototype cabinets were built and reworked taking into consideration the differences in the power supply, entry and routing of cables, and grounding versus the current setup, and any good ideas to ease maintenance.

Aside from the control system hardware, one has to consider whether to expand or upgrade the existing remote buildings and control room. The MOD™5 network is based on multi-mode fiber which is limited in transmission distance and has been in the ground for at least 20 years, so it may have to be replaced. Current capacity of uninterruptible-power-supplies (UPS) has to be assessed and new systems may be required as MOD™5 computers have internal batteries to ride through power upsets. Finally, the existing heat-ventilation-air conditioning (HVAC) systems are usually dated and of high maintenance so this may be an opportunity to procure new ones.

Instrumentation is another significant cost adder as some are simply obsolete and incompatible with the ABB 800xA system. Frequency-type meters such as vortex flow meters and densitometers may require new electronics to convert to conventional 4-20mA signals. Resistance-type temperature (RTD) sensors currently direct wired to MOD™5 need new transmitters as ABB does not offer redundant RTD input modules. A big benefit of migrating to the ABB platform is its integration to HART-enabled devices, allowing for advanced diagnostics and remote loop-check for these instruments.

Migration Planning

Each business was tasked to establish and maintain staged implementation plans with target completion date of 2030. With the high level of automation embedded in our existing MOD™5 control system, it is nearly impossible to replace the system safely without impacting plant operations. We have been successful with online migration of legacy and panel controls, but these are simple loop-based controls that can be easily managed one point at a time. Even this task took considerable planning and coordination between process automation (PA), operations, and maintenance. For continuous processes such as utilities, flare, compressors and distillation operations, the most opportune time to shut down the control system without impacting production is during a planned turnaround. As such, we planned the migration schedule around each facility turnaround cycle of approximately 8 years.

Figure 5 provides a quick overview of the migration plan of our Hydrocarbons business anticipating upcoming turnarounds and expected duration of each migration project. For a facility of approximately 7,500 I/O, the optimum duration is 3 years considering the resource loading for software work, and staged hardware production.

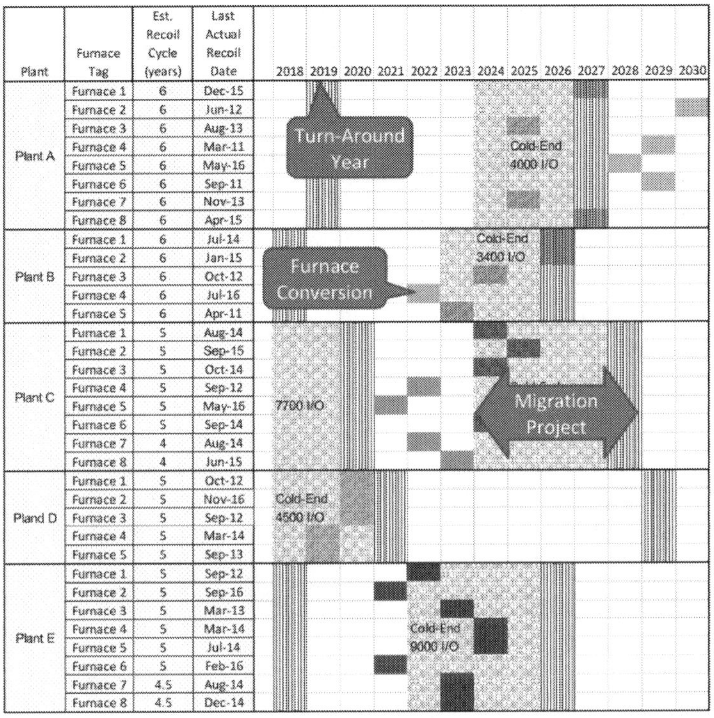

Figure 5 Migration Plan for Hydrocarbons Facilities

As cracking furnaces are stand-alone unit operations with shorter shutdown intervals of 4 to 6 years due to recoils, they allow for resource leveling and reduction of scope during the major turnarounds. With each recoil typically taking 2 weeks, this is the perfect opportunity for control system migration without impacting production from that furnace. In addition, we formed a Furnace PA team comprised of engineers from various sites and our high value center to take advantage of the many similarities between the 130 furnaces in our fleet that have to be migrated. They developed and leveraged application libraries for burner management and furnace control strategies from one furnace to another. Having to migrate around 10 furnaces a year will no doubt challenge the ability of this team to plan well and execute efficiently.

For sites which have multiple hydrocarbons facilities such as Bahia Blanca, Texas Operations, and Terneuzen, we will have a single project team per site to manage the whole scope in order to drive continuity and consistency of implementation. However, these projects may extend more than 5 years so attention to people is of utmost importance, such as providing sufficient growth opportunities and effectively managing turnover of personnel.

Other Challenges

Even with proper planning and proven migration strategies, we still face considerable risks associated with resourcing, false perceptions, and under-estimated scope.

PA resource gap

The most difficult challenge facing our migration program is the shortage of qualified PA resources. Unlike other companies, Dow has traditionally relied on chemical engineers to fill the PA role because they understand how the process should run and are able to devise innovative state-based control strategies to improve overall plant performance. Unfortunately, this has made it difficult for those not process-intuitive to understand the complex control logic. We believe that to be successful in these migration projects, we must select PA leads or technical coaches who have had prior MOD™5 experience in a hydrocarbons facility, for they are most familiar with our Hydrocarbons Most-Effective-Technology (MET) and Book-of-Conventions (BoC). The other skills such as application development in ABB 800xA using Dow libraries, and how to use Migration tools can be taught.

To supplement these plant knowledgeable resources, we select trained PA engineers from our high value center and from local control contractors. Some have previously worked on ABB projects or other control platforms, but few are familiar with Dowtran or our hydrocarbons processes. Since the hydrocarbons projects' scopes are relatively large, software development has to be distributed to at least 3 remote offices. The time zone difference of over 12 hours between the projects site and certain remote offices does impact team integration and productivity. Effective management of remote teams is essential for making and tracking progress.

Migration Projects are different

A migration project is not your typical capital project because software spending accounts for ½ of the budget. As described in the software migration strategy, software development can take considerable time and should be started as early as possible. Initial funding is necessary to secure the right talent and start training on the new control system. The other ½ of the budget is for the control hardware and associated work as there is practically zero steel and concrete work unless upgrades are required to building, HVAC, UPS, or cabling. Engagement, and thus engineering hours, for mechanical, electrical, civil or process engineering is minimal. Similarly construction and supervision hours are limited to replacement of the control system hardware.

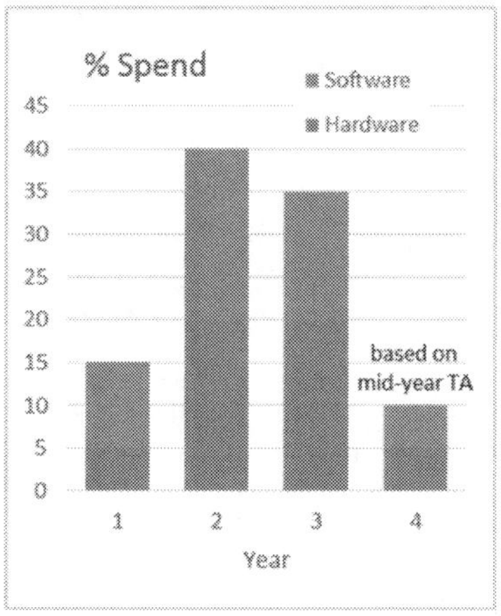

Figure 6 Spending Plan of Migration Budget

ABB ≠ MODTM5

Operators are so accustomed to the MOD™5 environment and its user friendly interfaces that they want to retain the same look and feel to the ABB system. The tendency to "*MOD-ify*" the ABB system is overwhelming. Early exposure, operator training, and leaders' commitment to common standards are ingredients for a successful transition.

Safety Instrumented System (SIS) Implementation

One of the great advantages of the MOD™5 system is its TÜV certification for up to SIL-3 [2]. This allows for us to combine Basic Process Control System (BPCS) control action with SIS protective functions in the same controller, making it easy to share I/O and avoid external communications. For hydrocarbons facilities built prior to 2001, we were able to apply Dow Good Engineering Practices (DowGEP) for our SIS design and implementation, based on the grandfathering clause in ISA84. However, we can no longer apply this clause when we change the control system and now must fully comply with IEC61511/ISA84. The design of the ABB 800xA system requires segregation of BPCS and SIS functions onto separate applications. The protection layers credited in the LOPA have to be reassessed based on the SIL certification of the new logic solver. Instrument sharing previously allowed per DowGEP and configured in the MOD™5 system now has to be reworked to meet the requirements of the ABB 800xA safety manual. For older facilities, it is not easy to add instrument taps or increase redundancy and there are bound to be obsolete or "non-smart" instruments, so additional engineering is required for fault tree analysis to determine if the probability of failure on demand meets the required SIL based on the revisions to the LOPA Independent Protection Layers (IPLs). Hence, SIS instrumentation work can be significant depending on the existing state of the facility and gaps in compliance.

The documentation work associated for each SIF loop cannot be under-estimated as well. Safety Requirement Specification forms have to be updated with sensor failure detection settings, valve leakage rates, process time constants, control logic and interlocks, based on the revisions to the LOPA IPLs. Each of the SIF loops will have to be validated after the control system is replaced. This activity is resource intensive and may be on critical path to plant startup. The only way to mitigate this risk is to have the right expertise to plan and execute this activity.

Conclusion

With every change comes great opportunities – a control system migration can help reinvigorate the organization, bring in new and better tools, offer a clean slate for future growth possibilities, and stimulate innovation to enhance plant performance. Our hydrocarbons business has had a good start on this massive migration program given clear direction from company leadership, a carefully planned strategy, an excellent collaborative relationship with ABB, and skilled resources to analyze risks, execute, and leverage good ideas.

Acknowledgement

Some of the information presented in this paper is prepared by the Core Engineering Disciplines Technology Center for the Corporate Migration Program. Special thanks to Program Director, John Kosel, and Technical Leader, Janette Brightwell, for taking the time to edit this paper. We also acknowledge the tremendous support from the Hydrocarbons business, especially Dan Rader, Trevor Boyce and Paul Broekaart of the Light Hydrocarbons Technology Center.

References

1. Walker, M., Sederlund, E., Gipson, J., and Cosman, E. (2006, February 10), "The MOD Squad: Process automation at Dow," Retrieved from https://www.controlglobal.com/articles/2006/029/
2. Walker, M., Sederlund, E., Gipson, J., and Cosman, E. (2006, May 11), "Process automation at Dow: Part 2," Retrieved from https://www.controlglobal.com/articles/2006/098/

AIChE Paper Number: 2a

Modernization of Existing Ethylene Crackers (Compressor & Turbine Footprint Replacement Options in Brownfield Locations)

Yuji Kuramoto
Manager, Revamp Projects
Mitsubishi Heavy Industries Compressor International Corporation

Satoru Yoshida
Application Engineering Manager
Mitsubishi Heavy Industries Compressor International Corporation

Clayton Jurica
Sales Manager
Mitsubishi Heavy Industries Compressor International Corporation

Bryan Barrington
Senior Advisor - Machinery Engineering
LyondellBasell

Todd Monroe
Machinery Consulting Engineer
LyondellBasell

Prepared for Presentation at the 2018 Spring National Meeting
Orlando, Florida, April 23-26, 2018

Modernization of Existing Ethylene Crackers (Compressor & Turbine Footprint Replacement Options in Brownfield Locations)

Yuji Kuramoto
Manager, Revamp Projects
Mitsubishi Heavy Industries Compressor International Corporation

Satoru Yoshida
Application Engineering Manager
Mitsubishi Heavy Industries Compressor International Corporation

Clayton Jurica
Sales Manager
Mitsubishi Heavy Industries Compressor International Corporation

Bryan Barrington
Senior Advisor - Machinery Engineering
LyondellBasell

Todd Monroe
Machinery Consulting Engineer
LyondellBasell

Abstract:
The recent drop in oil prices has had a negative impact on the Oil, Gas & Petrochemical industry and has emphasized the need for modernization and debottlenecking to reach ever higher levels of plant utilization and economies of scale. With that said, Turbomachinery manufacturers have been focusing on the application of state-of-the-art technologies to bring higher efficiency and reliability to the equipment market.

Traditionally, there are two options for compressor and/or turbine modification for plant revamps: one is re-rotoring of compressor and/or turbine; and the other is footprint replacement of the existing equipment. In the footprint replacement approach, the existing compressor / turbine can be replaced entirely to offer a more flexible and suitable design to maximize the plant capacity. The challenge for footprint replacements of equipment is matching the nozzle arrangement and physical footprint of the existing machine to avoid additional site work (modification of existing foundation, etc), extension of expected turnaround duration, and higher investment cost.

This paper presents MCO (Mitsubishi Heavy Industries Compressor Corp)'s latest footprint replacement technology to minimize facility modification by utilizing the same footprint and without extending the outage duration. Actual footprint replacement experiences are also referenced as a case study.

1. Introduction

Compressor & steam turbine installations in existing ethylene crackers are re-rotored or replaced, there are three reasons as follows:

(1) Capacity Upgrading --> To increase plant capacity
(2) Energy Saving --> To improve OPEX
(3) Plant Renewal --> To improve OPEX & Reliability

This paper focuses on the footprint replacement option to achieve the above requirements.

Figure.1 shows approaches for plant capacity upgrading; (1) changing process parameters, (2) revamping with internal remodeling, (3) equipment replacement or (4) establishing additional equipment in parallel.

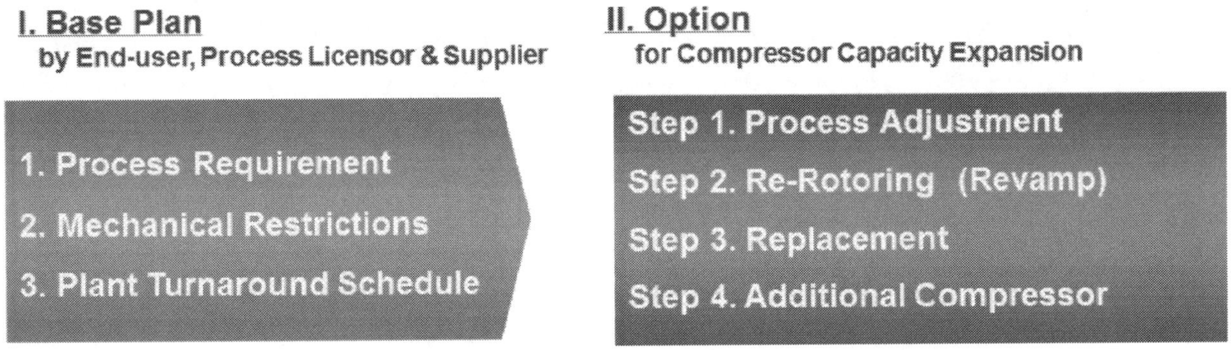

Figure.1 Basic Approaches for Plant Upgrading

As a first step, the end user should consider whether or not process improvements alone can satisfy their rerate targets. If so, this can be accomplished without equipment modifications. If process adjustments alone do not satisfy the requirements, the next step in this process is to consider the modification options of existing compressor & driver: one is re-rotoring the existing equipment; and the other is equipment replacement. When these modifications of the existing equipment will not achieve the plant target, the end user will need to consider parallel train applications.

2. Option for Plant Upgrading

The best approach adopted for plant capacity upgrading is determined in consideration of the mechanical limitation and the achievable increasing capacity rate with a budget accounting. The feature of each approach is summarized in Table.1.

"Changing process parameters" approach doesn't require remodeling of the existing equipment. The operating condition is limited by the existing design condition, and the capacity upgrading is small (less than 10%).

In "Revamping with internal remodeling" approach, an existing compressor casing is re-used. Only the internal parts (impeller and diaphragm, etc.) are revamped with new ones for the required larger flow. In this approach, even though there is an advantage that the existing casing can be re-used, capacity increase is limited to only 20 ~ 30%, and it is difficult to select the optimum impeller for the operating conditions because neither the impeller diameter nor bearing span of the existing compressor casing can be changed.

"Equipment Replacement" is the approach where an existing complete compressor assembly is replaced with a new larger compressor, and significantly increased production can be achieved up to +100% and optimum impeller selection with high efficiency can be achieved. The feature of footprint replacement is explained in detail in section 3.

"Parallel Train" approach establishes the need for a new compressor train. In this approach, the limitation is just site space for new construction, but this is also the highest cost approach.

In this paper, the focus is on the "Equipment Replacement" option.

Approach	Modification	Limitation
(1) Changing process parameters	Suction Pressure, Rotating Speed	-Design Pressure, -MCS, -Max. Output Power of Driver
(2) **Revamping** with internal modifications	Compressor Internal Parts (Impeller, Diaphragm, etc.)	Internal Diameter and Bearing Span of Existing Compressor
(3) **Equipment Replacement**	Compressor	Size of Existing Foundation, etc.
(4) **Parallel Train**	N/A New Construction	ex. Site Space for New Construction

Table.1 Feature of Approach for Plant Capacity Upgrading

3. Equipment Replacement for plant modification

For plant modification, equipment replacement is perceived as one of the preferred modification options, because of the significant benefit.

In terms of performance, "Equipment Replacement" allows for the selection of high efficiency impellers that best meet the required operation conditions. Performance improvement is greater with "Equipment Replacement" than "Revamping with internal remodeling" because impeller diameters & bearing spans can be changed and compatible with the new compressor casing.

In terms of reliability, the new replacement machines can be mechanically run-tested, and if desired, the machines performance can be verified with advance testing on the OEM's shop test stand.

With regard to the cost of "Equipment Replacement" vs. "Rerotoring", Table 2 shows a comparison of the initial investment cost at a 300 kiloton per annum ethylene plant. While it is assumed that the CAPEX for "Equipment Replacement" is higher than that for Rerotoring", what is less well-known fact is the overall total cost difference is not so large – on the order of 18% in our example. Furthermore, a 40% capacity increase could be achieved at this plant by replacement, whereas rerotoring would have only rendered a 20% capacity increase. So the initial investment cost difference may be offset by the higher profit differential generated by the higher production increase, as evidenced by the case at hand (300 KTA plant) where "Equipment Replacement" increased production by 120 kiloton per annum.

With regards to performance, the high efficiency impeller can be selected for required operating condition by "Equipment Replacement". Performance improvement by "Equipment replacement" is Revamping with internal remodeling" because impeller diameter & bearing span can be changed with new compressor casing.

Reliability is increased since-replaced new machines can be mechanically verified, and also, if required, machine's performance can be also verified at shop test in advance.

In terms of the cost side of revamping versus replacement, Table.2 shows a comparison of initial investment cost based on 300 kiloton per annum ethylene plant. It is very evident that equipment cost for replacement is higher than that of revamping. However, it is less evident that the total cost difference is not so large - only 18%. On the other hand, a 40% capacity increase could be achieved by replacement, while only a 20% capacity increase is achieved by revamping. So the initial investment cost difference may be able to absorb by the profit difference from the production increase between the two cases. For the case at hand (300 KTA plant), production increase by replacement is 120 kiloton per annum.

Example of Initial Investment Cost (300 KTA Ethylene Plant)

Item (Increasing Ratio)	Revamping (+20%)	Replacement (+40%)
Planning / Engineering	18 %	18 %
Equipment	70 %	86 %
Compressor & Driver	(20 %)	(28 %)
Furnace, Others	(50 %)	(58 %)
Field Installation / Start-up Work	12 %	14 %
Total Initial Investment Cost	**100 %**	**118 %**

Production Increase by Replacement
= 300 KTA x 0.4 = +120 KTA

Table.2 Comparison of Initial Investment Cost

Additional key factors to be considered in determining the feasibility of machine footprint replacement are minimizing installation costs and minimizing the turn-around period.

It is important that the new larger compressor be custom designed to allow the casing to be set on the existing foundation and to have the nozzles to fit the hole openings of the existing foundation. When a standard design compressor is applied to a replacement project, the nozzle orientations often will not match the existing main gas pipe positions and/or will not fit on the narrow openings of the existing foundation. Modification of foundations significantly increases the difficulty, time, and cost of the installation.

4. Technical Approach for Footprint Replacement

Figure.2 shows the equipment replacement for a plant capacity upgrade with new compressors. Though the new compressor size was much larger than existing, compressors were installed on the same footprint without foundation modification because of an OEM's custom-designed compressor. In this section, footprint replacement technology is introduced.

The custom-designed compressor made of a high quality casting enables the new compressor nozzle positions to fit on the existing process piping positions because MCO has foundry shop in their factory.

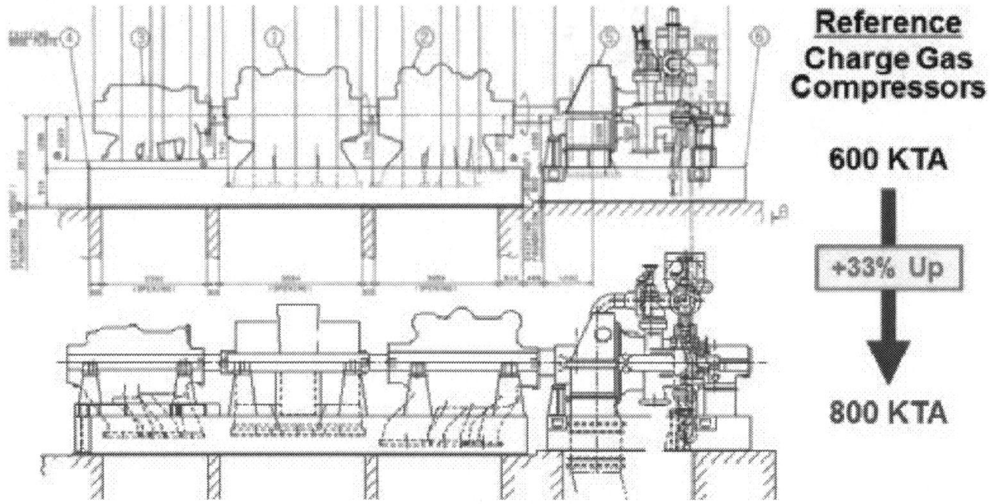

Figure.2 Replacement for Plant Capacity Increase

Also, for compact design, the new compressor is designed for as high of speed as possible to fit in the existing foundation. In order to increase the rotating speed for compact designs, the shaft rigidity needs to be higher for low vibration response and high stability. In addition, the bearing span needs to be shorter while the same number of stages remains the same. To realize both requirements, a new impeller standard series was developed and is in service in operational machines. Figure.3 shows the latest impeller design concept with high boss ratio and short span for high speed & compact selection.

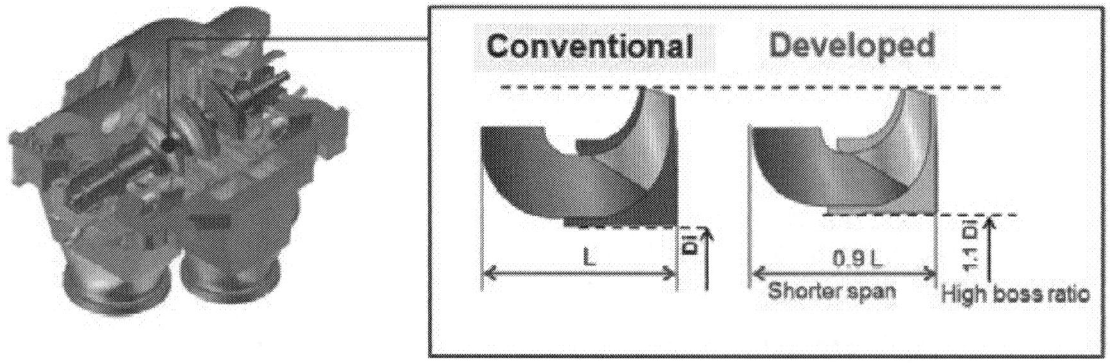

Figure.3 High Boss Ratio/Short Span Impeller Design

Therefore custom-designed compressor can remove the high site modification risk especially for the foundation and minimize the site work cost and period.

5. Performance Improvement by Footprint Replacement

The state-of-the-art technology is applied into the replaced new compressors, and higher efficiency and higher reliability are achieved and can be verified at shop tests in advance. For instance, one of those is a three-dimensional impeller.

Especially for replacement of existing old compressors with two-dimensional impellers manufactured before the first half of 1980's, its polytrophic efficiency can be raised on the average 5-10% by replacing it with new compressors that adopt higher efficiency three-dimension impellers. According to circumstance, any modification of an existing steam turbine may not be necessary as a result of this higher efficiency new compressor, even though new compressor handles higher flow than before. This higher efficiency three-dimensional impeller has been adopted at all stages of all new compressors manufactured, and is a contributing factor to a power decrease of the compressor.

The effect that the amount of a power decrease contributes to the plant utility cost and the effect that the CO2 emission gives to the global environment are demonstrated based on the example of a past replacement project. Figure.4 shows a case study of performance improvement. This case is the energy saving replacement project of charge gas compressor train in ethylene plant executed in 1991 (3 compressors, 5 sections, and steam turbine driven).

Thus, replacement of existing old-style compressors to advanced compressors will bring benefits as follows:

- Firstly, utility consumption as per plant capacity can be reduced and plant owner will save money by reducing OPEX of the plant.
- Secondly, decrease of CO2 emission can be achieved from utility consumption reduction. CO2 reduction will contribute not only to advancement of the company's brand image but also cash flow improvement with carbon credit.

	Existing	After Replacement	Improvement
Plant Capacity	300 KTA	300 KTA (with high efficiency 3-D impeller)	
Pressure	1.4 → 37.2 bara	1.4 → 37.2 bara	-
Measured Horsepower	25,516 kW @ 5,765 rpm	21,908 kW @ 5,634 rpm	3,608 kW Reduced (14% Reduced)
Steam Consumption (100 barg)	93 ton/hr	79 ton/hr	14 ton/hr Reduced (15% Reduced)

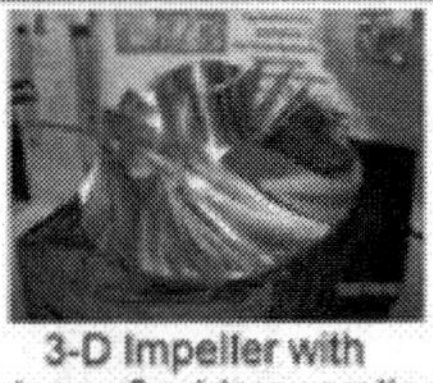
3-D Impeller with high efficiency & wide operating range

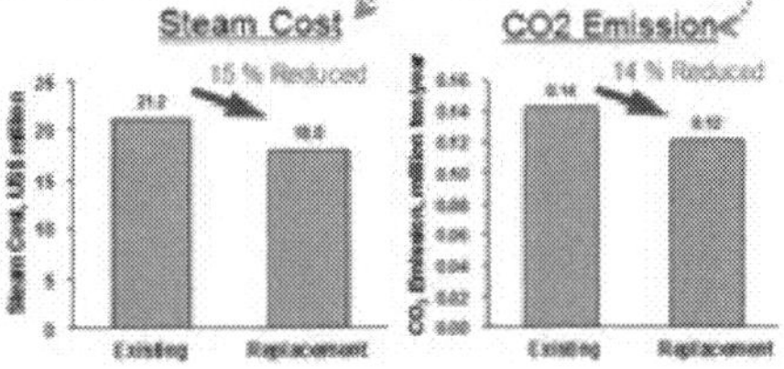

Figure.4 Performance Improvement by footprint replacement

6. Field Erection Work Schedule

To shorten the turn-around period, re-use of existing compressor base-plate is considered as much as possible.

For more detail investigation, an OEM can provide a site work schedule for each project during bidding stage during bidding stage and inform end users of required site work period and man-power for end user's information as shown in Figure.5.

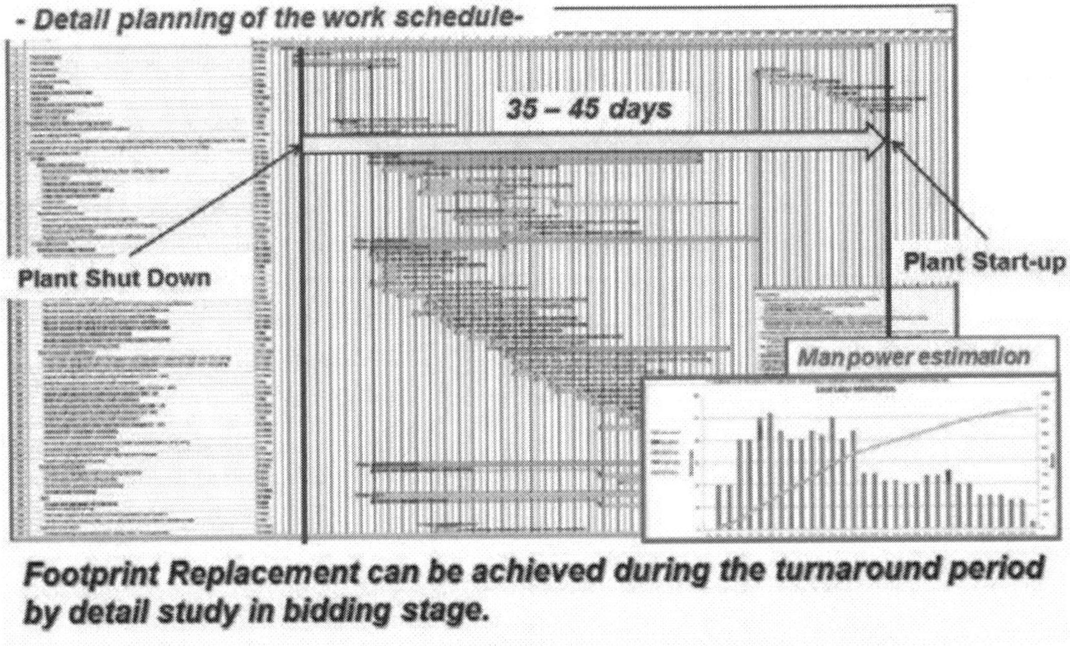

Figure.5 Sample Field Work Schedule

7. Case Study for Footprint Replacement

Figure.6 shows the two Case Studies of footprint replacement. One is Propylene Ref compressor train in La Porte, TX and the other is Ethylene Ref compressor train for Corpus Christi, TX.

Fig.6 shows the replacement project in La Porte, and that was plant capacity upgrading project from 875 kTA to 1,227 kTA (+40% increase) which meant upgrade to two new steam turbines, new steam turbine internals for a third turbine and one new compressor for this plant.

The compressor and steam turbines were replaced on the same footprint. All compressor nozzles fit in existing foundation openings due to the custom-designed compressor though 1st inlet nozzle size had to be increased due to the inlet flow increase. The exhaust duct centerline of new steam turbine was the same as existing and the surface condenser was re-used by apply a new connecting duct between the steam turbine exhaust flange and the existing condenser inlet flange. Baseplates were also replaced with new ones and existing foundation bolt locations for the steam turbine baseplate were re-used. At the customer's request, new anchor bolts were cored through the existing foundation as this was a better fit for the new compressor baseplate.

The existing foundation could be re-used with limited modifications due to the custom-designed compressor and steam turbines.

For this project, end user compared compressor replacement vs compressor re-rotoring.

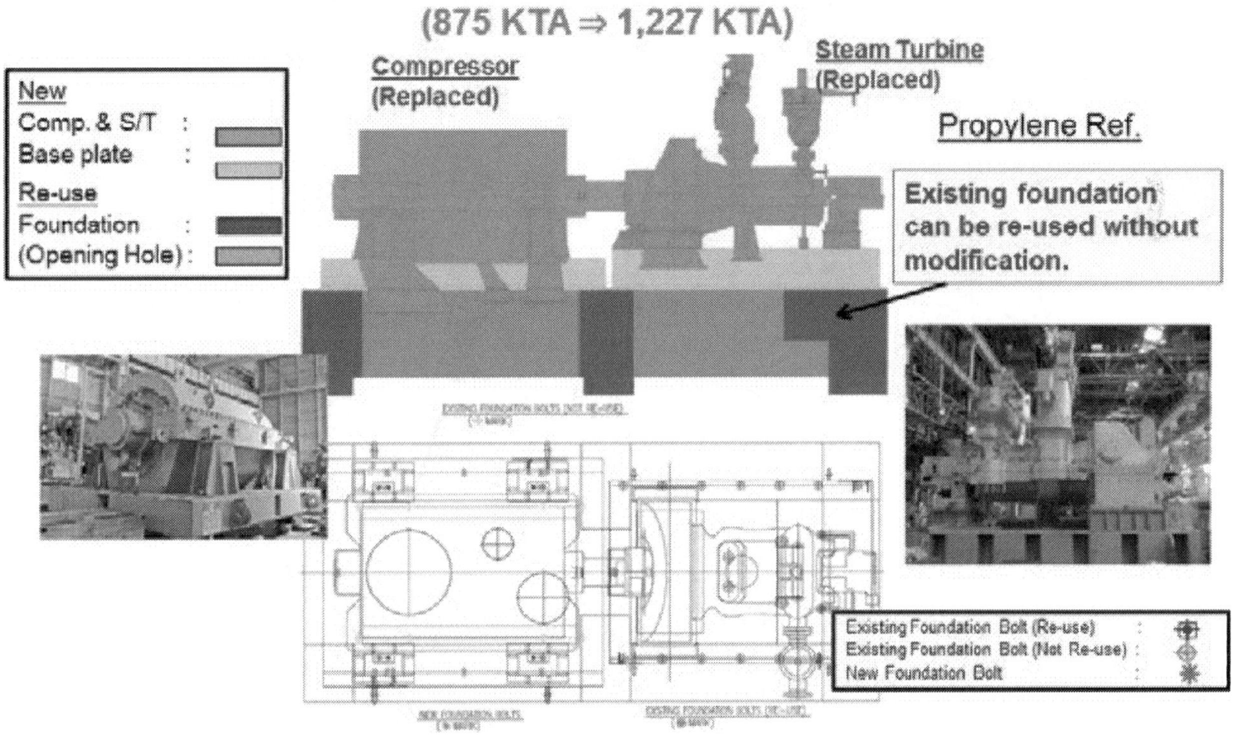

	Existing	Replacement	Re-rotoring
Plant Capacity	875 KTA	1,227 KTA	1,227 KTA
Certainty of Performance	-	High (verified at shop)	Medium (verified only through comparable machines and select shop tests)
Nozzle (Nozzle Flow Velocity)	1st: 48" (Within Criteria)	1st: 60" (Within Criteria)	1st: 48" (Exceeded Criteria)
Foundation	-	Re-Used (Minimal Modification)	Re-Used (No Modification)
Process Condition	-	Inlet Flow Increase: +80% Comp Power Increase: +60%	
Study Result		Applied	Not Feasible

Figure.6 Case Study1 – Footprint Replacement in US & Comparison Table between replacement and re-rotoring

As a result, the end user decided to replace the compressor because the nozzle size had to be increased. The new compressor had a higher efficiency (Improvement of OPEX), a higher certainty of performance, and minimal foundation modifications with essentially the same (or less) manhours for installation of the custom-designed compressor.

These new turbines and compressors were commissioned successfully and per the customer they meet or exceed the performance requirements.

Fig.7 shows the replacement project in Corpus Christi where that upgrade project increased plant capacity from 771 kTA to 1,100 kTA (+43% increase). In order to meeting this requirement, four new steam turbines and two new compressors were supplied for this plant.

The compressors and steam turbines were replaced on the same footprint. All compressor nozzles fit in the existing foundation openings and kept existing nozzle center line due to the custom-designed compressor, though one nozzle size (SS-IN) had to be increased due to flow increases. The turbine exhaust nozzle size was increased, but fit in the existing foundation opening. Common baseplate was also supplied that was nearly the same as the existing baseplate.

The existing foundation was re-used with no modifications (other than core drilling for anchor bolts) due to the custom-designed compressor and steam turbine.

For this project, end user compared compressor replacement vs compressor re-rotoring.

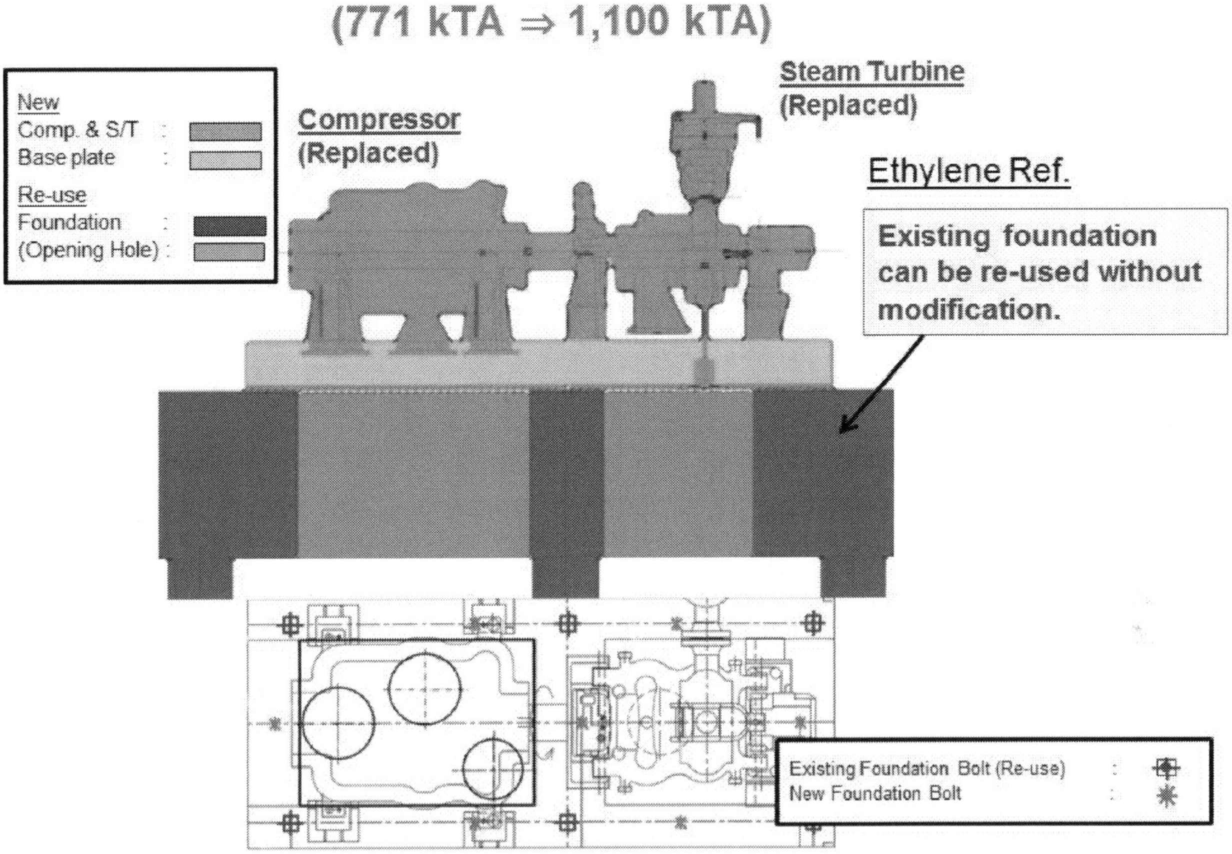

	Existing	Replacement	Re-rotoring
Plant Capacity	771 KTA	1,100 KTA	1,100 KTA
Certainty of Performance	-	High (verified at shop)	Medium (verified only through comparable machines and select shop tests)
Nozzle (Nozzle Flow Velocity)	SS-In: 12″ (WithinCriteria)	SS-In: 20″ (Within Criteria)	SS-In: 12″ (Upper limit of Criteria)
Foundation	-	Re-Used (No Modification)	Re-Used (No Modification)
Process Condition	-	Inlet Flow Increase: +23% SS-IN Flow Increase: +89% Comp Power Increase: +52%	
Study Result		Applied	Not Optimal

Figure.7 Case Study2 – Footprint Replacement in US
& Comparison Table between replacement and re-rotoring

As a result the end user decided to install a new compressor instead of re-rotoring because the new compressor provided future flexibility in altering loads on the compressor, had a higher efficiency (Improvement of OPEX), a higher certainty of performance, and minimal foundation modifications with essentially the same (or less) manhours for installation of the custom-designed compressor.

These new turbines and compressors were commissioned successfully and per the customer they meet or exceed the performance requirements.

Conclusion

For modernization (Capacity upgrading, performance improvement, etc) of existing ethylene crackers, footprint replacement is a proven method that can be achieved at a competitive cost and in a short duration. MCO makes good use of the technology that has been introduced in this paper, and offers not only the rotating machine replacements suitable for existing foundations but also achieves low OPEX and low CO2 emissions. Each Ethylene plant, however, has a different demand for plant modernization. An OEM however has to study the existing plant conditions for each project individually (including feasibility studies at very early project states) and suggest the best solution to meet each end user's requirements.

513176 The Soft-Side of the Control Valve, a New Approach for Control Valve Diagnostics

Abstract:

Control valves in the ethylene production process and its associated rotating equipment play a critical role as most of the variables are controlled by these throttling elements which are subject to wear and, in most cases are operating under continuous stress, changes in the control valve behavior will impact different aspects of the plant operation including; efficiency and safety.

MITIGATING FOULING in CRACKED GAS COMPRESSORS

Authors: Anahita Khanlari Ph.D.[1], Debby M. Rossana Ph.D.[2], Theodor C. Arnst Ph.D.[1]

1. Nalco Champion Research, Development, and Engineering (RD&E), 7705 Highway 90-A, Sugar Land, TX, USA 77478
2. Nalco Champion, North American Sales, 11177 S. Stadium Dr., Sugar Land, Texas 77478

Prepared for Presentation at the Year (2018)

Spring National Meeting, Orlando, FL, (April 23-26)

MITIGATING FOULING in CRACKED GAS COMPRESSORS

Authors: Anahita Khanlari Ph.D., Debby M. Rossana Ph.D., Theodor C. Arnst Ph.D.

Abstract:

Cracked gas compressors (CGC) in ethylene units are subject to fouling. Depending on the severity of cracking and feedstock, the extent of fouling in compressors varies. A correct combination of wash oil and treatment programs can maximize unit efficiency and gas throughput to make desirable run lengths of 6 years and more possible. The role of operational variables (feed, temperature, oxygen intrusion, water injection, and wash oil) have been discussed in previous papers. Over the past several years these operational variables have changed and increased fouling in compressors has been observed. With increased fouling comes a need to better understand the role of antifoulants (i.e. anti-polymerant, anti-oxidant, dispersants) to hinder or mitigate fouling and to protect the machine. This paper discusses some of the laboratory work done to develop the next generation of chemical treatment programs.

Keywords: Cracked gas compressor, fouling, treatment program

1 Introduction

Fouling in an ethylene unit is synonymous to increased power consumption, reduced throughput, increased vibrations, and ultimately reduced ethylene yield. As secondary effects, fouling leads to mechanical damage to equipment either due to vibrations or under-deposit corrosion and in severe cases unscheduled plant shut-downs. One ethylene licensor estimates that for an ethylene plant with a production capacity of 800,000 Tonne/year (1.76 billion lb/year), every 1% lost in efficiency can cost as much as $200,000 per year in energy consumption. Based on two surveys that Nalco Champion conducted globally in 2005 and in 2015, more producers are relying on a continuous treatment program in recent years (Figure 1). The percentage of treated CGC's has increased from 32 to 43%. With target run-lengths of 8-10 years while maintaining profitability and preventing un-expected shut-downs and maintenance, the need for a reliable continuous CGC treatment is further emphasized.

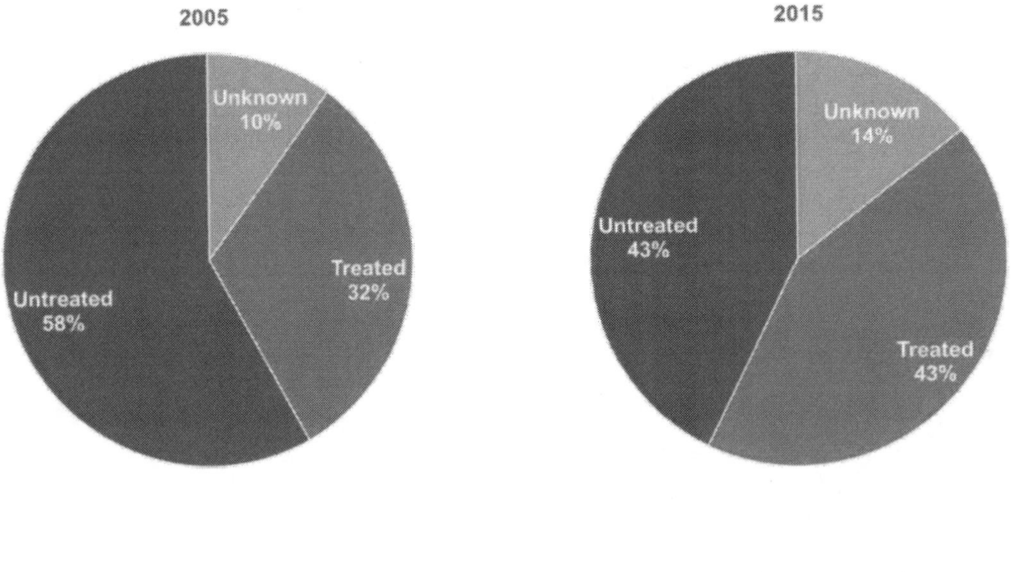

Figure 1- Percentage of CGC's treated globally in 2005 vs. 2015 [1].

The ethylene industry relies on a combination of treatment and mitigation techniques to address the fouling issues. These techniques range from lowering the severity of cracking (the ratio of propylene to ethylene), lowering the outlet temperature of the machine by using attemperation water, coating internal parts of the machine, and using a proper wash oil. If not chosen properly, each mitigation option, by itself, can result in undesirable effects. For example, when injecting attemperation water, poor water quality can introduce salts, oxygen and other impurities and free water can increase corrosion and erosion in the system. Developing a holistic strategy to treat fouling in the compression train can significantly add to the run-length of the unit and ensure smooth operation. In this paper, after reviewing the underlying mechanisms of fouling in CGCs, we describe how to scrutinize and choose a treatment program that is compatible with unit operations. Then, individual performance criterium such as: reactivity, effect on emulsion and the dispersion power are discussed and examples are provided.

2 General considerations in operation of the cracked gas compressors

2.1 Chemical Mechanisms of fouling formation

Production of polymers that cause fouling is dependent upon concentration of reactive monomers, concentration of initiators, temperature and time. Reactive monomer (Figure 2) concentration is a function of feedstock and cracking severity, both of which are dictated by plant economics. Concentration of initiators is very small and typically beyond control. Temperature is a function of compressor operating rates and conditions along with interstage cooling water conditions; these variables are optimized as best as possible. Over the years, multiple papers have been written that discuss fouling mechanisms occurring in process gas compressors [2]. Most agree on three mechanisms; free radical polymerization, Diels-Alder condensation, and the thermal degradation to coke-like solids.

1,3-Butadiene
(B.P. @ -4C)

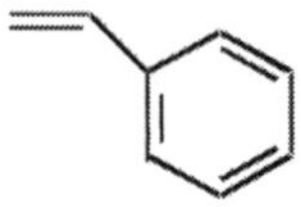
Styrene
(B.P. @ 145C)

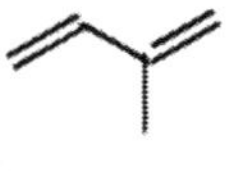
Isoprene
(B.P. @ 34C)

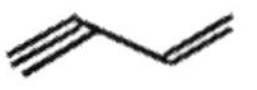
Vinyl acetylene

Fig 2- Example of reactive monomers that undergo free-radical polymerization

2.2 Wash Oil

Wash oil is an important part of a compressor treatment program. Wash oil acts as a solvent for low molecular weight polymers that can form in the compressor and interstage coolers. It also provides some level of physical scouring of exposed surfaces on the compressor. It forms a thin film on the metal surfaces which helps deter attack by corrosive species. Finally, it is an excellent carrier for treatment chemicals for even distribution throughout the machine. Wash oil quality is critically important. Low quality wash oil not only will perform poorly as a solvent for polymeric deposits, it may contribute to fouling. We recommend reviewing the following guidelines to producers as what their wash oil quality should be:

- High solvency for butadiene and styrene polymers
- Low volatility
    - Ensure scouring and polymer solvency
    - Minimize re-deposition of entrained solids
- Low soluble polymer / suspended solids content
    - Distilled product is best
- Low reactive monomer (BD, styrene, etc.) content

As the industry has evolved, the importance of an adequate wash oil monitoring program has become even more imperative [3]. Over time we've noticed an increase in wash-oil testing demands, a decrease in wash-oil supply (due to excess gas cracking), and an increase in cost. Due to high cost and low availability of wash oil, some producers have turned to attemperation water to control fouling in their operation of CGC.

2.3 Attemperation Water

Water injection is another attempt by ethylene producers to minimize fouling [4]. When a proper amount of water is injected, it evaporates in the compressor and due to the heat of vaporization, lowers the compressor discharge temperature. As polymerization is a function of temperature, a lower temperature means lower polymerization and fouling rates.

2.4 Oxygen:

Presence of oxygen promotes fouling and corrosion in compressors. Oxygen centered radicals form when cracking products react with free oxygen in the environment. Oxygen can enter the system either through direct intrusion or via water or recycle streams. If possible, it is best practice to avoid routing recycle streams that contain oxygen to the compressor. Even parts per million of oxygen or peroxides in the process gas can greatly increase polymerization rates. If recycling a stream to the compressor cannot be avoided, consideration of a compressor antifoulant to mitigate the effect oxygen has on fouling rates can benefit the system and the run length.

In an example to simulate the effect of oxygen on the soluble polymer number (*i.e.* gums) of hydrotreater feed and de-butanizer bottoms, we designed an experiment [5]. A sample of the above-mentioned streams was obtained and divided into 3 individual samples. First, the existing gums content of the original sample was measured. In a second experiment, the sample was heated at 100 °C under nitrogen blanket for 1 hr. Finally, the third sample was heated to 100 °C this time under an air blanket. Figure 2 shows the measured gum contents in each case. The samples that were heated while exposed to oxygen had the highest amount of gum formation, several thousand times higher than the original sample or the one heated under nitrogen. *Note that the levels of oxygen in this test far exceed those found in a compressor.*

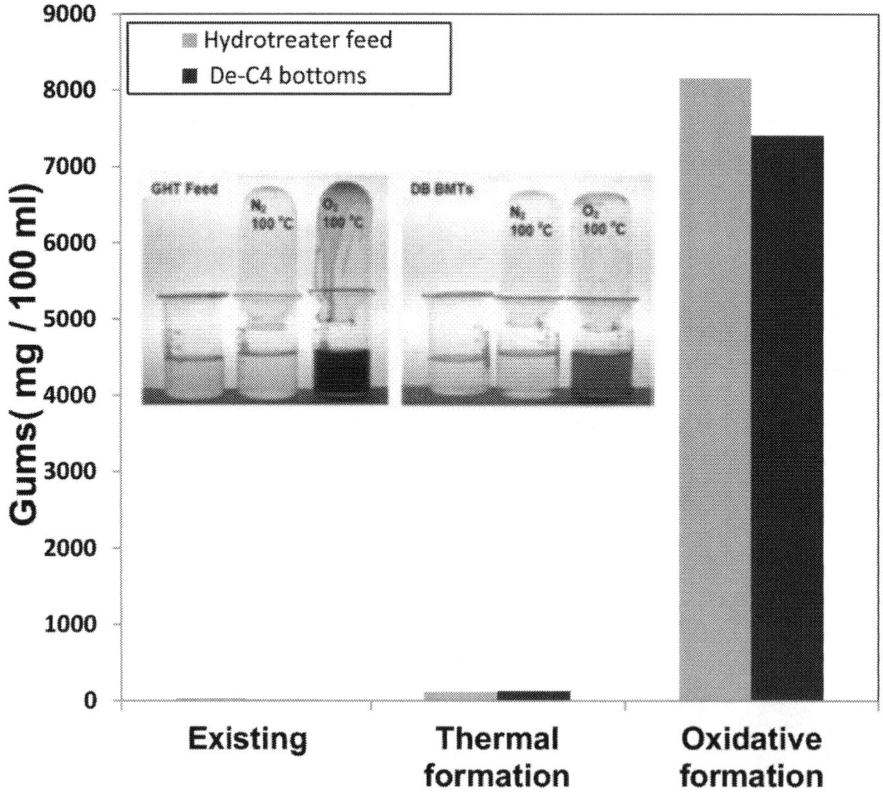

Figure 2- The effect of presence of absence of oxygen on increasing the soluble polymer number in hydrotreater feed and de-butanizer bottoms [5].

3 Key parameters of designing a reliable chemical treatment program

3.1 Dispersant:

Surface active agents (surfactants) used in controlling fouling in process gas compressors are organic molecules with building blocks of polar/non-polar (less polar) origin. The differences in charge in different sections of the molecule leads to different electrostatic profile and hence different interactions with the foulant. The polar heads have a higher affinity for polar sections of the foulant and non-polar heads for non-polar sections. Polar sections of the dispersant can physically bind to oxidated species on the foulant. Non-polar sections of the dispersant can interact with aliphatic/aromatic chains of the foulant via van der Waals interactions; ultimately resulting in solvating these chains. Here, it should be emphasized that there are no chemical reactions between the foulant and dispersant, *i.e.* no covalent bonding. With the help of wash oil (or the solvent present in the antifoulant program), dispersants can effectively pull deposited foulant into solvent/wash and remove it from the machine. When used continuously (and assuming an ideal distribution on the surface), the surfactants can effectively prevent deposition of foulant in the compressor. Figure 3 shows how an effective surfactant can suspend the majority of a foulant and at the same time control the particle size distribution. The third sample (from left) has the most dispersed particulates with ultra-fine size distribution. When developing a treatment program, the finer the dispersion the more desirable.

Figure 3- Three test tubes with different dispersants at equal dosage (2000 ppm) disperse a CGC foulant in a non-solvent environment. The foulant was obtained from a North American ethylene producer's machine (3rd stage).

3.2 Onset of crosslinking (solid foulant formation) in a py-gas mixture

Polymerization reactions in cracked gas compressors take place in liquid phase and on the metal surface with the help of higher pressures and temperatures of the environment. Presence of free radicals, freely moving monomers, and conjugated monomers (*e.g.* isoprene and butadiene) provide an ideal medium for formation of long polymer chains. The chains can then get crosslinked and form solid seeds that now can accelerate the fouling process. Understanding this mechanism and finding most effective species to battle the onset of solid formation is the key to design a truly effective antifoulant program.

Thereby, we designed an experiment to mimic reactions and molecular interactions that lead to the onset of foulant formation inside the compressors and intercoolers over several years. A simulated-pygas mixture was used to perform this experiment. The mixture consisted of conjugated diene, a vinyl monomer, dimer of cyclopentadiene and a crosslinker. The experiment was performed at 130 °C in a temperature controlled oven environment. Each tube was dosed with 500 ppm of the studied chemical. Then, each tube was purged with 100% nitrogen for 1 minute and immediately capped. Each formulation was done at minimum in triplicates and the results were averaged. After placing the tubes in the oven, time to gelation was measured. After the first tube gelled, the samples were checked every 5 minutes (error of ±5 min). The gelation time was determined visually and recorded when the liquid became viscous and no longer freely flowed in the tube.

Table 1- Composition of samples (S) tested to study individual component's effect on hindering solid polymer formation

	S1	S2	S3	S4	S5	S6	S7
Antioxidant (AO)			×	×	×		×
Antipolymeranat (AP)			×		×	×	×
Dispersant type I (D1)		×		×		×	×
Dispersant type II (D2)					×		
Solvent (S)		×	×	×	×	×	×

In this experiment, each formulation was designed so the effect of each component in hindering polymerization could be independently verified. The components of each formulation are listed in Table 1. The time to gelation would reflect the effectiveness of the formulation to prevent free radical and condensation reactions. The longer it would take for the mixture to polymerize, the more effective the antifoulant program. As shown in Figure 4, the shortest time to gelation belonged to S1 (control group). The addition of dispersant, antipolymerant, and antioxidant increased the time to gelation as shown by S2, S3, and S4 groups. The difference between S5 and S7 directly is related to the dispersant type (type II vs. type I). The longest time to gel (170 min) belonged to the S7 case. Sample S7 demonstrates the cumulative effects of anti-polymerant, anti-oxidant, and dispersant. This result verifies the existence of certain synergy among not only the antioxidant and antipolymerant components, but also the particular dispersant (type I) in the formulation. The effect of dispersant (type I) and the cumulative improvement on gelation time is the novelty of present work and has not been previously studied or used.

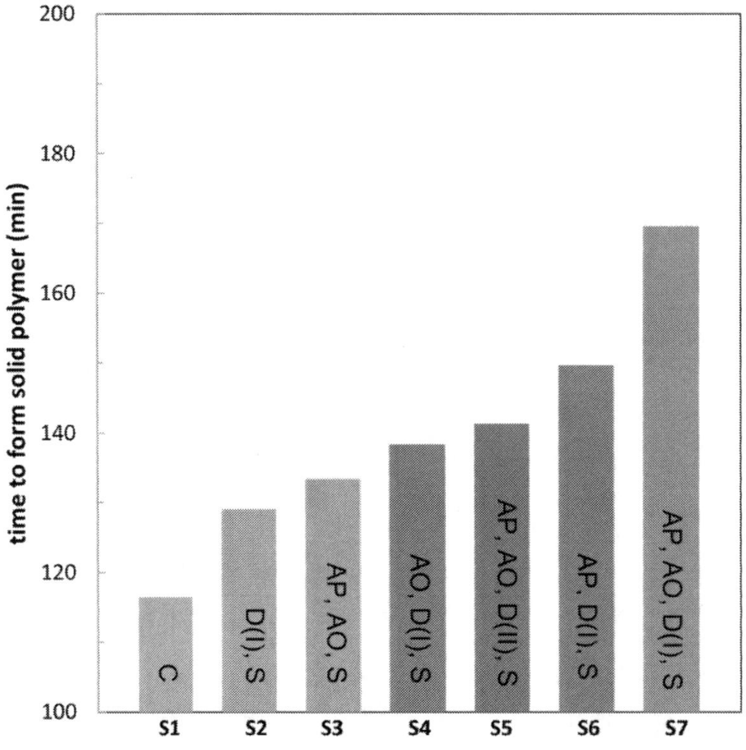

Figure 4 - The synergistic effects among S7 components resulted in delayed polymerization of the simulated pygas mixture. The reaction was carried out at 130 °C in sealed tubes. Each tube was purged with nitrogen for 1 minute and then tightly capped. In this experiment, the type of AP, AO, and S were kept the same to showcase the effects of presence or type of dispersant on fouling. S1 is the control group (simulated py-gas + crosslinker).

3.3 Emulsion

A variety of factors can lead to emulsion formation in compressor knock out drums and quench settlers. Mixing energy, residence time in the knockout drum or quench water settler and impurities that are knocked out into the drum are a few of the factors that affect emulsions. Some types of compressor antifoulants (containing surfactants) can also stabilize emulsions. Therefore, it is important to first conduct a holistic evaluation of a knockout drum emulsion before taking corrective measures. However, in this section, we show a few examples that only focus on the role that a surfactant (dispersant) can have on an emulsion.

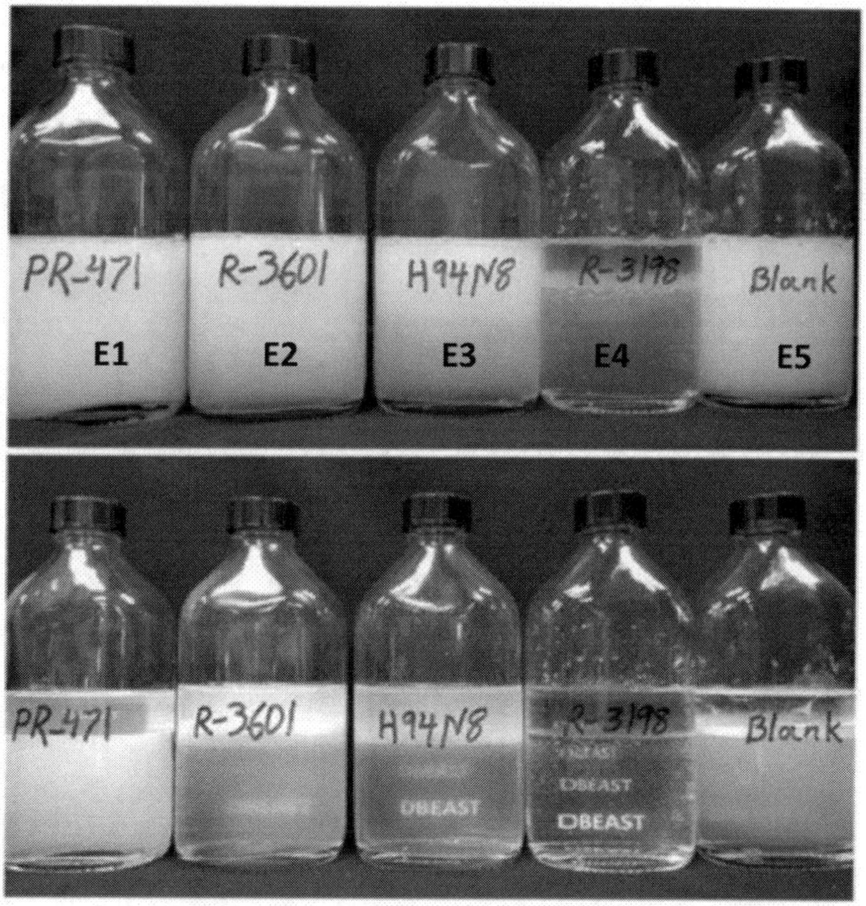

Figure 5- The role of dispersant in causing emulsion. In each case 100 ppm of dispersant was dosed to the mixture. Water (de-ionized)-pygas (simulated) were mixed at a ratio of 75:25. mixture. At t=0 and t=3 min the top and bottom photos were taken respectively. **Note**: the names on the jars do not represent any commercial products and are simply used for identification.

In this study, a mixture of de-ionized water:simulated py-gas (75:25) was used to provide a baseline for the emulsion (blank sample, E5). After dosing 100 ppm of dispersant in each case, the jars were shaken vigorously and left to settle. Photos were taken when the jars were left to settle (t=0) and 3 minutes later. After settling, E1 was emulsified in the water phase while E2 and E3 were mostly emulsified in the hydrocarbon phase. The clarity of water between E2 and E3 was different emphasizing the effect of dispersant type on defining emulsions. E4 got cleared in both phases almost immediately. The clarify of the water phase in E4 was significantly better than E5 (blank) implying that the surfactant in this case acts to break the emulsion.

4 Conclusions:

The emphasis of ethylene industry on more efficient and longer run-lengths drives new research on better chemistries to treat the compressors and intercoolers forward. After reviewing the main factors involved in fouling, based on research data, we showcased a new method to mimic fouling reactions. By breaking down the performance of the antifoulant to the performance of individual components, comprehensive programs can be designed to best suit a particular operation.

References:

1- S. Snider, Ethylene Plant Cracked Gas Compressor Fouling, *AIChE-EPC proceedings, Orlando*, FL, April 2006.

2- H Richter, J.B Howard, Formation of polycyclic aromatic hydrocarbons and their growth to soot—a review of chemical reaction pathways, *Progress in Energy and Combustion Science*, Volume 26, Issues 4–6, 2000, pp 565-608.

3- D Dutschmann, M Harrison, Y Yang, Compressor Wash Oil in a Light Feedslate Environment, *AIChE-EPC proceedings*, Houston, TX April 2016.

4- D G. Puchan, M A. Gerty Boiler Technology to Monitor and Control Corrosion in Ethylene Plant Steam Generation Process, *AIChE-EPC proceedings*, San Antonio, TX, March 2010.

5- J Hancock, Determination of Gums Loading on GHT Reactor, *Technical reports*, Nalco Champion, Sugar Land, TX, 2009.

AIChE Paper Number: 32a

NOx-polymer based fouling in cold box

Helene Launay, *R&D Manager, Dow Chemical Company*

Stefan Korf, *Principle R&D Scientist, Dow Chemical Company*

Mark Brayden, *R&D Fellow, Dow Chemical Company*

Pieter Vijfhuizen, *Production Engineer, Dow Chemical Company*

Bart Dervaux, *Analytical Specialist, Dow Chemical Company*

Ron Bassie, *Analytical Technologist, Dow Chemical Company*

Klaas van Gelder, *Reactive Chemicals EMEAI Leader, Dow Chemical Company*

Peter Feld, *Technology Center Leader, Dow Chemical Company*

Prepared for Presentation at the 2018 Spring National Meeting

Orlando, Florida, April 23, 2018

NOx-polymer based fouling in cold box

Helene Launay, *R&D Manager, Dow Chemical Company*

Stefan Korf, *Principle R&D Scientist, Dow Chemical Company*

Mark Brayden, *R&D Fellow, Dow Chemical Company*

Pieter Vijfhuizen, *Production Engineer, Dow Chemical Company*

Bart Dervaux, *Analytical Specialist, Dow Chemical Company*

Ron Bassie, *Analytical Technologist, Dow Chemical Company*

Klaas van Gelder, *Reactive Chemicals EMEAI Leader, Dow Chemical Company*

Peter Feld, *Technology Center Leader, Dow Chemical Company*

Abstract:
Serious fouling of a coldbox led to high pressure drop increase over the first cracked gas pass heat exchanger. Pressure drop eventually limited the plant capacity and production rate had to be reduced. Root cause investigations were initiated including extensive analysis of the fouling material. The main contributing root cause for the cold box plugging was identified as the co-cracking of a C4-recycle stream from the Butadiene plants containing NO containing inhibitor residuals leading to NO_x formation in the furnaces. These NO_x species are assumed to react in gas phase with dienes to form a non-explosive, non-shock sensitive NO_x-polymer material. The non-explosive, non-shock sensitive NO_x-polymer material found in the heat exchanger is a completely new phenomenon in Dow and in the ethylene industry, different from the known NO_x-gums formed in the low temperature zone of the cold box reported in Dow and ethylene industry. The intent of this paper is to share the current knowledge on the composition and mechanism of formation of such polymer and recommendations to understand its reactivity properties which is essential for the development of efficient and safe cleaning procedures.

Introduction

NO_x compounds that are formed or introduced into olefins units can represent a serious risk, due to the potential for formation of explosive nitrated gums and ammonium salts in the cold section of the unit. Several incidents, of different degrees of impact, have occurred in olefins units in the past years. Some of them have been reported in the EPC NO_x safety subcommittee like the incident that happened in the BASF Port Arthur naphtha cracker which was discovered to be caused by decomposition-triggered detonation of NO_x gums in a 1-inch drain line in the coldest portion of the cold box.[1]

Nitric Oxide (NO), which might be contained in the feed or formed during cracking of certain feedstocks, can pass unhindered through strippers and the pre-cooling section into the low temperature section due to its very low boiling point. [2] In presence of oxygen, nitric oxide converts to nitrogen dioxide (NO_2), whose equilibrium toward NO_2 is favored at low temperatures (below -110 °C) and high pressures. Low temperatures and high pressures also favor the reaction of NO_2 with the remaining NO to form di-nitrogen trioxide (N_2O_3), also referred to as "blue ice". NO_2 and N_2O_3 are less volatile than NO and freeze out and deposit typically in the lowest temperature sections in cold boxes. They can react with ammonia to form ammonium nitrite (NH_4NO_2) and ammonium nitrate salts (NH_4NO_3) and with olefinic and diolefinic hydrocarbons to form nitrated gums. NO_x salts and NO_x gums compounds are highly unstable and sensitive to both heat and impact. Safety hazards are especially introduced on temperature changes (warming-up): ammonium nitrite is known to decompose explosively at temperatures around 60 °C, with energy release equivalent to TNT. [2-5]

In 2013 a serious pressure drop increase over the first pass of the cold box occurred in one of the Dow crackers. Because of the impact on the plant capacity and production rate, the cold box had to be replaced. Sampling could then take place and a root cause investigation was initiated. The results of the fouling material analysis and of the investigations are described below, and how these results were used to determine cleaning methods to recover full functionality of the fouled cold box.

Fouling event description and location

The particular Dow cracker related to the event reported here is a front-end demethanizer cracker with a known history of NO_x salts formation in the cryogenic area of the plant. In 2013 a serious pressure drop increase was observed over the first pass of the cold box (Figure 1), which corresponds to the first heat exchanger located downstream of the cracked gas dryers (Figure 2). As depicted on Figure 2, excess liquids in the cracked gas are removed before water is adsorbed on the dryer guard beds. Downstream of the guard beds and upstream of the cold box, a high efficiency cartridge filter removes solids and liquids. This first pass heat exchanger of the cold box is a brazed aluminum plate-fin heat exchanger which cools down the cracked gas from around 15 °C to around 8 °C.

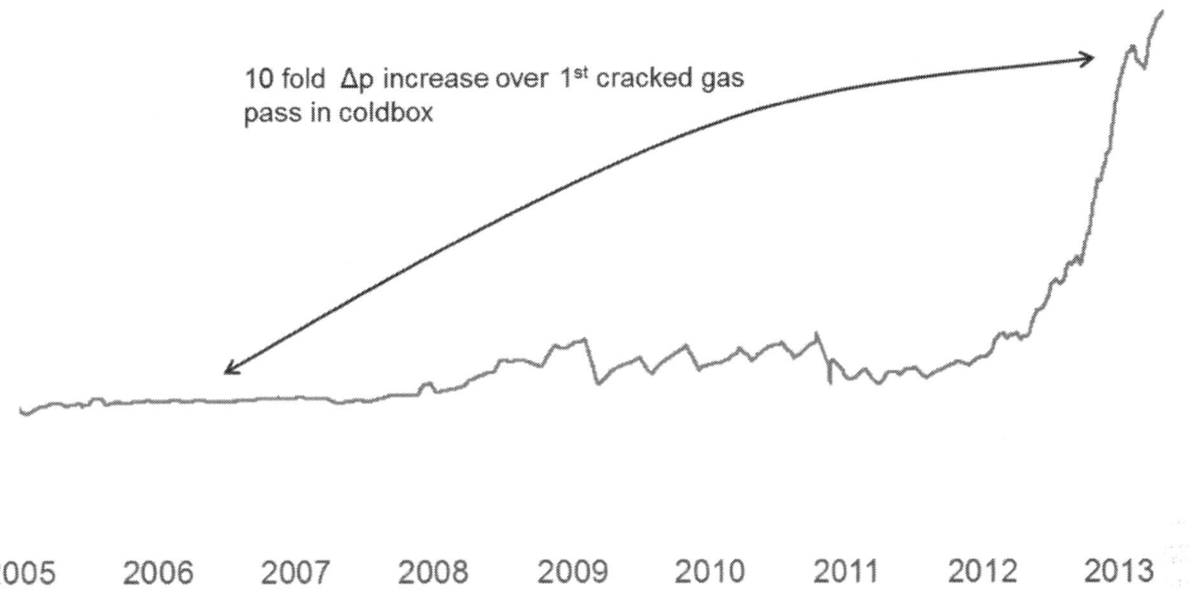

Figure 1: Observed flow corrected pressure drop over the first pass of the cold box.

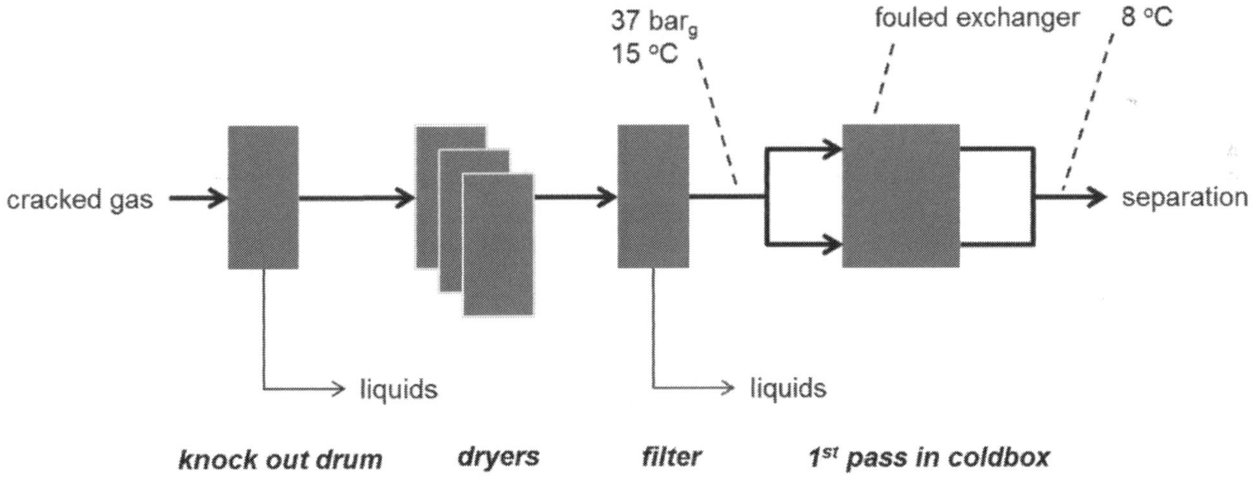

Figure 2: Plant lay-out

When the pressure drop limited the plant capacity, the cold box was removed from the process. Brown particles were found in the header to the heat exchanger and on the distributor fins and were sampled wherever possible for analysis by scraping the walls of the header and inlet pipes. Plugging by this brown material is clearly visible on the pictures of the distributor fins (Figure 3).

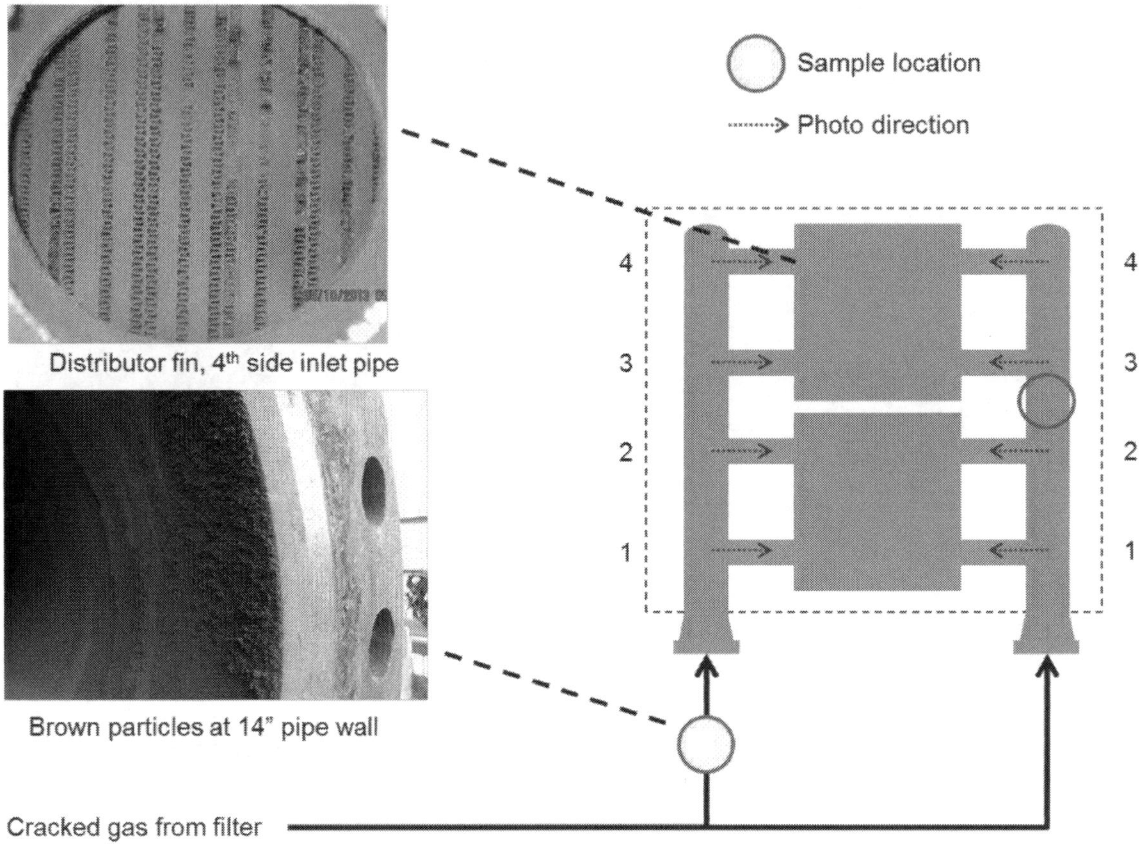

Figure 3: Schematic view of the heat exchanger with location of the fouling solid sampling and of the photos

Fouling material characterization

Analytical techniques

CHNSO

Approximately 5 mg of sample is packed in a silver foil and subsequently analyzed with a Vario EL Cube CHNS elemental analyzer with a thermal conductivity detector (TCD). The temperature of the combustion tube and reduction tube are at 1150 °C and 850 °C respectively. Calibration is performed with sulfanilamide. The procedure is performed in triple.

X-Ray Fluorescence (XRF)

The samples are dried in an oven at 75 °C and ground in a mortar prior to XRF analysis on a PANalytical Axios X-ray spectrometer (wavelength-dispersive) in P1 cup with 4 µm prolene film under helium atmosphere. Quantification is done with Uniquant. The data are semi quantitative, but give the correct order of magnitude. The detected elements (Na and higher) are listed as the oxides (this is an assumption; it was not measured) and expressed as a percentage of the total sample. Elements with a concentration below 0.02 wt% and sodium below 1000 mg/kg are not reported. The amount of organic rest is obtained from TGA.

Thermogravimetric Analysis (TGA)

The TGA experiments are performed via a TA Instruments model Q500 TGA with a purge gas flow (N_2 or Air) of 90 mL/min. Data are collected using the Thermal Advantage for Q series (version 2.7.0.380) software package and reduced using version 4.4A of the Universal Analysis 2000 software. The samples are placed on a Pt-pan accurately weighed by the calibrated TGA balance. The TGA oven is allowed to cool back to 90 °C before the purge gas is switched to air and the second heating run is performed. Details on the method used can be found in Table 1.

Table 1: TGA method

Temperature Scan Rate	10	°C/Min	Sample Container Material	Pt	Open
Minimum Test Temperature	20/90	°C	Sample Weight	60	mg
Maximum Test Temperature	750	°C			

Fourier Transform Infrared spectroscopy (FT-IR)

IR spectra are acquired using a diamond ATR accessory (single bounce) installed on a Nicolet 5700 spectrometer. 16 scans at 4 cm^{-1} resolution are co-added. Additionally spectra in transmission mode are acquired using a Hyperion IR microscope installed on a Vertex 70 spectrometer. In this case 64 scans at 4 cm-1 resolution are co-added.

Anionic Ion Chromatography (IC)

The deposits (25 mg) are extracted with H_2O (2 mL) or a KOH solution (0.1 M; 2 mL) by shaking the solutions for 5 minutes and incubation for 5 days without stirring. The solutions are filtered and the filtrates are diluted 50x, or 100x with demi water before IC analysis. Also a hot extraction with water (3.5 g sample in 15 mL H_2O) is performed by heating the sample for 8 hours at about 90 °C under reflux. Quantification is carried out with external standards in demineralized water. Analysis is performed on a Dionex Ion Chromatograph ICS-5000 with KOH eluent generator and conductivity detection. An IonPac AG11-HC + AS11-HC is used for separation.

Cationic Ion Chromatography (IC)

The deposits (35 mg) are extracted with H_2O (3 mL) by shaking the solutions overnight. The solutions are filtered and the filtrates are diluted with factor 10-20 with demin water before IC analysis. After extraction additionally several aliquots are spiked with known amounts of analytes of interest for peak verification and for recovery determination. Quantification is carried out with external standards in demin water. Analysis is performed on a Dionex Ion Chromatograph ICS-5000+ with methanesulfonic acid 1.5 mM as eluent and conductivity detection. An IonPac CG17 + CS17 is used for separation.

Inductive Coupled Plasma – Optical Emission Spectroscopy (ICP-OES)

The extraction solutions, prepared using the same method as for the Cationic Ion Chromatography analysis, are diluted with factor 10-20 with demin water prior to ICP analysis. All aliquots are measured by using a Thermo Scientific™ iCAP 7600 ICP-OES equipped with a spray chamber with baffle and burgener Mira Mist nebulizer. An external calibration is used for all elements of interest.

Pyrolysis-GC

The deposit is analyzed as it is. A few milligrams of sample is introduced into the pyrolysis module. The same sample is pyrolyzed first at 350 °C and a second pyrolysis is carried out at 600 °C. Experiments are carried out on an Agilent 7890 GC (Agilent Technologies), which is equipped with a Gerstel pyrolyis module (Gerstel PM-1) and Agilent 5975C MSD mass spectrometer. A HP-Ultra 2 column (50m x 0.32mm x0.52um) is used with He as carrier gas.

Solubility test

Solubility of the particles collected in the inlet pipes to the heat exchanger are tested in various solvents: methanol, toluene, acetone, chloroform. Hot extraction in water and methanol are also performed, as well as an extra test with methanol at 50 °C. In all tests, weight of non-dissolved particles are measured to determine the dissolution rate.

Shock sensitivity test or "hammer" test

An unofficial qualitative shock sensitivity test, or "hammer" test, is performed in which a small amount of sample is attached to the head of a hammer with tape and whacked against a solid object (like a concrete floor or tile). The tests is done in a dark area and visual and noise observations, that is, a flash or a "boom", are noted.

Differential Scanning Calorimetry (DSC)

DSC tests are conducted under nitrogen, oxygen or air with a temperature rate increase of 10 °C/min till 400 °C. An inert sample is used as a reference allowing the thermal effect to be measured and quantified. Glass capillaries or ampoules are used as a sample container. Both the capillaries and the ampoules are sealed before the test to obtain a closed system. This prevents evaporation of volatile components, an endothermic process that can interfere with exothermic behavior.

Thermal Screening Unit (TSU)

The TSU uses an ARC sphere as a sample container and the oven temperature is ramped up at a specific rate. During the run both the sample temperature and the pressure in the sample container are monitored. Because no reference is used and the test is not adiabatic, the thermal events cannot be quantified and the results cannot be used to derive kinetic data like the activation energy. The main purpose of the TSU test is to look for pressure generation. In combination with a DSC test, both thermal and pressure effects can usually be sufficiently quantified to identify potential hazards and define safe operating conditions.

Analytical results

Solubility tests indicate a rather poor dissolution rate (up to 20% maximum) in methanol, toluene, acetone, chloroform. Hot extraction in water or methanol allows to increase the dissolution rate to 35 and 75% respectively. With methanol at 50 °C a solubility of about 75% is obtained. After filtration on a 25µm PTFE filter and cooling down of the filtrate, a turbid solution is obtained which remains stable after cooling down (no precipitation is observed).

The TGA analysis of the solid particles collected in the heat exchanger inlet header and pipes reveals that the solids are composed of organic material at about 90 % (Figure 4). The inorganic part is mainly iron and aluminum with traces of silicon, sodium and potassium as shown by the semi-quantitative XRF analysis reported in Table 2. While iron most probably comes from rust present in the pipes and equipment, aluminum and the other trace elements may be explained by carry over of guard bed adsorbent present in the upstream dryers.

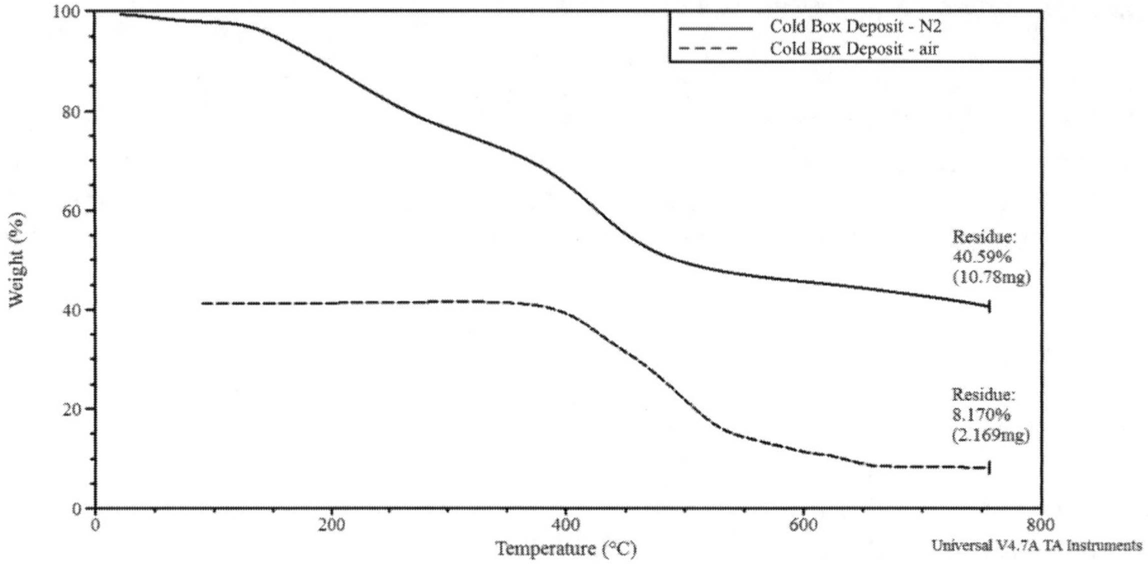

Figure 4: TGA plot of heat exchanger solid particles (first heating in nitrogen and second heating in air)

Table 2: Elemental analysis by XRF of heat exchanger solid particles

Element	Fe	Al	Si	Na	K	S	Mg	Ca	P	Zn	Mn
Wt %	1.89	1.28	0.68	0.29	0.29	0.31	0.06	0.06	0.03	0.03	0.03

With the help of a pyrolysis-GC (not reported here), the organic fraction was identified as being composed of cyclopentadiene, with the presence of aromatics (benzene, toluene) and N-heterocyclic aromatic groups being detected.

An infra-red spectrum of the solid particles can be found in Figure 5 with a tentative assignment. Bands in the 3000 cm^{-1} region are typical of unsaturated, branched hydrocarbons. Specific bands attributed to organic nitrates are depicted with arrows according to assignments reported elsewhere [4, 6]. Organic nitro species (C-NO_2) are also tentatively assigned while organic nitrite species (C-O-NO) are not detected.

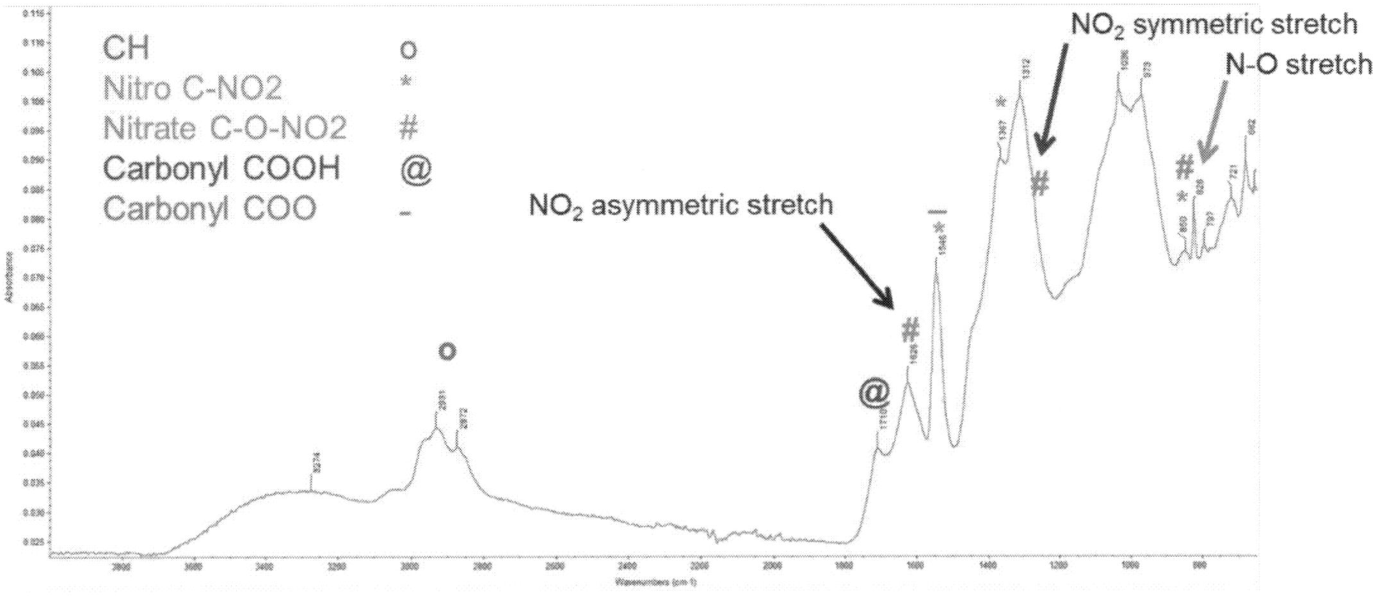

Figure 5: IR spectrum of solid particles found with tentative assignment

By CHNSO analysis, it is shown that the fouling material contains these elements at the following concentrations: 7.1 wt% N, 57.7 wt% C, 6.1 wt% H, 0.3 wt% S and 20.8 wt% O. The N/O ratio gives NO_x with x=2.6, indicating the presence of a mix of nitrite and nitrate species with a predominance of nitrate species. Those species are organic as shown by IR but the presence of salts is also seen by ion chromatography of the aqueous extract which detected 1.6 wt % NO_3^- and 0.1 wt% NO_2^-. In the hot aqueous extract, 3.25 wt % NO_3^- was detected. Cationic IC reveals NH_4^+ and Na^+ as the main counter ions (Table 3). Other cations were determined via ICP-OES. In total 3.35 wt% nitrate and nitrite salts are measured: these species represent thus about 10% of the total N present in the solid.

Table 3: Cationic ions determination

Element	Average concentration (wt%)	Corresponding nitrate	Nitrate content of water soluble fraction (*calculated from compounds*) (wt%)
NH_4^+ (IC)	0.074	**NH_4NO_3**	0.25
Na (IC)	0.10	**$NaNO_3$**	0.28
K (IC)	0.11	**KNO_3**	0.17
Al (ICP)	0.090	**$Al(NO_3)_3$**	0.62
Ca (ICP)	0.077	**$Ca(NO_3)_2$**	0.24
Fe (ICP)	0.015	**$Fe(NO_3)_3$**	0.048
Mg (ICP)	0.011	**$Mg(NO_3)_2$**	0.055
Zn (ICP)	0.005	**$Zn(NO_3)_2$**	0.009
		Total	**1.67**

In summary, the results of the various analyses carried out on the solid samples collected in the inlet header and pipes to the heat exchanger show that the material is a NO_x-containing, CPD-based polymer. The polymer is cross-linked and contains aromatic and N-heterocyclic aromatic groups. There is, as well, the presence of a minor amount of NO_x salts and of adsorbent material. The NO_x groups, in the form of organic species or salts, seem to be predominantly nitrate NO_3.

Further investigations to determine the safety hazards associated with such polymer in terms of thermal and mechanical stability are presented below. The "hammer" test evidenced that the sample is not shock-sensitive: no flash is observed, nor sound heard. The tape remains intact, that is, not scorched. The TSU reveals a significant, but gradual pressure rise with temperature and not an explosive pressure rise (Figure 6), showing that the material is not explosive.

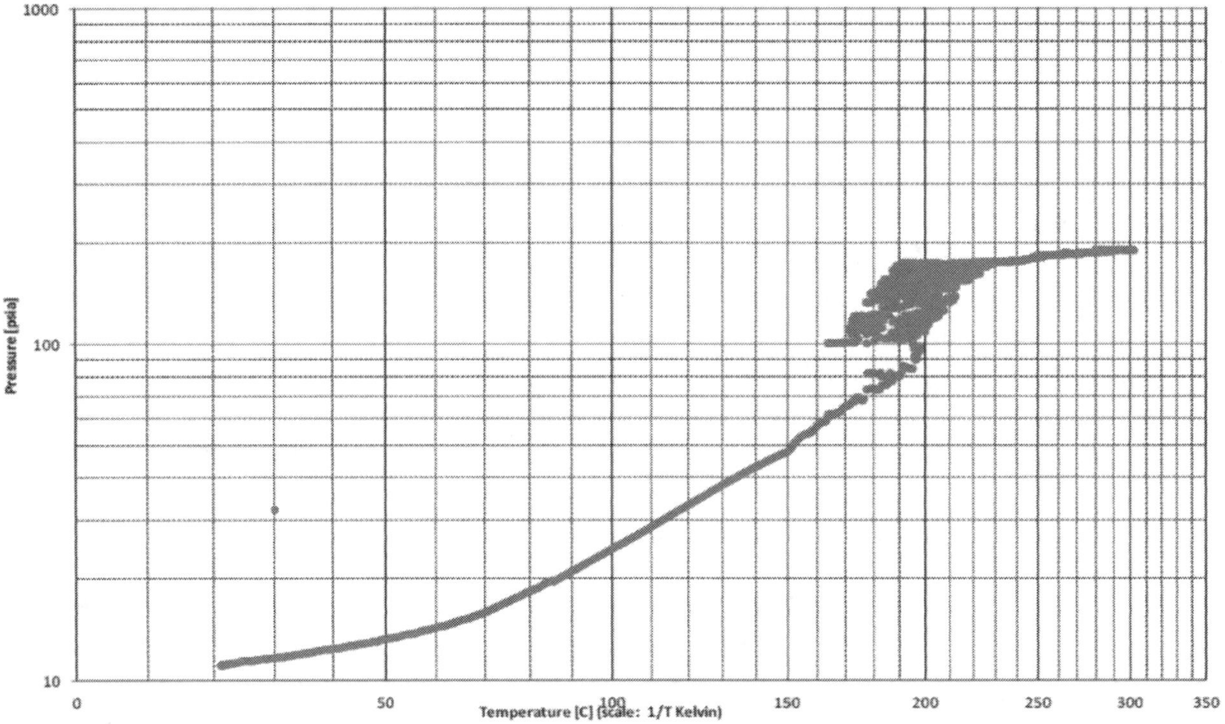

Figure 6: TSU pressure vs temperature

The TSU observations are confirmed by the DSC analyses (Figure 7 and 8). Under nitrogen (Figure 7), an exotherm is observed with an onset temperature of 65 °C. The energy release is medium (-755 J/g) and spread over a very wide temperature range (up to 310 °C). An explosive sample would show a high, small, intense peak instead of such a widely spread out peak. Similar exotherms are observed under oxygen and under air (Figure 8) with a mild and widely spread energy release. The total energy release is not different from the one obtained under nitrogen, which is expected because the amount of oxygen present in the headspace is limited. The onset temperatures under oxygen and under air are around 82 °C and 85 °C respectively, which is well above ambient temperature and would indicate a safe handling of the material under air at ambient conditions.

Based on these results, it can be concluded that the material is not a NOx-gum, which is shock-sensitive and explosive, and typically found in the low temperature zone of the cold box as reported in the ethylene industry.

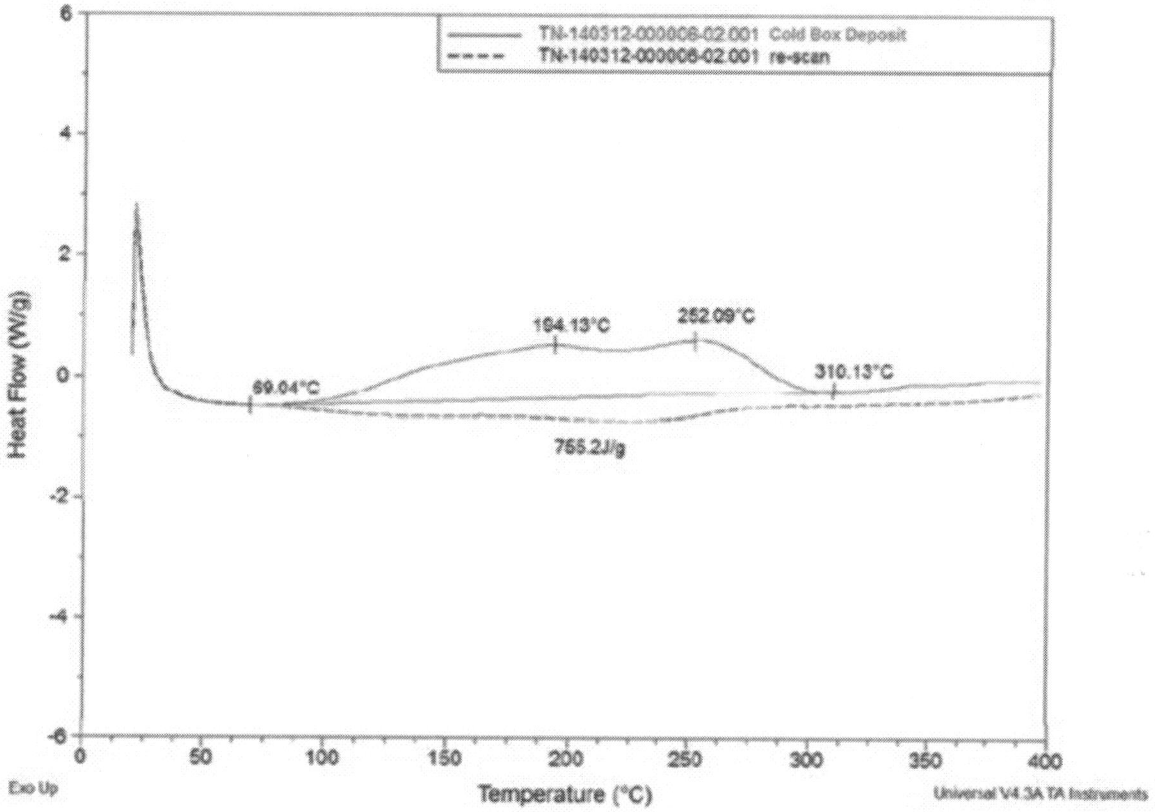

Figure 7: DSC under nitrogen

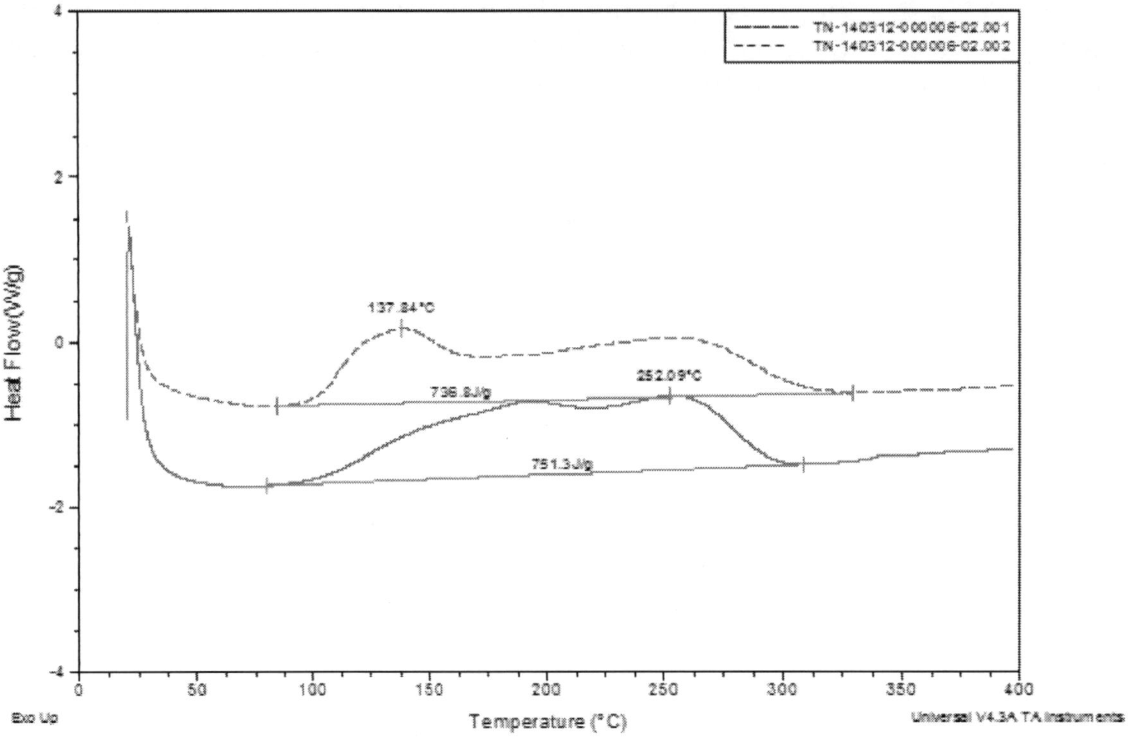

Figure 8: DSC under oxygen (top, dotted line) and under air (bottom, plain line)

Cold box cleaning methods

Based on the characterization results of the fouling material, two cleaning methods for the fouled heat exchanger were developed. The thermal and mechanical stability was an essential finding that allowed the successful implementation of the methods without safety incidents.

The first method is an on-line and off-line methanol wash. The efficiency of the on-line wash in dissolving the fouling material is directly monitored by the pressure drop measurement over the heat exchanger, for which an example can be found below in Figure 13.

For the off-line wash, based on the solubility tests results, it was decided to use warm methanol. The off-line wash was carried out by recirculating methanol at 50 °C until saturation of the solvent. This was monitored by regular methanol sampling and total nitrogen analysis. Completely dry methanol is not recommended as this is corrosive to aluminum, the material of construction of the heat exchanger. To ensure the methanol was not dry, water was added at a concentration of about 0.2 wt%. Two methanol batches were recirculated for respectively 2 days and one day. Based on the total N analysis and on the known N composition of the fouling material, an estimated amount of about 2.5 kg of polymer was dissolved.

The second method applied is a mechanical method recommended by Alpema, the brazed aluminum plate-fin heat exchanger manufacturers' association, for the removal of plugging particles and is based on a "back-blow" of the plugged stream. [7] For this method, a bursting disc is installed at the inlet and nitrogen is pressurized up to the rupture of the disc. This operation is repeated until no more particles are observed being discharged. This method was applied before and after off-line warm methanol cleaning.

The efficiency of the cleaning was measured by off-line measurements of the pressure drop under known nitrogen mass flow rate and temperature. The combination of both chemical and mechanical methods allowed the recovery of the cold box functionality.

Mechanism of formation

Several hypotheses on the mechanism of formation of such polymer have been developed. It must be noted than an extensive sampling plan in various locations upstream the cold box (liquid drains from knock-out drums) was carried out in order to identify whether the polymer was formed upstream and carried over to the cold box where it deposited. Neither NO_x-polymer nor NO_x-oligomers were found upstream of the dryers.

The regeneration system of the dryers has also been investigated. A possible leak path from the process side to the stand-by dryer during regeneration, which could initiate polymerization at high temperature, was hypothesized but eventually rejected as neither evidence of leaking valves nor presence of NO_x containing materials in the regeneration area could be found.

NO_x-oligomers were found in the liquid drain of the filter located downstream of the dryers and upstream of the cold-box (Figure 9). The reaction could thus take place in the liquid phase formed by condensation of the heavier hydrocarbons present in the cracked gas stream (C4's and C5's). The deposit was however also observed on the top of the cartridges at a level far above any liquid level. Also, plugging events, indicated by an increase of the heat exchanger pressure drop, were observed in the summer time when there was no liquid level in the filter: these observations would rather support a gas phase reaction between dienes and NO_x species.

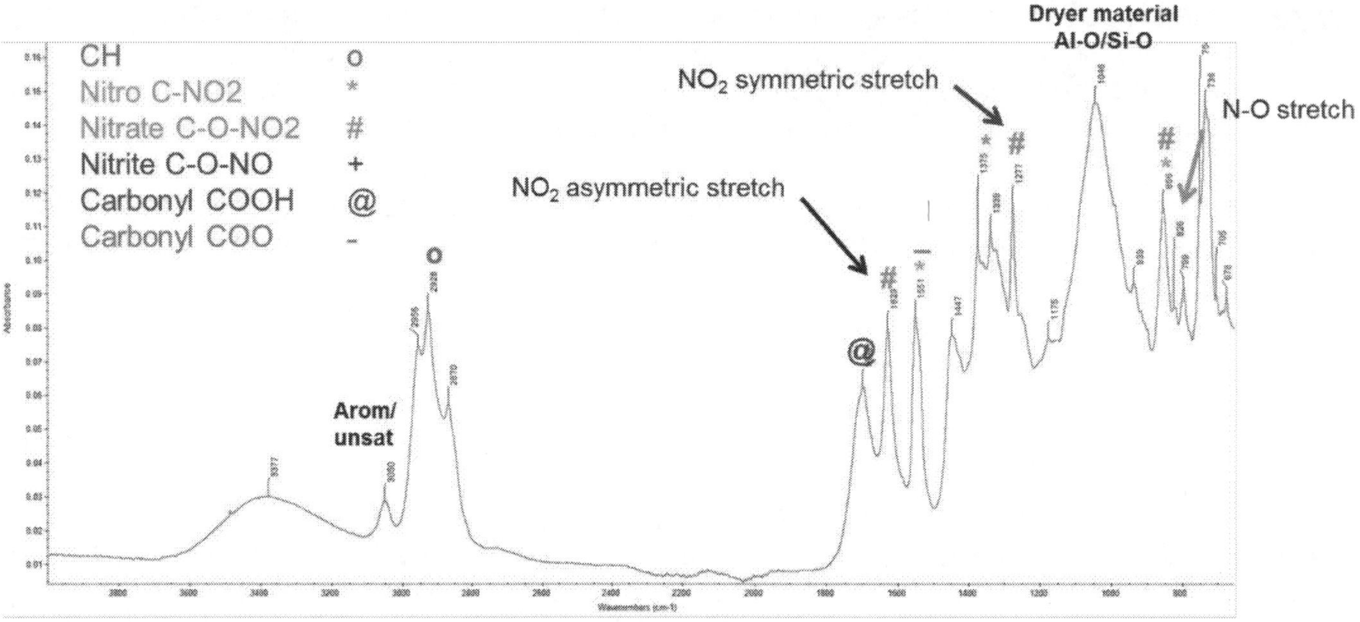

Figure 9: IR spectrum of the liquid drain sample from the upper stream filter, with tentative assignment

The formation of nitrogenous "vapor-phase gum" has been reported in the early 20th century town gas (or coke-oven gas) systems literature. [8-10] It was shown that NO_x reacted with unsaturated hydrocarbons in gas phase to form dark brown particles. The gum was thermally unstable and its decomposition under heat became violent around 175 °C. It contained a high amount of nitrogen and oxygen (about 3 to 7 wt% N and 22 to 32 wt% O). Based on thermal decomposition of the material and analysis of the products generated, it was found that the gum was mainly composed of NO, indene, aromatics and probably cyclopentadiene, amines and N-heterocyclic aromatic compounds and phenols. [8] Formation of gum particles was reproduced in the laboratory by adding NO_2 to various hydrocarbons gas mixtures. In presence of NO, particles were also formed but to a lesser extent and it was not excluded that this could have been caused by NO_2 formed from NO and traces of oxygen. [8]

In an attempt to validate the gas phase mechanism hypothesis, a nitric oxide addition experiment in cracked gas was carried out in the laboratory. Cracked gas was sampled upstream of the dryers by using a sample cylinder whose internal steel walls were passivated with an inert silica layer. Prior to use, it was checked that the cracked gas samples did not contain NO_x by GC-NCD (detection limit 1 vol ppm). Nitrogen Chemiluminescence Detector (NCD) allows the detection of only nitrogen containing species present in the sample. With NCD detection, all non-nitrogen containing species are invisible to the detector.

In the laboratory, Tedlar® sample bags were filled with cracked gas to which NO was added at 2000 ppm. For the first addition test, the sample bag became brown. The content of the sample bag was extracted with chloroform. A brown, clear solution was obtained and analyzed by GC-NCD (Figure 10, b). The following day, a deposit was observed in the chloroform solution, which was analyzed by FT-IR (Figure 11,b). As this first test was done with an already used bag, it was decided to repeat the test with a new, unused sample bag. For this second test, another cracked gas sample had to be taken in the plant (on a different day than the previous one). The chloroform extract was also analyzed by GC-NCD (Figure 10, c) and by FT-IR after evaporation of the solvent (Figure 11, c).

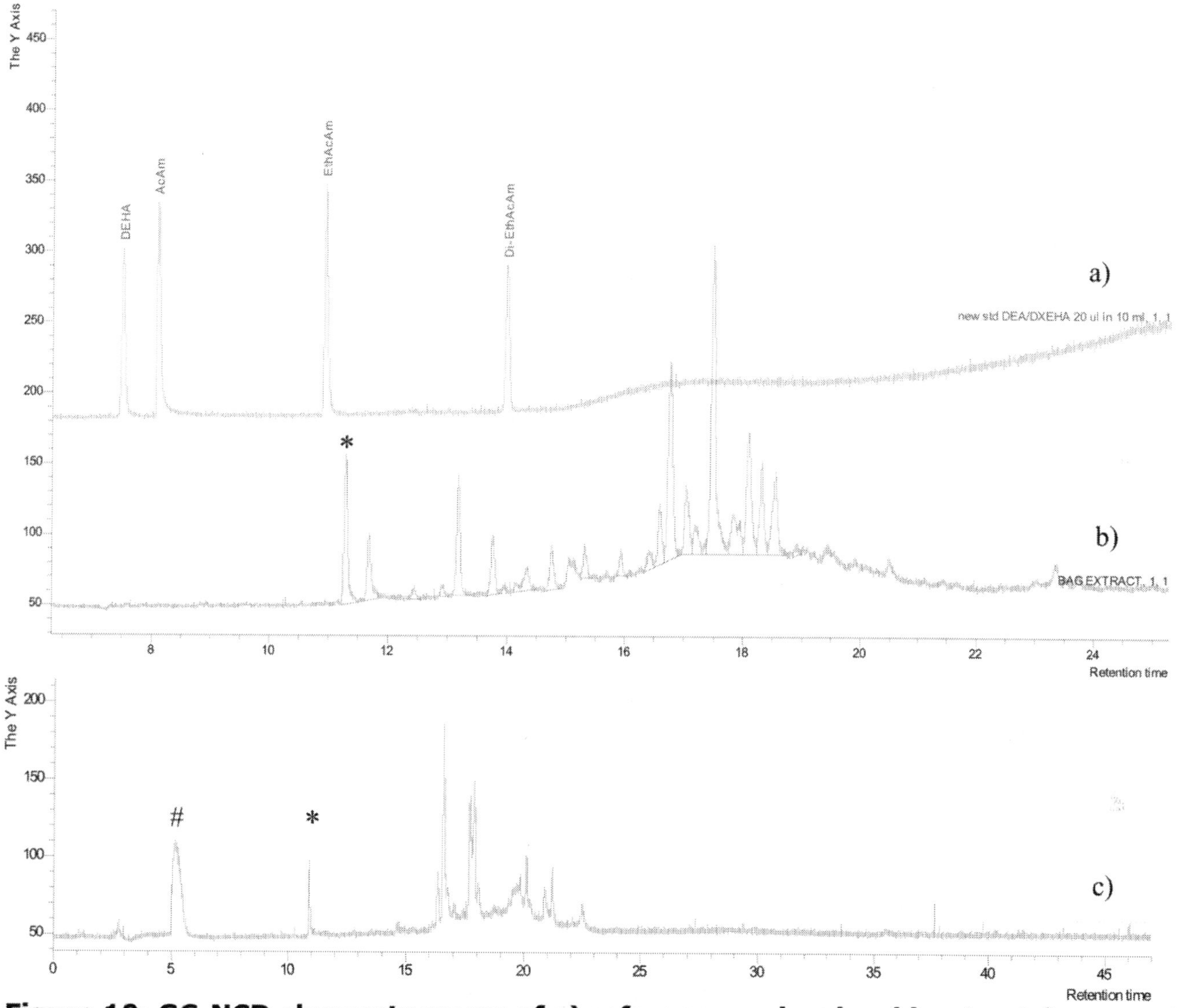

Figure 10: GC-NCD chromatograms of a) reference molecules, b) extract from used sample bag, c) extract from unused sample bag

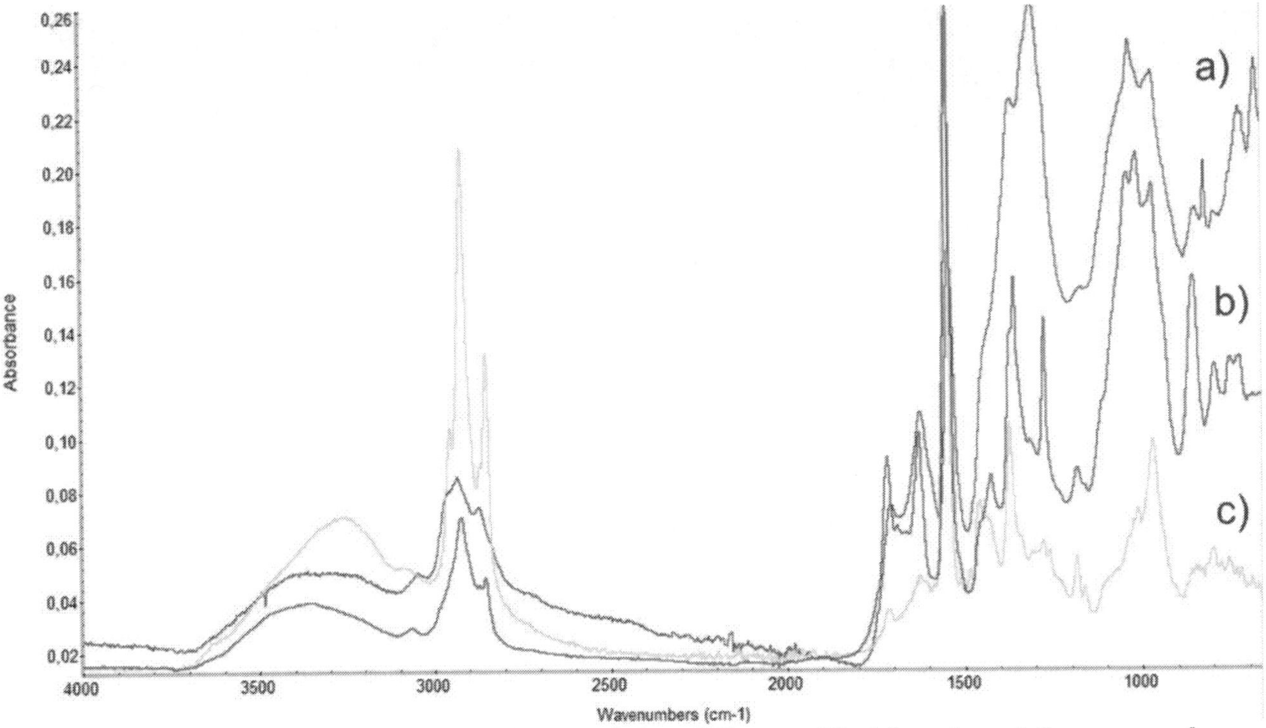

Figure 11: IR spectrum of a) solid particles (Figure 5), b) extract from used sample bag, c) extract from unused sample bag

In Figure 10, the chromatogram a) indicates the retention time of various N containing molecules as references. The chromatogram b) shows that the non-polymeric, soluble in chloroform fraction of the material formed upon addition of NO in the used sample bag contains various N-containing components in the range of C4 and higher as most of these components elute around and at higher retention times than ethylacetamide (C_4H_9NO) and diethylacetamide ($C_6H_{13}NO$). With the new sample bag test (chromatogram c) in Figure 10), N-containing components present in the extract elute in the same region between 15 and 25 min, which would indicate that similar species were formed during the second experiment. By running blank analyses (chloroform alone and chloroform extract of an unused sample bag), one interference with the solvent and one impurity in the sample bag are seen and depicted respectively by the symbols # and * on Figure 10.

On Figure 11 the IR spectra b) and c) are compared with spectrum a) corresponding to the IR spectrum of the solids collected in the cold box (reported from Figure 5). The solids formed during the first NO addition experiment is very similar to the fouling solid of the cold box. A similar pattern is also observed with the extract of the second experiment with however a higher CH ratio which is especially visible in the region of 3000 cm^{-1}.

Complementary analyses of the extract from the unused sample bag were performed by GC-MS and TOF-MS (not reported here). They demonstrated the presence of CPD and of CPD-NO fragments.

By carrying out successive GC-NCD analyses on the same sample after NO addition, it was observed that NO added at the level of around 10 ppm was stable for about 10 min and for more than one hour when added at about 2000 ppm. The experiment was repeated with cracked gas samples collected at other times in the plant, these results could however not be reproduced. For few samples NO was stable for several hours. To explain these differences of reactivity between samples, the operating conditions of the condensate stripper at the various times of sampling and the cyclopentadiene content of the samples were checked. The variations of these parameters could not explained the differences in reactivity. Presence of oxygen may have played a role but this parameter was not controlled during sampling and handling of the samples in the laboratory.

These exploratory experiments demonstrate that a similar NOx-containing polymer to the one found in the cold box can be formed in ambient conditions, in gas phase and in presence of NO_x. Questions related to the mechanism of formation, the nature of the oxide(s) of nitrogen as reactant and the particular location of the solid deposition in the plant remain however unanswered.

Root Cause Identification and Validation

Investigations were initiated to identify potential sources of NO_x. In particular recycle streams were inventoried and their composition analyzed. One co-crack C4 recycle stream from the Butadiene plants was identified as a potential source of NO_x as it contained a high diethylhydroxylamine (DEHA) residual concentration. DEHA is used as a polymerization inhibitor in the Butadiene plants. Previous studies demonstrate that DEHA mainly decomposes into ammonia during cracking, but NO is also formed. [4, 11]

As a test to validate the root cause, the filter upstream of the cold box was by-passed and during that time the C4-recycle co-cracking was stopped and started again (Figure 13). While co-cracking the C4-recycle stream, the pressure drop of the first pass heat exchanger increased dramatically. Then the pressure drop stabilized when "pumping" the C4-recycle stream to the cracker furnaces was stopped. Initial pressure drop could be recovered by dissolving the freshly made plugging material by immediate on-line injection of methanol to the heat exchanger.

Based on this results, it was decided to re-route the C4-recycle stream. Since then the pressure drop of the heat exchanger has been stable and the cycle time of the filter cartridges has been extended by a factor 6, which validates this stream as the main source of NO_x.

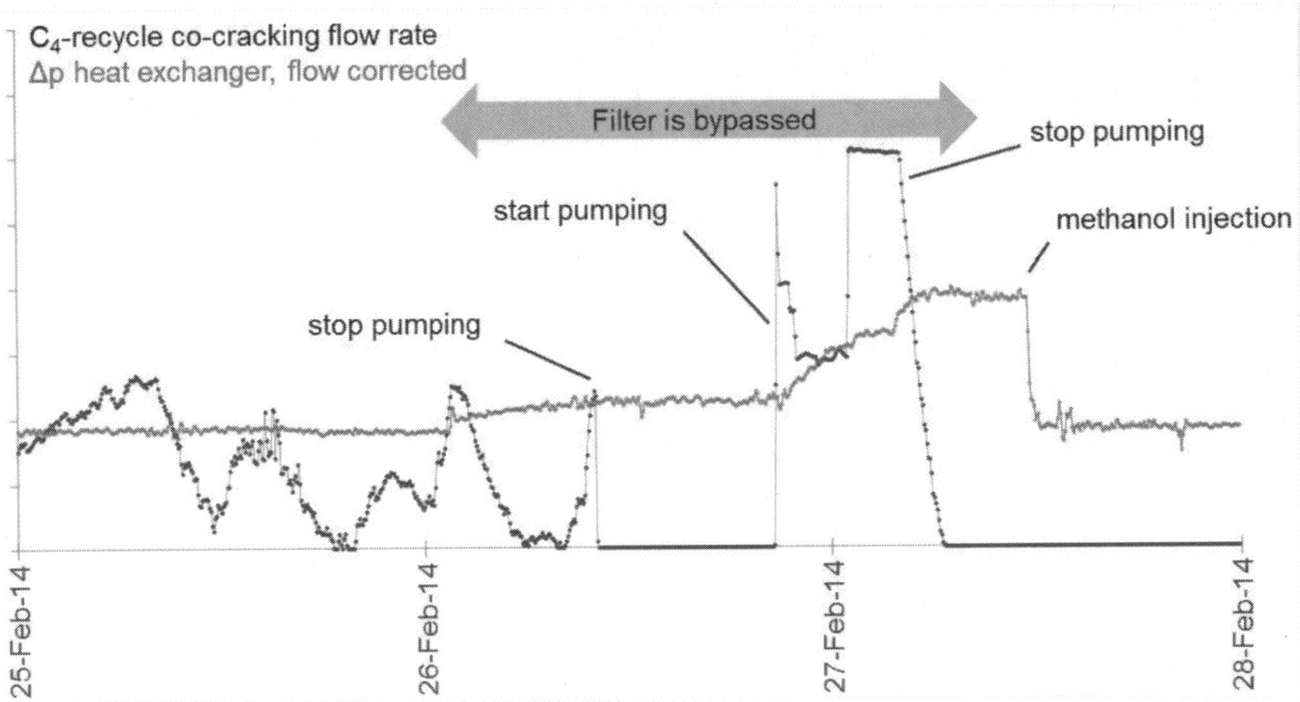

Figure 13: C4-recycle stream flow rate and pressure drop (Δp) of the heat exchanger during the filter by-pass test

Conclusions

A NO_x-containing, cyclopentadiene based polymer plugged the first pass heat exchanger of a cold box. This polymer is non-shock-sensitive and non-explosive, and is assumed to be formed in the gas phase and as such is different from the known explosive and shock-sensitive NO_x-gums formed in the low temperature zone of the cold box reported in the ethylene industry.

Chemical and mechanical cleaning methods were applied to the fouled cold box to recover its functionality. Such methods could be developed based on the fouling material characterization results that demonstrated the thermal and mechanical stability of the polymer and that identified a suitable solvent.

Similar material could be made in the laboratory by adding NO into cracked gas sample at ambient conditions. Experimental parameter like the presence of oxygen was however not controlled and NO reactivity experiments could not be reproduced. The mechanism of formation and the particular location of the polymer deposition in the high temperature section of the cold box remain unexplained.

The most probable and main contributing root cause for the cold box plugging was identified as the co-cracking of a C4-recycle stream from the Butadiene plants containing DEHA residuals. Such a compound is known to decompose into NO in the cracking furnaces, which could be thus the source of NOx. This hypothesis was validated by removing the proposed root cause which led to longer cycle times of the upstream filter and to stable operation of the heat exchanger.

References

[1] Hack, W., R. Copeland, and D. Hines, *Investigation of NO_x Related Cold Box Incident.*, Paper presented at the Ethylene Producers Conference at the 2009 Spring National AIChE Meeting, 2009.

[2] Nalco/Exxon Energy Chemiclas, *Potentiental for NO_x Formation from Process Additives*, CPI Chemicals Marketing Letter No. 97-05, May 20, 1997

[3] Henstock, W.H., *NO_x in the Cryogenic Hydrogen Recovery Section of an Olefins Production Unit*, Plant/Operations Progress, Vol 5, No 4, page 232, 1986

[4] Halle, R., *Potential hazards of nitrogen oxide compound accumulations in cryogenic ethylene recovery facilities*, Paper presented at the Ethylene Producers Conference at the 1993 Spring National AIChE Meeting, 1993.

[5] Simms, D.R., *Update on the activities of the NO_x taskgroup*, Proceedings of the 9th Ethylene Producers Conference, Vol.6 AIChE, 1997, page 480

[6] Moran, M., *2012 Baytown Olefins Plant NO_x Washing Findings*, Presentation given at September 26th 2012 EPC NO_x Subcommittee Meeting, 2012

[7] ALPEMA, *The standards of the brazed aluminum plate-fin heat exchanger manufacturers' association*, Third edition, 2010

[8] Jordan, C.W, Ward, A.L., Fulweiler, W.H., *Gum Deposit in Gas Distribution Systems, Vapor-Phase Gum*, Industrial and engineering chemistry, Vol 26, No 9, page 947, 1934

[9] Jordan, C.W, Ward, A.L., Fulweiler, W.H., *Gum Deposit in Gas Distribution Systems, Vapor-Phase Gum (Continued)*, Industrial and engineering chemistry, Vol 26, No 10, page 1028, 1934

[10] Jordan, C.W, Ward, A.L., Fulweiler, W.H., *Gum Deposit in Gas Distribution Systems, Vapor-Phase Gum (Concluded)*, Industrial and engineering chemistry, Vol 27, No 10, page 1180, 1934

[11] Wang, H., et al., *Kinetic Study of Thermal Decomposition of N,N-Diethylhydroxylamine (DEHA) in Steam Crackers*, Paper presented at the Ethylene Producers Conference at the 2017 Spring National AIChE Meeting, 2017.

AICHE PAPER NUMBER 32B

32b: Zero waste/polymer disposal from caustic tower saves 1(€)MM in Versalis ethylene unit

Patrick Lucas
Global Technical Director, CPI
SUEZ Water Technologies & Solutions

Fabrizio Lucchese
Technical Manager Marghera
Versalis S.p.A.

Oreste Vassillo
HPI/CPI Product Application
SUEZ Water Technologies & Solutions

Prepared for Presentation at the 2018 Spring National Meeting
Orlando, FL, April 23 – 26, 2018

*Trademark of SUEZ; may be registered in one or more countries.

Zero waste/polymer disposal from caustic tower saves 1(€)MM in Versalis ethylene unit

Patrick Lucas
Global Technical Director, CPI
SUEZ Water Technologies & Solutions

Fabrizio Lucchese
Technical Manager Marghera
Versalis S.p.A.

Oreste Vassillo
HPI/CPI Product Application
SUEZ Water Technologies & Solutions

Abstract:

The process gas in ethylene plants contains acid gases, namely carbon dioxide and hydrogen sulphide, due to the cracking and coking/decoking reactions in the cracking furnaces. These acid gases are typically removed in a caustic tower. However, aldehydes and ketones, contaminants that are also produced in the cracking furnaces, can polymerize under basic conditions to form aldol condensation polymer. This polymer can deposit on the internals of the caustic tower and in the downstream spent caustic handling facilities (tankage, strippers, etc.). This fouling can lead to serious operating problems that impact the plant's overall efficiency and reliability and generate high polymer disposal costs.

Today, ethylene producers are often faced with more stringent environmental, health and safety regulations. This is particularly true in Western Europe and certain countries in Asia Pacific. Environmentally friendly process treatment technology is, therefore, of great interest and sometimes even mandatory. SUEZ Water Technologies & Solutions has developed a patented, non-toxic aldol polymerization inhibitor program, which can provide reliable protection for both the caustic scrubber and the spent caustic system. This paper will review an application case in Versalis' ethylene unit in Porto Marghera, Italy. The innovative, environmentally-friendly technology provided excellent fouling protection and reduced the waste/polymer disposal costs by 1(€)MM (380 k€/yr in three years contract).

1. Process Overview

The cracked or process gas in ethylene plants contains acid gases, such as carbon dioxide and hydrogen sulphide. The carbon dioxide is present because of the coking/decoking reactions which occur in the cracking furnaces. Hydrogen sulphide is also formed in the furnaces from sulphur compounds in the feed-stock and sulphur compounds injected into the furnaces to minimize coking and CO formation. These contaminants in the process gas must be removed from the gas to prevent contamination of the acetylene converter catalyst by hydrogen sulphide, and to meet product specifications for polymer grade ethylene.

The acid gases are removed by either of two methods: an alkanol amine (MEA, DEA) system followed by a caustic tower, or by a caustic tower alone. The use of both systems is generally needed for treating process gas with high acid gas. Where acid gas levels are low (<500 ppm), a caustic scrubber alone provides economical removal of the contaminants. In many older plants, the amine systems have been shut down or converted to additional stage(s) of the caustic tower. The acid gas removal system is usually located between the 3rd and 4th stages of compression or the 4th and 5th stages of compression or after the 5th stage.

To enhance plant economics, some ethylene plants process refinery off gases to recover primarily the ethylene and propylene. The refinery gases are usually from the FCC or Coker units and contain high levels of hydrogen sulphide. The refinery gases are usually introduced into the ethylene plant in the process gas compressor. Therefore, the acid gas introduced with the refinery gas directly impacts the load on the acid gas removal system. Some plants have separate caustic systems to handle the streams recovered from the refinery.

Caustic Tower

Acid gases are removed from the cracked gas by counter current contact and scrubbing with recirculating caustic solutions. The caustic system may include one or more towers which are divided into several sections. The sections are divided according to the concentration of free caustic and there is a water wash before the cracked gas leaves the system. A system may have two or three caustic scrubbing sections or stages. The 1st stage (or bottoms or weak caustic section) contains a circulating solution of caustic with 1 -3 % free caustic. This is the section where the cracked gas enters the tower. The spent caustic is removed from this section, from the column bottoms and/or through a skim system. To keep caustic usage low, the plant tries to run the caustic strength as low as possible in this section. The next caustic loop may be called the 2nd stage (or middle or strong caustic section). The free caustic level in this section may vary from 5-10%. If a plant has a third caustic stage, it will operate at the highest free caustic level of the three sections. The water wash section is used to ensure that no caustic is carried out of the tower to downstream areas. If there is a compressor section downstream, the caustic can foul the compressor internals causing a decrease in performance. Make-up caustic is added to the strong section of the tower to maintain the desired caustic concentrations. The plant receives concentrated caustic into the plant and dilutes it down to 10% or 20% for use in the caustic tower. The make-up caustic rate is roughly equal to the spent caustic rate. Figure 1 shows the typical configuration of a caustic scrubber.

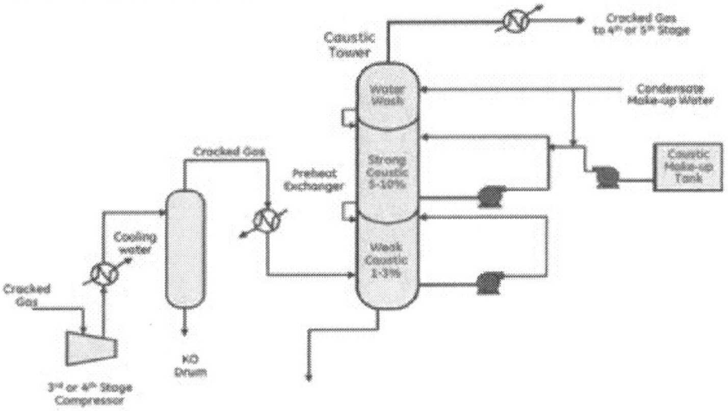

Figure 1

Caustic towers are normally operated in a temperature range of 100 – 125 °F (38 – 52 °C). The temperature is controlled by the temperature of the inlet cracked gas. A cooler temperature will increase the amount of hydrocarbon that condenses in the caustic tower. Hydrocarbons can promote fouling and foaming in the caustic tower. Skimming the caustic in the first stage is an effective way to remove the hydrocarbons. To avoid/minimise condensation of hydrocarbons, it is important that the temperature of the gas is higher (5 °C/10 °F) than the temperature of the caustic. It is also important, however, to avoid the temperature exceeding 50 °C. Although that will lower the concentration of fouling precursors, the rate of the polymerization reactions will increase significantly. It has also to be considered that the physical absorption of the acid gas and its chemical reaction with caustic is favoured by low temperature. Therefore, too high a temperature might limit acid gas removal.

Handling and disposing of the spent caustic has become increasingly expensive as environmental regulations have put limitations on the plants. In response, many plants have been installing equipment to handle and treat the spent caustic locally. The spent caustic also contains benzene and a steam or gas stripper can be used to remove this. Other approaches are to neutralize the spent caustic and then treat it in the plant wastewater system, the Zimpro unit, which is a form of wet air oxidation. Figure 2 shows a typical spent caustic treatment scheme.

*Trademark of SUEZ; may be registered in one or more countries.

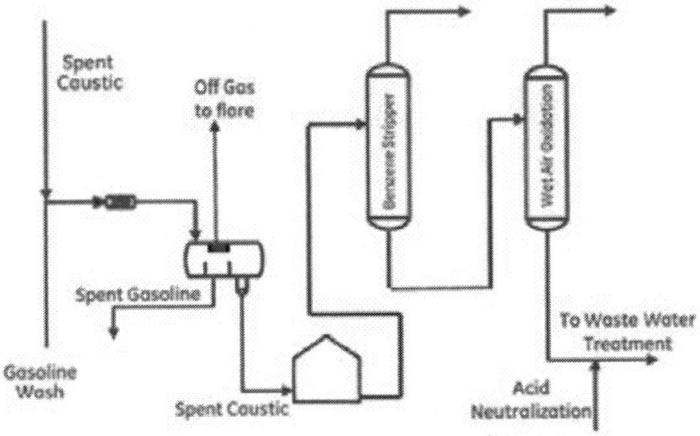

Figure 2

2. Areas of concern

Fouling, emulsification and foaming are the three major problems encountered in the caustic system. If these problems are severe and lead to shutting down the caustic tower, the entire ethylene plant will have to be shut down.

Fouling in the caustic system is characterized by increase in pressure drop, a decrease in acid gas removal efficiency or a build-up of solids in the spent caustic. Fouling is due mainly to aldol condensation reactions, but also in some cases due to Free Radical Polymerization if the temperature is higher than 50 °C.

Foaming is also characterized by pressure drop increase in the tower (more correctly by fluctuating pressure drop), and by an increase in the caustic level in the water wash section and a corresponding increase in caustic make-up requirements.

Emulsification may occur if the loading of hydrocarbons to the system is too high. This may be due to operating the tower too cold, or to inadequate separation of the hydrocarbons in the compressor knock out drum upstream.

Influence of Feed-stock Contaminants

Contaminants in the feed-stock, such as oxygen, alcohols and ethers, can lead to the formation of constituents that cause fouling and corrosion problems downstream. Compounds, such as ethanol and methanol found in naphtha and NGL feed-stocks, can convert to ketones and aldehydes. Ketone hydrolyses to acetic anhydride and acetic acid. Ketones can also react with amine or caustic to form stable compounds, which cannot be regenerated. This increases caustic demand. Aldehydes can lead to polymer formation or pass through product purification to the crude butadiene. In the crude butadiene, aldehydes may cause the product specification to be exceeded. Oxygen may be introduced via naphtha, gas oil, refinery gas or flare gas recycle. Oxygen can form Ketones and aldehydes thus affecting mainly the carbonyl specification on the C_4 cut. MTBE in the feed can crack to methanol and aldehydes giving the same problems listed before. Also, higher butadiene content could be detrimental and increase the fouling rate because of Free Radical Polymerization. This type of polymerization occurs at higher rate if the temperature increases and it could be critical at temperatures higher than 50 °C. An example for rate of butadiene dimer formation is showed in Figure 3.

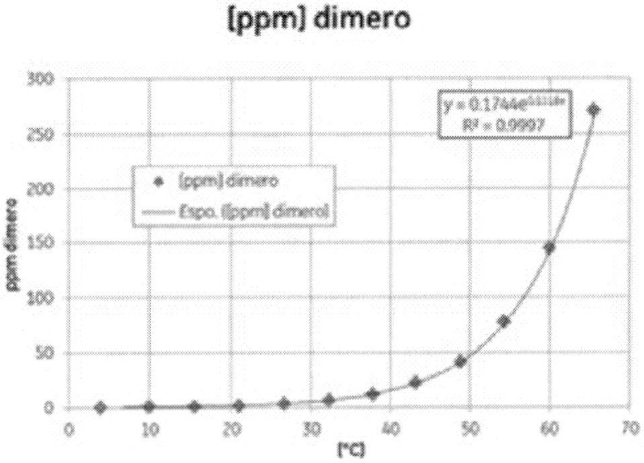

Figure 3

Fouling Mechanisms

Organic fouling of amine systems and caustic systems can be caused by two mechanisms, free radical polymerization and aldol condensation. A combination of the two mechanisms may also occur. Free radical polymerization is the less common of the two mechanisms. Both mechanisms occur in the liquid phase where hydrocarbons have been knocked out of the cracked gas.

Free radical polymerization occurs according to the traditional steps of initiation, propagation, and termination. The reaction involves dienes or unsaturated compounds, which have condensed from the cracked gas. In the amine system, this mechanism may occur in the hotter areas of the process, such as the regenerator or stripper reboiler, and the lean side of the lean/rich exchanger. Free radical polymerization products are darker in color and may be rubbery.

The treatment program for free radical polymerization uses chain-stopper type antifoulant chemistry. The antifoulant terminates active free radicals.

The Aldol Condensation mechanism is the prevalent mechanism causing fouling in amine and caustic systems. The conditions suggesting Aldol Condensation include:

- process caustic sample coloration typically orange, brown-red to red
- red hydrocarbon liquid layer on the spent caustic
- granular orange to brown-orange deposits
- the presence of yellow tacky polymer in stage or skim samples

The red liquid is commonly referred to as red oil. Further polymerization of the red oil yields the tacky, yellow polymer. The waste oil from the caustic tower is typically classified as hazardous waste The Aldol Condensation reaction occurs when carbonyls, aldehydes and ketones, condense from the cracked gas. In the basic environment of the system, the molecules of aldehyde or ketone combine to form a B-hydroxyaldehyde or B-hydroxyketone. In every case the product results from addition of one molecule of aldehyde or ketone to a second molecule in such a way that the alpha carbon of the first becomes attached to the carbonyl carbon of the second. The reactions continue due to the on-going presence of the C=O. The reaction product dehydrates to form a structure containing double bonds. The polymer formed will have a red color when six or more molecules are reacted and there are alternating double bonds in the structure. Ketones react slower due to the steric hindrance in their structure.

Basic or acid environment

$$CH_3CHO + CH_3CHO \Leftrightarrow CH_3-CHOH-CH_2-CHO$$

In the cracked gas, a variety of carbonyl compounds will be present and will react via the aldol condensation mechanism. Testing has indicated that acetaldehyde is the most prevalent and will most directly form red oil. Types of carbonyls which are typically found in the cracked gas include: acetone, acrolein, acetaldehyde, propionaldehyde and formaldehyde. Some plants process a polyethylene plant off-gas stream through the compressor to recover ethylene. This recycle gas can contain vinyl acetate used in the polyethylene process. Vinyl acetate will condense into the caustic and hydrolyze to form acetaldehyde. If an aldehyde or ketone does not contain the a-hydrogen, a simple aldol condensation cannot take place. This is the case for formaldehyde, where formaldehyde will not react with itself. However, in the presence of aldehydes or ketones with the alpha hydrogen, formaldehyde will react via aldol condensation. Thus, it is assumed that in the process system, the true structure of the polymer formed cannot be assumed since a variety of carbonyls will participate in the reactions to form polymer.

The aldol condensation mechanism can be determined by analyzing the level of carbonyls in the gas, deposit analysis, sludge analysis, the color of the caustic on each loop, and a laboratory fouling simulation. The carbonyl level in the cracked gas can be determined using a field test called Aldol-Trac. This test uses a Sensidyne pump with a GasTech acetaldehyde detector tube. While the detector tube is specific for acetaldehyde, the detection principle is such that all carbonyls will be detected. The result from the test is the vppm of carbonyls in the gas, which must be converted to a wppm using the molecular weight of the gas and composite molecular weight of the carbonyls. The molecular weight for acetaldehyde is 44.

3. Fouling Control

SUEZ Water Technologies & Solutions has recently developed a patented, non-toxic, nitrogen-free aldol polymerization inhibitor program, which can provide reliable protection for both the caustic scrubber and the spent caustic system. This approach

*Trademark of SUEZ; may be registered in one or more countries.

can be used alone or in combination with other SUEZ's carbonyl scavengers, depending on the severity of the problem.

The technology is a combination of three functionalities: carbonyl scavenger, oxygen scavenger and dispersant. It is composed of a second-generation new carbonyl scavenger and polymer solubilizing agent for caustic system. The carbonyl scavenger includes multiple hydroxyl groups on the incorporated carbohydrates and helps to solvate the otherwise insoluble, hydrophobic aldol condensation products as well.

The non-toxic aldol polymerization inhibitor hinders aldol polymer condensation by reaction on the carbonyl group, reducing polymer formation induced with peroxides, reducing oxygen and cleaning the system through the dispersant. It is a technology that enables provision of a full protection.

4. Monitoring the effectiveness of antifouling program

The monitoring to determine the performance of a caustic tower antifoulant is designed based on the goals for the program. Typically, the goals are to achieve no polymer in the caustic tower stage samples, no red oil in the spent caustic, no fouling in the towers which would result in lost production or unscheduled shutdown, and no loss of caustic tower acid gas removal efficiency.

The service support for a caustic tower should include routine monitoring for performance related criteria, process conditions, program optimization, predictive tools and characterization/quantification of the sludge. Specific service techniques include: Aldol-Trac gas analysis, spent caustic color before and after treatment, red oil/polymer content in spent caustic, analysis of non-volatile residue in spent caustic, laboratory inhibition studies, tower pressure drop trending, prediction using computer modelling and Multiple Regression Analysis & Statistical Process Control (MRA/SPC)

5. Case Study: Versalis Porto Marghera (Italy)

a) Description of the circuits

The ethylene plant in Porto Marghera is a steam cracking plant, CBI LUMMUS designed, and fed mainly by Virgin Naphtha. Production capacity for Marghera Steam Cracking plant is 460,000 ktons/year of ethylene. This plant has two caustic scrubbers, named C-202 & C-285, that work in parallel. The spent caustic stream goes, after degassing, in a settling tank and to the Waste Water Treatment plant. Marghera industrial site has no other pretreatment plants before the biological plant (no de-oil system, no oxidation furnace). For this reason, it is very important to have low level of toxicity/pollutants coming from the spent caustic stream.

*Trademark of SUEZ; may be registered in one or more countries.

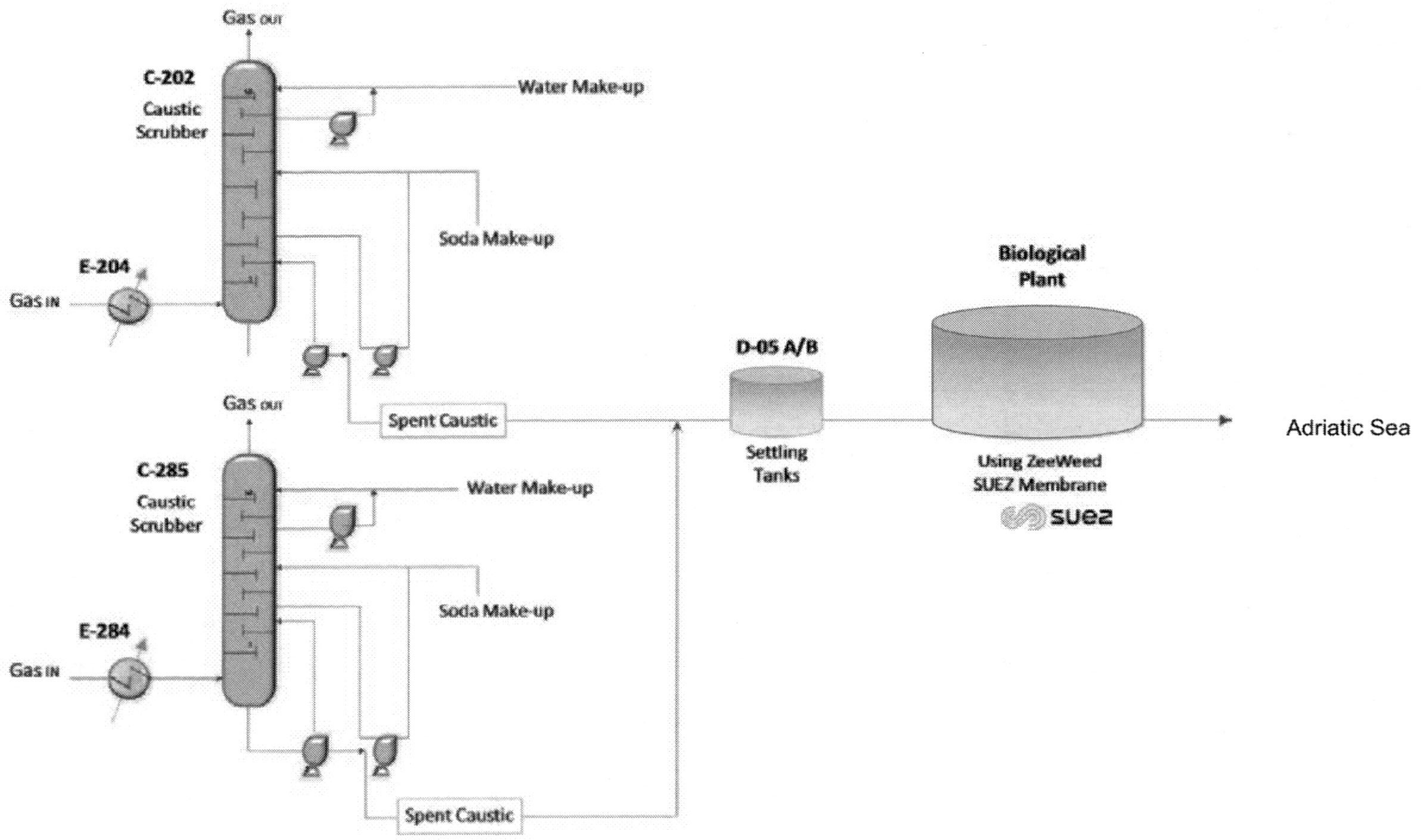

Figure 4

*Trademark of SUEZ; may be registered in one or more countries.

b) Background

The two caustic scrubbers, named C-202 & C-285, were treated for several years with aldol scavenger technology to avoid fouling in caustic scrubber with a dosage rate of 60÷70 ppm based on the throughput to the plant. The aldol scavenger was injected into the fresh caustic makeup.

The ethylene plant in Porto Marghera experienced severe fouling of the caustic scrubbers C-202 & C-285 and quality issues in the spent caustic during the last quarter of 2015. Due to high amounts of polymers, the two scrubbers were stopped and cleaned.

A deposit analysis was performed and it was identified as:

- 35% aldol polymer due to aldol condensation
- 65% polymer due to free radical polymerization (FRP) containing reactive materials such as propadiene, butadiene, isoprene, cyclopentadiene, etc.

During the diagnosis phase, other evaluations were carried out, including:

- MTBE concentration in the feedstock. More than 100 ppmwt was measured causing a high generation of ketones and aldehydes during the cracking process.
- 1,3 butadiene content in the Mix C_4. A level of 55%wt was measured compared to 40-45%wt from other ethylene plants in Italy.

c) Lab Tests and Results

To evaluate the effectiveness of SUEZ's non-toxic aldol polymerization inhibitor, some comparative evaluation tests were performed at the Versalis site.

The following procedure was used in the lab to simulate plant conditions:

1. Two liters of a fresh sample of spent caustic were taken using glass bottles, to check with visual inspection the effectiveness of the treatment
2. The samples were then separated into three samples and treated as follows;
    a. "Blank" or untreated sample
    b. Treated sample with 500 ppm of non-toxic Petroflo* aldol polymerization inhibitor
    c. Treated sample with 1500 ppm of non-toxic Petroflo aldol polymerization inhibitor
3. The samples were warmed up to 40 ° C and stirred for 30 minutes
4. The samples were then stored at room temperature for 48 hours and every 12 hours a visual inspection was performed.

*Trademark of SUEZ; may be registered in one or more countries.

The pictures below (figure 5) show the results of the lab tests. The untreated sample bottle has some sticky polymer attached to the bottom. The treated samples do not have any visual deposits, indicating effective fouling control due to the treatment.

Figure 5 – Spent caustic with SUEZ non-toxic Petroflo – lab. test

d) Plant Tests & Results

Based on the laboratory results, the technology was applied in the field.

The chemical program started in May 2016. To have a smooth transition between the aldol scavenger treatment and the treatment using the non-toxic aldol polymerization inhibitor, the dosage strategy was to start to inject, the non-toxic technology, in the bottom of only one of the columns to the spent caustic stream and maintaining active the aldol scavenger injection in the caustic make-up.

*Trademark of SUEZ; may be registered in one or more countries.

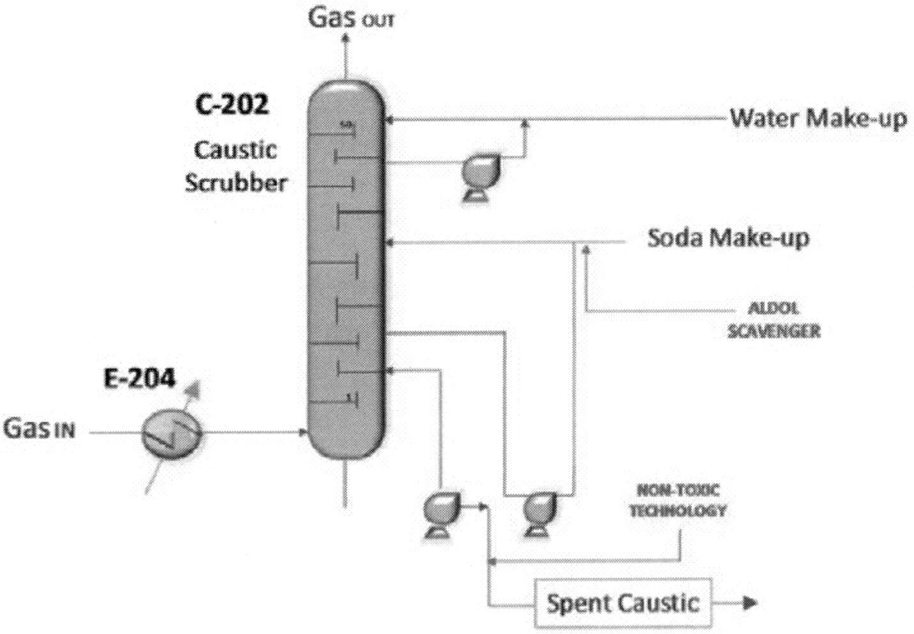

Figure 6 – First Step of dosage

The initial dosage was 3 ppm based on the throughput to the plant. Based on results of the samples (visual inspection, gum contents), this was increased in steps of 2 to 3 ppm, to a dosage of 40 ppm. After one month the injection point was changed from the spent caustic draw off to the strong caustic loop of the column and so on until the aldol scavenger was completely substituted by the new non-toxic technology in the caustic make up.

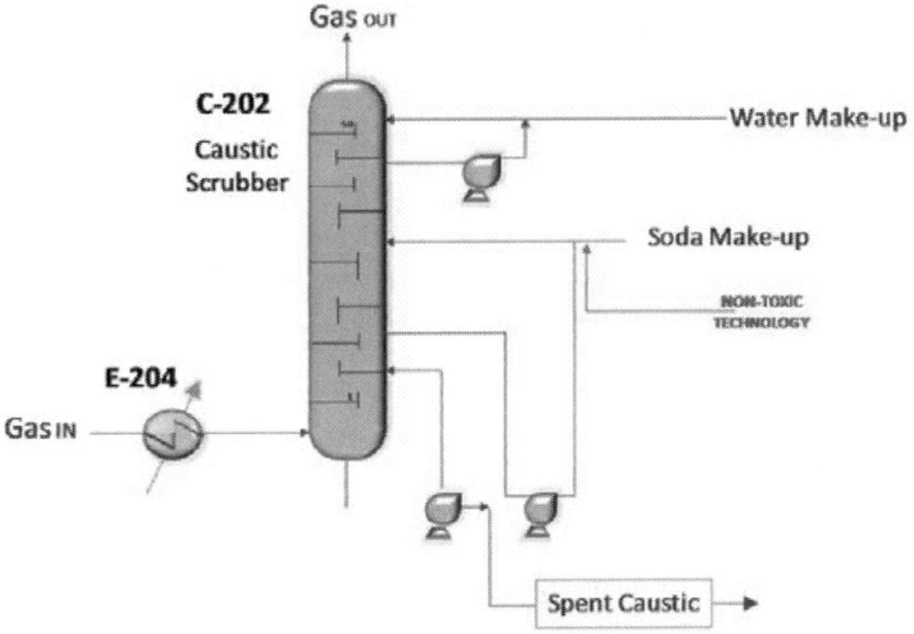

*Trademark of SUEZ; may be registered in one or more countries.

Figure 7 – Last Step of dosage

After almost one year the normal dosage was confirmed at 30 to 50 ppm based on throughput of the plant. This dosage ensured good results in all conditions tested.
As showed in below pictures comparing some spent caustic treated with Aldol scavenger (Figure 8) with spent caustic treated with the non-toxic Petroflo aldol polymerization inhibitor (Figure 9), and comparing the volume of the deposits taken from each sample, it was possible to estimate that, for those specific samples, the product resulted in **10 times less "solid" formation** and higher performance vs. other carbonyl scavengers. From an analytical point of view, the polymer content did not vary compared to the use of aldol scavenger. And indeed, the non-toxic technology ensures a better dispersion of the produced polymer.

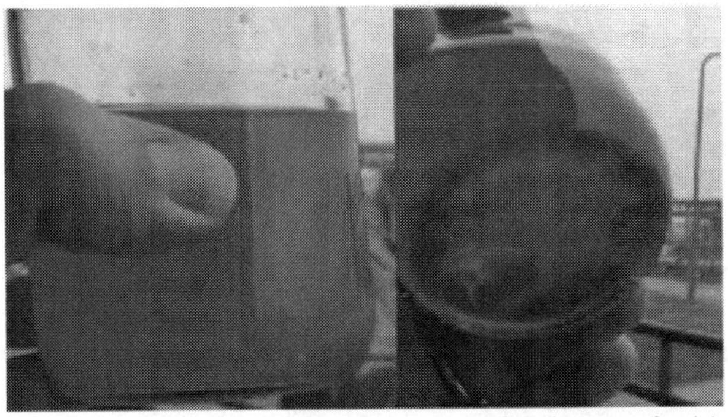

Figure 8 – spent caustic with previous technology – plant test

Figure 9 – spent caustic with the non-toxic Petroflo aldol polymerization inhibitor– plant test

SUEZ's non-toxic aldol polymerization inhibitor ensured, besides the easier handling of the product, a significant reduction in fouling rate of the system, with a consequent reduction of the disposal cost evaluated at 1 MM€ generated during three years of contract with SUEZ Water Technologies & Solutions non-toxic.

*Trademark of SUEZ; may be registered in one or more countries.

During the test period, SUEZ Water Technologies & Solutions suggested also to run the column with some different operational conditions. In particular:

1. Decrease the operating temperature of the Caustic Column to reduce the impact of the polymerization rate as explained in the previous paragraph and shown in the following figure.

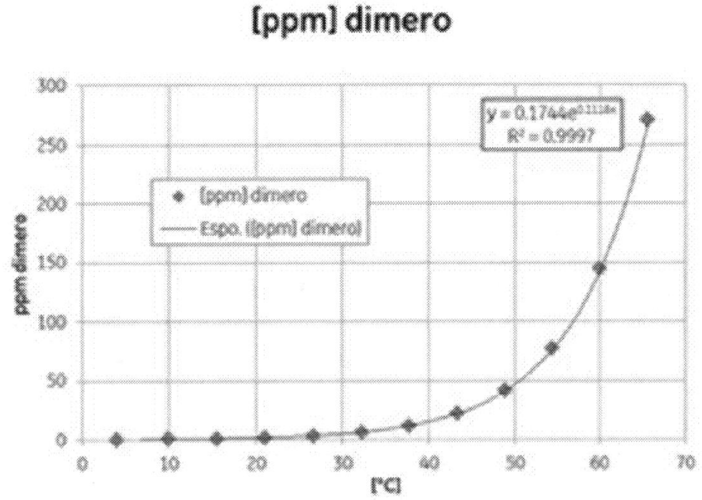

Figure 10 – Vinilcicloesene monomer formation rate

The customer agreed, and a balance was found to reduce polymerization rate whilst ensuring that the temperature remained high enough to avoid hydrocarbon condensation. Figure 11 shows a trend of the caustic scrubber inlet temperature

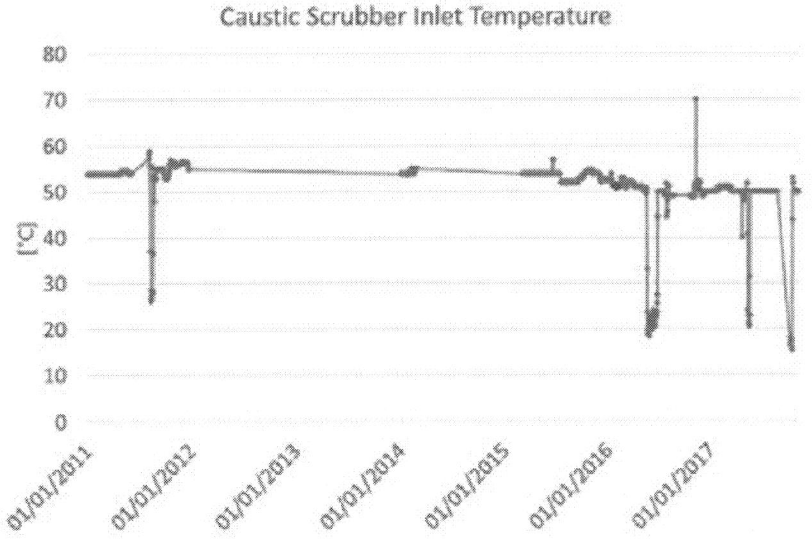

Figure 11

*Trademark of SUEZ; may be registered in one or more countries.

2. Decrease the ΔT on the preheat exchanger of the cracking gas to the caustic scrubber for the same reasons explained in the previous point.

The following pictures is self explaining about effectiveness of the treatment program. It show the difference of cleanliness after almost one year of treatment with the SUEZ non-toxic Petroflo aldol polymerization inhibitor, compared to the previous program.

Figure 12

In summary the implementation of the SUEZ non-toxic Petroflo aldol polymerization inhibitor helps to achieve;

1. Targeted Versalis' KPI of maintaining clean conditions in the caustic scrubbers as well as the spent caustic loops, thereby reducing the plugging of the system ensuring a continuous flow of the spent caustic to the waste water and biological section
2. The resulting saving was calculated of 380 k€/y and a part of it (100k€/y) was due to maintenance cost reduction minimizing the equipment fouling of the spent caustic circuit.
3. EHS Improvement: improving the handling of the non-toxic technology instead of aldol scavenger.

*Trademark of SUEZ; may be registered in one or more countries.

STUDY OF ALDOL POLYMERIZATION IN ETHYLENE PLANT CAUSTIC TOWERS

Satheesh Sambandam, Ph.D.
Lead Chemist
Nalco Champion, An Ecolab Company

Bassam Alnasleh, Ph.D.
Lead Chemist
Nalco Champion, An Ecolab Company

Debby Rossana, Ph.D.
Senior District Account Manager
Nalco Champion, An Ecolab Company

Prepared for Presentation at the 2018 Spring National Meeting
Orlando, Florida, April, 23, 2018

STUDY OF ALDOL POLYMERIZATION IN ETHYLENE PLANT CAUSTIC TOWERS

Satheesh Sambandam, Ph.D.
Lead Chemist
Nalco Champion, An Ecolab Company

Bassam Alnasleh, Ph.D.
Lead Chemist
Nalco Champion, An Ecolab Company

Debby Rossana, Ph.D.
Senior District Account Manager
Nalco Champion, An Ecolab Company

Abstract:

The caustic tower in an ethylene plant removes acid gases H_2S and CO_2 from cracked the gas. The cracked gas flows upward while a caustic solution flows downward in a liquid/gas absorption tower that operates at 110+/- 10 F. The acidic species are absorbed into the caustic where they react to form water and sodium salts. The primary concern in caustic towers is fouling and the most common mechanism of polymer formation is the aldol condensation of Acetaldehyde. Aldol polymer is formed by the base catalyzed polymerization of Acetaldehyde which continually grows in molecular weight if left unchecked. As the polymer increases in molecular weight it precipitates from solution leading to fouling. Nalco Champion has developed chemistries that react with Acetaldehyde and its oligomers to prevent polymer formation and they are typically aqueous solutions of amines or hydrides from various sources including amino acids.

Historically we have always studied aldol polymerization and its inhibition (qualitatively) in the lab at room temperature in 2 oz bottles, using Vinyl Acetate as the Acetaldehyde source, since Acetaldehyde is a gas at room temperature. But as we venture into developing new products there is a desire to understand the reaction mechanisms under a laboratory set-up that closely mimics the caustic tower conditions. When the lab evaluation method was moved from batch to semi-batch and then to continuous mode, greater insights on the mechanism of inhibition that reflects field performance was obtained and this work summarizes the results quantitatively. The use of ELSD technique proved to be a powerful method in quantifying aldol polymer as a function of time.

INTRODUCTION

Fouling in caustic towers is caused primarily by polymer formation; the most common mechanism of polymer formation in these units is the aldol condensation of Acetaldehyde. Acetaldehyde is formed in the furnaces and enters the caustic tower with the cracked gas feed. Aldol polymerization is an ionic mechanism, and it does not involve a free radical intermediate. Specifically, the polymer is formed through the reaction of Acetaldehyde with itself as it is absorbed into the circulating caustic solution and deprotonated. Vinyl Acetate, usually present from a recycle stream, will also be absorbed into the caustic solution where it readily decomposes to Acetaldehyde.

In the initial polymer formation steps, Sodium hydroxide (caustic) deprotonates Acetaldehyde and generates an anion. This anion attacks the carbonyl group of a second molecule of Acetaldehyde. Caustic initiated dehydration generates an unsaturated aldehyde that can continue to react with more Acetaldehyde and form aldol polymer. If the polymer possesses an active carbonyl site and is in the presence of Acetaldehyde and caustic, it can continue to grow (Figure 1). Polymer formation under these conditions is commonly referred to as Red Oil.

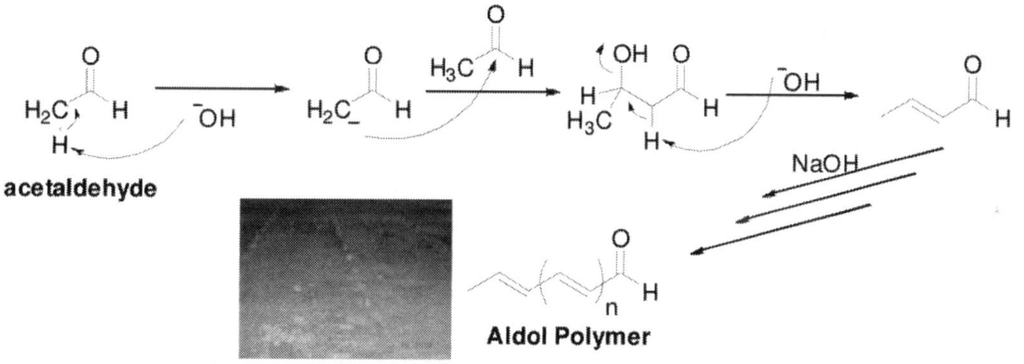

Figure 1. Acetaldehyde aldol polymerization mechanism, and photograph of aldol polymer fouling.

Vinyl Acetate monomer is typically used in all lab aldol polymerization inhibition studies in part due to its relative low volatility when compared with Acetaldehyde. Aldol polymerization and its inhibition have been studied in the lab at room temperature in glass bottles. The reaction progression or effectiveness of carbonyl scavenging was monitored visually at different time intervals or by measuring turbidity. Many times, the experiments did not reveal the true performance of the scavengers as the experimental conditions sometimes did not favor chemistries whose kinetics was faster and performed better at higher temperature. To understand the reaction mechanisms under a system that closely mimics the caustic tower conditions a quantitative method to quantify the polymer formation with time is needed. When the lab evaluation method was moved from batch to semi-batch and then to continuous mode, greater insights on the mechanism of inhibition that reflects field performance was obtained and this work summarizes the results. Evaporative Light Scattering Detector (ELSD) was used in quantifying aldol polymer as a function of time.

EXPERIMENTAL

Batch Testing

Vinyl Acetate and 2M NaOH was used for batch aldol polymerization/inhibition studies. Experiments involved 4 oz glass bottles. The tests were done at 25 °C and with a water bath maintained at 40°C. Intermittent stirring was undertaken to ensure good mixing of reactants. Carbonyl scavengers were dosed before the addition of Vinyl acetate. Estimation of aldol polymer by gravimetry, similar to the polymer extraction method used in the field, was used to determine gums contents in the test solutions. The contents were washed thoroughly with a solvent capable of dissolving aldol polymer and then the solvent was removed using a rotary evaporator operating under pressure.

Semi Batch Testing

In a semi batch mode, a syringe pump was used to inject Vinyl acetate in to a round bottom flask with caustic and carbonyl scavenger. The mixture was maintained at 40°C using an external control while keeping the mixture stirred all the times. An aliquot of sample was taken from the reaction mixture and diluted in a solvent and then injected in Evaporative Light Scattering Detector (ELSD) to estimate the polymer. A calibration curve of the aldol polymer was then used to estimate the concentration of the injected sample.

Continuous Stirred Reactor

A set up similar to the semi-batch system, under same conditions was used in a continuous mode with two syringe pumps each sending Vinyl Acetate and 2M Caustic mixed with carbonyl scavengers and a third syringe pump was used for withdrawing samples for testing. The pump rates were ensured that the residence time is 30 minutes and that there is no build-up of reaction mixture.

RESULTS

Aldol polymerization and inhibition studies can be performed quickly in the lab by reacting Vinyl Acetate and caustic and because the polymer formed is colored, the reaction progress can be visually tracked. Therefore, the performance of different carbonyl scavengers can be analyzed without the need for any analytical tool. Recording the reaction progress by time lapse pictures can be used as the conclusive proof of one chemistry performing better than the rest. In Figure 2, it is clear in 30 minutes that the Blank is turbid with Aldol polymer concentration being the highest while carbonyl scavenger C is clear with no visual evidence of polymer formation. A and B are nearly identical but show presence of Aldol polymer albeit lower than the Blank. In addition to color, turbidity was also used either as a visual or analytical tool in analyzing the performance of chemistries. The limitations of this study are many 1) Scavengers themselves can be colored or add color to the mixture during the reaction (example Scavenger A turns dark orange color during reaction with time), 2) scavengers with fast kinetics may not show performance over longer periods and 3) quantitative results in shorter time scale is not possible.

Figure 2. Aldol polymer inhibition studies using three Carbonyl Scavengers at 25°C.

A quick experiment to compare the effectiveness of the scavengers when the starting monomer is Vinyl acetate and Acetaldehyde was conducted. When the starting monomer is Vinyl Acetate, it hydrolyzes in caustic solution to form Acetaldehyde which then undergoes aldol condensation reaction. The goal of the experiment was to find out if aldol polymerization inhibition mechanism has any effect on the monomer used. The moles of Acetaldehyde and Vinyl Acetate was kept constant in these experiments. Results in Figure 3 indicated that the inhibition trends pretty much followed the same profile whether it was Vinyl Acetate or Acetaldehyde at both room temperature and 40°C. Vinyl Acetate is used by Nalco Champion to simulate caustic tower aldol fouling to circumvent the handling issues due to low boiling point (20°C) of Acetaldehyde.

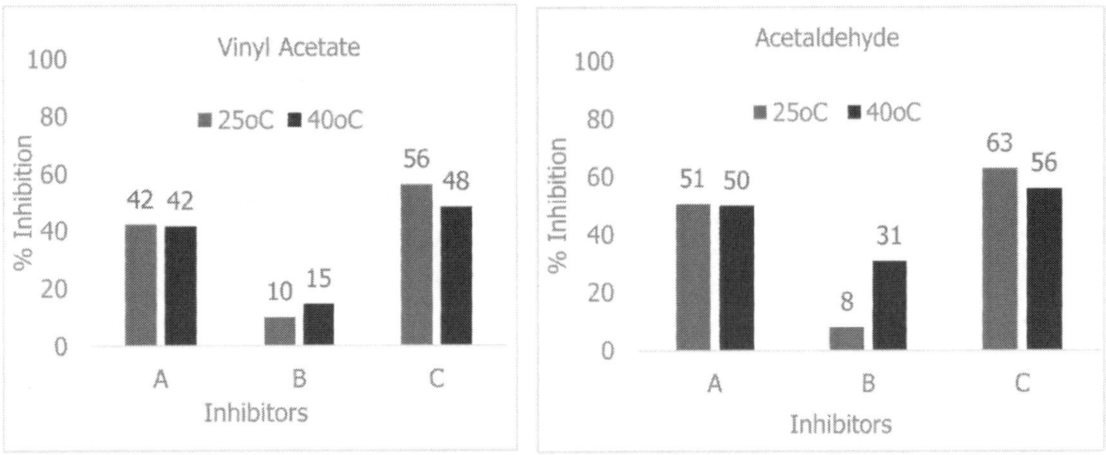

Figure 3. Effectiveness of Carbonyl scavengers as a function of Temperature and Monomer. % Inhibition equals ratio of performance with respect to blank.

The effect of temperature on the performance of carbonyl scavengers was also studied. The two common temperatures most of the aldol polymer inhibition experiments are carried out are room temperature and 40-45°C. Carbonyl scavenger A's performance did not change with temperature for both the monomers while B's performance improved when increasing the reaction temperature to 40°C for Vinyl Acetate and Acetaldehyde. The difference was higher in the case of Acetaldehyde as starting monomer. If the size of the scavenger molecule is larger, then higher temperature can help in scavenging. There was only a slight difference in performance for carbonyl scavenger C.

The experiments involving Vinyl Acetate and caustic were run for different reaction times starting from 15 min to 3 hours to determine rate of the reaction under a specific set of lab conditions. Figure 5 indicates the mass of aldol polymer for a blank sample under various reaction times. The data shows that the rate of the reaction is indeed faster and a measurable (by gravimetry) amount of polymer is formed within the first 15-30 minutes of reaction. Since the reaction is faster it would be wise to study the performance of carbonyl scavengers under shorter time scale rather than longer time periods. Therefore, new experiments were designed to study the inhibition efficiency of carbonyl scavengers. But the challenge now lies in measuring low concentrations of the polymer. Gravimetric method that was used thus far has limitations in measuring low polymer concentrations and/or larger error associated with smaller weight data in addition to errors such as solvent trapped in polymer after drying etc.

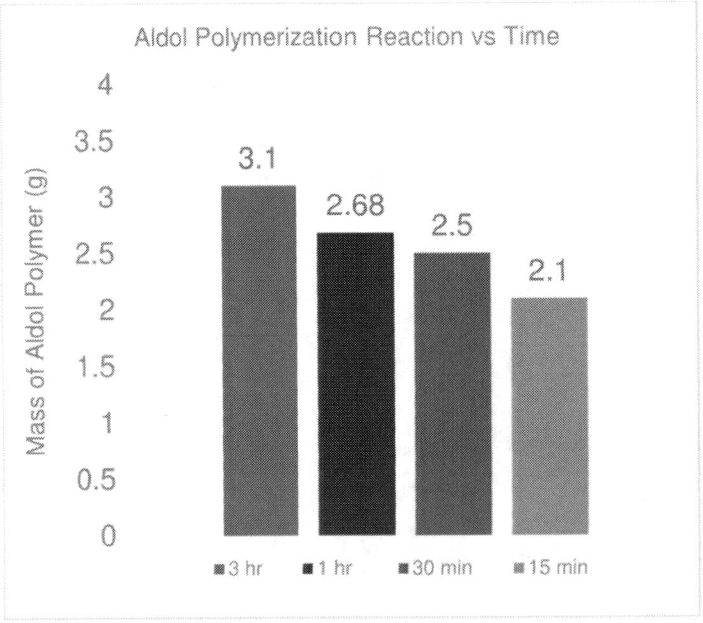

Figure 5. Mass of aldol polymer formed as a function of reaction time under blank condition (no scavenger)

Evaporative light scattering detector (ELSD) was used to measure the amount of polymer in the samples. This is the first instance of any attempt to measure aldol polymer using ELSD. The only challenge that prevented the use of this method in the past is caustic present in the reaction. Caustic will remain even after solvent evaporation and hence it will show up along with the polymer during detection. Therefore, to work around this issue, the reaction mixture of caustic and aldol polymer sampled at different time intervals was added to a vial containing solvent that can dissolve aldol polymer. The resultant solvent was then injected in to ELSD which then helped quantify the aldol polymer present in the sample.

A semi batch experiment with Vinyl Acetate monomer flowing into a flask containing 2M caustic dosed with carbonyl scavenger was conducted with the objective of sampling the reaction vessel at different time intervals and tracking the progress of the reaction under inhibited and uninhibited conditions. Aldol polymer concentration in ppm was measured using ELSD and the experiment was stopped after 30 min to get at least 5 data points to plot the reaction profile. Figure 6 shows the polymer profile comparing blank with carbonyl scavengers A, B and C. Scavenger C showed the best performance and the result is in line with the earlier study. Scavenger B was inhibiting the formation of aldol polymer for the first 15 minutes and then it was acting like a retarder. Scavenger A's performance was not at par probably due to its slow kinetics. Although the semi-batch reactor refined the test conditions, the experiment still lacked true continuous mode operation that mimics the real world conditions.

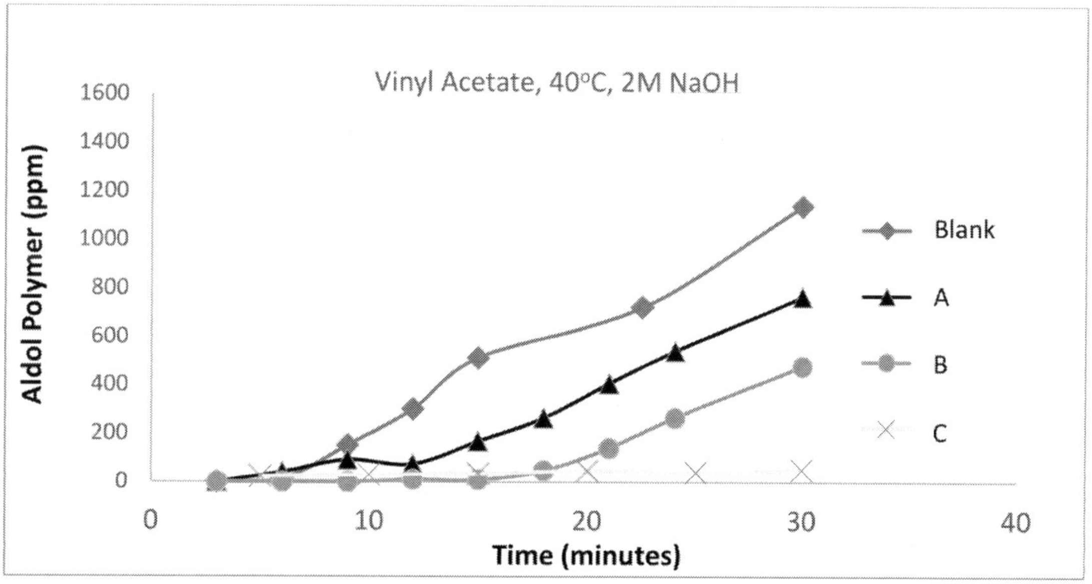

Figure 6. Aldol polymerization under Semi batch mode

An experimental set up that mimics a Continuous Stirred Tank Reactor was used to study the reaction under continuous operation mode. A syringe pump was used to pump both the reactants to a flask and a third syringe pump was set to withdraw the reaction product from the flask at a flow rate that did not change the volume of the mixture in the flask. The residence time was set to 30 minutes. Since it's a CSTR, the reaction set up is expected to achieve steady state in 2 or 3 residence times. Hence was decided to pick the 90 minute data as steady state for each run. Figure 7 shows steady state concentrations of aldol polymer under various treatments including the blank. The system was sampled every 30 minutes to measure aldol polymer. The blank achieved steady state fast at ~500 ppm of aldol polymer. The sample treated with carbonyl scavenger A had a slow transition to steady state, possibly due to its poor kinetics, and this was evident in the semi batch experiment as well. Performance of B and C was in the range of 100-200 ppm of aldol polymer.

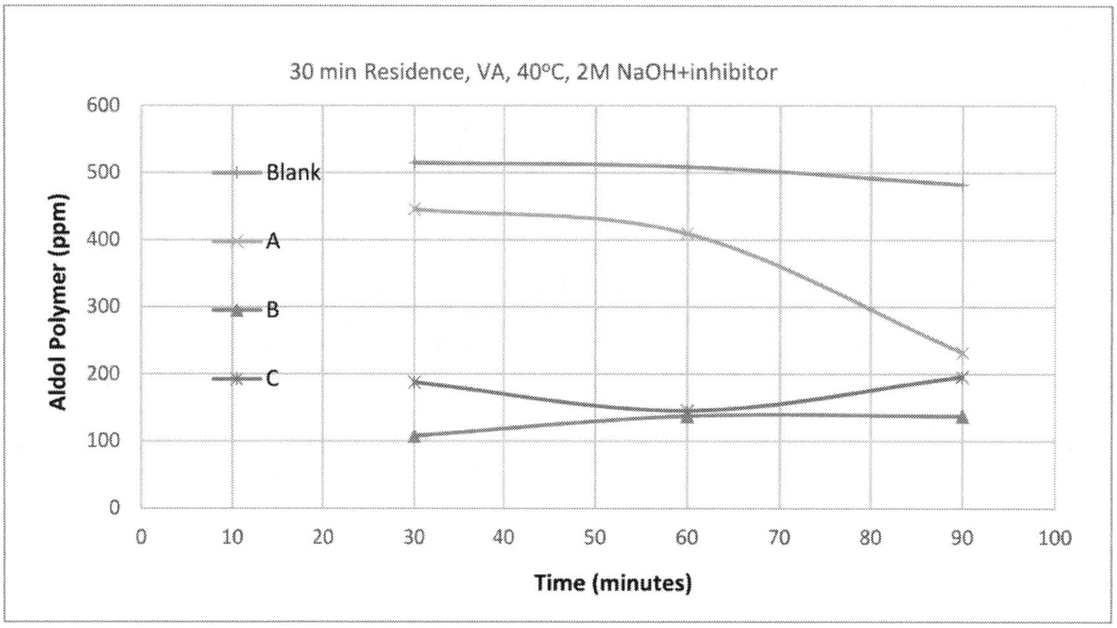

Figure 7. Aldol polymerization under CSTR (Continuous Stirred Tank Reactor) mode

CONCLUSIONS

Performance testing of carbonyl scavengers using bottle testing may not be the right method even though it's the easiest method. Experiments should be conducted for shorter time scale at temperatures close to the caustic tower operation. Some of the scavengers show better performance at caustic tower temperature than at room temperature. There are two different methods of performing the experiment based on the order of dosage – pre-dosing the aldol scavenger and then forming the polymer or allowing the polymer to form and then dose the chemistry. Interestingly, pre-dosing and post dosing of aldehyde scavenger can change the results of the batch test. In the batch method using glass bottles, the chemistry with faster kinetics will get consumed quickly and performance over extended periods will be poor. However, this problem can be averted in the case of CSTR which runs continually. One should be cognizance of kinetics and moles of active when evaluating different carbonyl scavengers. Every effort should be made to quantify the aldol polymer formed rather than use pictures or turbidity as scavenger can affect both parameters. Some inhibitors are colored, and pictures should not be used to select the appropriate chemical. The use of ELSD technique gave us a method to quantify aldol polymer at low concentrations and determine the efficacy at shorter time scales.

AIChE Paper Number: 32d

HOW MUCH IS TOO MUCH?

FEED CONTAMINANTS AND THEIR CONSEQUENCES

Kandasamy M. Sundaram, PhD.
Distinguished Technologist
CB&I

Barbara Stancato
Manager, Aftermarket Services
CB&I

Prepared for Presentation at 2018 Spring National Meeting
Orlando, Florida, April 22-26, 2018

HOW MUCH IS TOO MUCH?
FEED CONTAMINANTS AND THEIR CONSEQUENCES

Kandasamy M. Sundaram, PhD.
Distinguished Technologist
CB&I
Barbara Stancato
Manager, Aftermarket Services
CB&I

Abstract

Olefin producers use economically attractive feedstocks from different sources to be competitive. Sometimes this introduces contaminants that, untreated, can produce adverse effects in plant operation. Some contaminants cause cracking heater problems ranging from short run length to coil damage. Other contaminants cause problems in the recovery section such as fouling of equipment and catalyst poisoning. Contaminants in cracker feedstocks have been discussed by many authors in previous ethylene producer's conferences [DeHaan, 2006; Bernard and Snider, 2008; Mohundro, 2013; Brayden et al., 2017; Graham 1993 for example]. This paper discusses some known contaminants in the hydrocarbon feeds and dilution steam and their effects on plant operation. Conventional gaseous and liquid feeds are reviewed. Some uncommon contaminant sources such as external feed streams that enter the plant downstream of the heaters are also considered. The recommended maximum limit to avoid known problems for each contaminant is provided at the end of the paper.

Introduction

Ethylene plants are designed to crack a wide range of feedstocks from gaseous feeds, like ethane, to very heavy liquid feeds, like vacuum gasoil. The operating feeds are often much different from the design feeds for economic reasons. Since the feeds come from a variety of sources, they can contain contaminants that cause serious operational problems. Figure 1 shows the results of a recent industry survey of sixty plants and shows the percentage of plants that reported a measurable amount of contaminants in their heater feeds. Typical contaminants routinely analyzed are shown in the X-axis of Figure 1. Only a few plants exceeded the acceptable limits.

"Contaminant" is a general term often used in the industry. Other terms, such as impurity and pollutant, are also interchangeably used to describe the same effect. Contaminant is the presence of an unwanted specie in another substance. Pollutant is something that causes the contamination of the natural environment. Impurity results from insufficient purification. Methane in the ethane feed is an impurity. However, carbonyl sulfide (COS) in ethane feed is a contaminant. The ethylene plant can tolerate a high concentration of methane in ethane feed but even a few ppm COS is undesirable in a cracker feed.

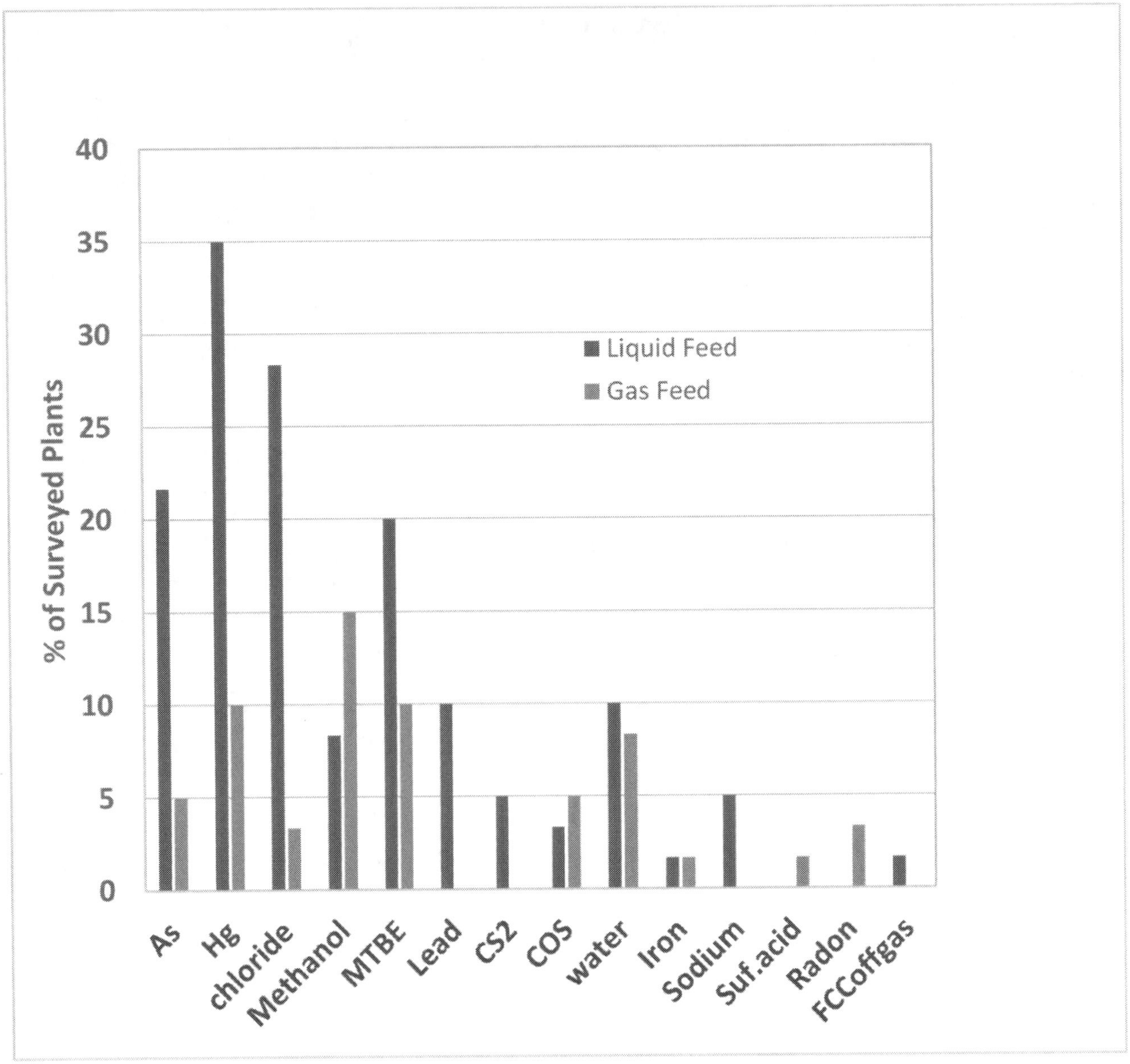

Figure 1. Reported Contaminants in Plants Surveyed

Feed source contaminants can be classified primarily as gas feed contaminants and liquid feed contaminants. Both gas and liquid feeds may also contain solid contaminants present as fine particles in gas feeds and finely dispersed in liquid feeds. In this paper the term "cracker feeds" are those feeds that pass through the cracking heaters. In some plants, olefin rich streams are directly sent to recovery section. These are denoted as external feeds (or streams). In addition to feeds, contaminants can be present in dilution steam, chemical additives and catalysts. When certain contaminants are detected immediate action is required. Other contaminants may affect the plant operation slowly over time.

This paper focusses on carbonyl sulfide (COS), mercury (Hg) and sodium (Na) as examples of the contaminants reported in the industry, frequently the subject of technical service requests and their impact on olefin production.

Gas Feed Contaminants

Notable contaminants in gas feedstocks are COS, oxygenated compounds (e.g. methanol), chlorides, dissolved salts and traces of mercury. Mercury is usually present in extremely small concentration (not detectable to <120 µg/Nm^3) [Corvini et al., 2014]. Ethylene gas crackers very rarely require a mercury removal system. Naturally occurring radioactive materials are also found [Taylor and Cleghorn, 1996].

Liquid Feed Contaminants

Many contaminants are reported in liquid feeds. Metals such as arsenic, mercury and vanadium, oxygenated compounds such as methanol, sulfur compounds such as carbon disulfide, and nitrogen compounds such as amines are reported. In addition, high molecular weight hydrocarbons may be present in low molecular weight liquid feeds and in that respect, those high molecular weight species will be considered as contaminants. For example, sometimes crude transporting ships are used to transport naphtha feeds. If the vessels are not adequately cleaned some residues (very high boiling material) can be present in the naphtha shipment and cause problems.

External Feed Contaminants

External feeds are generally sent directly to the recovery section. Some external streams (vents from other processes) are mixed with fresh cracker feeds and sent to heaters. These external feeds are previously processed streams such as FCC off gas, polyethylene vents, etc. These feeds sometimes contain impurities in significant levels that require removal. Typically, oxygenates, catalysts and some inorganic materials are present.

Other Modes of Contaminants

Cracker hydrocarbon feed is always mixed with dilution steam before being cracked in the heaters. There are instances where contaminants in the dilution steam have resulted in adverse effects. Various fuels are used for supplying the heat duty necessary for the reaction. Contaminants in both gaseous and liquid fuels have affected the performance of the heater.

To illustrate the potential impact of contaminants on olefin production a few of the contaminants are discussed individually. The table at the end is a general summary of the contaminants, recommended limits and their consequences.

Carbonyl Sulfide (COS)

Carbonyl sulfide (COS) can be present in gas feeds mainly in ethane and propane. During coal gasification, a portion of sulfur in the coal is converted to COS. The natural gas liquids from some fields contain COS. A typical composition from one field is shown in Table 1 before treatment.

Table 1. Typical Ethane and Propane Feed Compositions from Gas Separation Plant

Component	Units	Ethane	Propane
Methane	Mol%	1.8	0.
Ethane	Mol%	96.4	.5
Propane	Mol%	1.5	98.2
i-Butane	Mol%	0.	1.0
n-Butane	Mol%	0.	0.3
CO2	ppmv	50	0.
H2S	ppmv	7	0.
Mercaptans	ppmv		183
COS	ppmv	76	2
H2O	ppmv	2856	

COS in the feed is not completely converted to products (H_2S) in the cracking heater. In fact, in almost all cracking heaters COS is formed in trace quantities by the following reactions:

$$CO2+H_2S \rightarrow COS + H_2O \quad (1)$$

In trace quantities COS is not a problem in the recovery section. However, when it is present in significant quantities (see Table 8) though small, it has caused catalyst deactivation and off spec propylene product. The properties of carbonyl sulfide are shown in Table 2.

Reactions

The decomposition of COS has been discussed by many authors [Preston and Cvetonovic, 1972; Schecker and Wagner, 1969; Karan et al., 1999]. Under pyrolysis conditions there are many reactions involving COS. COS is formed in the reactor and is also destroyed. Some of them are reversible reactions. The main ones are:

The reverse of reaction (1)

$COS + H_2O \rightarrow CO_2 + H_2S$ (2)

Similarly, the reaction of CO and H_2S is important.

$COS + H_2 \leftrightarrow CO + H_2S$ (3)

Elemental sulfur is involved in the formation of COS.

$H_2S + M \leftrightarrow H_2 + \frac{1}{2} S_2 + M$ (4)

$CO + \frac{1}{2} S_2 \leftrightarrow COS$ (5)

(M= metal)

Table 2. Properties of Carbonyl Sulfide (COS) and Propylene

Parameter	**Carbonyl Sulfide**	**Propylene**
Molecular Formula	COS (S=C=O)	C3H6
Molecular Weight	60.075	42.08
Boiling Point	-50.2°C	-47.6°C
Melting Point	-138.8°C	-185.2C
Vapor Pressure at 0°C	6.53 bar	13.69 bar

Karan et al. (1999) in fact claim that COS formation by reaction (1) is extremely small and the decomposition appears to be equilibrium limited. They concluded that hydrogen sulfide decomposes to sulfur and hydrogen and the elemental sulfur plays a critical role in the formation of COS. At higher temperature, COS formation is limited by thermodynamic equilibrium. Though simple molecular reactions are shown, the actual reaction mechanism is more complex and involves free radicals. The equilibrium constants for these reactions are shown in Table 3.

Table 3. Equilibrium Constants for Some COS Forming Reactions

Temperature, °K(°C)	1000(727)	1100(827)	1200(927)
$CO_2 + H_2S = COS + H_2O$	0.0237	0.0339	0.0341
$CO + H_2S = COS + H_2$	0.0345	0.0340	0.0340
$H_2S = H_2 + \frac{1}{2} S_2$	0.0072	0.0193	0.0439
$CO + \frac{1}{2} S_2 = COS$	4.7757	1.7580	0.7752

Under pyrolysis conditions, most organic sulfur compounds are converted to hydrogen sulfide [Brayden et al. 2008]. The remaining sulfur appears in the gasoline and fuel oil products. There are also some reforming reactions producing CO and CO2, paving the way for COS formation in cracking heater. Equilibrium COS formation as a function of temperature is shown in Figure 2 for a typical ethane cracking heater. Actual measured values for COS are much lower than the equilibrium values shown. When COS is present in the feed all COS is not destroyed [Karan et al. 1999]. COS tracks with propylene and can cause the product to become off spec (see propylene properties compared with COS in Table 2). In naphtha cracking COS can be formed when carbon disulfide (CS_2) is present and has caused significant reduction in pyrolysis gasoline hydrogenation catalyst activity [Coenen, 2016]. The propylene specification is typically less than 50 ppb of COS.

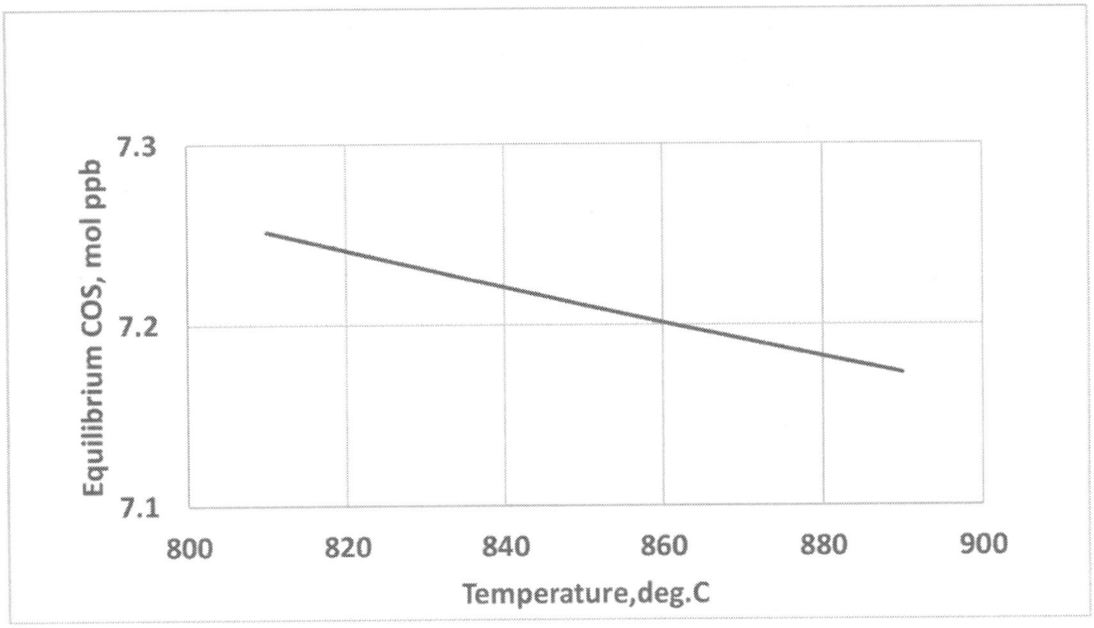

Figure 2. Equilibrium COS Formation in Ethane Cracking

COS Removal

COS in the feed can be reduced significantly by amine absorbers [Katz and Giesen, 2009; Hugo and Wagner, 2000; Huntsmann, 2003 and Zhao et al., 2013]. Even trace quantities of COS can be removed by adsorbents [Huntsmann, 2003]. Kinetics of COS removal with primary and secondary amines in aqueous solutions are discussed by Little et al. [1992]. Often absorbers are used in gas liquid separation plants since CO2 has to be removed anyway. When additional removal is required, it is handled in the ethylene plant. Lummus has used both approaches successfully. For example, using a guard bed, over 1000 ppb of COS in the feed was reduced to below 30 ppb (detection limit) in one plant.

Mercury and Other Metals in Liquid Feed

As mentioned above there are many liquid feed contaminants. As an example, mercury in liquid feed will be considered. Generally, naphtha feeds do not contain mercury. Some condensate feeds contain metals like arsenic, mercury and other heavy metals. In natural gas mercury is present in trace quantities (< 200 μg/m^3) [Shafawi, 1999] but this can be enough to cause problems. Though elemental mercury boils at 357°C, mercury derivatives boil at lower temperatures. Typical mercury concentrations in natural gas and condensates found in various regions are shown in Table 4 [Shafawi, 1999]. Mercury found in a typical South East Asian condensate as a function of boiling point is shown in Figure 3. Acceptable limit for mercury is less than 5 ppb in the feed.

Table 4. Typical Mercury Concentrations Found in Condensates and Natural Gas

Location	**Mercury Concentration**	
	Gas(μg/m^3)	Liquids(μg/kg),ppb
Europe	100-150	-
South America	50-120	50-100
Gulf of Thailand	100-400	400-1200
Africa	0-100	500-1000
Gulf of Mexico (USA)	0.02-0.4	-
Overthrust Belt (USA)	5-15	1-5
North Africa	50-80	20-50
Malaysia	1-200	10-100
Indonesia	200-300	10-500

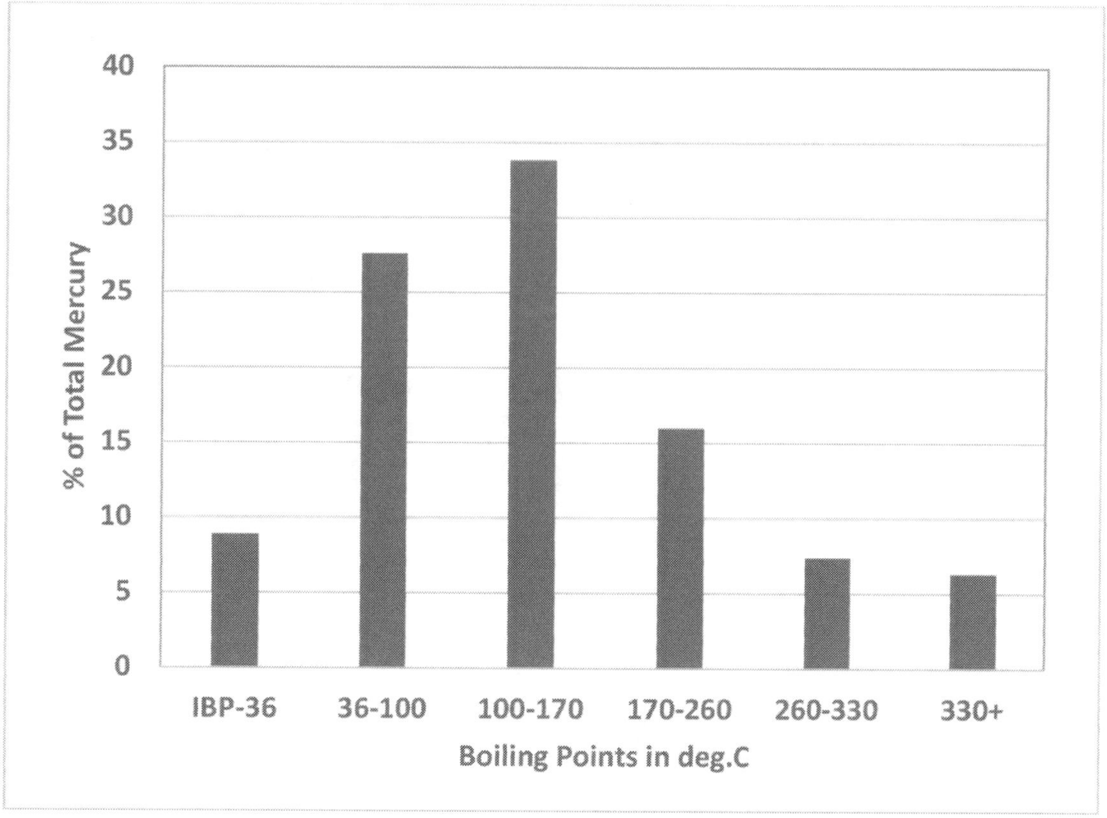

Figure 3. Mercury Distribution in a Condensate

Figure 3 clearly shows more than 50% of mercury compounds are concentrated in naphtha boiling range (IBP to 200 C) in this condensate. Typical compounds found in natural gas and liquids are shown below (Table 5).

Table 5. Compounds of Mercury

Species	**BP,deg.C**
Dimethyl mercury ($(CH_3)_2Hg$)	96
Diethyl mercury ($(C_2H_5)_2Hg$)	159
Diisopropyl mercury ($(i.C_3H_7)_2Hg$)	170
Dipropyl mercury ($(C_3H_7)_2Hg$)	190
Dibutyl mercury ($(C_4H_9)_2Hg$)	206
Mercury (Hg)	357

Ezzeldin et al. [2016] have given a detailed description of mercury affected units in natural gas processing plants and the effects of not removing the mercury. Mercury causes corrosion and affects brazed aluminum exchangers in the recovery section. Mercury is a toxic chemical and hence there are severe restrictions in emission standards, which is a big concern during turnaround maintenance and human exposure. Mercury has to be removed when it exceeds a critical value. Often detection of mercury compounds is difficult. Induced plasma mass spectrometry (ICP-MS) and other methods are used to measure even trace quantities [Bouyssiere et al., 2000]. Cold vapor fluorescence can detect quantities as low as 0.0001 nanogram [Shafawi, 1999].

Metals like Al, Cu, Cd, Ag, and Zn react with water and diffuse into mercury and this promotes corrosion. Mercury reacts with aluminum forming amalgam. This reaction affects all brazed aluminum exchangers, and is discussed in detail in previous ethylene producers' conferences [English,1989; Baumgartner and Coleman,1996]. Mercury induced stress cracking occurs in brazed aluminum exchangers at the aluminum metal grain boundaries.

$$4\ Hg + Al_3Mg_2 \rightarrow 2MgHg_2 + 3\ Al \qquad (6)$$

Mercury also catalyzes reaction with water.

$$Al_x\ (s) + Hg_y\ (l) \rightarrow Al_{x-2}\ (s) + Hg_yAl_2\ (l) \qquad (7)$$
$$Hg_yAl_2(l) + 3H_2O(l,v) \rightarrow Hg_y(l) + Al_2O_3\ (s) + 3H_2(g) \qquad (8)$$

Aluminum oxide protects further corrosion of aluminum.

Removal Techniques

In ethylene plants, mercury compounds are removed either in the feed or in the cracked products. Unless the type of mercury compound is known, the efficiency of removal in the liquid feed may be low. Liquid phase removal technology using sulfide containing resins has been used in some plants [Franken and Hubbers, 1991]. Vapor phase technologies have also been commercialized for mercury removal [Sarazzin et al., 1992; Coleman et al., 2004; Corvini et al 2014.] Vapor phase removal generally employs activated charcoal adsorbents [Corvini et al., 2014]. Mercury compounds are generally removed prior to cryogenic recovery. Sainal et al. [2007] have used sulfur containing resin based adsorption technology to remove mercury from natural gas condensates and found better than 95% mercury removal for Petronas condensate.

As mentioned above both liquid phase and vapor phase removal technologies are used. Liquid phase technologies generally use catalysts that demetallizes. Vapor phase technologies use adsorption type technology. Ionic solid adsorbents are also used to remove mercury from natural gas [Abai et al., 2011]. Activated carbon is typically used for adsorption.

Sodium

Sodium found in liquid feeds, may be the residues left from desalting of crudes. Sweetening of gasoline (catalytic conversion of mercaptans to disulfides in presence of air and caustic) and storage of feeds in salt caverns can also add sodium to feeds [Devakotai et al., 2011]. Sodium can also be naturally present as an organic molecule in the feed. Brayden et al. [2006] mentions that the contained water in naphtha feeds had as much as 13000 ppm of sodium. However, the main source of sodium entering the heaters for most plants is the dilution steam. Caustic is used sometimes to control the pH of quench water and for acid gas removal. Occasionally, due to some upset, sodium can carry over into the dilution steam and can cause significant damage to the coil surface in the cracking heaters. As a result, some units have switched to amines for pH control. Caustic carry over is an upset condition and is therefore a short term phenomenon. Therefore, after the incident the analysis of dilution steam shows no caustic contamination. Even if the heater has experienced a short burst of sodium carry over, subsequent operations typically show significantly reduced radiant coil run length. A typical run length impact is shown in Figure 4. Cycle 1 is before sodium attack with 60 days run length. Subsequent run lengths are less than 40 days and never were restored to the original values. Sodium carry over can be confirmed only by metallurgical analysis of the affected metal. Typical metallurgical analyses of a fresh coil and a coil that has suffered sodium attack are shown in

Table 5. Devakotai et al. [2011] have discussed the effect of caustic contamination in gasoline feed. Sodium attack can be from sodium chloride or from caustic. Sodium attack is from inorganic sources rather than from organically bound sodium in a naphtha fraction. Sodium attack on metals is very similar whether it is from dilution steam or inorganic sodium present in liquid feeds. Inorganic sources affect the metallurgy and reduce the run length.

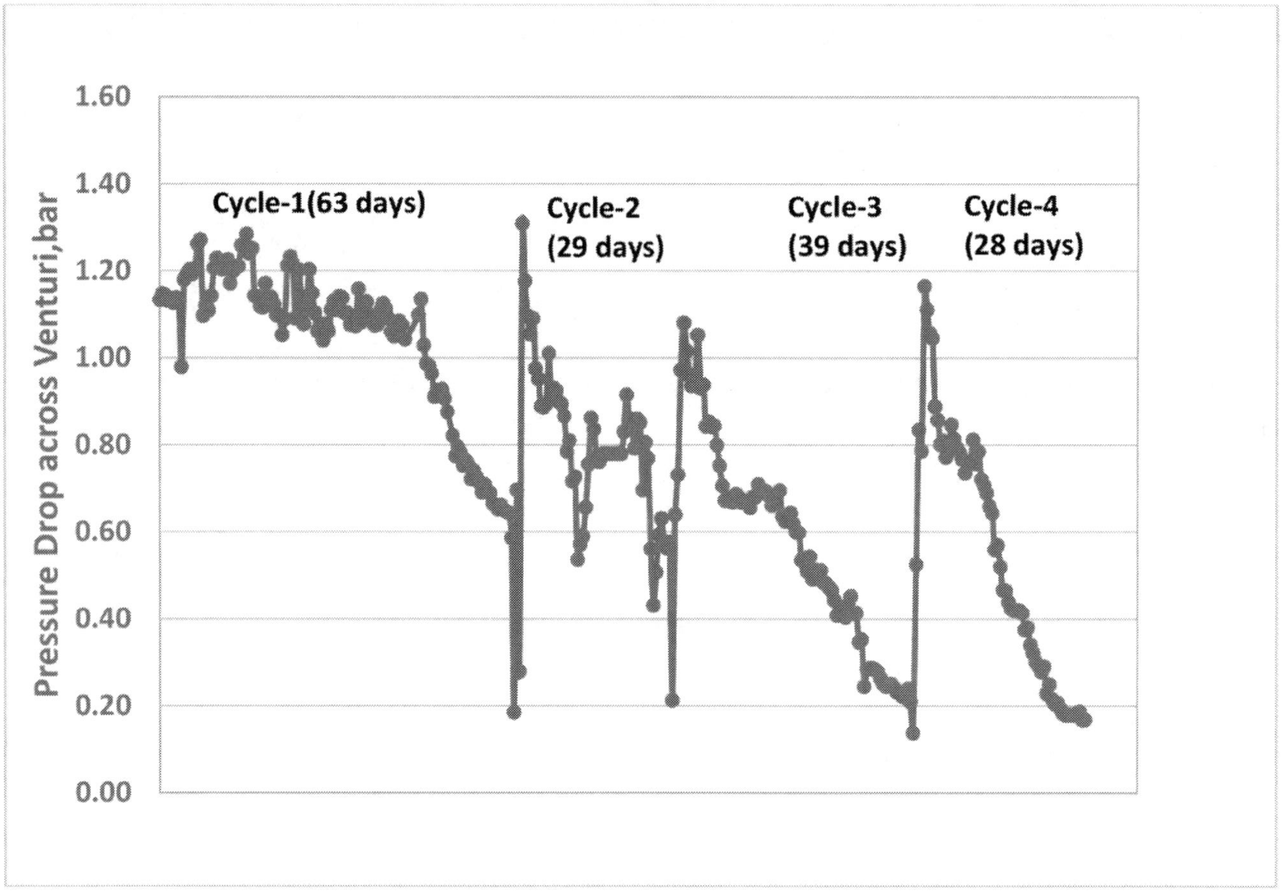

Figure 4. Effect of Sodium Carry Over on Ethane Cracking Run length

Sodium reacts with Cr in the coil metallurgy and forms compounds which have low melting points. For example, sodium chromite ($NaCrO_2$) has a melting point of 762°C and sodium chromate (Na_2CrO_4) has a melting point of 792°C. This affects the surface properties by harming the chromium oxide layer and depleting the Cr content of the coil. The surface chromium oxide acts a protective layer for diffusion of carbon and protects against catalytic coking. This protective chromium oxide layer is not easily restored with sodium attacked radiant coil tubes even after many repeated controlled decokings. Some of the sodium and chlorine reactions on metals are shown below. Hasan and Mobin [2013] concluded that at high temperatures sodium chloride is converted to hydrogen chloride, which also attacks the metal. Both sodium oxide and chlorine attack oxide scales to form corrosion products ($Na_2O.M_2O_x$).

$2 NaCl + 0.5 O_2 \rightarrow Na_2O + Cl_2$ (9)
$Na + CrO_2 \rightarrow NaCrO_2$ (10)
$Na_2O + Cr_2O_3 \rightarrow Na_2CrO_4$ (11)
$Na_2O + Al_2O_3 \rightarrow 2NaAl + 2O_2$ (12)
$Na_2O + Cr_2O_3 \rightarrow 2NaCrO_2$ (13)
$Al_2O_3 + 3Cl_2 \rightarrow 2AlCl_3 + 3/2 O_2$ (14)
$Cr_2O_3 + 3Cl_2 \rightarrow 2CrCl_3 + 3/2 O_2$ (15)
$Cr_2O_3 + 3C + 3Cl_2 \rightarrow 2CrCl_3 + 3CO$ (16)
$2Cr_2O_3 + 3C + 6Cl_2 \rightarrow 4CrCl_3 + 3CO_2$ (17)
$Cr_2O_3 + CH_4 + 3Cl_2 \rightarrow 2CrCl_3 + CO + 2H_2O$ (18)
$2NaCl + H_2S \rightarrow Na_2S + 2HCl$ (19)
$Cr_2O_3 + 6HCl \rightarrow 2CrCl_3 + 3H_2O$ (20)

Since the radiant coil surface is affected usually it is necessary to replace the coil in order to restore the run length. Attempts to clean the coil with BFW or other chemical agents are typically not successful. These methods yield some improvements but do not generally even restore 50% of the lost run length. However, if the chromium depletion is not too severe, several cycles of aggressive decoking may be able to re-establish a reasonable chromium oxide layer. On the other hand, naturally occurring (organic?) sodium in naphtha feeds has caused some increase in CO and CO2 in the effluent only. In such cases significant reduction in run length was not noticed. A sketch of sodium affected alloy coil is shown in Figure 5. This is also dicussed by K.Enqvist [2008],which contains a photograph of a damaged coil. The metallurgical analysis of the alloy is in Table 6.

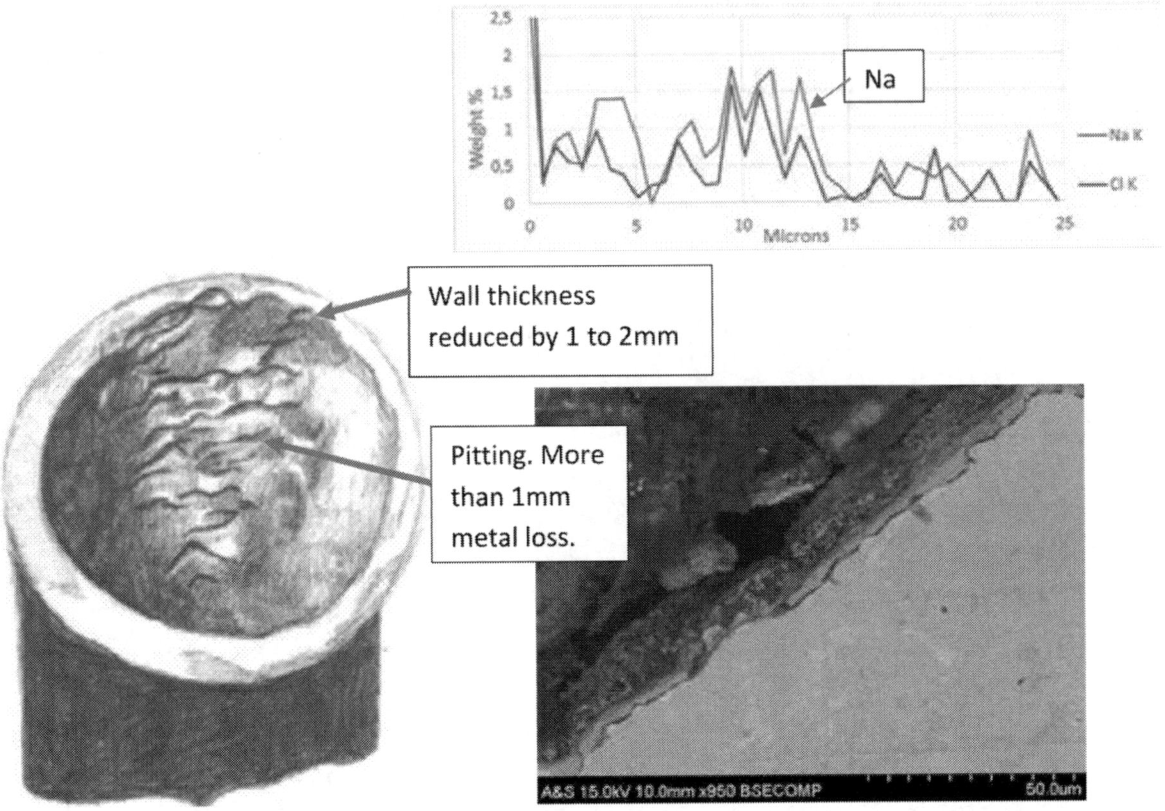

Figure 5. Sketch of Sodium Affected Alloy Coil (Caustic Carry Over in Dilution Steam) and EDS Profile analysis

Table 6. Metallurgical Analysis of Fresh and Na Affected Alloy (Surface Particles Collected)

	New Alloy,wt%	**Surface Deposit (wt%)**
Nickel	**46.2**	0.6
Sodium		**34.7**
Iron	**16.2**	1.0
Chromium	**35.6**	**2.0**
Sulphate		**10.4**
Sulphur		**3.4**
Silica		**1.2**
Cobalt		0.06
Manganese	1.1	0.03
Others	0.9	46.6

Oxygenates In External Feeds

Some ethylene plants use external streams derived from other processes. FCC produces a significant amount of ethylene, but the concentration of ethylene in the dry gas is low and typically not removed in the FCC unit. These ethylene-rich streams are usually sent to nearby ethylene plants, reacted as dilute ethylene or used as fuel. The FCC off-gas contains many species including some contaminants in addition to ethylene. Typical FCC off-gas composition is shown below (Table 7).

Table 7. FCC Off Gas Contaminant Levels Before and After Treatment

	Before			**After**	
	amount	units		amount	units
H_2S	6.3	Mol%		<20	wppb
H_2	5.8	Mol%			
N_2	10.4	Mol%			
O_2	0.1	Mol%		<0.1	Mol%
CO	1.1	Mol%			
CO_2	4.4	Mol%		<0.5	wppm
CH_4	31.8	Mol%			
C_2H_4	20.4	Mol%			
C_2H_6	13.8	Mol%			
C3s	5.1	Mol%			
C4+	0.5	Mol%			
Water	0.3	Mol%		<0.1	wppm
Ammonia	1	wppm		<0.1	wppm
NitrogenOxide	2	wppm		<0.1	wppm
Nitriles	10	wppm		<0.5	wppm
COS	5	wppm		<20	wppb
SO_x	5	wppm		<0.1	wppm
HCN	35	wppm		<0.5	wppm
Phosphines	1	wppm		<5	wppb
Arsine	1	wppm		<5	wppb
Mercury	0.7	wppm		<5	wppb

FCC off gas contains inerts like nitrogen, CO_2 and also some elemental oxygen and NO_x. This stream is not fed to the heaters, but sent directly to the recovery section. NOx will form explosive gums in the chilling train and is discussed by Grenoble [2012]. Oxygen is a poison to some catalysts and also affects the olefin purity. Methanol-to-Olefin (MTO) effluents often contain significant amounts of oxygen and oxygenated compounds. Free oxygen has to be removed. Often copper catalysts are used for oxygen removal. Other oxygenates are removed by adsorption and absorption methods. Some oxygenates like alcohols and phenols are formed in trace quantities in pyrolysis reactors [Baur, 1997]. Hence, the concentration of these species have to be higher than those typically found in crackers to merit removal. Some removal methods are given by Chuang et al. [2008].

High Molecular Weight Hydrocarbons and Solid Particles

Cracking heaters are designed to crack many different type of hydrocarbons to produce olefins. Each heater is designed to crack a specific range of feedstocks. For example, a heater that is designed to crack only ethane feed will not have a vaporization coil in the convection section. When a liquid heater is designed for a specific liquid feed boiling range, feeds heavier than the design feed range can cause problems. Therefore, sometimes liquids in a gaseous feed or a heavy liquid in a furnace that was designed for a lighter liquid feed range has caused some operational problems. When liquid feeds like naphtha are transported in barges, the tanks may not be cleaned from the previous shipment. Sometimes these ships carry crude oil on one route and naphtha in the return route. The residues left in the tank of crude oil transport has caused problems. Since these are present in extremely small amounts (<0.1 wt. %), distillation curves will not show any abnormal behavior. In some cases, ASTM distillation may show significant residue. Presence of such materials has caused significant fouling in the convection section, radiant coil, TLE and /or transferline.

Figure 6 shows the level of residue found in a naphtha feed. Over several years of operation the residue content resulted in coke production and deposition in the transferline and almost blocked the transferline!

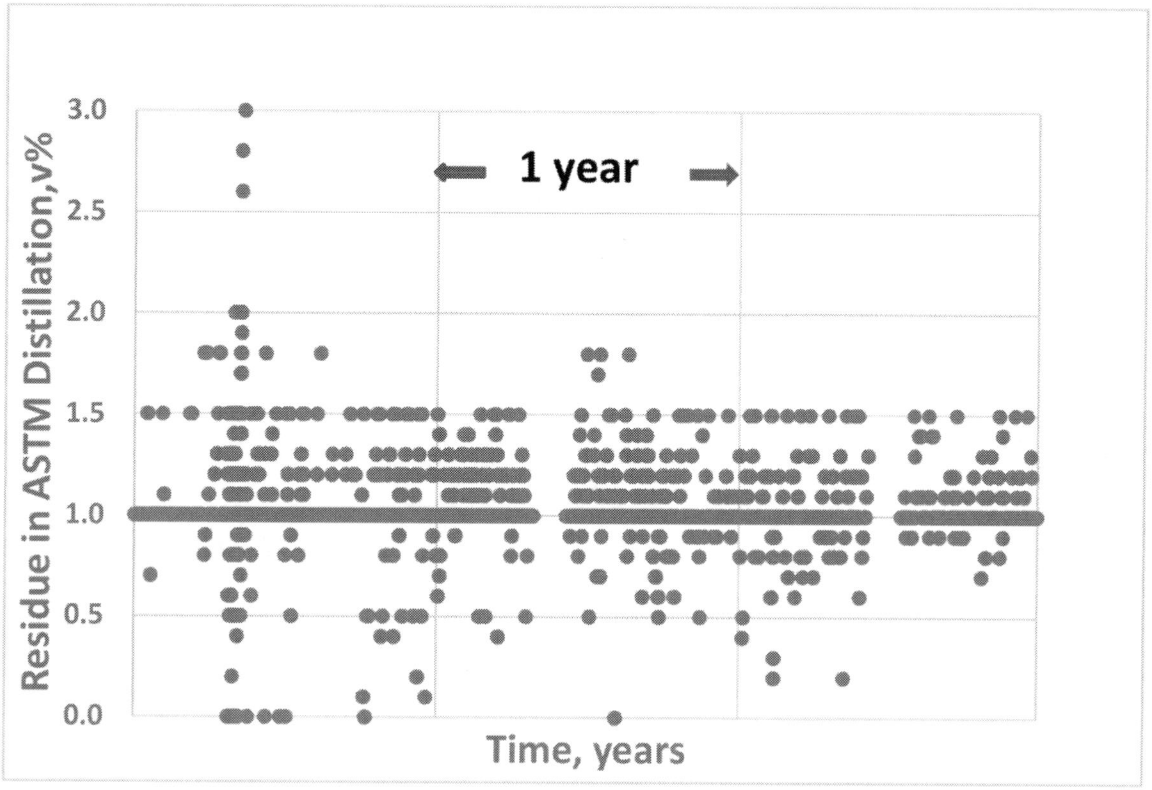

Figure 6. Residue Naphtha Distillation Curve

In another plant, significant coronenes and ovalenes were present in hydrocracked vacuum gasoil [Fernandez-Baujin et al., 1990]. These compounds have chemical structures with closely bonded 6 ring and 10 ring aromatic molecules (pericondensed) respectively.

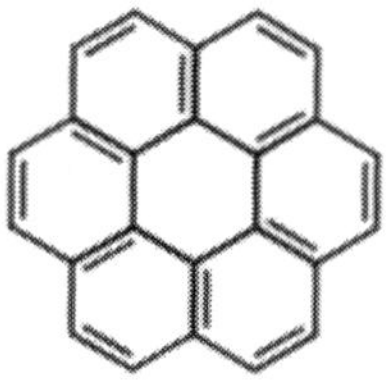

Coronene($C_{24}H_{12}$)

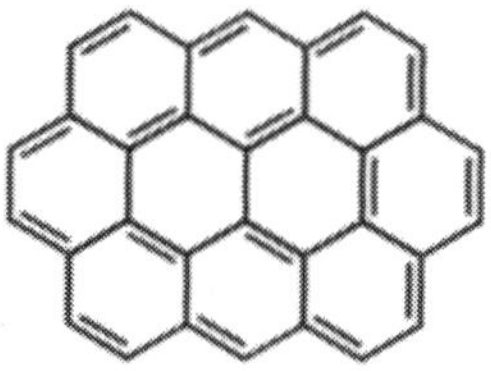

Ovalene($C_{32}H_{14}$)

At room temperature, they are solid and they sublime at high temperatures. There are no issues in vaporization of this feed in the convection section and there is no significant increase in coking rate in the radiant coil. However, massive coke deposition in the transferline exchanger (TLE) was reported within one day of oil in! The TLE was opened and it was observed that coke was growing on the outlet tubesheet as a tube resembling an extension of the TLE tube! More than 40% of the tubes had this "coke tube". The length of the extended coke tube varied up to 9 inches as shown in Figure 7. There were two shell and tube exchangers per heater and both exchangers showed similar pattern. Reducing the severity or increasing the steam to oil ratio did not improve the TLE fouling.

Another situation is when the light liquid feed contains a high boiling tail composed of condensed ring molecules. In this case, contamination of some heavy molecules in light liquid feeds can be removed using HOPS (heavy oil processing scheme) [Maddock, 1992].

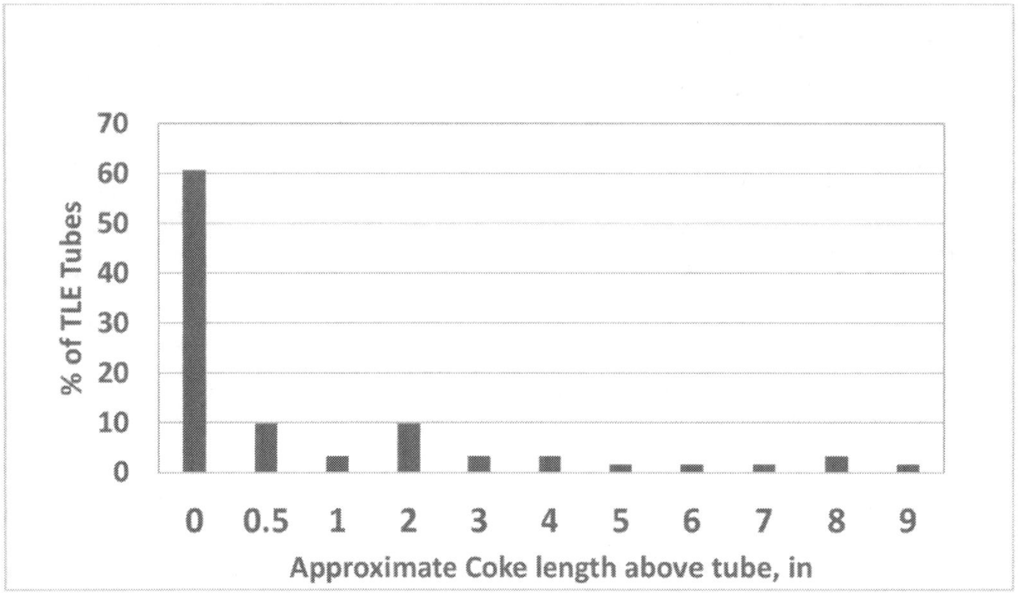

Figure 7. Effect of Contaminant in HVGO Feed Cracking

Typically, a feed is only analyzed to detect contaminants that are considered foreign materials. Sometimes, asphaltenes and polycyclics in the feed have to be considered as contaminants. Often liquid feeds contain suspended particles either in the form of rust and/or inorganic matter. Significant radiant coil run length reduction has been reported when heaters are operated without feed filters [Robinson, 2009]. Brayden et al. [2006] recommend filtration combined with a coalescer.

Summary of Contaminants

Table 8 summarizes some of the known contaminants, their consequences and typical recommended limits. The limits are based on available literature data and in-house experience.

Table 8. Summary of Contaminants

Contaminant	Limit in feed	Impact	Cleaning Method	Comments
Acetone	10 wppm	Forms aldehydes and other oxygenated compounds that lead to downstream fouling		
Amines	ND (not determined)	Can form NH3 and HCN		NH3 will react with N2O3 and other NOx compounds to form ammonium nitrite in the cold box
Ammonia	<0.1 ppm in effluent	C2H4 purity, catalyst poison, formation of ammonium compounds in chilling train	Adsorbents	See amines. Limit included produced ammonia so limit is based on effluent content.
Arsenic	5 wppb	Catalyst poison in C2H2 and MAPD and pygas reactors	Adsorbents; selective hydrogenation	>50 wppb has been reported to increase coking rate by a factor of 3.
Ash	Nil	Fouling in convection		Indicative of other metallic contaminants
Calcium	<0.5 wppm	Fouling and corrosion		
Carbonyl sulfide	<2 wppm	MAPD catalyst poison and C3H6 purity	Absorption/ adsorption	Alumina
Carbondisulfide	<1 wppm	Poisoning of DPG catalyst	Partial removal with absorbents	
Caustic	ND	Corrosion		Rapid coking
Chlorides	<5 wppm	Stress cracking of stainless steel; fouling in convection	Adsorbent	
Chlorates	ND			
Coronenes	<15 wppm			Possibly mal operation of hydrocracker; rapid coking
Dimethyl ether	ND			
Ethanol	ND	Forms CO, ketenes, carbonyls and potential fouling of C4 catalysts.		
Fluorides	<2 wppm	Catalyst poison	Activated alumina	
Green oil	ND	Excessive coking and fouling		
HCl	<2 wppm	Increased corrosion and lowers pH of Quench water		
HCN	ND	Catalyst poison		
Hydroxylamines	ND	NOx producer; explosion risk		Can form explosive gums
Iron	<1 wppm	Increased coke formation and fouling	Passivate equipment to reduce rust initiated fouling.	
Lead	<50 wppb	Corrosion; catalyst poison		

Table 8. Summary of Contaminants cont.

Contaminant	Limit in feed	Impact	Cleaning Method	Comments
Metals	<1 wppm	Convection section fouling		
Mercury	<5 wppb	Corrosion of brazed aluminum exchangers and catalyst poison	Activated carbon; selective hydrogenation	
Methanol	<50 wppm	Propylene product purity		
MTBE	<50 wppm	Produces methanol and affects C3H6 purity	Waterwash	
NOx	nil	Explosion risk	Selective hydrogenation	
Oxygen	<1 wppm	Fouling,NOx and carbonyl formation; corrosion		
Oxygenates	ND	Generally catalyst poison and affects product purity	Some (e.g.MeOH) can be removed by water wash. As carbonyls, sodium bisulfite is used.	
Ovalenes	<40 ppm	Rapid coking		Hydrocracker operation
Phenol	ND	Increased tendency to foam in the caustic tower		
Phosphine	ND	Poison to hydrogenation catalyst	Selective hydrogenation	
Phosphorous	ND	Corrosion		
Potassium	<0.5 wppm	Radiant coil corrosion,fouling	Some claim as coke inhibitor	
Red or Yellow oil	ND	Rapid coking		
Rust/sediments/ particles	<1 wppm	Rapid coking		
Silica	<1 wppm	Poisoning of DPG catalyst		
Sodium	<125 wppb	Radiant coil damage; higher coking rate; PFO contamination	Desalting, high efficiency coalescers to remove water and water soluble salts.	
Vanadium	<1 wppm	Radiant coil corrosion; PFO contamination; Hydrogenation reactors		
Wash oil	ND			
Water	No free water	Corrosion	Adsorbents	
ND: Not determined				
Note: Consult the Ethylene Producers feed contaminants database; these are typical limits and will vary with other conditions.				

Summary and Conclusion

Contaminants in cracker feedstocks are unavoidable. Identifying those contaminants and their effects on olefin production is of prime importance. Removal of contaminants is a challenge to the engineering community. However, with changes in technology the limits on contaminant level will change. New contaminants may be added. Filtering feedstocks, analyzing for contaminants, and careful monitoring of the plant is the best policy to minimize the impact of contaminants.

Acknowledgement

The authors are thankful to CB&I for granting permission to present this paper and to all our colleagues and clients with whom we work to continually learn more about this subject.

Disclaimer

References (listed in alphabetical order)

1. M. Abai, M. P. Atkins, A. Hassan, J. D. Holbrey, Y. Kuah, P. Nockemann, A. A. Olifrenko, N. V. Plechkova, S. Rafeen, A. A. Rhaman, K. R. Seddon, S. M. Shariff, G. Srinivasan, Y. Zou (2011), "Mercury Removal from Gas Streams using new Sold Adsorbents" International Gas Union Research Conf., Korea.

2. A. Baumgartner and S. Coleman (1996),"Principal Feedstock Contaminants in Ethylene Plants", AIChE National Spring Meeting.

3. K. Baur (1997),"Formation and Behavior of Oxygen Containing Organic Compounds in Ethylene Plants", Linde Reports on Science & Technology, 59, p19.

4. A. Bernard and S. Snider (2008), "Ethylene Plant Contaminants: A Database and Interactive Tool", AIChE Spring National Meeting, New Orleans, LA.

5. B. Bouyssiere, F. Baco,L. Savary and R. Lobinski (2000),"Analytical Methods for Separation of Mercury in Gas Condensates", Oil & Gas Sci. Tech-Rev. IFP, 55,639.

6. M. Brayden, A. Baumgartner, S. Coleman, D. Smith, L. A. Nairn, J. Lally, A. Bernard (2017), "Feedstock Contaminants in Ethylene Plants-2017 Update", AIChE 2017 Spring National Meeting, San Antonio, TX.

7. M. Brayden, D. Hines, J. Graham and T. Prickett (2008), "Top 5 Contaminants in Ethylene Production Unit Feedstocks", AIChE Spring National Meeting.

8. M. Brayden, T. H. Wines and K. D. Giudice (2006),"Improve Steam Cracking Furnace Productivity and Emissions Control Through Filtration and Coalescence", AIChE Spring National Meeting, Orlando, Florida.

9. S. Y. Chuang, R. Nanda, J. Rizopoulos (2008), " Contaminants Key to Refinery Offgas Treatment Unit Design", Oil & Gas Journal.

10. P. Coenen (2016),"Carbon disulfide (CS2): A New Challenge for Steam Cracker Complexes?", AIChE Spring National Meeting, Houston.

11. S. Coleman, J. Ries, D. Campbell and K. R. Clark (2004),"Mercury Removal from Cracked Gas and Liquid Streams", AIChE Spring National Meeting.

12. G. Corvini, J. Stiltner and K. Clark (2014),"Mercury Removal from Natural Gas and Liquid Streams", UOP Publication.

13. S. De Haan (2006), "An Overview of Self Inflicted Contaminants", AIChE Spring National Meeting, Orlando, FL.

14. B. S. Devakotai, P. S. Misttry and R. Pritchett (2011),"Feed Contamination with Sodium and its Impact on Ethylene Furnaces", AIChE Spring National Meeting, Chicago, IL.

15. J. J. English (1989),"A Mercury Induced Aluminum Alloy Piping Failure in an Ethylene Manufacturing Plant", AIChE National Spring Meeting.

16. Kari Enqvist,"Significant Sodium Carryover To Cracking Furnaces", EEPC Ethylene Seminar, 15-17 October, 2008 ,Madrid, Spain

17. M. F. Ezzeldin, Z. Gajdosechova, M. B. Masod, T. Zaki, J. Feldmann, E. M. Krupp (2016),"Mercury Specification and Distribution in an Egyptian Natural Gas Processing Plant", Energy Fuels, 30, 10236.

18. J. M. Fernandez-Baujin, M. J. Maddock and K. M. Sundaram (1990)," A Case Study Based on Commercial Experience: Vacuum Gasoil to Petrochemicals", AIChE Spring National Meeting, Orlando, FL.

19. P. Franken and T. Hubbers (1991), AIChE Spring National Meeting, Houston.

20. M. A. Graham (1993),"Ethylene Producers Committee-Feedstock Issues Subcommitte Questionnaire/Survey Data: Selected Feedstock Impurities", AIChE Spring National Meeting, Houston.

21. D. Grenoble (2012),"History of NOx in the Ethylene Industry", AIChE National Spring Meeting.

22. S. K. Hasan and M. Mobin (2013)," Hot Corrosion Behavior of Protective Oxide Scales with Sodium Chloride in Chlorine Gas Environment", Nace Corrosion, paper#2125.

23. R. S. Hugo and R. Wagner (2000),"Deep COS Removal with new Formulated MDEA Solvents", Paper presented at the Lawrence Reid Gas Conditioning Conference, University of Oklahoma, OK.

24. T. Huntsman (2003),"Study Compares COS-Removal Processes", Oil & Gas Journal.

25. K. Karan, A. K. Mehrotra and L. A. Behie (1999), "A High Temperature Experimental and Modeling Study of Homogeneous Gas-Phase COS Reactions Applied to Claus Plants", Chem. Eng. Sci., 54, 2999-3006.

26. T. Katz and V. Giesen (2009),"The Challenges of Deep COS Removal: Which Options do we Have?" Gas Processors Association Europe.

27. R. J. Littel, G. F. Versteeg and W. P. M. van Swaaij (1992). ,"Kinetics of COS with Primary and Secondary Amines in Aqueous Solutions", AIChE Journal, 38, 244.

28. M. J. Maddock (1992),"Improved Field Condensate Cracking Trends in Industrial Application", paper presented at Refining, LNG and Petrochem Asia 92., Singapore.

29. E. Mohundro (2013),"A 'look –back' at EPC Contributions to Dealing with Ethylene Plant Contaminants", AIChE Spring Meeting, San Antonio.

30. G. Nowowiejski and J. A. Reid (2003)," An Overview of Oxygenates in Olefin Units in Relation to Corrosion, Fouling, Product Specification and Safety", AIChE Spring National Meeting, New Orleans, LA.

31. C. Polito (1996),"Arsine and Carbonyl Sulfide Contaminants in Ethylene Plant Feedstocks", AIChE Spring National Meeting, New Orleans.

32. K. F. Preston and R. J. Cvetonovic (1972), "The Decomposition of Inorganic Oxides and Sulfides" in "Decomposition of Inorganic and Organometallic Compounds" Ed. C.H.Bamford, R.G Compton & CF. H. Tipper, Elsevier.

33. J. Robinson (2009),"Naphtha Contamination and Furnace Coil Plugging", AIChE Spring National Meeting, Tampa, FL.

34. M. R. Sainal, A. Shafawi, A. J. H. Mohamad (2007),"Mercury Removal System for Upstream Application: Experience in Treating Mercury from Raw Condensate", E & P Environmental and Safety Conference, p. 154, Galveston, Texas, USA.35. P. Sarrazin, C. J. Cameron, Y. Barthel and M. E. Morrison (1992),"Arsenic and Mercury Removal from Hydrocarbon Streams", AIChE Spring National Meeting, New Orleans.

36. H. G. Schecker and H. G. Wagner (1969), "On the Thermal Decomposition of COS", Int. J. Chem. Kinetics, 1, 541.

37. A. B. Shafawi (1999),"Mercury Species in Natural Gas Condensate", PhD thesis, University of Plymouth, U.K.

38. G. Taylor and M. Cleghorn (1996),"Naturally Occurring Radioactive Materials: Contamination and Control in a World Scale Ethylene Manufacturing Complex", AIChE Spring National Meeting.

39. S. Zhao, H. Yi, S. Jiang, F. Gao, B. Zhang, Y. Zuo and Z. Wang (2013)," The Hydrolysis of Carbonyl Sulfide at Low Temperatures: A Review", The Scientific World Journal, , ID739501.

AIChE Paper Number: 32E

A NEW TECHNOLOGY FOR REDUCTION OF SULFUR CONTAMINATION IN PETROCHEMICAL FEED STOCKS

Sai Reddy Pinappu

Baker Hughes, a GE Company

Hua Mo

Baker Hughes, a GE Company

Jerry Weers

Baker Hughes, a GE Company

Waynn Morgan

Baker Hughes, a GE Company

Prepared for Presentation at the 2018 Spring National Meeting

Orlando, FL, April 22-26, 2018

A NEW TECHNOLOGY FOR REDUCTION OF SULFUR CONTAMINATION IN PETROCHEMICAL FEED STOCKS

Sai Reddy Pinappu[1], Hua Mo[1], Jerry Weers[1], Waynn Morgan[1]

[1] Baker Hughes, a GE Company.

ABSTRACT

Contamination removal in ethylene cracker feed stock is always necessary. Sulfur species are one of the potential contaminants that need to be removed. The Sulfur compounds in petrochemical grade naphtha (PCN) and gas feeds have been linked to many downstream process issues such as corrosion, contamination of products, catalyst poisoning, etc. Managing these sulfur compounds in PCN is a challenge for most producers because most of the sulfur species in the feedstock could create Hydrogen Sulfide (H_2S), Carbonyl Sulfide (COS) and other species through the cracking furnace. Although some of the sulfur species may be removed by quench water, condensation at the compressor and caustic tower, some species would be hard to remove because of high volatility and short residence time in the process. As a result, it would be preferable to remove the sulfur species before they enter the cracking furnace.

Currently, there are multiple technologies for effective management of sulfur species in PCN. The technologies include mechanical, chemical, or a combination. One such treatment methodology is a chemical treatment program based on an extraction process. An early form of sulfur management in PCN was treating the process streams with inexpensive materials such as sodium hydroxide solution. The water insoluble sulfur species will convert to the water soluble material and be extracted into the sodium hydroxide solution. Although liquid-liquid extraction processes are still prevalent in today's refining and petrochemical operations, there are limits to the degree of sulfur removal and separation that can be accomplished.. The efficiency of removal and separation are decreased with increasing molecular weight of the sulfur species and the pKa of the sulfur species. This led to the development of a new technology which focuses on improving the chemical treatment of scavenger-insensitive and higher molecular weight molecules such as CS_2, C4+ Mercaptans. The technology is designed to increase scavenger reaction products' solubility in the aqueous phase to maximize extraction efficiency. This paper will review the capabilities of this new technology. Several case studies will also be presented to discuss the efficiency of this technology.

Keywords

Sulfidation, Sulfur reduction, complex sulfur compounds, sulfur extraction, desulfurization

INTRODUCTION

Increased competition and rising costs of raw materials have narrowed margins in ethylene production which promoted producers to look for various feed stocks. Options include feedstocks such as naphtha, ethane, propane, butane and gas oils. While producers have choices to reduce raw material costs they do come with new challenges depending on the source. The impurities in the feed could affect the further downstream process in the ethylene process. One of the important contaminations are the sulfur species. Some of these sulfur species are naturally present in the feed stream and some are intentionally added to minimize coking in the furnace. When these sulfur species burn in the ethylene cracker, different sulfur compounds (H_2S, COS, etc.) are formed. If these sulfur species are not removed in the process, they will cause downstream processing issues. These issues include the contamination of the product, corrosion of the process equipment and poisoning the catalyst in the py gas hydrogenation.

To avoid the process issues caused by the sulfur species, chemical or physical treatments are preferable to treat the feed stream. Traditional processing technologies for sulfur removal from hydrocarbon streams in petroleum refining industries mainly include catalytic-hydrodesulphurization (HDS or Hydrotreater), caustic treating and amine scrubbing. Designed in alignment with refinery capacity and are usually capable of processing large volumes. These technologies also provide a certain level of operation flexibility for constantly changing feedstocks regarding sulfur species and sulfur levels. With modern day demands for higher desulfurization, throughputs and sulfur removal efficiencies, these processes possess fundamental gaps. Increasing throughput and severity in the hydrotreating process can lead to high hydrogen consumption and frequent fouling in hydrotreating equipment[1]. On the other hand a caustic scrubber is generally used for washing petrochemical grade naphtha (PCN) and gas feed with caustic solution. During the caustic wash, some sulfur species, such as H_2S and low molecular weight sulfur species,. These caustic reacted sulfur species can be removed by separation because of their affinity towards aqueous phase. However caustic wash may not remove all the sulfur species in the feed streams due to its limited reactivity and partition between the oil / water phase. These species include thiophenes, carbonyl sulfides (COS), higher chain Mercaptans and disulfides of the nature of RSR, RSSR[2].

Realizing that there is a gap in sulfur removal technologies within the industry, a new sulfur extraction technology (EXALT™) was developed. The EXALT™ technology combining with mechanical equipment can achieve an effective way to remove sulfur species from light hydrocarbons. The EXALT™ technology can now remove sulfur species that were not able to be removed with the predecessor extraction technologies at a fraction of the cost of hydrotreating. Although the lab experiments and trials were successfully carried out in the refinery, the same technology can be used to treat petrochemical feed streams to remove sulfur species. This paper shares some lab and case studies of EXALT™ technology.

METHODOLOGY

The basic approach of EXALT™ technology consists of two steps

- Step 1: Identifying the effective sulfur removal chemistries
- Step 2: Designing a modular mechanical process with optimal removal efficiency and design flexibility.

Unlike the conventional mechanism of a H_2S or mercaptan scavenger, EXALT™ sulfur removal chemistry forms hydrophilic product with the sulfur species in the hydrocarbon upon liquid phase contact. The hydrophilic sulfur-containing products retain sulfur in the water phase while phase separation occurs, thus achieving sulfur extraction from the hydrocarbon to the hydrocarbon (Figure 3).

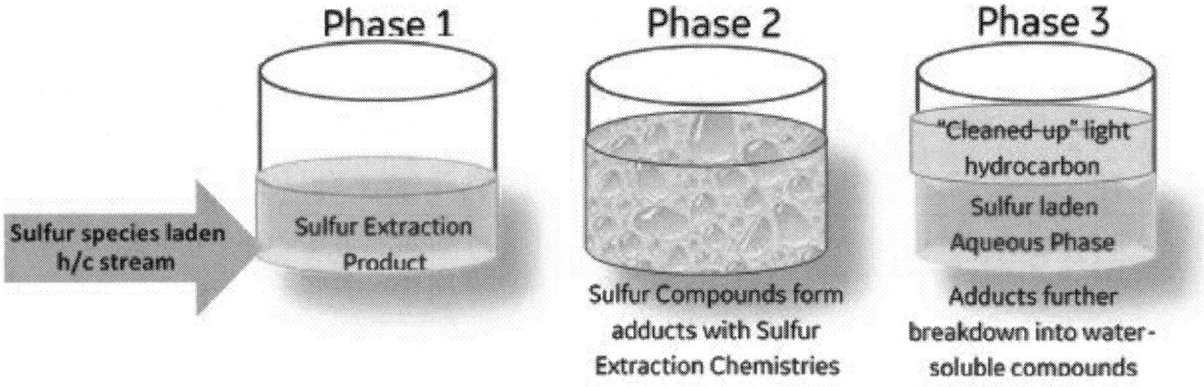

Figure 3 Function Mechanism: Sulfur Extraction Chemistries

Sulfur Containing Reagents

To determine the representative target sulfur compounds for evaluation, sulfur speciation (per ASTM D5623) results from multiple light hydrocarbon streams were studied. Then a list of model sulfur compounds were generated for lab sulfur removal testing. The species included carbon disulfide, diethyl sulfide, dimethyl disulfide and n-butyl mercaptan.

Simulation Sulfur Removal Process by Lab Test

A liquid-liquid extraction procedure was developed to simulate the static sulfur removal process on a lab bench. 50 ml of the light hydrocarbon sample with different pre-dosed sulfur species to be treated was added to a 500 ml glass separatory funnel, to which 50 ml of the extraction solution with / or without EXALT™ was added. The mixture was agitated for 1 – 2 minutes to allow adequate contacting (with frequent pressure release to avoid vapor pressure buildup). The mixture was then allowed to settle undisturbed to enable complete phase separation.

The sulfur species in the light hydrocarbon phase was determined by ASTM D7039, using a Sulfur Analyzer (XOS SINDIE 2622-10) with LOD of 0.4 ppm at 300 s. The efficiency of sulfur removal can be calculated by comparing the sulfur content in the hydrocarbon phase before and after the extraction procedure.

RESULTS AND DISCUSSION

I. Efficiency of Sulfur Removal (Lab Test)

The lab test results are shown in Table 1. And Figure 4. The results indicated that both EXALT™ 1000 and EXALT™ 1100 have high sulfur extraction efficiency on the main sulfur compounds seen in light hydrocarbons from refinery. Additionally, EXALT™ 1100 showed removal efficiency on some sulfur compounds such as diethyl sulfide and n-butyl mercaptan. These compounds are not effectively removed by the conventional removal method.

Table 1 Lab Results: Efficiency of the Sulfur Extraction Chemistries

Model Sulfur Compound Type	Model Sulfur Compound conc. (ppm)	Sulfur Extraction Chemistry	Sulfur content (ppm)	S Removal Efficacy, %
Carbon Disulfide	1000	EXALT™ 1000	135.4	86%
Carbon Disulfide	1000	EXALT™ 1100	112.5	89%
Dimethyl Disulfide	1000	EXALT™ 1000	37.9	96%
Dimethyl Disulfide	1000	EXALT™ 1100	34.5	97%
Mixed Sulfur Compounds**	4000	EXALT™ 1000	2020	50%
Mixed Sulfur Compounds**	4000	EXALT™ 1100	1090	73%

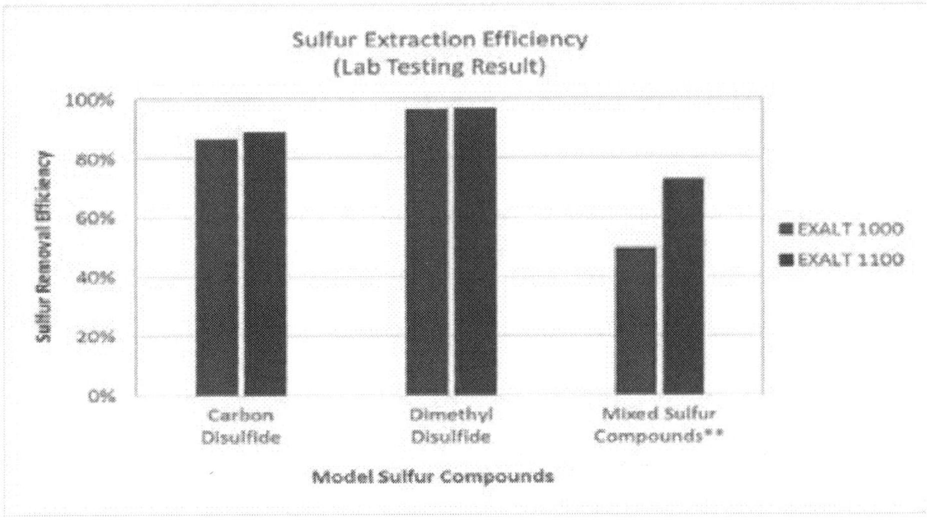

Figure 4 Sulfur Extraction Chemistries Lab Evaluation Results

**Mixed sulfur compounds include 1000 ppm each carbon disulfide, diethyl sulfide, dimethyl disulfide and n-butyl mercaptan

II. Equipment Design with Optimal Removal Efficiency

Another aspect of the EXALT™ technoloy is to increase the sulfur removal efficiency by mechanical equipment. To fill the gap in sulfur removal efficiency, several flexible, a modular system with multiple equipment elements were designed in response to specific application requirement.

An example pilot sulfur removal system is illustrated in Figure 5. The core components are the two-stage extraction vessels. The first stage vessel contains different chemistry formulations to target H_2S and C1-C4 mercaptans in the first stage, and C4+ mercaptans. The other sulfur species will be treated and removed in the second stage. Additional components including particulate filters, recirculation filters, recirculation pumps, and coalescers can be arranged in various sequence and locations to maximize the flexibility of the overall system. This design will help to meet quality and spec requirements of the hydrocarbon stream to be processed.

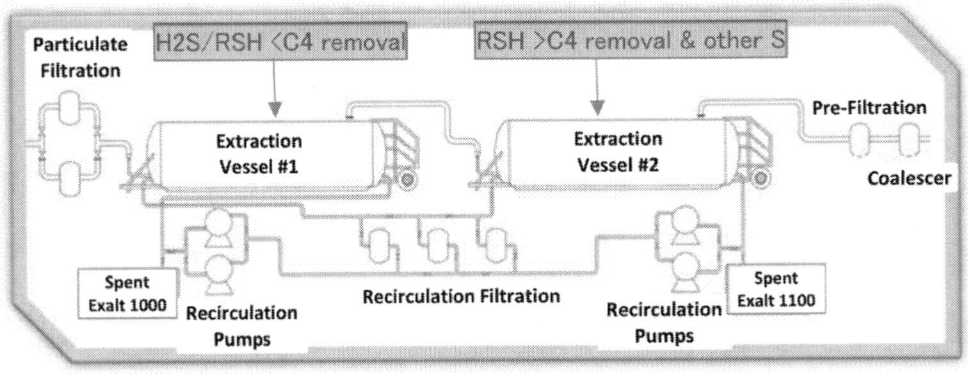

Figure 5 Example Layout of the Sulfur Extraction System

Another example of the modular sulfur removal system that is used in the field includes a vertical contact tower to maximize contact between the liquid phases, and a horizontal vessel that allows optimal phase separation to happen. A series of specialty filters downstream of the second vessel are designed for removing residual contaminants before tankage (Figure 6). By performing upfront testing and analytics to understand the specific system and the process streams being processed, an optimal treatment program can be customized and developed. It can also include specialty chemical additive injection before the extraction vessels as pretreatment option (Figure 6). A full-size unit of the two-vessel design (vertical plus horizontal) was built as a mobile system for field evaluation (Figure 7).

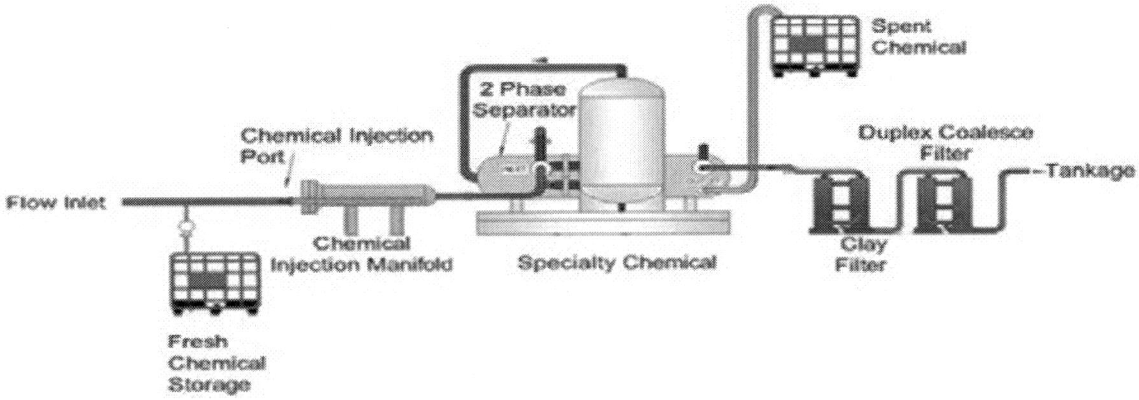

Figure 6 Modular Extraction System Example

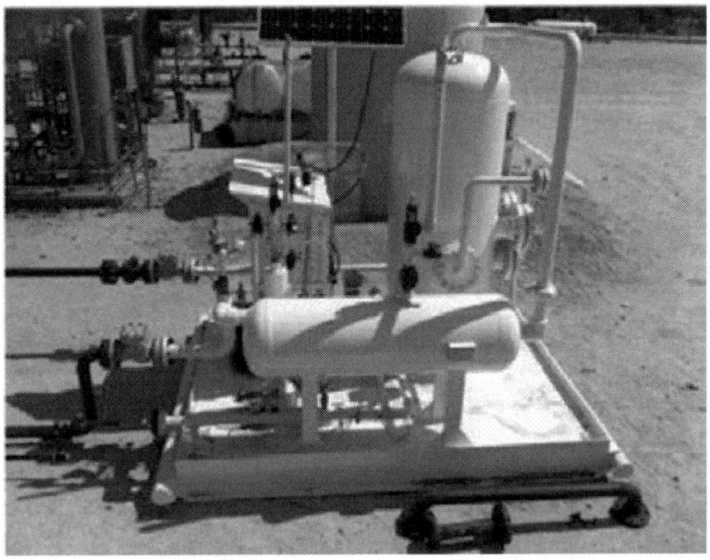
Figure 7 Pilot Sulfur Removal Modular System

II. Efficiency of Sulfur Removal (Field Trial)

1. Case Study I

A US Gulf Coast refinery faced several challenges in producing on-spec naphtha and propane as a result of reactive sulfur contamination. Due to the presence of reactive sulfur, the naphtha failed the ASTM D-1838 copper corrosion specification. Therefore the propane had to be sold at a discount, because In addition, valuable naphtha was being lost into the propane stream, causing the light naphtha to be downgraded. The refinery considered installing a hydrotreater. However a hydrotreater for this type of operation would require large capital investment, as well as additional operating expenses.

After surveying the system, the sulfur extraction services were recommended to the refinery. It delivered a turnkey solution that removes the sulfur compounds and brings the streams back on spec. The treatment included a combination of sulfur removal chemistries paired with proprietary equipment for optimal extraction efficiency. The extraction system was installed on the Y grade propane stream to separate the light naphtha and heavier sulfur compounds, generating a new liquid light naphtha stream. The treatment results showed more than 98% removal of the total sulfur compounds (Figure 8), including reduced levels of methyl mercaptan, carbonyl sulfide, and carbon disulfide, which could not be removed by caustic treatment alone. It is important to note that this treatment did not affect the characteristics of the product (color, volatility, etc.) and the propane successfully passed the copper corrosion tests. This solution also helped reduce other extraneous contamination, limiting future catalyst and corrosion issues. This novel technical approach helped the refiner to bring the propane stream back on-spec, and recovered significant revenues that were lost due to off-spec propane. This on-spec product combined with the new naphtha stream allowed the refinery to debottleneck operations, realizing a 20% increase in overall process efficiencies and valuable products with high profit. The efficiency of sulfur removal is shown in Fig. 8.

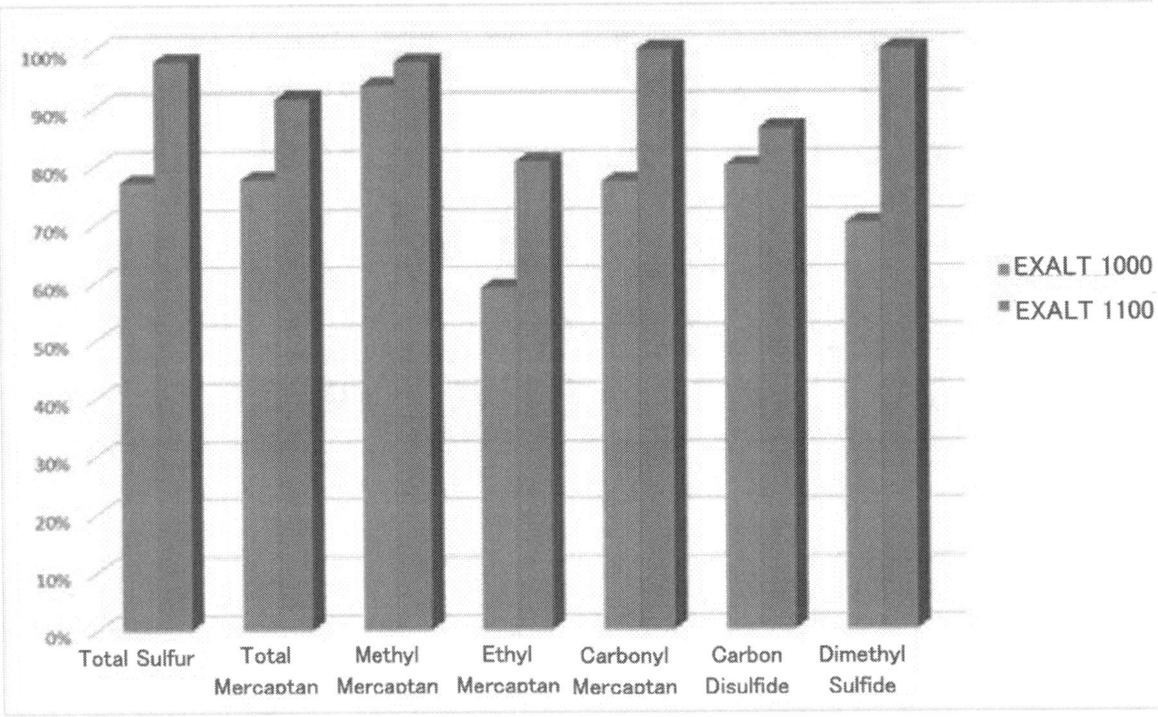

Figure 8 Sulfur Compounds Removal Efficacy

2. Case Study II

A petroleum refiner experienced a typically high level of mercaptans (over 1000 ppm) in its shale oil feedstock. The traditional mercaptan scavenger approach was able to reduce the mercaptan content to 190 ppm. However, this mercaptan level is still higher than operation requirement (< 20 ppm). Lab tests indicated that EXALT™ 1000 and EXALT™ 1100 were able to bring down the final mercaptan level to less than 10 ppm (Table 2). A full scale system was developed to apply technology at field scale on-site

Table 2 Mercaptan Reduction from Shale Feedstock

Sample	Treatment Chemistries	Mercaptan (ppm)	% Mercaptan Removal
Shale Feedstock	-	1020	-
Shale Feedstock	Traditional Mercaptan Scavenger	190	81.4
Shale Feedstock	EXALT™ 1000	9	99.1
Shale Feedstock	EXALT™ 1100	46	95.5
Shale Feedstock	1:1 EXALT™ 1000 : EXALT™ 1100	7.7	99.2

3. Case Study III

A gulf coast refiner was buying off spec Isobutane from a nearby gas plant. When processing this fuel in their alky unit, they were having some major issues with respect to corrosion and high acid consumption. After a thorough investigation, sulfur was one of the major contaminants that cause issues in the unit. Sulfur in the feed reacts acid to form light acid soluble oils/polymer and contributed to acid loss. These polymers were difficult to remove in the acid regenerator and contribute to the corrosion. The sulfur compounds most commonly found in the feed are H_2S, Mercaptans and small quantities of carbonyl sulfide (COS). To simplify the analysis, the total sulfur content of the feed was used in the analysis. The customer would like to have total sulfur <25 ppm. However in some cases, total sulfur species <40 ppm was also acceptable.

Isobutane was arriving onsite in batches of truck loads (4 to 5 trucks per day) offloading into a butane sphere. EXALT™ technology in conjunction with proprietary mechanical equipment was used to remove sulfur in the butane before it being sent to butane sphere for storage. The field test data is shown in the Fig. 9. The result indicated that EXALT™ could successfully control the total sulfur level below the 25 ppm at most time. It control the total sulfur level below 40 ppm all the time.

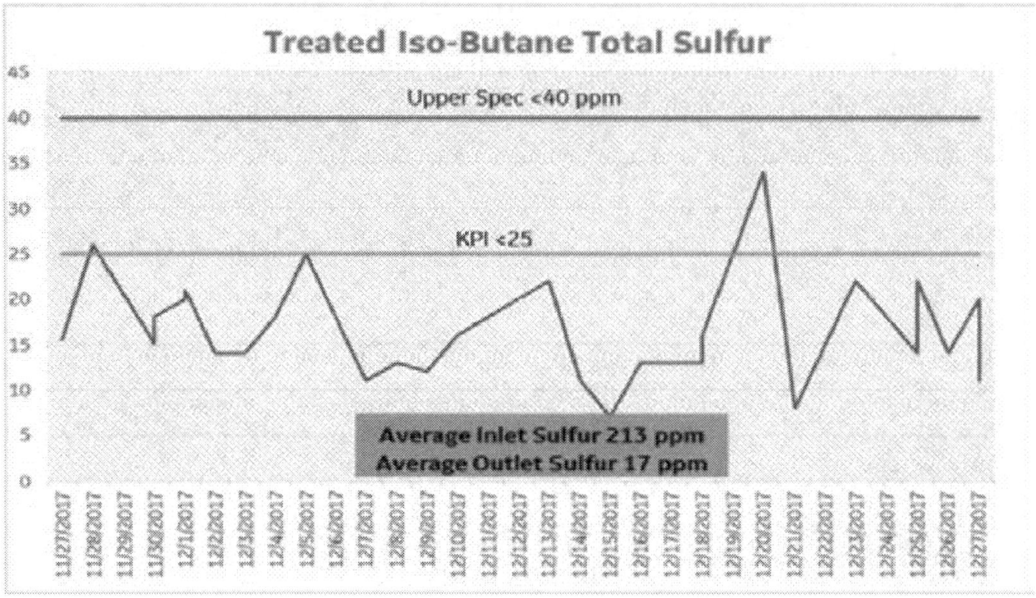

Figure 9 Sulfur Removal Efficacy on an Isobutane Stream

CONCLUSIONS

A new approach of sulfur removal technology (EXALT™) was developed. The technology includes patent pending sulfur removal chemistries and modular mechanical designs. This technology can bring flexibility and customizable solution to meet sulfur spec for many feedstocks. Continuous development from both chemistry and mechanical aspect will expand the capability of the technology to wider range of sulfur removal challenges. Although the technology is currently apply in the refinery streams. We believe that this technology can also be used to treat the feedstocks for the ethylene cracker, which contained trace amount of sulfur species.

REFERENCES

1. G. Medine and B.E. Wright. (2008) "Distillate Hydrotreater Fouling" 2008 AIChE Spring National Meeting
2. Blaschke, Marilyn W. (2003), "Caustic tower fouling: Identifying the causes" 15th AIChE Ethylene Producers Conference.

AICHE PAPER NUMBER: 32F

CARBON DIOXIDE: AN OVERVIEW ON REMOVAL TECHNOLOGY

Dr David Slim

Technology Advisor

Evonik Corporation

Dr Phillip James

Technology Director

Evonik Performance Materials GmbH

Joseph Lally

Technical Service Manager

Evonik Corporation

Prepared for Presentation at the 2018 Spring National Meeting

Orlando, FL, April 22-26, 2018

CARBON DIOXIDE: AN OVERVIEW ON REMOVAL TECHNOLOGY

Dr David Slim

Technology Advisor

Evonik Corporation

Dr Phillip James

Technology Director

Evonik Performance Materials GmbH

Joseph Lally

Technical Service Manager

Evonik Corporation

Abstract

Carbon dioxide, which falls into the category of acid gases, is commonly found in natural gas streams at high percentage levels. In combination with water, it is highly corrosive and can harm pipelines and operating equipment unless it is partially removed. Carbon dioxide reduces the heating value of natural gas, wastes pipeline capacity and can adversely impact ethylene plant operation A wide variety of upstream acid gas removal technologies are available. Each process has its own advantages and disadvantages. Gas scrubbers using sodium hydroxide or amines are commonly used. Emerging technologies include Solid Physical Adsorption, Low Temperature Distillation and Membranes Separation. This paper examines the dominant technologies with limits and advantages.

Summary

Carbon dioxide (CO_2) has been identified as a top 5 EPC feedstock contaminant. It is commonly found in natural gas streams in high percentage levels. In combination with water, CO_2 is highly corrosive and can damage pipelines and operating equipment. CO_2 is normally removed from ethane or ethane/propane mixes to <500ppm mainly by amine treating. At the ethylene plant, unremoved CO_2 has the potential to cause operating problems and adversely impact product quality. All ethylene plants have an acid gas removal system to ensure that product quality specifications are met,

Introduction

In the 2003 and 2016 AIChE Ethylene Producers' Committee (EPC) feedstock contaminants surveys, carbon dioxide was the most problematic contaminant for crackers fed with ethane and ethane/propane mix (ref 1). The main areas of concern for ethylene producers are corrosion, freezing and contaminating the pure ethylene product. This is compounded by the variability that can be expected in natural gas, meaning that carbon dioxide levels in the ethylene plant may not be constant. In order to avoid upsets, it has to be carefully and constantly managed.

This paper provides information that addresses these concerns.

Discussion

Properties of carbon dioxide

Carbon dioxide is a colorless gas and is found in the atmosphere. The current concentration is about 400 ppm having risen from pre-industrial levels of 280 ppm. It is present in natural gas at widely varying concentrations.

CO_2 dissolves in water (1.5 gm/litre). In water CO_2 forms the weak acid known as carbonic acid since its ionization in water is incomplete. The majority of the carbon dioxide is not converted into carbonic acid, but remains as dissolved CO_2 molecules. The pH of carbonated water is about 4.

CO_2 sublimes at minus 78°C. Since it has no dipole moment, it tends to fractionate ideally. In the Table 1, the boiling point of CO_2 is compared to those of other hydrocarbons. As can be seen, it tends to concentrate in C2 streams. Thus in furnace feed, it is present in ethane and in steam cracker products it will be found predominantly in ethylene. CO_2 is more of an issue for ethane than propane feedstocks because of its boiling point being closer to C2s. For heavier feedstocks, CO_2 is a minor issue.

Table 1: Boiling Points of Select Light Hydrocarbons

Compound	Boiling point, deg C
Hydrogen	-253
Methane	-162
Ethylene	-104
Ethane	-89
Acetylene	-84
Carbon Dioxide	-78
Propylene	-48
Propane	-42

Ethane Supply and Purification

Ethane is a component of natural gas. Natural gas consists primarily of methane and can contain a significant amounts of ethane and heavier hydrocarbons but also non-hydrocarbons e.g. carbon dioxide. It is estimated that there are about 900 000 km^3 of "unconventional" gas such as shale gas, of which 180 000 km^3 may be recoverable (ref 2). Shale gas is natural gas produced from shale. Almost all shale gas wells today require fractures artificially created by hydraulic fracturing. Shale gas is the major source of natural gas in the United States.

Natural gas is gathered from several wellheads and conditioned to make residue gas, a pipeline quality material with less than 4% inert material. During this conditioning CO_2 is typically removed via amine absorption.

This residue gas can go on to local markets and/or interstate pipelines. Additionally, residue gas can be further processed at gas processing plants to raw natural gas liquids (NGL), loosely referred to as x grades, y grades or NGL. Methane is removed at this point. CO_2 is further removed at this point to make grades with 0.35LV% CO_2/C2

Higher hydrocarbons, including ethane, go on to fractionation to serve specific industries. CO_2 is further removed here to prevent plugging in the cryogenic process and typically reaches values below 500 ppm in C2 streams.

The process overview is outlined in Figure 1

Figure 1 Ethane Supply Process

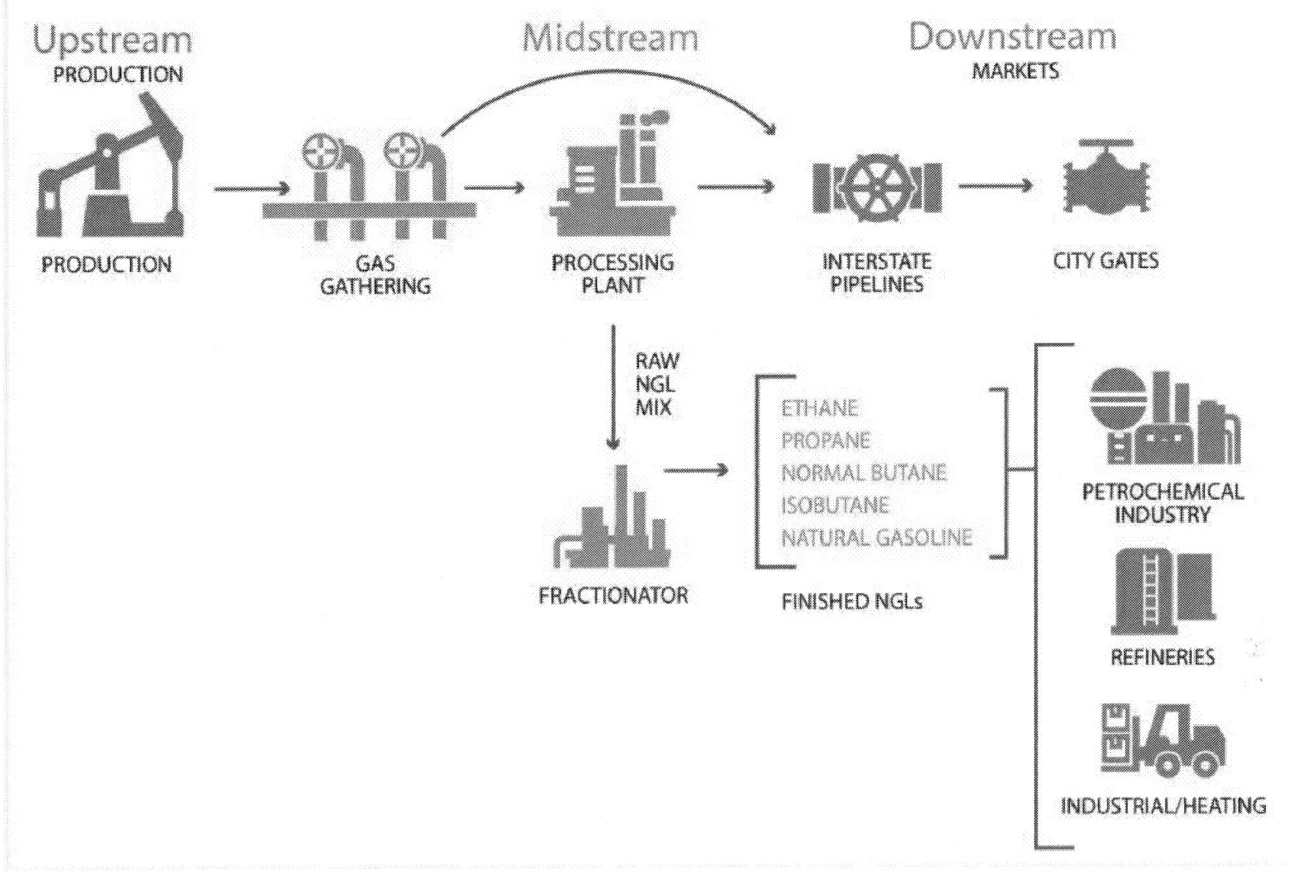

Source: Gas Processing Association Brochure

Carbon Dioxide Removal Steps

Since CO_2 is present in natural gas at percentage levels, it must be removed to avoid corrosion, reduce volumes in pipelines and to meet ethylene plant target specifications. Typical ethane pipeline values are <500 ppm, although ethane distributors can reduce the CO_2 level to 100 ppm with only minor excursions to 300-400 ppm. In rare cases, an ethylene plant may use a local crude NGL supply, in which case a dedicated feed purification unit may be needed. To meet the <500 ppm target, CO_2 is removed by a combination of fractionation and gas sweetening.

The main technology for the bulk removal of carbon dioxide is via amine treating. Both CO_2 and H_2S react readily with amines. In amine treating, the feed gas is sweetened in the absorber with the sweet gas leaving the top of the tower. Amine is regenerated in stripper and acid gases e.g. H_2S and CO_2 are purged. A variety of amines are used commercially. These include: MDEA (methyldiethanolamine), MEA (monoethanolamine), DEA (diethanolamine), DGA (diglycolamine) and MDEA + piperazine. A number of potentially improved amines are currently under development. These include CAPLUS from Evonik. A typical amine treating flow-plan is shown in Figure 2.

Figure 2 Acid Gas Sweetening Unit

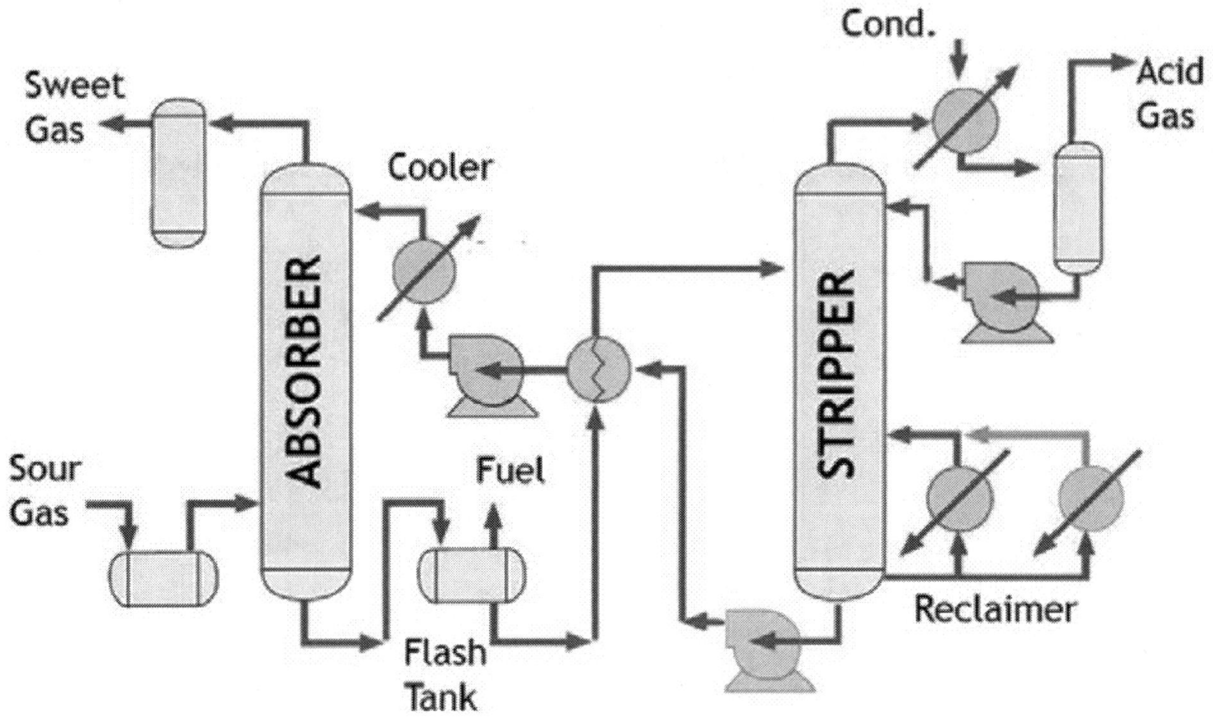

Source: Gas Processors Suppliers Assoc. Eng. Data Book, Vol. 2, sec. 21 (1987)

Although amine treating is the commonest CO_2 removal technology used, there are several emerging technologies that are being developed. These include membranes, cryogenics and Pressure Swing Absorption. However it seems at this time that there has been very little, if any, adoption of these techniques in the ethane supply chain as they currently lack the capacity to competitively perform as well as amines. Developments, like enhanced oil recovery which involved injecting large amounts of CO_2 to stimulate wells may change how gas gathering plants tackle CO_2 recovery.

Amine-based gas sweetening is a well-proven technique with many installations in service, and conveniently for ethylene producers very little risk of contamination caused by the technique itself. Because of the configuration of the plants it is unlikely that the amine extraction fluid is able to break through with the sweetened gas, and even if it would, the amine at low levels would be highly unlikely to impact steam cracking operations. All of the amines in commercial use would likely crack to form ammonia and short chain hydrocarbons.

Impact of CO_2 on ethylene plant operation

Carbon dioxide can have a number of impacts on an ethylene plant operation. The key ones are as follows:

- In furnace feed can take up furnace capacity. However this is only an issue if the level is significantly greater than the <500 ppm feed specification i.e. in the percentage levels.

- Carbon dioxide is a weak acid when dissolved in water. Thus it can influence quench water pH and corrosion tendency and a neutralizer may be required

- Ethylene has CO_2 specifications that are generally <0.1 ppm

- When CO_2 condenses, it condenses as a solid, which could be an issue in the cold box. As well as a risk of plugging, flushing out CO_2 could take a long time as it sublimes to equilibrium.

Carbon dioxide removal within the cracker

All ethylene plants have acid gas removal facilities in the process sequence between stages of compression. Amine treating can remove approximately 95% of acid gases. Caustic (aqueous NaOH) reacts irreversibly with acid gases and can remove >99.9%. However spent caustic is produced and must disposed of at a cost. As a general rule economics suggest the removal system consists of just caustic treating when $CO_2 + H_2S$ <1000 ppm and in a few cases amine treating followed by caustic treating when $CO_2 + H_2S$ >1000 ppm. Systems are designed such that the CO_2 and H_2S exiting the caustic tower ensure ethylene product quality is met

There are several potential issues with caustic treating. These include: fouling due to red oil formation, excess CO_2 results in sodium carbonate formation in the caustic tower which may exceed its solubility limit and it can break through the caustic which can possibly result is precipitation and plugging in the chill train. The latter two factors are normally not an issue if feed ethane specifications are met.

CO_2 and product quality

Approximately 62% of ethylene goes to polyethylene production (ref 3). Impurities in ethylene can be a problem for polyethylene producers because they are catalyst poisons. For example, CO_2 can react with triethyl aluminium to produce a less active species (ref 4) Because of these factors, ethylene specifications are tight. Other uses for ethylene include styrene, ethylene oxide and vinyl chloride. These processes are less sensitive to impurities and have looser specifications

The sensitivity of polyethylene systems to impurities is a function of the polyethylene technology. In the 1996 EPC meeting the following specifications were presented, see Table 2

Table 2 CO_2 Specifications for Polymer Grade Ethylene

Technology	CO_2 specification, ppm
Solution	5
Polymer grade	2
Slurry	1
Gas Phase	0.1
Metallocene	0.1

Generally producers tend to make one grade of ethylene, rather than different grades for different end uses. The reasons for this are to avoid increasing plant complexity, complex logistics, it may be difficult to sell excess product and it may not be compatible with ethylene pipeline network. Hence almost all ethylene is produced with <0.1ppm CO_2. In some cases, a portion of the product ethylene may be sent to a dedicated customer. In this case, the ethylene quality is such that it meets the lower specification.

Conclusion

Carbon dioxide has been identified as a top 5 EPC feedstock contaminant. It is commonly found in natural gas streams in high percentage levels. In combination with water, CO_2 is highly corrosive and can damage pipelines and operating equipment. CO_2 is normally removed from ethane or ethane/propane mixes to <500ppm mainly by amine treating. At the ethylene plant, unremoved CO_2 has the potential to cause operating problems and adversely impact product quality. All ethylene plants have an acid gas removal system to ensure that product quality specifications are met,

References

1. [EPC feedstock contaminant survey 2016]
2. Helen Knight (12 June 2010). "Wonderfuel: Welcome to the age of unconventional gas". *New Scientist*. pp. 44–47.
3. Ethylene Chemical Economics Handbook, HIS Chemical, 2017
4. Introduction to Industrial Polyethylene: Properties, Catalysts, and Processes, page x, D. Malpass, page 50, 2010

AIChE Paper Number:

Compressor Wash Oil Impact on Process Gas Compressor Efficiency and Vibration

Michael J. Player
Fluids Marketing Technical Support Advisor
ExxonMobil Chemical Company

Yang Yang
Fluids Marketing Technical Support Advisor
ExxonMobil Chemical Company

Marie Conrad
Fluids Regional Marketing Manager
ExxonMobil Chemical Company

David Dutschmann
Recovery Engineering Associate, Basic Chemicals
ExxonMobil Chemical Company

Prepared for Presentation at the 2018 Spring National Meeting
Orlando, Florida
April 23-26, 2018

Compressor Wash Oil Impact on Process Gas Compressor Efficiency and Vibration

Michael J. Player
Fluids Marketing Technical Support Advisor
ExxonMobil Chemical Company

Yang Yang
Fluids Marketing Technical Support Advisor
ExxonMobil Chemical Company

Marie Conrad
Fluids Regional Marketing Manager
ExxonMobil Chemical Company

David Dutschmann
Recovery Engineering Associate, Basic Chemicals
ExxonMobil Chemical Company

Abstract:

Organic fouling in process gas compressors (PGC) is an important issue that many olefin plants must manage. When it occurs PGC efficiency drops and compressor vibration increases. Minimizing fouling is critical to limiting production losses and high maintenance costs in existing or new steam crackers. The introduction of lighter feed stocks from shale gas is expected to increase the probability of fouling in PGCs. There are numerous factors to consider when managing PGC organic fouling and compressor wash oil (CWO) plays a vital role in compressor reliability. In this paper, key wash oil properties, general operational concepts, and two cases relevant to PGC efficiency and vibration will be discussed.

Shale Gas Contribution

The dominance of ethane as a cracking feedstock for ethylene production has been growing and is expected to continue growing (Figures 1 and 2). This higher ethane consumption has translated to a greater need for compressor wash oils.

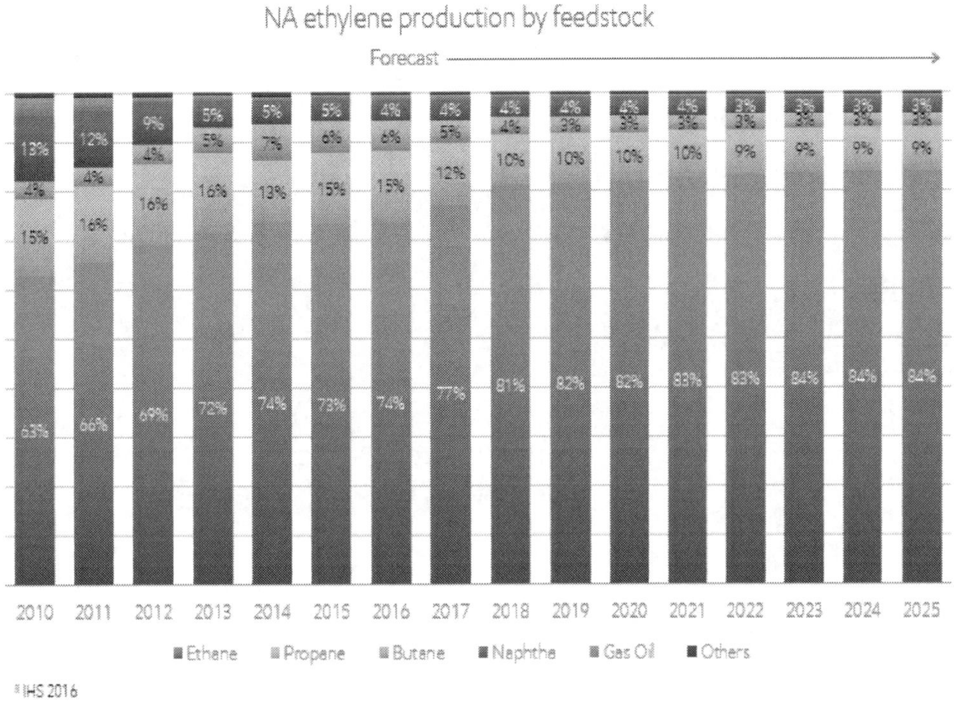

Figure 1

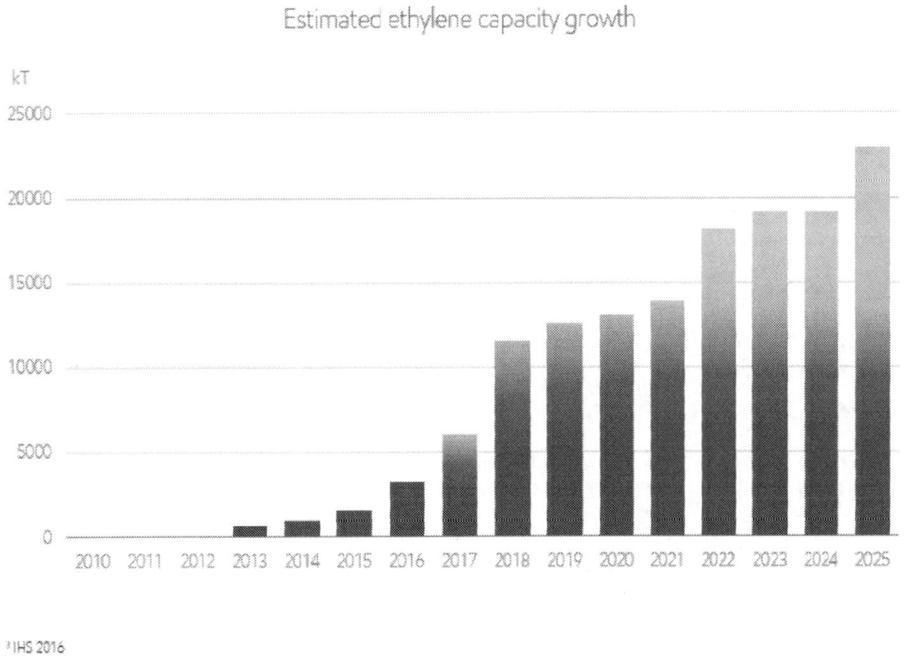

Figure 2

Wash Oil Necessity: Why and how it works

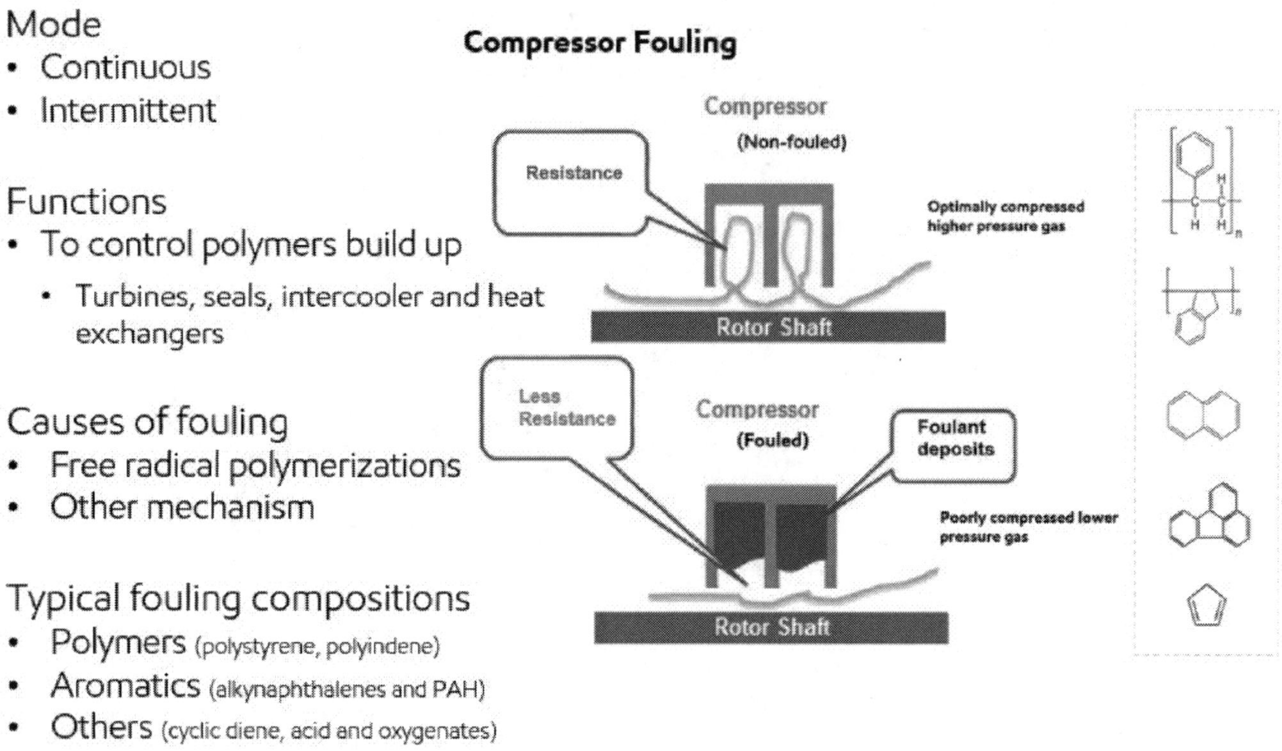

Figure 3

Producing olefins by steam cracking or pyrolysis brings with it unwanted, reactive precursors. These precursors readily polymerize by a free radical or Diels Alder mechanisms and if left unchecked will rapidly decrease runtime for a PGC. Wash oil, whether continuously or intermittently injected, will help dissolve low molecular weight (MW) polymer and facilitate removal of its resistive buildup along the various compressor stages. There are other key characteristics for a wash oil that will be covered next.

Wash Oil Quality is Important

Using Wash oil should not contribute to compressor complications with regard to quality. Quality wash oil is essentially defined by three pillars: solvency, purity and volatility. Strong wash oil solvency promotes effective foulant removal which is its primary function. Kauri-Butanol (KB) and Mixed Aniline Point values measure solvency power:

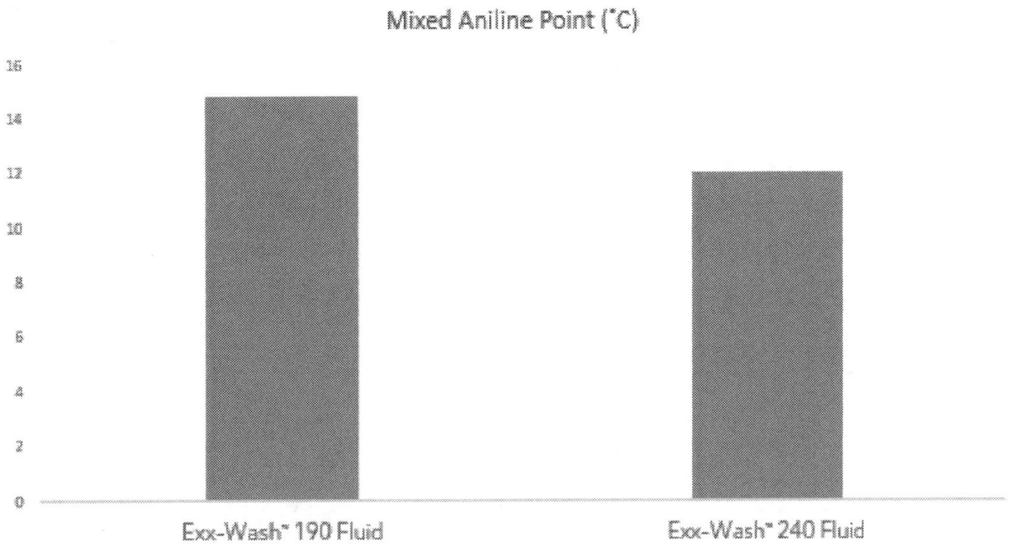

Property	Exx-Wash™ 190 Fluid	Exx-Wash™ 240 Fluid	Unit	Test Method
Mixed Aniline Point in C	14.8	12.0	C	ASTM D611-A
Mixed Aniline Point in F	58.6	53.6	F	ASTM D611-A

Figure 4

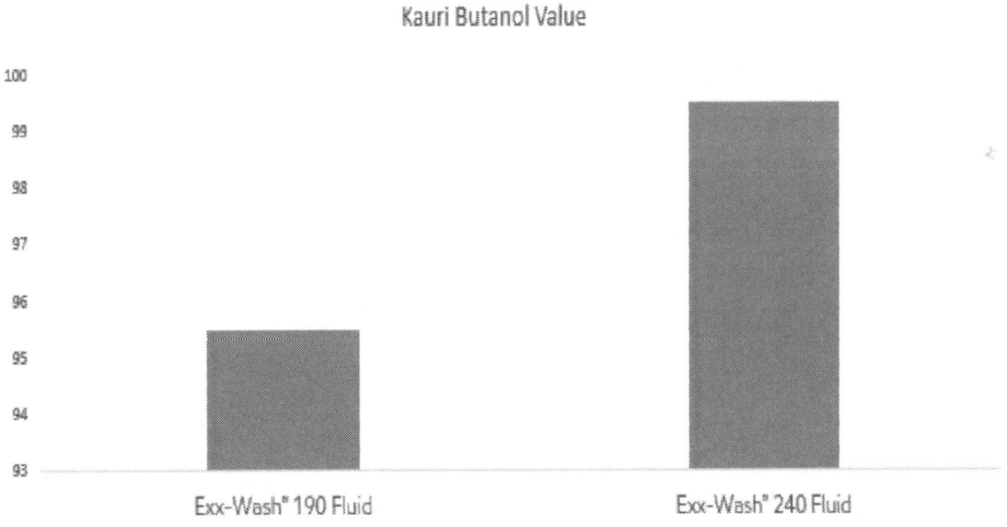

Property	Exx-Wash™ 190 Fluid	Exx-Wash™ 240 Fluid	Unit	Test Method
Kauri-Butanol Value	95.5	99.5	-	ASTM D1133

Figure 5

Impurities consist of a number of important items such as gum, organic chloride, sediment, sulfur, and olefin content (defined by Bromine number). Gum content needs to be low so that it does not provide an avenue of resistance nor a platform for additional foulants to build upon. Organic sulfur and chlorides contribute to corrosion which will weaken compressor internals and lead to foulant formation. Sulfur and chloride can aid crosslinking which enhance polymer growth which impedes solvation. The effect of sediments is akin to sand blasting the compressor train. Some sediments like rust will provide a boost to foulant formation. Low trace metals content should not be discounted either. Key spot property data provided in table below:

Table 1

Property	Fluid 190[1]	Fluid 240[2]	Unit	Test Method
Total Organic Chloride	0.3	<0.3	ppm	UOP 779
Sediment and Water	<0.01	<0.01	vol%	ASTM 2709
Total Sulfur	<3	<3	wppm	ASTM D2622
Bromine Number	1.3	0.8	gBr2/100g	ASTM D1159
Washed Gum Content	<0.5	<0.5	mg/hml	ASTM D381
Cloud Point	-10	-14	°C	ASTM D2500

High cloud point indicates the presence of heavy wax, asphaltenes and other polymer contributions.

Consistency in a wash oil is an often overlooked and under-valued property. There can be significant variation in key properties, say aromatic content for example. Some wash oils can have wide ranges for key properties. This inconsistency can lead to more adjustments and consumption to keep the process running smoothly and efficiently.

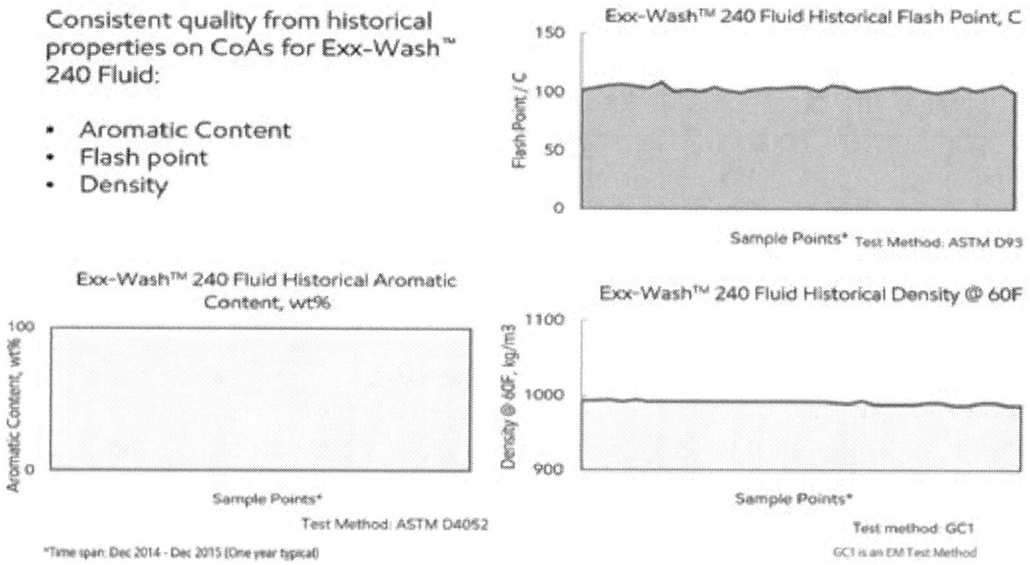

Figure 6
Property Consistency

One might believe that the heavier wash oil is the better it will be from a volatility stand point. In theory that would be case, but oils with final boiling points (FBP) greater than 300 °C can introduce heavier molecules that are more prone to forming carbonaceous deposits and contain more polycyclic aromatic hydrocarbons (PAH). These heavier molecules can contribute to carbonaceous deposits which can be more difficult to remove.

A comparison of wash oils in Table 2 below highlights several key quality items that have been discussed- aromatic content and volatility. The aromatic content of the competitive materials is significantly lower than wash Fluid 240. Not only is the distillation range wider, the FBPs are also above 300 °C. Wider cut wash oils will have more inconsistency.

Table 2

Typical Values Properties	Fluid 240	Similar Product[3]	Catalytic Lite Heating Oil*	Unit	Test Method	Comments
Aromatic Content	>99	82	78	wt% vol%***	GC1** ASTM D1319	- Better Solvency to remove the foulants
Flash Point	102	82	72	°C	ASTM D93A	- Safer handling
Initial Boiling Point	229	204	162	°C	ASTM D86	- Narrower distillation range ensures
50% Distillation Point	243	246	274	°C	ASTM D86	higher quality consistency
Final Boiling Point	285	316	351	°C	ASTM D86	- High IBP reduces carry-over
Specific Gravity @60°F	0.99	0.96	0.95	-	ASTM D4052	

* Data for single sample only GC1-EM test method ***Vol% assumed based on published data

Water-Oil Separation

Water washes can be a concern for several reasons when using wash oils. Having emulsions or foam forming in a PGC is undesirable as it will drive efficiency and pressures down. The compressor will need to work harder to maintain throughput. A wash oil must provide good separation from water under operating conditions. With increasing temperature the density difference between water and Fluid 240 provides a margin of safety against forming emulsified material, as illustrated by Figure 7.

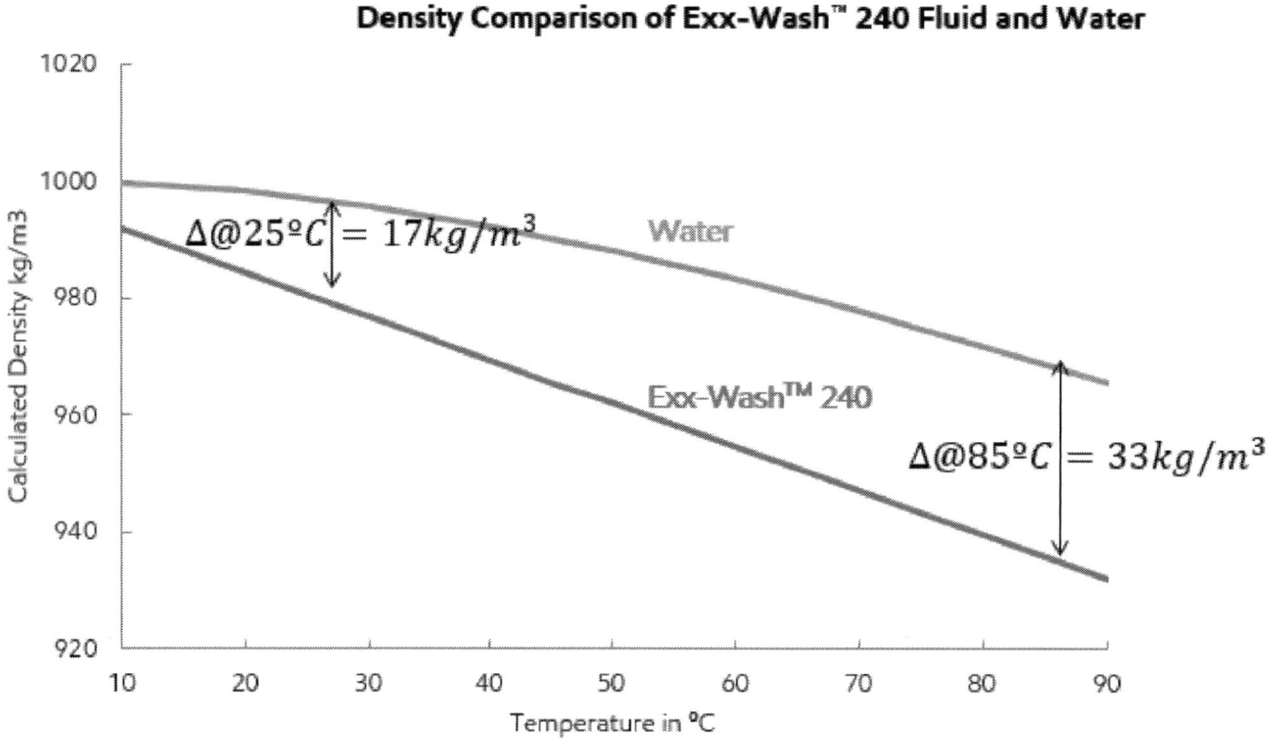

Figure 7
Wash Oil/Water Density Comparison

Wash Oil and Compressor Vibration

Excluding mechanical compressor issues, vibration will increase as foulant continues to deposit on rotors and impellers which lead to imbalances as well as build- up in the stator blades and casing areas. Sudden releases of polymer deposits will hasten mechanical wear and or failure. Injecting wash oil will lower vibration and soften the impact of polymer deposits within the compressor internals.

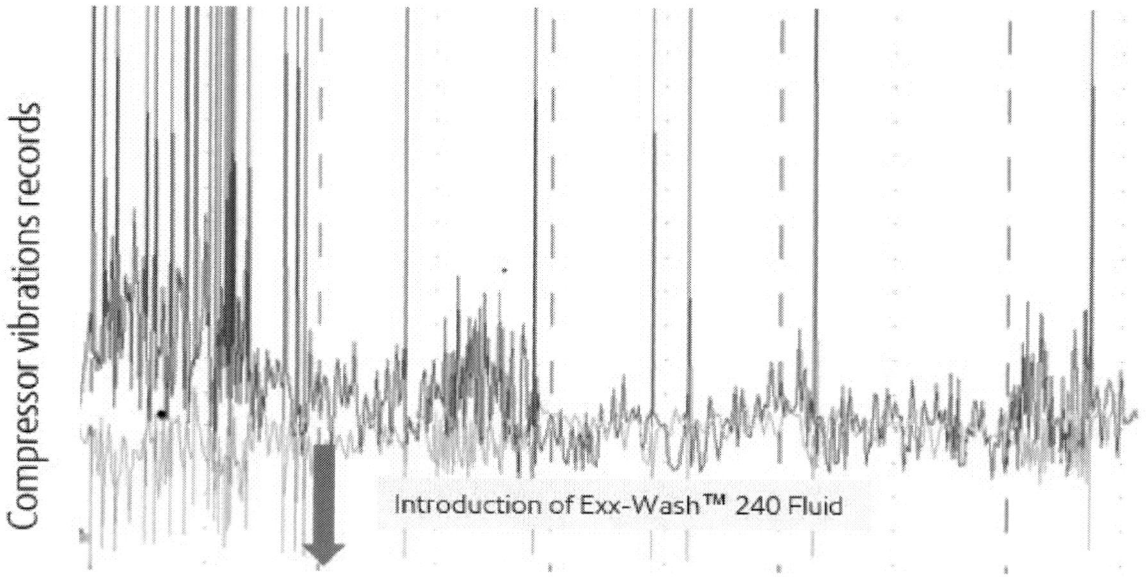

Figure 8
Vibrational Decrease with Wash Oil Use

Maintaining Efficiency

Polytropic efficiency computation provides an assessment of compressor performance. The efficiency plots below show positive impact from wash Fluid 240 injection relative to the previous period.

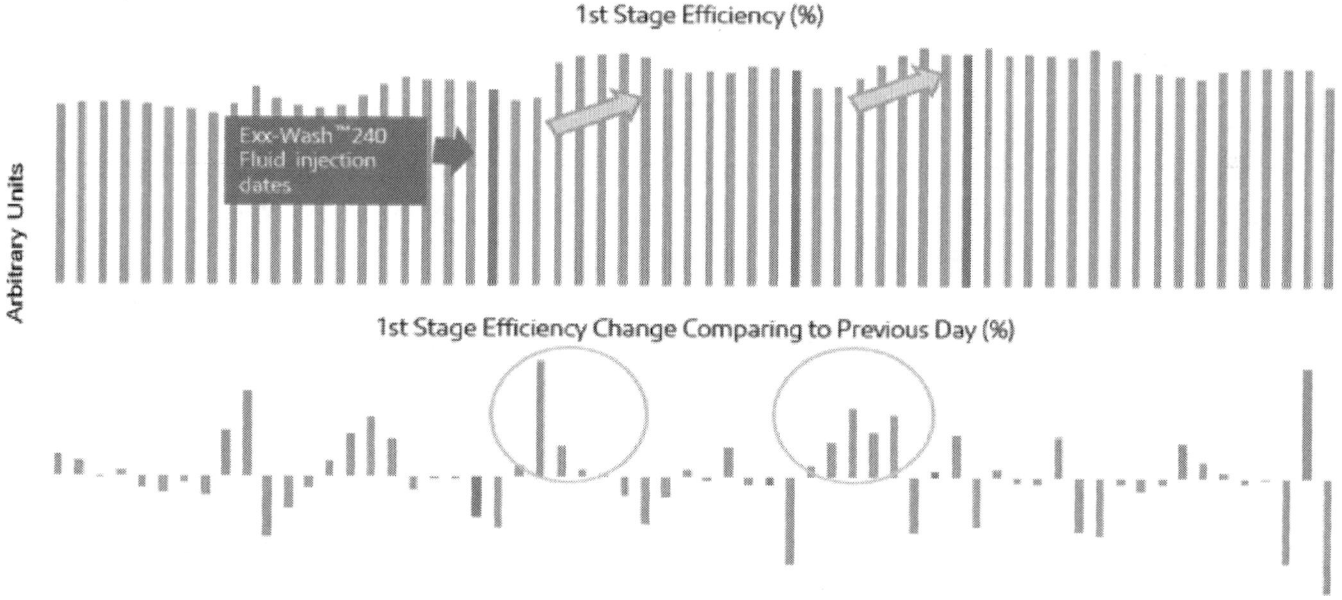

Exx-Wash™ 240 Wash Dates (highlighted in red): Dec 05, 2012, Apr 03, Apr 17 and Apr 25, 2013
Data collected: over 1 year

Figure 9
Efficiencies Before and After Wash Oil Use

Over time the intermittent injections of Fluid 240 facilitate in keeping overall throughput higher and energy consumption lower.

Operational Guidance

Wash oil can be implemented in an intermittent or continuous fashion. In the base case, excluding state-of-the-art compressors, high volume periodic rinses at typical levels 1-3% can be performed. Although lower in material cost, there is more risk with periodic washes due to polymer building up and if large chunks of polymer shift and suddenly break free resulting in severe vibrations. Continuous wash oil injection at 0.1-0.3% levels offer the prospect of keeping compressor internals "slick" so that foulant does not become stationary and surfaces are more resistant to other forms of attack from corrosive impurities. Of course this treatment mode is more expensive, but maximizing run time on the compressor without having to completely shut down and open the casing has immense appeal. ExxonMobil ran compressors for over ten years without having to throttle down and remove casings.

Measuring PGC vitals such as pressures, effects of feed swings, temperatures, flow rates, and wash injections frequency, is paramount when a treatment regimen is deployed. Monitoring these parameters permits better assessment of expected compressor design efficiency in addition to computing pressure drop limits within the system. Higher temperatures can dramatically fuel foulant formation, so keen attention must be given to intercooler and quench tower vitals to keep heat removal optimal. Knock-out drum content and trace (ppm), low levels of either oxygen or peroxides intrusion must be avoided since they readily promote radical polymerization.

Conclusion

Keeping a compressor running smoothly and efficiently requires careful consideration of foulant management. Wash oil use is an effective means of mitigating deleterious effects of organic foulant formation and accumulation within a PGC, particularly in the era of light cracking feeds where probability for fouling is higher. Wash oil quality is important because impurities and lack consistency can complicate compressor operation. Evaluating key PGC parameters when using wash oil is necessary to achieve best results.

References:

1. Fluid 190 is Exx-Wash™ 190 Fluid
2. Fluid 240 is Exx-Wash™ 190 Fluid
3. Similar Product is Advasol™ 150
4. Snider, Sheri "Ethylene Plant Cracker Gas Compressor Fouling" AIChE Spring National Meeting, April 2006
5. Yang, Yang; Harrison, Megan; Dutschmann, David "Compressor Wash Oil Supply in a light Feedslate Environment", AIChe Spring Meeting, April 2016
6. "World Analysis – Ethylene." November 2015. *IHS Chemical.* 2016 Edition
7. Snider, Sheri "Olefins plant cracker gas compression fouling" April 2007
8. HP Special Report: "A chemist's perspective on organic fouling in ethylene operations" *Hydrocarbon Processing*, June 2014

Permission to publish:

Disclaimers

512748 Refrigeration Compressors in Ethylene Plants

Abstract:

The separation side or cold end of an ethylene plant typically utilizes two levels of refrigeration. A -40F (-40C) level utilizing a propane or propylene refrigeration compressor (PRC) and a -150F (-101C) level using an ethylene refrigerant compressor (ERC). Some process designs utilize a binary refrigerant (BRC) as well for this coldest level. We will discuss what makes refrigeration compressors different from process gas compressors such as Cracked Gas Compressors (CGC)

Topics of discussion include:

Refrigeration Cycle basic overview

Sideloads and Extractions: impact on machinery design and performance

Materials of constructions

Process and surge control

Ethylene Producers' Committee

AIChE Paper Number: 513117

HIGH EFFICIENT AND COMPACT SOLUTIONS FOR NEXT GENERATION TURBOCOMPRESSORS IN ETHYLENE PLANT

Lorenzo Cosi

BHGE Steam Turbine Eng Manager

Marco Fiori

BHGE Steam Turbine Senior Product Manager

Sergio Ghezzi

BHGE Centrifugal Compressor Senior Product Manager

Giuseppe Iurisci

BHGE Centrifugal Compressor Eng Manager

Francesca Monti

BHGE Petrochemicals Platform Manager

Prepared for Presentation at the 2018 Spring National Meeting

Orlando, FL, April 23rd 2018

HIGH EFFICIENT AND COMPACT SOLUTIONS FOR NEXT GENERATION TURBOCOMPRESSORS IN ETHYLENE PLANT

Lorenzo Cosi

BHGE Steam Turbine Eng Manager

Marco Fiori

BHGE Steam Turbine Senior Product Manager

Sergio Ghezzi

BHGE Centrifugal Compressor Senior Product Manager

Giuseppe Iurisci

BHGE Centrifugal Compressor Eng Manager

Francesca Monti

BHGE Petrochemicals Platform Manager

ABSTRACT

In the continuously evolving global petrochemical industry all the companies are looking more and more for the highest level of technology that can guarantee lower operating costs coupled with increased reliability.

Compressor and steam turbine performances are key to reach top level production of the ethylene plant through efficiency. Power density technology applied to the charge gas train and refrigerants is able to increase several points of efficiency while optimizing CAPEX investment thanks to compact design and weight reduction.

New stages with modern design have been developed and completed with Computational Fluid Dynamics and model test. The new high-efficiency stages can be used in different configuration to reach the compressor compactness needed to minimize CAPEX and installation costs, up to stacked configuration.

Real examples of these technologies and relevant results are presented in this paper.

NOMENCLATURE

AF	Amplification factor
SM	Separation Margin
HP/IP	High Pressure / Intermediate Pressure
LP	Low pressure
HPRC	High Pressure Ratio Compressor
2MCL	GE Oil&Gas 2 section Horizontally split compressor
DMCL	GE Oil&Gas Double Flow Horizontally split compressor
3MCL	GE Oil&Gas 3 section Horizontally split compressor
LMF	Liquid Mass Fraction
τ	Impeller work coefficient
γ	Specific Heats ratio (C_p/C_v)
μ_p	compressor polytropic efficiency
N	Number of Impellers
N_{ST}	Steam turbine rotating speed
s	Sound Speed
TTT	Tip Timing Test
U	Impeller tip speed
WBT	Wheel Box Test

ETHYLENE MARKET ENVIRONMENT

While the whole petrochemical industry is experiencing a great momentum, plans for ethylene expansions are clear; this market is going to grow fast in the next years.

A long list of new plants is continuously under revision in terms of start date, due to changes in potential feed, cost of labor, schedule overlapping, delays, but it is still increasing and very promising.

This market revitalization is linked to different parameters worldwide. All the variables influencing industry mapping, exports and imports, short and long-time schedules will continue to impact producers' margins, both at present and in the future, during the whole plant life.

The probability of all these factors to be predictable and manageable are clearly low, and in a such competitive panorama, the need of optimization and cost reduction of plant equipment is stringent.

The importance of building a new steam cracker with the best in class equipment will certainly remain a priority, that is the reason why key rotating equipment for the ethylene industry are continuously improved by manufacturers.

Producers need new and efficient technologies that can reduce capital expenditure, installation cost and contemporarily optimize reliability and availability. Compression efficiency is a key element to reach these targets and coupled with the efficient use of space and minimized weight, can bring greater value to the overall performance of ethylene plants, as analyzed in the following chapters.

DESIGN DEVELOPMENT AND VALIDATION

In the last 15 years the OEM has been investing significantly to improve Turbo-Compression technology, to deliver cost effective compression stations that could reduce Capex, Opex and maintenance cost.

The major area of development has been:

1. Improvement of the compressor and steam turbine efficiency to reduce absorbed power and steam consumption
2. Improvement of rotodynamic predictability to provide reliable solution when moving to higher speed and over or beyond critical speeds
3. Increase the compactness of the compression station, reducing equipment size and number of compressors for higher reliability, availability and lower cost

STEAM TURBINE DEVELOPMENTS

Steam turbine efficiency translates directly in plant energy efficiency. In this regard OEM has invested in the development of stages and seals technology aimed to increase turbine efficiency.

Steam turbines applied in mechanical drive industry must provide an optimized combination of thermodynamic and operating flexibility to match the different process conditions.

When looking for chances of overall turbine efficiency improvement, it is important to bear in mind that overall turbine power output can be typically split into two main contributions: the 60% coming from HP/IP stages (including control stage) and 40% from the LP section. Therefore, both sections matter and this is also reflected in available literature where papers covering stage and seals improvements of both HP/IP and LP sections can be retrieved. As far as concern HP/IP sections, Maughan et al. [5] illustrates how to increase efficiency through degree of reaction, the aspect ratio, and the reduced stage load. Mindok et al. [6] and Segawa et al. [7] illustrates their work aimed to improve the profile robustness to incidence and also the beneficial effect of nozzle endwall contouring. In some cases, the performance improvement was obtained just improving the airfoil shape of prismatic blades (Sakai et al. [8]), in others by replacing the prismatic blades with twisted blades, and applying also bowing and leaning (Schaarschmidt et al. [9]). Remarkable examples of LP section development increasing annulus area to reduce the exhaust losses can be found in (J. D. Denton and A. Spurr [3], Yoshinori et al. [4]).

Steam turbine efficiency have been improved by working on these major areas:

- High and Intermediate Pressure stages
- Low pressure stages
- Seals technology

HP/IP STAGES

In 2010 HP/IP stage technology development program was launched with the objective to develop a set of master stages with higher efficiency and wider flow/work coefficients range usually required in mechanical drive application. Development process was based on a closed loop design method that combines CFD optimization, stage performance validation, stage master model definition, turbine application and field testing.

CF D Optimization

Many optimization techniques are nowadays applied to redesign the aerodynamic components of turbomachines. The industrial interest pushes for a continuous improvement of these tools towards automatic and effective procedures for aerodynamic and mechanical design. The emphasis is justified by the capability of optimization algorithms to exploit the great amount of information provided by numerical simulations, and correlate them to the geometrical parameters to find optimal combinations. OEM adopted the same approach proposed by Pierret and Van den Braembussche [13], which consists of coupling CFD analyses with a neural-network-based optimization method. Bellucci and others [14] can be referenced for more information about the optimization process followed in this program.

Stage Performance Validation

Stage validation process is key to anchor the prediction models and ensure performance predictability. A two steps approach was adopted. Firstly, the new airfoils profile losses were initially validated in a cascade testing; secondly, an air turbine test was conducted to validate secondary losses, leakage contribution eventually providing an overall validation of the performance.

The air turbine test facility (Figure 1) consists of a subsonic closed loop wind tunnel for turbine stages. The test section can house up to two full turbine stages.

The test section consists of one and a half axial turbine stage preceded by an IGV row. The rotation of the casing over 360 degrees is obtained by a rotatable gear ring, allowing the investigation of the flow-field over the entire section. In this rig stage performance were validated through local and integral measurements; by isolating individual losses and eventually calibrating the correlation models used on a project basis, was possible to predict the efficiency and the flow functions of the HP/IP stages sections.

Figure 1: Air turbine validation facility

Stage Master Model Definition

Thanks to this development programs, 4 new master stages were added to the available HP/IP masters' family. New masters are optimized for different combinations of flow and load coefficients to cope with mechanical drive application flexibility requirements. New masters feature 3D airfoil (tapered and twisted) to minimize incidence losses and optimize blade solidity. Platforms and covers leading edges have been optimized to minimize the losses associated with secondary flows incursions.

New master stages allow up to 2 points of efficiency increase versus traditional reaction stage technology design and thanks to the higher nominal loading a reduction in stage count can be achieved. Stage count reduction allows axial span reduction which is a key advantage in ST where rotodynamic design is challenged by power density machines or multiple extractions.

Projects Application

New HP/IP stages have been applied on 20+ turbines (see Figure 2) employed for different Power Generation and Mechanical drive services (Ammonia drivers, Ethylene drivers, Methanol divers). On those turbines subjected to field performance testing it has been possible to verify consistency of the new stages predicted efficiency with overall turbine performance.

Figure 2: New HP/P stages application

LOW PRESSURE STAGES

LP section rotating speed capability is typically the main bottleneck to the increase of steam turbine rotating speed capability of the turbine because last stage blades have the highest AN^2 (i.e. last stage rotating blade annulus area multiplied by the rotating speed squared).

Power density centrifugal compressors requires the usage of steam turbine with same volume flow capability at a higher rotating speed; in this scenario the traditional LP stages may represent a bottleneck.

To overcome this limit, OEM has developed a new family of low pressure blades (i.e. HS Family) for Mechanical Drive and Power Generation applications. The new design features a 60% increase in AN^2 and 6 points increase in section efficiency with respect to the traditional design.

This new generation of blades has been scaled to produce a given number of sizes and thus to cover a wide range of exit annulus areas. The overall family is ranging from the 8" (height of the last rotating blades) running at 11250 rpm to the 31" last stage blade running at 3000 rpm and suitable for the 50 Hz power gen market.

The new LP stages feature 3D airfoils (see Figure 3)

The design of the L0 stage blade consists of an axial entry dovetail skewed in the axial-tangential plane. The blade is shrouded at the tip with an integral cover, providing a rigid form of continuous coupling for robust operation. No coupling was included within the steam path to minimize performance losses. The L1 (next to the last), stage blade is composed of a 3D airfoil, a two hooks axial entry dovetail, and an integral V-shape cover. L2 and L3 blades (third and fourth to the last), blades are composed of a 3D vane, a tangential entry T shape root and a V shape integral cover.

Aero Design

The LP stages are designed aerodynamically to work together as a system using a combination of streamline curvature design methods, 2D cascade analysis, and state-of-the-art 3D computational fluid dynamics analysis techniques. This four stages group has been designed to meet up with an intermediate pressure section whose inner diameter is considerably lower than the exit diameter of the last stage blade. This design strategy allows the retrofitting of existing LP rotors, while providing maximum work output from the LP section. Since power density is an important feature, the stage spacings were limited to maintain close wheel spacing while allowing for a high recovery exhaust diffuser configuration.

Figure 3: HS LP Rotor

Aerodynamics and Aeromechanics validation

Verification of the predicted aeromechanical behavior and aerodynamic efficiency was conducted by mean of a full scale full load instrumented test performed in an experimental test turbine (Figure 4). The test turbine facility is used to carry out aeromechanical and aero performance tests of low pressure steam turbine stages. It consists of a steam source (with steam conditions appropriate for low pressure turbine testing), a low-pressure stage steam turbine and two water-brakes that absorb the energy that the LP turbine produces. Stages are instrumented with strain gages and thermocouples on the blades and with static temperature, total temperature, static pressure and total temperature probes in the flow path and in the diffuser.

Figure 4: Experimental Test Turbine

The aeromechanics test space is a three-dimensional space with the three critical variables being mass flow, exhaust pressure and turbine speed. Test points are established by fixing the mass flow first. With the mass flow constant, the exhaust pressure assumes distinct values between the minimum and maximum exhaust pressure (in this case between 0.05 and 0.7 bara). At each of those combinations of mass flow and exhaust pressure, the speed of the turbine is varied between 50% of Maximum Continuous Speed 100% of Maximum Continuous Speed while acquiring the strain gages data.

Main result of the aeromechanics validation consists in the mapping of blades vibratory response over the entire operating range. Vibratory response is post-processed on a statistical basis to take into account blade to blade variability and build to build variability. Eventually the experimental Goodman diagrams of the blades are built for any operating condition. A broader description of the test set-up and results is presented in Piraccini et al. [15].

Turbine performance is a function of several operating parameters, namely: mass flow (Reynolds number), rotating speed, condensing pressure and wetness content. For this reason, the performance test is conducted in a set of operating points which are selected in order to map the dependence of the efficiency upon the above parameters.

For each point the operating conditions are set and maintained until all the temperatures stabilize and eventually all the pressure probes are purged to remove water lags and the data are recorded.

Data are then post-processed to measure all the performance characteristics of the turbine like: section efficiency, individual stages efficiency, diffuser recovery factor and moisture loss contribution.

Projects Application

New LP stages have been applied on 50+ turbines (see Figure 5) employed for different Power Generation and Mechanical drive services (Ammonia drivers, Ethylene drivers, Methanol divers). On those turbines subjected to field performance testing it has been possible to verify consistency of the new stages predicted efficiency with overall turbine performance. Fleet leader is in operation since 2009 and has been successfully inspected early 2017.

Figure 5: LP stages application

SEALS TECHNOLOGY

New seals technology has been developed for stage and balancing piston sealing. Sealing technology development has been built upon the basis multi-year multi-millions sealing technology investment carried out in the last 20 years in support of Aviation and Gas Turbine sealing technology. Out of the multiple available sealing solutions abradable and brush seals technology have been selected, developed and validated to address steam turbine specific operating needs. Also traditional back spring solutions have been developed to offer an alternative to the traditional reaction turbine strip seals solution. As a result of this effort, advanced sealing technology are available as an option to improve performance and maintainability of OEM steam turbines.

Abradable seal design are applied on balancing piston seal which is one of the major contributor to overall turbine efficiency. As shown in picture zz, it consists of a back spring packing ring coated with an abradable proprietary material. Its main advantage is the possibility to reduce clearances without risk of rubbing and its superior long-term effectiveness. Even in terms of maintainability abradable packing can be easy replaced without di-installing the inner casing.

Brush seals are applied for low delta P on balancing piston or on rotating blade tip. A proprietary brush seals design has been developed and characterized in terms of efficiency and durability. Brush seals enables a superior leakage reduction compared to traditional labyrinth (i.e. 2-3 times lower flow). Brush application on blade tip ensures a significant efficiency increase particularly in HP drum operating at low volume flow condition where tip leakage has a significant role in overall stage losses (i.e. 1-3 points).

Back spring labyrinth stage tip seals has been developed to offer a blade carrier solution with improved maintainability and more sustained efficiency compared to traditional reaction machines strips seals.

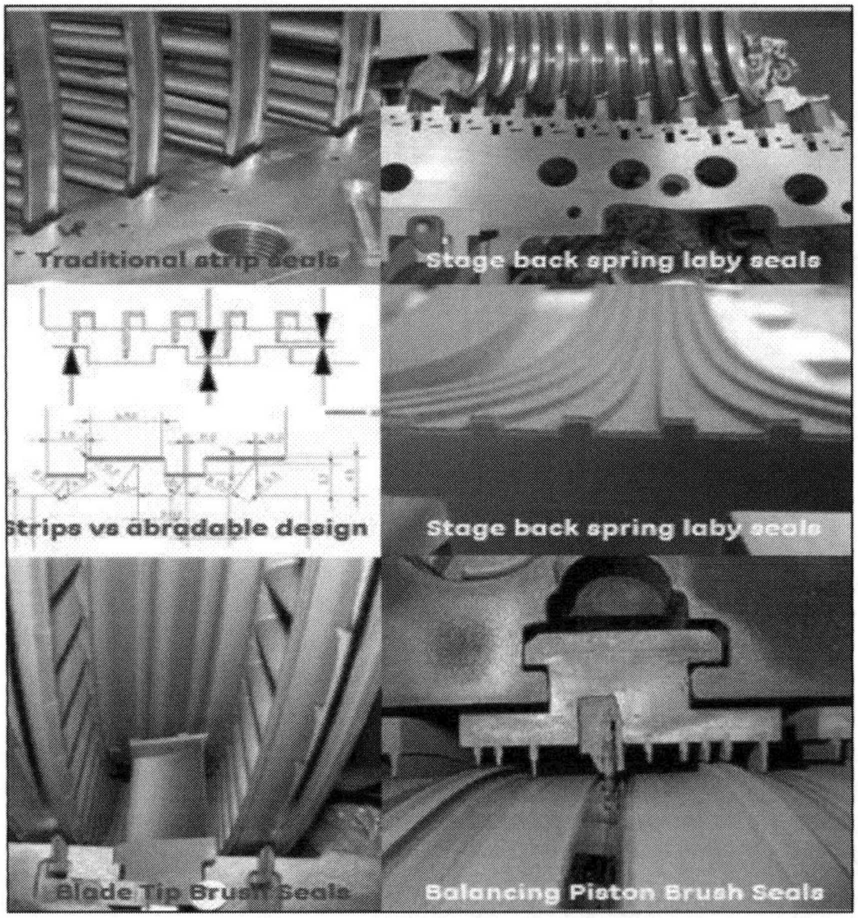

Figure 6: seals technologies

CENTRIFUGAL COMPRESSOR DEVELOPMENTS

COMPRESSION EFFICIENCY

Compression efficiency is one of the primary target for the OEM and the major areas of development have been:

a. Aerodynamic design optimization: flow-path components

b. Seal design

c. Higher speed and higher flow coefficients

a. Aerodynamic design optimization: flow-path components

Machine customization has been maximized with a dedicated stage design interpolated among pre-engineered, optimized and tested geometries (Masters) defined in a range of flow coefficient, Peripheral Mach Number and diameter.

These master stages are extensively optimized, constantly improved and then tested in scale model condition in dedicated test facilities. In this way, impellers are tailored for each specific project to improve efficiency and are designed with a standard approach to keep high levels of reliability and predictability.

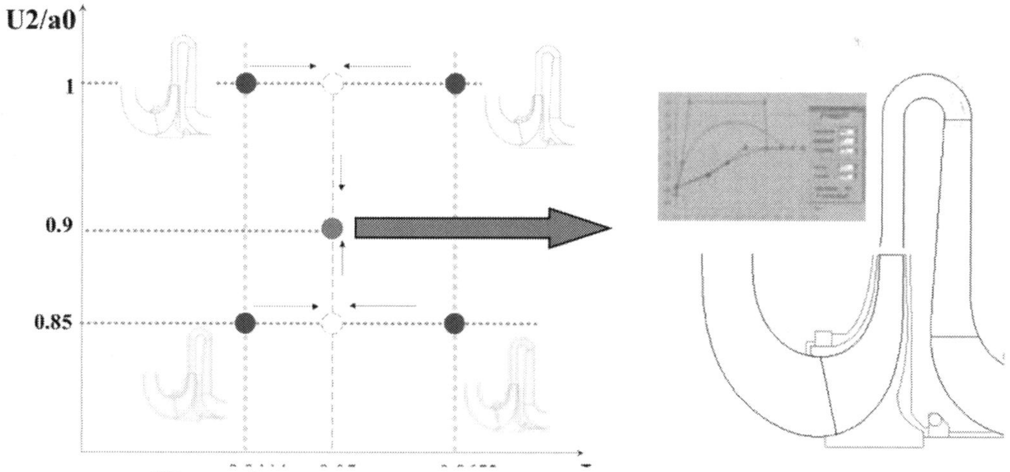

Figure 7: Stage Technology Process

The design is carried out through an **integrated** aero design system aided by advanced Fluid Dynamic Computation capabilities and validated with structured internal test campaigns.

The major highlights of the integrated design system are:

- Integrated, modular & extensible environment of design tools
- Full modeling capability up to 3D CFD
- Interactive Flow-path & Airfoil design and Performance data matching
- Interactive and dynamic visualization and geometric analysis capabilities using CAD systems
- Geometry verification and analysis in a very short amount of time
- Coupling with different optimization algorithms

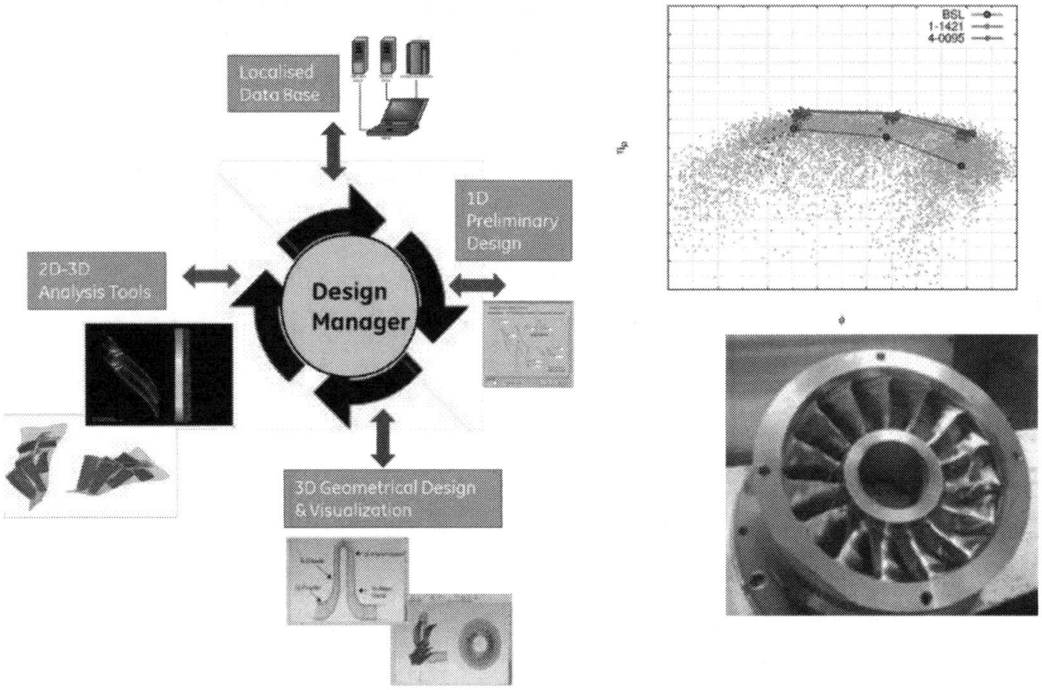

Figure 8 – Stage Technology Process

The Computation Fluid Dynamic (CFD) improvement helped a lot on the single geometry optimization, becoming gradually a simulation describing a flow behavior very close to the reality. Some of the improvement on the CFD computation for impeller stage design are:

- Additional and improved numerical models: streamline curvature correction, real gas, transitional laminar-turbulent, boundary layer capturing
- Fidelity geometrical modeling with cavity addition, fillets, grid resolution
- Steady and Unsteady simulations for time averaged and time accurate flow field simulations
- Both in-house and commercial codes used (well established codes with many validations in open literature)

The "High Fidelity" CFD provided significant improvements extending from simple main flow path modelling to full geometry (including secondary flow leakages) and physics.

In this way physical phenomena not correctly reproduced by standard CFD can now be better captured and characterized.

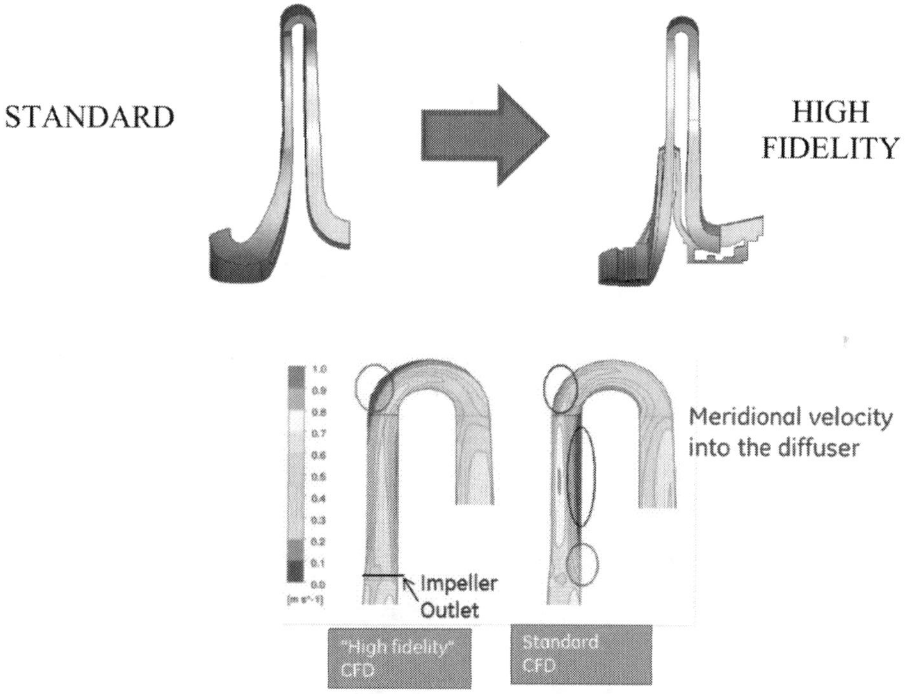

Figure 9 – CFD Results

In parallel with design optimization improvement, testing capabilities have been also increased for a deep performance diagnostic with several internal instrumentations allowing either to derive impeller, diffuser and return channel detailed performances than to fine-tune the mathematic model for higher predictability.

Figure 10 – Aerodynamic Test Rig

These advanced tests are done using the best available sensors. One of those is for instance the so called FRAPP Fast Response Aerodynamic Pressure Probe that allowed to measure the complete channel flow field in complex aerodynamic area as the impeller exit, showing good agreement between CFD prediction and test.

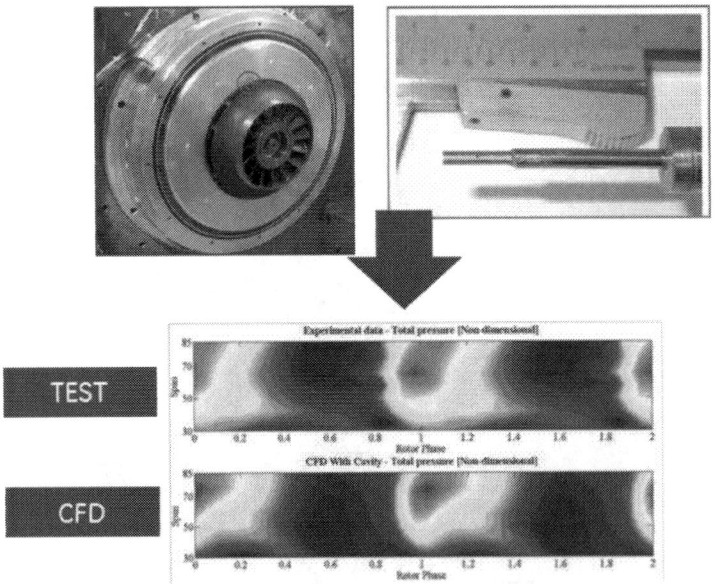

Figure 11 – Test vs. CFD comparison

This design method, the advanced computation and tested capabilities allowed to gradually improve impeller stage performance, improving blade aerodynamic profiles and exploiting more effective static pressure recovery element as vaned diffusers.

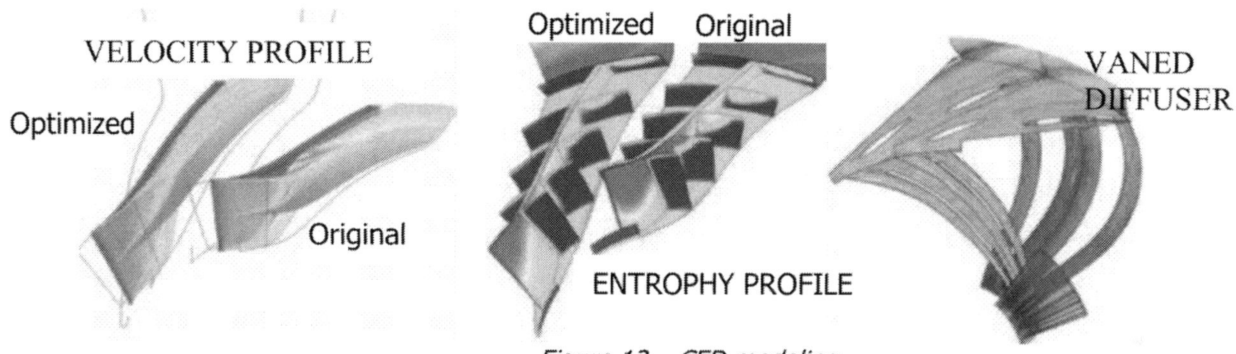

Figure 12 – CFD modeling

In addition to the impeller stage design, all other stator components have been improved by adopting same approach with High Fidelity CFD and tests.

Figure 13

By combining all the above, today it is possible to rely on compressors with an improved efficiency of around **5%** compared to old designs

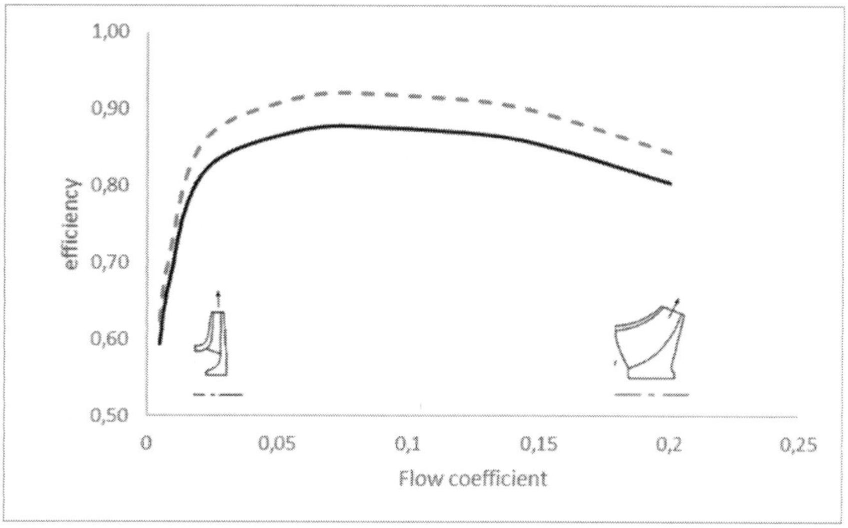

Figure 14

b. Seal design

Dedicated development has been done to improve seals technology focused on internal leakages reduction, forces modeling as well as their effect on rotor stiffness and damping characteristics. Such seal innovations include new optimized type, shape and materials to reduce clearance and improve fouling resistance.

Advanced computation allowed a more accurate seal design and a better understanding of secondary flows behavior. Together with the computation, dedicated test rig was built to have seal characterization, both from an aerodynamic point of view but also to characterize the stiffness and damping for different flow condition (density and swirl) fundamental for a proper stability rotor-dynamic analysis.

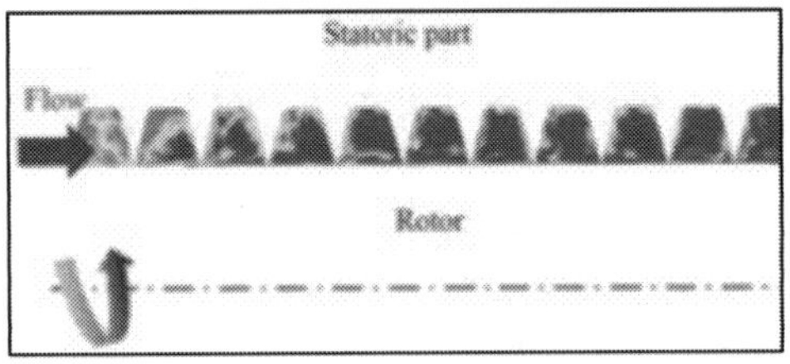

thermoplastic laby seals

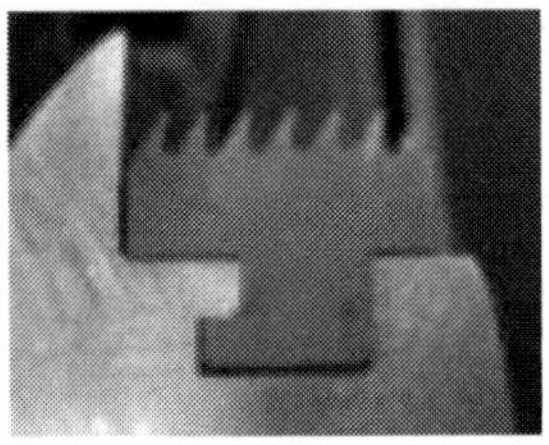

thermoplastic laby seals

leakage rig to characterize the seals performance

Figure 15

c. Higher speed and higher flow coefficients

The curve envelope of the impeller efficiency vs flow coefficient shows that there is an efficiency trend with flow coefficient and a best efficiency region. According to flow coefficient definition, by increasing RPM and reducing the impeller diameter it is possible to design more "power density" machines.

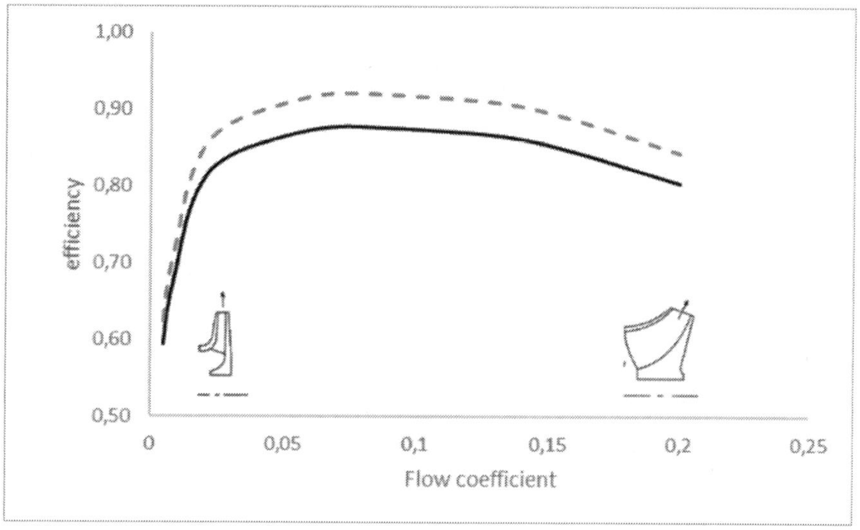

Figure 16

The effect of the speed increase ("power density" effect) and improved stage component aerodynamic, in addition to the above mentioned up to 5% efficiency gain, allow also a reduction of compressor size from 15 to 30%.

From mechanical perspective all these developments lead to higher rotor flexibility and more challenging dynamic conditions; an accurate rotor-dynamic predictability is critical to assure a robust and reliable compressor.

IMPROVEMENT OF ROTORDYNAMIC PREDICTABILITY

Rotor dynamic predictability level is linked to the precision modelling of the rotor stiffness, inertias, bearings and seals in all possible operating conditions. Dedicated test-rigs and a full scale high speed prototype compressor have been built to validate and calibrate prediction tools.

The same approach is used also for journal bearings, to validate stiffness and damping coefficients, power losses, oil consumption, speed and bearing design (flooded, direct lube, different pivot design).

Large scale Journal bearing test rigs up and above 280 mm

Figure 17

Specific ping tests are used to match the correct rotor stiffness and inertia distribution and obtain an accurate rotor model.

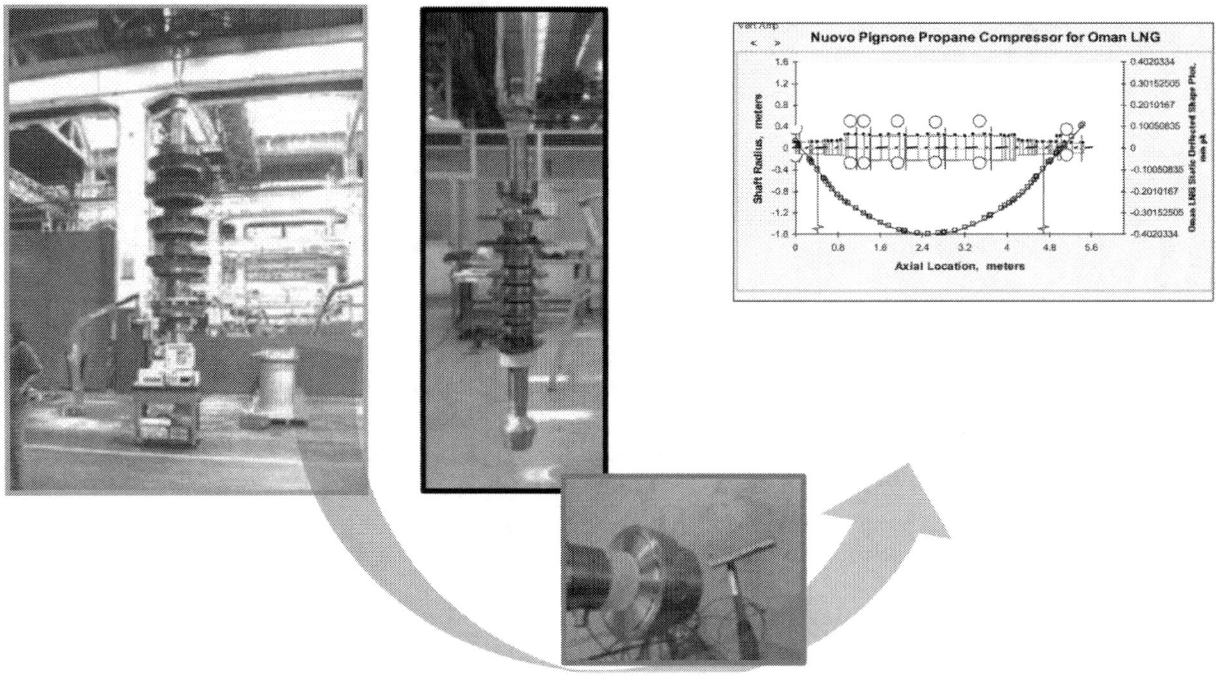

Figure 18

INCREASE THE COMPACTNESS OF THE COMPRESSION STATION. FROM TWO COMPRESSOR CASINGS TO ONE CASING

An important outcome of the above-mentioned programs is the significant increase of the head capability of each unit, allowing to even reduce the number of compressor casings while keeping same level of robustness and reliability required by ethylene applications.

The reduction in number of compressor casings from two to one has several advantages:

- Elimination of a complete dry gas seal system;
- Reduction of baseplate footprint and weight: reducing the number of components to be supported not only the baseplate dimensions will be reduced, but also main beams' size could be decreased and the sub-structure to the baseplate (for top mounted shaft lines);
- Reduction of lube oil consumption and therefore size of the Lube Oil Console (LOC), with size reduction for both tank and main components.
- Improved availability and maintenance

Case study

Below is reported a case study made for a 3 sections charge gas compressor, with high volumetric flow (200000 m3/h) and high-pressure ratio (10), to compare how it would look like a new compact and efficient BHGE solution versus a conventional solution

Mol Weight g/mol	Q m^3/h	PR	Stage #	STANDARD solution			HPRC solution		
				Train composition	RPM	Casing #	Train composition	RPM	Casing #
18.5	200000	10	3	SAC1-10 + DMCL1006 + 2MCL1006	4869	2	SAC1-10 + 3MCL1407	5460	1

Table 1. Standard Configuration vs compact and efficient for Charge Gas

Charge gas compressor design and configurations

The design conditions for the compressor train are:

- Volumetric flow rate of 200,000 m3/h

- Gas mixture (major components are Hydrogen, Methane, Ethylene, Ethane, Butadiene, Steam Water...) with a molecular weight or around 18.5 g/mol
- Inlet pressure of 1.75 bar and an Outlet pressure of around 175 bar.

In a standard configuration, the large flow warrants a first casing with a double flow configuration and a following casing with 2 sections, driven by a steam turbine with large controlled extraction.

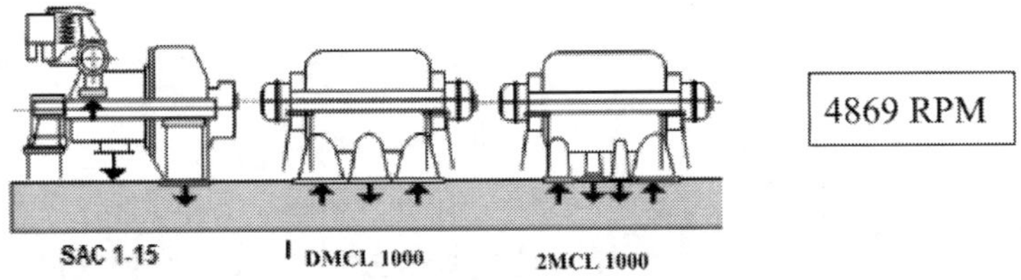

Figure 19 - Charge Gas Standard Configuration Schematic

The new configuration allowed by the above-mentioned technology improvements can allow to have a single casing 3 section compressor, more efficient and providing a more compact overall general arrangement

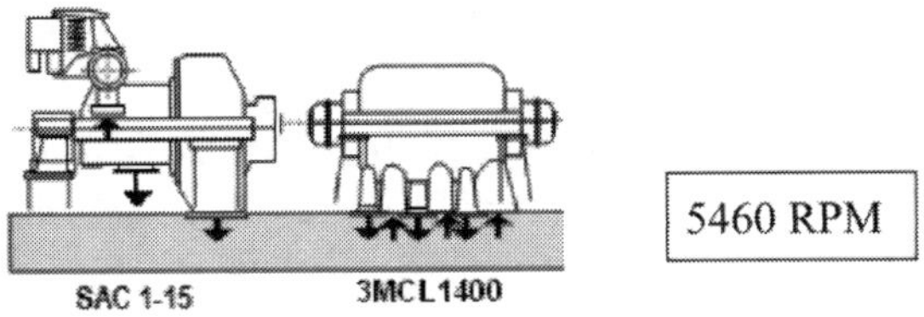

Figure 20 - Charge Gas HPRC Configuration Schematic

The overall absorbed power is reduced of about 3% and the speed increase allows also a Steam Turbine improved efficiency, so the overall benefit in steam consumption is around 5%.

The compressor casing contains 7 impellers, all 3D optimized blade shape. The shaft ends maintain a standard configuration with standard oil bearings and DGS.

Even increasing the RPM, the compactness of the solution maintains a rotor L/D and flexibility ratio within acceptable limits, yielding typical rotordynamic behavior, and full API compliance.

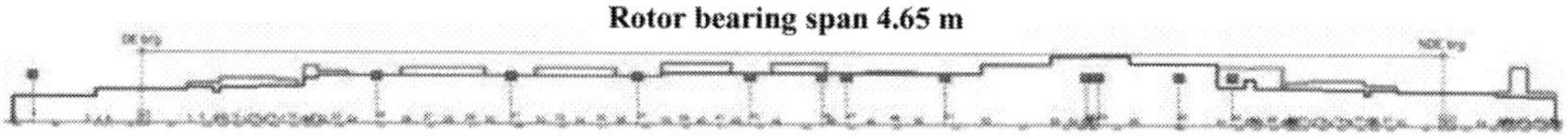

Figure 21 - Charge Gas single unit rotor scheme

Below is the undamped critical speed map, which clearly shows the unit is operating between the first and second critical modes.

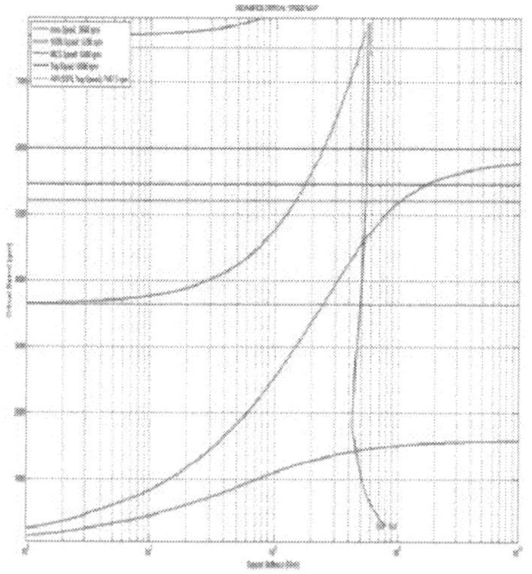

Figure 22 - Charge Gas single unit Critical Speed map

The machine operates with sufficient margin from the first and second critical speeds and is critically damped due to proper selection of the oil bearings. The graphs and table below provide evidence of the Separation Margin and Amplification Factor from the damped response analysis, confirming API compliance.

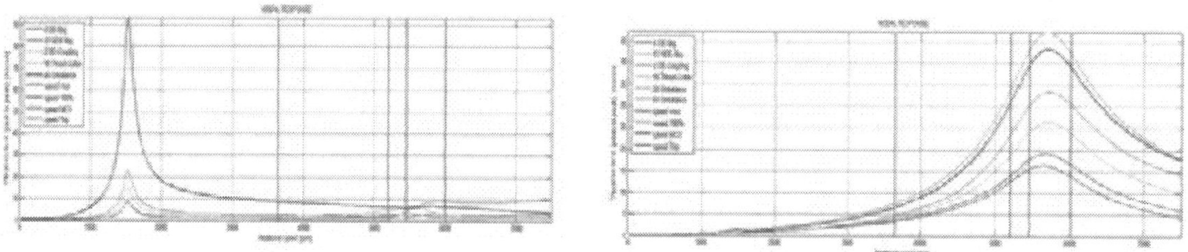

Figure 23 - Damped response

		ROTOR RESPONSE		
		Amplification factor	Separation Margin	Separation Margin required by API
1st MODE	**Avg bearing clearange**	3.8	61	≥10
2st MODE	**Min bearing clearange**	1.5	-	≥0

Figure 24 - Amplification Factor and Separation margins

From a stability point of view, the machine is very robust, having a log decrement on the 1st mode with minimum bearing clearance, and all seal destabilizing effects above 1 vs. the API minimum requirement of >0.1

CONCLUSION

After over 30 years OEM has been able to renew the charge gas train configuration with tangible benefits in terms of CAPEX and OPEX while preserving the confidence of industry standard compliance.

The test campaign is carried out as companion for all conceptual analysis to fully validate the studies and fine tune the predictive model. In this way, a high predictability and minimized risk profile can be successfully reached since the first introduction. This process represents the OEM standard path for novelty validation.

In addition to the efficiency increase on the steam turbine and compressor stages, the high-pressure ratio technology brings benefits also to the equipment operating life, installation, train reliability as well as footprint, maintenance time and capital spares reduction.

At present more than forty steam turbines and six compressors designed with the technology described in this paper have been sold to oil and gas companies.

REFERENCES

[1] Havakechian, S., and Greim, R., 1999. "Aerodynamic design of 50 per cent reaction steam turbine stages". In IMechE, J. of Mechanical Engineering Science, Vol. 213, pp. 1–25. part C.

[2] Yoon, S., Curtis, E., Denton, J. D., and Longley, J., 2010. "The effect of clearance on shrouded and unshrouded turbines at two levels of reaction". In ASME Turbo Expo 2010, June 14-18, Glasgow, United Kingdom, no. GT2010-22541.

[3] Denton, J. D., and Spurr, A., 1979. "Aerodynamic factors in the design of the final stage of large high-speed steam turbines". In Desing Conference: Steam Turbines for the 1980s, October 9-12, London, United Kingdom, no. I Mech E 1979-12.

[4] Tanaka, Y., Yokota, H., Ohyama, H., Nakano, T., and Koeneke, C. E., 2003. "Advanced design of Mitsubishi large steam turbines". In PowerGen Europe, May 6-8, Dusseldorf, German.

[5] Maughan, J. R., Hill, J. M., and Goel, S., 2000. "Development of the dense pack steam turbine: A new design methodology for increased efficiency". In International Joint Power Generation Conference & Exposition, June 23- 26, Miami Beach, Florida, USA.

[6] Mindock, M., DiOrio, J., and Golinkin, S., 2002. "Applying new and improved technology to an existing generator-drive steam turbine". In International Joint Power Generation Conference, June 24-26, Phoenix, AZ, USA, no. IJPGC2002-26099.

[7] Segawa, K., Shikano, Y., and Takano, T., 2004. "A high performance optimized reaction blade for high pressure steam turbines". In ASME Power Conference, March 30-April 1, Baltimore, Maryland, USA, no. POWER2004-52110, pp. 307–314.

[8] Sakai, Y., Nakamura, K., and Shiokawa, K., 2005. "Recent technologies for geothermal steam turbines". In Fujii Electric Review, no. 3, Vol. 51. Shinagawa-ku, Tokyo, Japan.

[9] Schaarschmidt, A., an G. Nitch, E. J., and Michels, B., 2005. "Performance increase through world class technology and implementation". In PowerGen Europe, June 28-30, Milan, Italy.

[10] Simon, V., Stephan, I., Bell, R. M., Capelle, U., Deckers, M., Schnaus, J., and Simkine, M., 1997. "Axial steam turbine with variable reaction blading, advances in turbine material, design and manufacturing". In Proceedings of the 4th International Charles Parsons Conference, pp. 46–60. Londra.

[11] Littlewood, D. M., and Hogg, S. I., 2008. "Compact diaphragm turbine technology and its first application in a turbine retrofit". In ASME Power Conference, July 22-24, Orlando, USA, no. POWER2008-60141, pp. 317–323.

[12] Cotroneo, J. A., Cole, T. A., and Hofer, D. C., 2007. "Aerodynamic design and prototype testing of a new line of high efficiency, high pressure, 50% reaction steam turbines". In ASME Turbo Expo 2007, May 14-17, Montreal, Canada, no. GT2007-27315.

[13] Pierret, S., and den Braembussche, R. A. V., 1999. "Turbomachinery blade design using a navier-stokes solver and artificial neural network". ASME Journal of Turbomachinery, 121, pp. 326–332.

[14] Bellucci, J., Sazzini, F., Rubechini, F., Arnone, A., Arcangeli L., Maceli, N., 2013, "Using cfd to enhance the preliminary design of high-pressure" Proceedings of ASME Turbo Expo 2013

[15] Piraccini F., Cosi L., Biondi R., "Aeromechanical Validation of a New Steam Turbine LP Section: Test Major Outcomes", Proceedings of ASME Turbo Expo 2010, Glasgow, UK.

AICHE PAPER NUMBER: 512744
DESIGN CHALLENGES TO APPLY MIXED FLOW IMPELLER TECHNOLOGY TO 3000 KTA ETHANE CRACKER

Daniel Jones
Director of Quality
Technip FMC

Ronald Josefczyk
Director of Engineering
Elliott Group

Prepared for Presentation at the 2018 Spring National Meeting
Orlando, Florida, April 23, 2018

DESIGN CHALLENGES TO APPLY MIXED FLOW IMPELLER TECHNOLOGY TO 3000 KTA ETHANE CRACKER

Daniel Jones
Director of Quality
Technip FMC

Ronald Josefczyk
Director of Engineering
Elliott Group

Abstract:
World scale ethylene (MEGA) plants continue to increase in scale. Current installed plant technology ranges from 1500 to 2000 total olefins capacity, with larger plants under consideration. This presentation/paper addresses some of the equipment design challenges associated with the application of mixed flow thermal turbocompressor technology to the cracked gas service of a 3000 kta ethane cracker. Solutions proposed are based on a comprehensive design considering aerodynamic development and testing, operating range and efficiency, rotor dynamics and other mechanical challenges. The resulting hardware solution presented can achieve an equipment risk profile consistent with currently applied equipment.

Introduction:

From the first on purpose ethylene plant constructed approximately 100 years ago to modern mega ethylene crackers (>1000 kta ethylene capacity) there has been a consistent growth trend in worldwide ethylene demand (recently 4% per annum). The upper end of individual new olefins plant capacity also continues to grow as the limits of economy of scale have not yet been reached. In recent years the plant growth trend has increased dramatically (See Figure 1).

Year over year olefins production increased approximately 20 KTA/year from 1980 through 2000, and increased again to approximately 35 KTA/year in the following decade.
Since 2010 to the present total capacity increase per year has averaged 43 KTA/year. Plants in excess of 1000 KTA that were previously referenced as world scale have become a baseline with modern world scale plants commonly reaching 1500 kta and with 2000 KTA plant being commissioned in North America.

Based upon these trend projections, 2500 KTA plants are within available rotating equipment manufacturer capabilities. In fact, machinery solutions for 3000 KTA total olefins production for both liquids and gas feeds have been vetted.

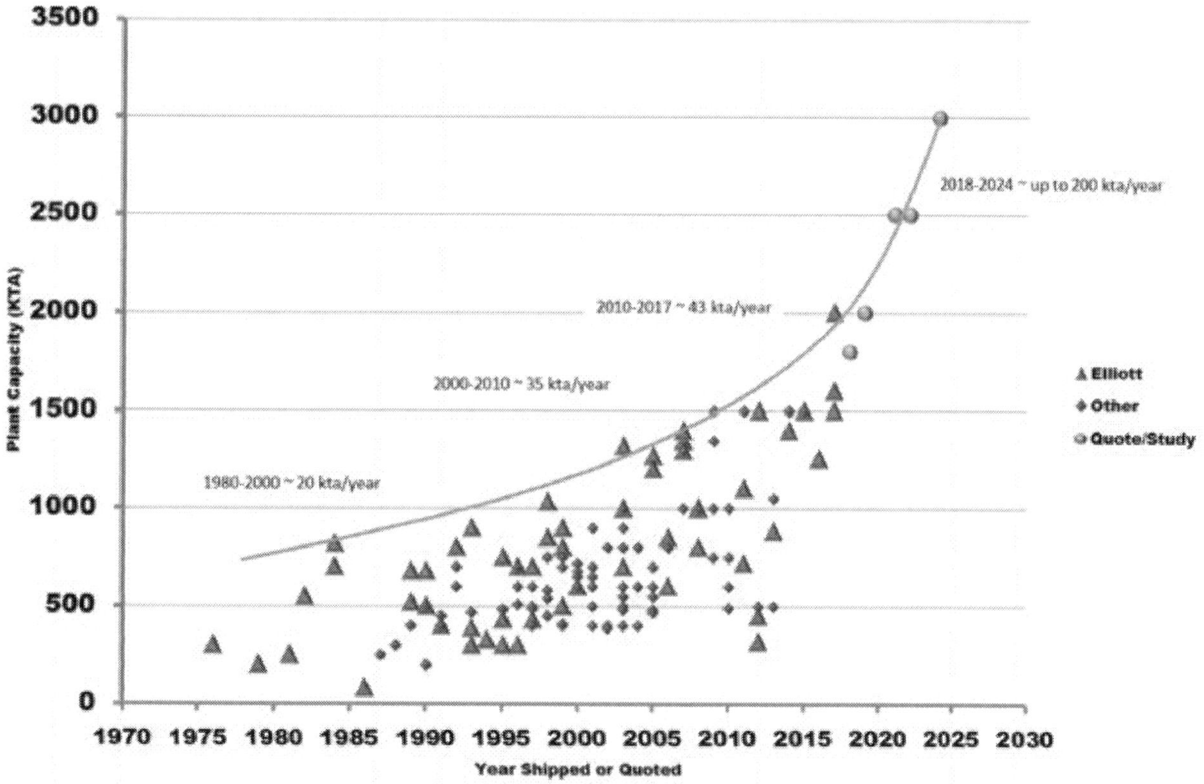

Figure 1: Ethylene plant capacity historical trend and future projection

Turbomachinery manufacturers have met the requirements of historical plant capacity growth by increasing scale of the radial in line compressors and steam turbines. These methods proved robust for plant capacity increases applied to date. As the market, plant capacity and equipment demands continue to grow, conventional increase in scale may be a less viable solution than alternative equipment design options available to realize highly reliable systems beyond current mega ethylene capacity.

Challenges of Mega Compressors:

There are multiple options available to the equipment designer to meet the challenge of increasing plant capacity. The single train/body configuration has been the historical lower cost solution selected by the market. This allows the end user to take advantage of "economy of scale" and is time and again the preferred solution over parallel bodies or multiple strings of equipment in the same service.

From a capital expenditure point of view there is an economy of scale associated with application of increased frame vs parallel bodies as shown in the following graph, see Figure 2. Elliott frame sizes are roughly 15% scale from the adjacent frame which equates to a >30% increase in volumetric flow per frame. The nominal cost per compressor body has been normalized to the smaller casing size applied to mega ethylene plant cracked gas applications. Casing cost and casing capacity per unit cost is plotted relative to the other available frames, with the same number of impellers per respective frame.

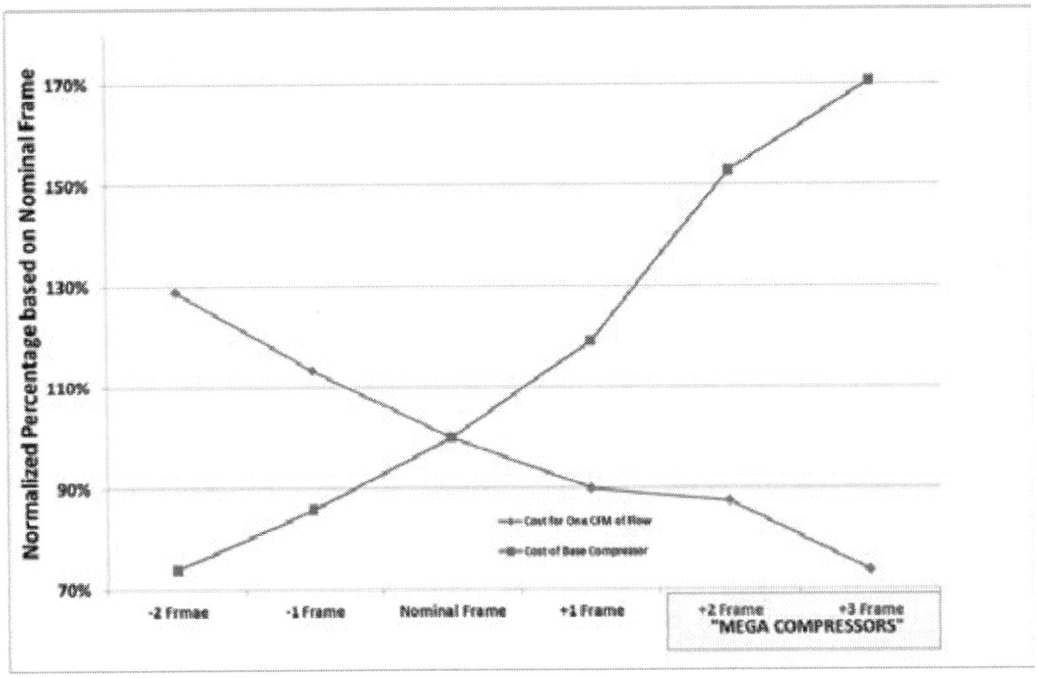

Figure 2: Normalized unit cost and cost per volume flow trend

There is an increase in cost as we go up in frame but there is a noted step change in the escalation for "mega" compressors at the upper end of the scale. Factors driving this step change in cost with scale include the methods of manufacture, material availability, logistics and general increased difficulty handling extremely large compressors associated with mega ethylene applications.

Shown in blue is the ratio of normalized flow vs cost. Flow increase is a function of the square of compressor frame linear dimension. Considering the increased costs associated with the larger frames, there remains an economy of scale capital advantage to increased unit frame size.

While there is a capital and performance advantage with increased frame sizes there are also significant challenges that need to be addressed. The compressor weight increase is a function of the cube of compressor frame linear dimension. As larger lower pressure machines typical of olefins and LNG manufacturing grow with plant size, unit mass increase invites more challenges. Recent experience requires large frame compressors weighing greater than 400,000 lb (181,500 kg) with largest compressors capable of reaching the 600,000 lb (272,200 kg) mark.

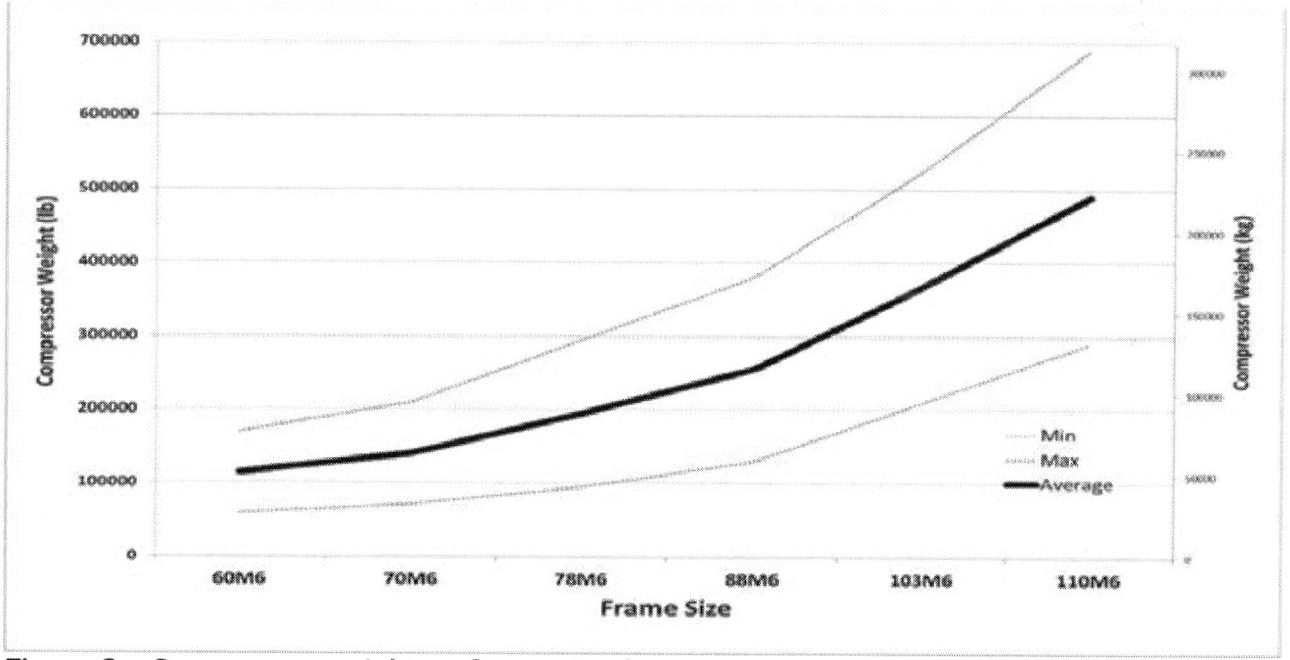

Figure 3: Compressor weight vs frame size for six stage compressor with single interstage cooling

As ethylene licensors have increased plant capacity, system operating pressures have been more stable, with modest changes depending on specific process and site requirements. Therefore, as the volumetric flow has increased the process pressure required remains *relatively* unchanged. This aspect of scale increase presents pressure containment challenges. Historical frame scale increases go hand in hand with casing working pressure decreases. However, as process pressure levels remain relatively constant, axial split line, bolting and casing wall thickness must be redesigned to accommodate the out of scale pressure requirements.

The massive dimensions and weight of larger frame units increase both the complexity and associated cost for logistics. Logistics, shipping permits, and methods of delivery play a greater role with mega compressors. The larger units require disassembly for shipment, and associated costs for field assembly (Figure 4).

Figure 4: Top half of large cracked gas compressor being shipped to customer site.

Mixed Flow Technology Solution:

The alternative to increasing frame size is to increase the volumetric flow capacity for given frame size of compressor. This can be accomplished by one of two primary methods.

- Increasing operational speed of the compressor while applying existing aeropath components
- Increase the flow coefficient of aeropath components design.

An increase in operational speed is accompanied by multiple challenges that provide limited benefit beyond the increase in flow. A 10% increase in rotating speed proportionally increases the flow 10% but also increases the rotating component stress by approximately 21%. The greater speed also increases head and heat rise per stage which coupled with smaller passage width increases the likelihood of polymerization and reduced efficiency. Speed increase alone presents challenges in the form of reduced operating range, rotordynamic stability, increased bearing metal temperatures, seal face velocity and increased driver speeds.

To address these concerns Elliott can offer an increased impeller flow coefficient solution. Using modifications to proven and well referenced stage designs the compressors can meet the challenges of increasing volumetric flow and plant capacity without the corresponding increase in frame size and associated costs and difficulties outlined previous.
A solution to the high flow coefficient need was the development of a moderate head, high efficiency mixed flow stage. The flow coefficient is defined as $\Phi_0 = \frac{4\,Q_0}{U_2 \pi\, D_{2m}^2}$ where

Q_0 = Suction volume flow rate

Ω= Rotor shaft speed and $U_2 = \frac{\Omega\, D_{2m}}{2}$

D_{2m} = Impeller mean diameter

The practical limitation of fully radial stages diminishes at flow coefficients approaching 0.2. At and above this limit both the head and efficiency ratings fall off rather steeply. Application of mixed flow stage technology allows this performance limitation to be extended to an approximate flow coefficient of 0.3.

A mixed flow stage falls between a full radial centrifugal stage and axial flow stage and combines positive aspects from each. The exit flow angle is between ~20 and 90 degrees from the shaft centerline. The stage design is optimum for applications requiring high flow coefficients, upper-level head coefficients and high efficiency while maintaining a radially compact and robust design.

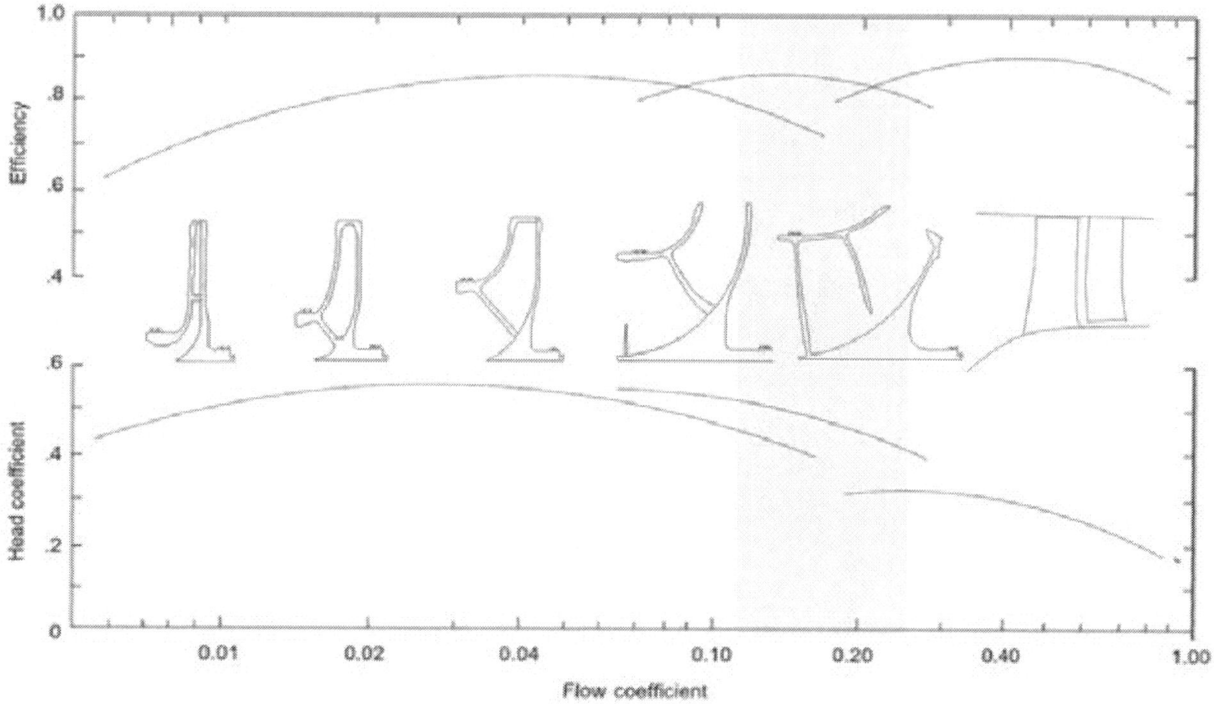

Figure 5: Typical Efficiency and Head Coefficients vs flow coefficient and stage configuration

Mixed flow technology increases design flow coefficients beyond Φ = 0.25 while maintaining a head and efficiency levels consistent equivalent to modern radial flow aerodynamic stages.

This allows each frame of compressor to produce a throughput increase of approximately 30% while maintaining the original operating speed. The increase in capacity is directly comparable to flow achieved by an increase in frame size without the negative impact of increased size and weight and lower operating speed. Additional benefit of this design is that the large through flow area of the high flow coefficient mixed flow stages are less prone to fouling than lower flow coefficient stages. The reduced wetted surface to flow ratio reduces the impact of fouling on efficiency.

Figure 6: Modern high flow coefficient mixed flow stage.

The increase in high flow performance of a mixed flow stage does come with some tradeoffs. In particular axial stage length for mixed flow stages is longer than traditional radial stages, in some cases by as much as 20%. This challenge is generally manageable for ethylene applications, as the gases compressed have medium to higher molecular weight, and typically require only one mixed flow stage to feed into a traditional radial stage.

One part of the solution to accommodate increased length was the compaction of Elliott's full radial stages. Using tested, field proven aero technology the critical parameters that define aero performance were preserved while identifying and reducing the axial length of non-critical locations.

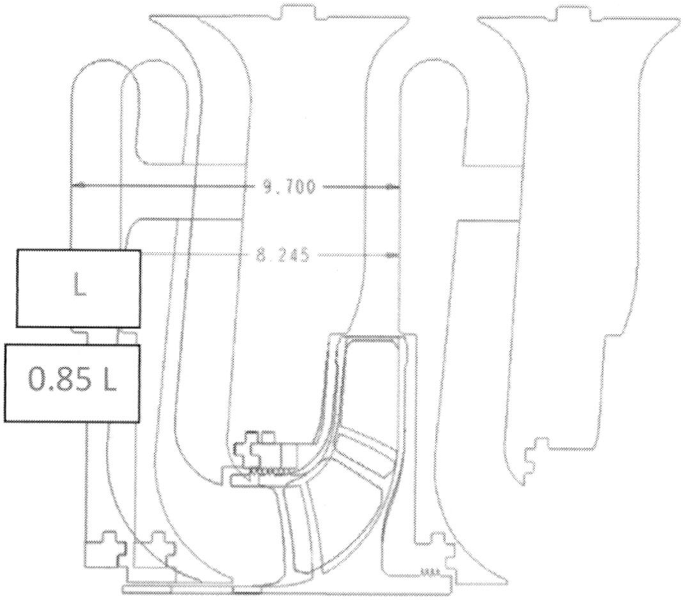

Figure 7: Overlay of axially compacted stage and preservation of defining stage geometry

Extensive CFD and bench performance testing validated the reduced stage length performance, with no negative impact throughout the operating range. Overall the average stage length reduction achieved was 15% which when applied to downstream stages offset the length addition required for mixed flow inlet stages.

Increased capacity and volumetric flow require greater process nozzle area to maintain velocity levels at an acceptable range to minimize flow separation and losses.

Inlet and discharge nozzles have been redesigned such that the increased volume flow is evenly distributed around the casing opening allowing a shorter axial space all while improving distribution into the first stage.

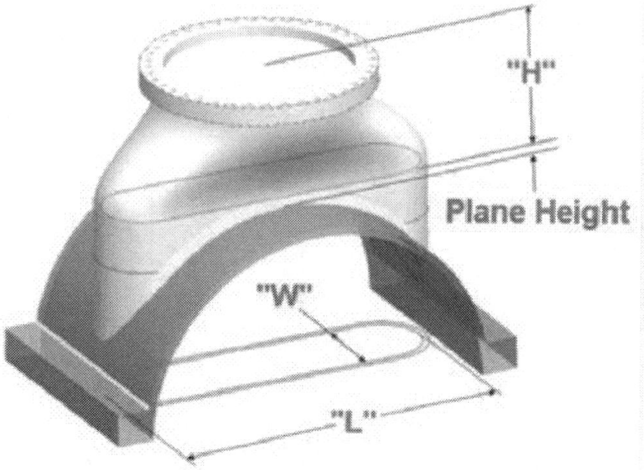

Figure 8: High volume flow, reduced axial space inlet with increased projected area.

Flow turning losses and choke are minimized by maintaining critical relationship and similitude of inlet parameters to operating nozzle and stage designs which have a large and well referenced experience list.

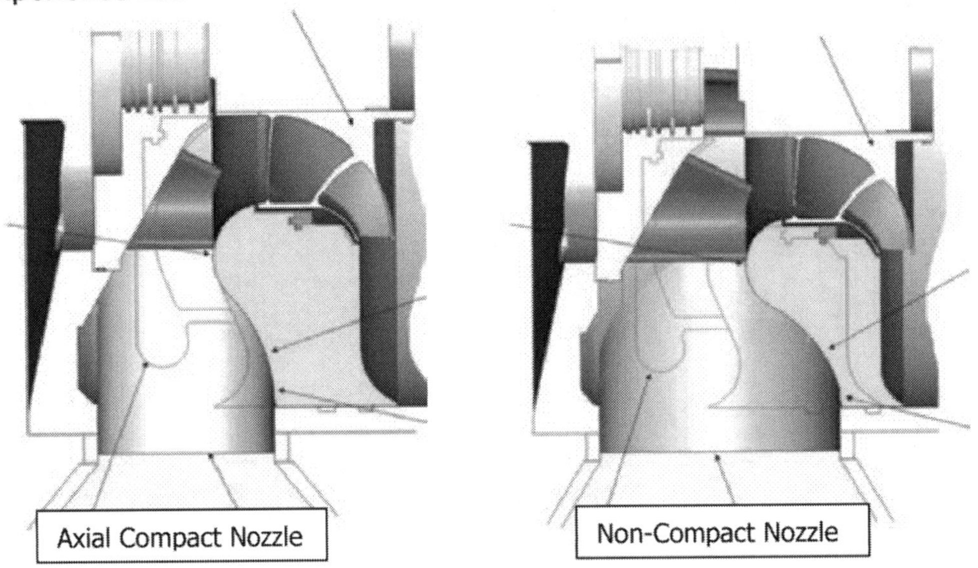

Figure 9: Axially compact high volume flow inlet vs predecessor of equal volumetric flow. Flow path requirements overlaid.

Once flow enters the casing a low count profiled flow catching guide vanes direct flow into the impeller inlet with minimal blockage. This is required for higher flow applications to prevent choke as well as resonant excitation.

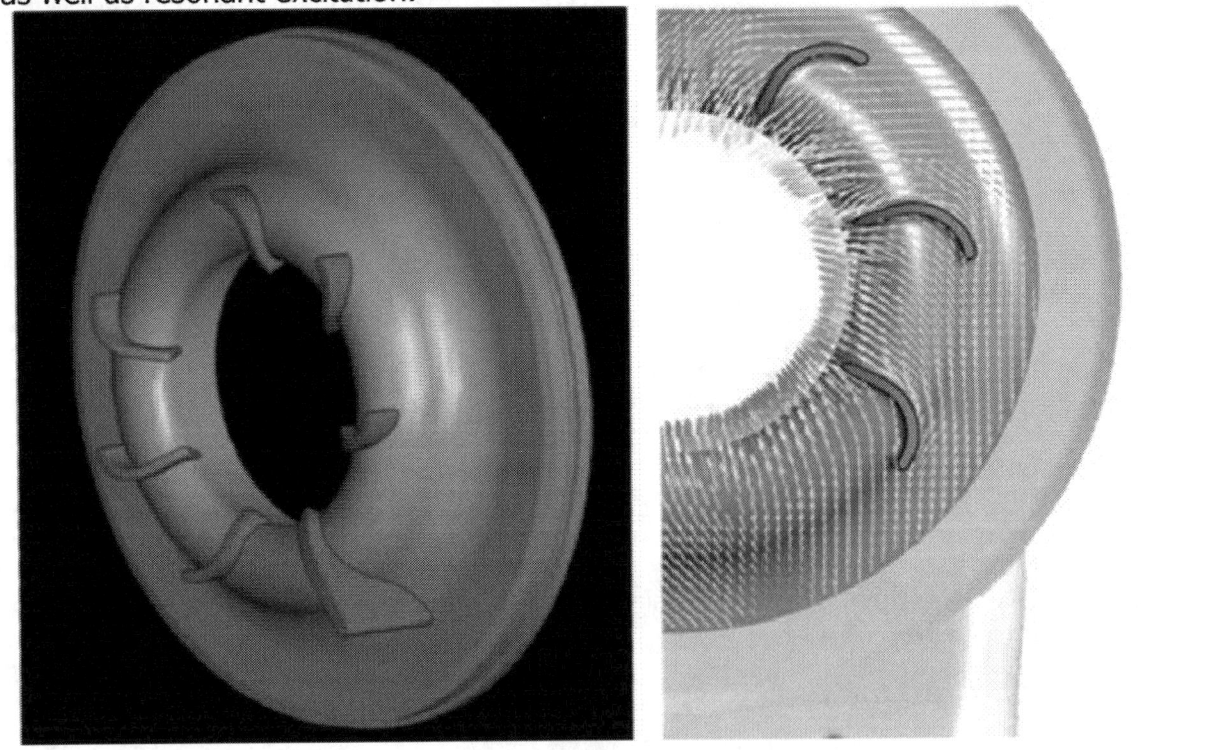

Figure 10: State of the art low blockage - high volume inlet guide vanes. CFD results showing even pressure/velocity distribution

Conclusion:

With the application of these major components the typical compressor flow range can be increased by approximately 32% with no impact on overall efficiency, rotor dynamics or machine reliability.

Application of new technology that is well anchored to proven reliable components and extensive testing enables users to expand plant capacity to unprecedented levels.

Legacy compressor and turbine designs have had the capability to reach 3000 kta with increased scale and compressor frame sizes. Increased volumetric flow capacity and power density can achieve these larger capacities without the associated challenges and cost associated with increasing frame size.

These options allow end users to take advantage of both improved capital expenditure of smaller frames sizes as well as benefits traditionally associated with economy of scale.

AIChE Paper Number: 513188

Supercritical Ethylene Export – Process Solutions and Fundamental System Design Issues for Trans-critical Ethylene Compression

Torben Hoefel

Process Engineer Petrochemical Plants

Linde AG, Engineering Division

Martin Kamann

Senior Expert Rotating Equipment

Linde AG, Engineering Division

Gunther Kracker

General Manager Process Engineering Petrochemical Plants

Linde AG, Engineering Division

Prepared for Presentation at the

2018 Spring Meeting and 14th Global Congress on Process Safety

Orlando, Florida, April 23, 2018

Supercritical Ethylene Export – Process Solutions and Fundamental System Design Issues for Trans-critical Ethylene Compression

Torben Hoefel

Process Engineer Petrochemical Plants

Linde AG, Engineering Division

Martin Kamann

Senior Expert Rotating Equipment

Linde AG, Engineering Division

Gunther Kracker

General Manager Process Engineering Petrochemical Plants

Linde AG, Engineering Division

Abstract: Some ethylene plants require the product delivery at supercritical conditions. This can be achieved in two ways: Pumping of liquid ethylene or compression of vapor ethylene. The trans-critical product compression can be combined with the open loop subcritical refrigeration cycle resulting in one ethylene compressor serving multiple tasks. However, the combination of the different tasks in one machine results in fundamental design issues particularly induced by physical properties of trans-critical ethylene.

This paper outlines the principal compressor design features to be considered, describes the selection of an adequate control scheme and finally addresses the verification process for transient operation to assure highest level of system safety and reliability. The process design for trans-critical ethylene product compression is further compared to pumping of liquid ethylene considering costs and other constraints including operability, possible liquefaction requirements and intermediate BL pressure specifications.

1. Introduction and Process

Ethylene production is mainly performed by steam cracking, a high temperature pyrolysis in the presence of steam, where hydrocarbon feed stocks are converted to lower olefins [1]. The reaction effluent consisting of ethylene as well as various other main- and by-products is routed through a complex product recovery section. Some separation units in the recovery section require refrigeration on a low temperature level between - 50 to -100 °C [1, 2]. These temperatures are typically generated in a conventional refrigeration cycle operated with ethylene. To supply refrigerant on different temperature levels, liquid ethylene is expanded to different pressures. After vaporization in various consumers the ethylene vapors are fed to a compressor with multiple process stages according to their respective pressure levels as shown in Figure 1. The ethylene gas with the lowest pressure enters the first stage of the compressor and the ethylene vapors on higher pressure levels are fed to compression stages on higher pressure levels, respectively. The ethylene is compressed to approximately 20 bar a at which it can be liquefied by means of C3 refrigerant and other process media.

In the final separation step for the recovery of ethylene from the reaction effluent, it needs to be separated from ethane, which is typically performed by distillation [1, 2]. Some plants utilize a so-called heat-pumped low-pressure C2 fractionation, where the distillation column is operated at approximately 8 bar a. This process has some advantages with respect to investment and energy consumption and requires the compression of the vapor ethylene overhead product to a pressure level around 20 bar a, at which it is liquefied in the re-boiler of the distillation column providing heat duty for re-boiling. Part of the liquid ethylene is then recycled as reflux to the top of the column.

Both compression tasks can be integrated in one ethylene compressor using the high-pressure refrigerant stage also for the heat pump as shown in Figure 1, resulting in an open loop refrigerant cycle. [1, 2].

Ethylene product is typically discharged to battery limits as a gas. For pressures up to 20 bar a, the compressor described above can be utilized as well. If the required pressures are significantly higher or even supercritical, the ethylene needs to be either liquefied and pumped to the respective pressure level or it can be directly compressed. For supercritical product delivery, pumping of liquefied ethylene is the typical approach applied in industry. If the compression of vapor ethylene is considered, the product compressor can also be integrated into the ethylene compressor train as shown in Figure 1 thus combining three process tasks in one machine. The number of required process stages is defined by the required final discharge pressure and the maximum allowed ethylene temperature.

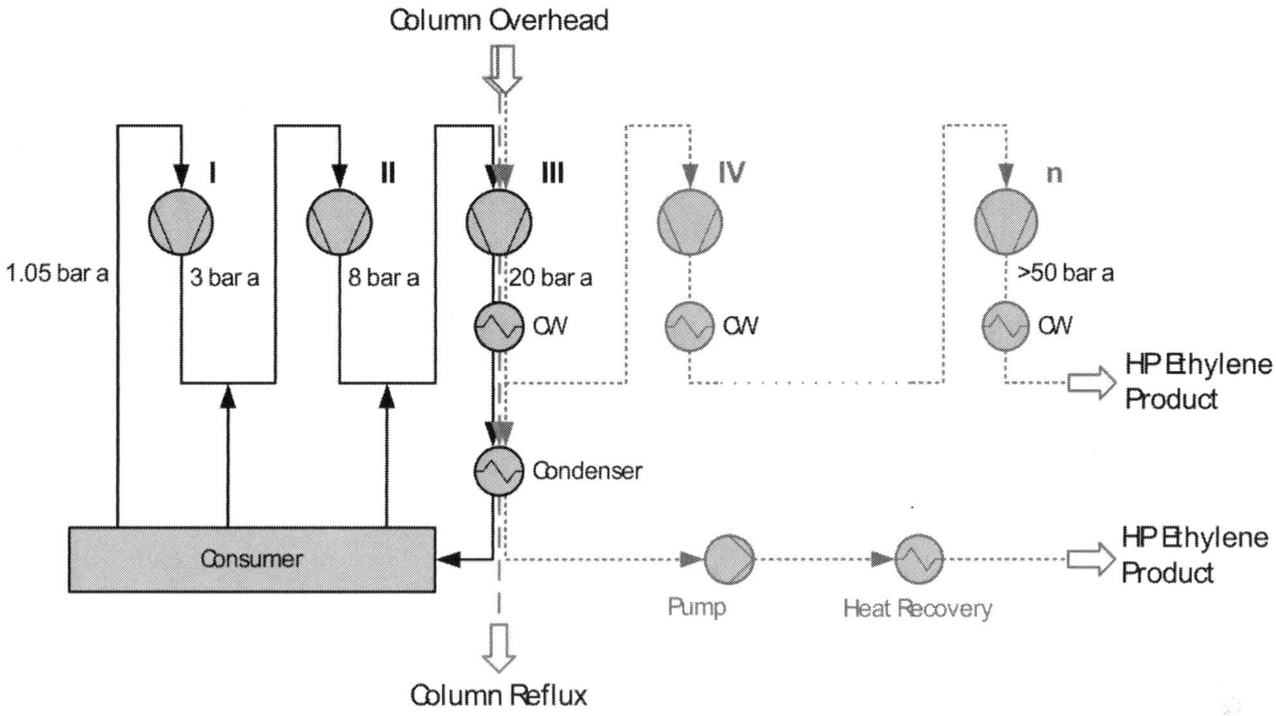

Figure 1: *Ethylene compressor designed as open loop refrigerant system and ethylene product delivery. The compression Stages I-III are used for the refrigerant system (black solid lines). Liquid ethylene is utilized as refrigerant on different temperature levels in the plant. The resulting ethylene vapors on different pressure levels are compressed to approximately 20 bar a, cooled against cooling water (CW) and liquefied in one or more condenser. Ethylene vapor from the low-pressure C2 fractionation column overhead is also compressed in Stage III, cooled, liquefied and sent back to the column as reflux (red dashed lines). The product fraction from the column overhead (green dotted lines) can be exported as low-pressure ethylene product (not shown). If high pressure (HP) ethylene product is required, the ethylene product can either be liquefied and pumped to the required pressure or it is directly compressed from the vapor phase. For this purpose, further compression stages are required, which can be integrated into the ethylene compression train as well.*

The essential safety review of a trans-critical compression system turns out fundamental design issues induced by physical properties of trans-critical ethylene. Compressor operation in dense phase is challenging for aerodynamic and rotor-dynamic design. Temperature-induced compressibility changes and Joule-Thompson effect are significant for compressor control synthesis and proper design of its anti-surge and performance control. The ability of liquid formation during expansion and supercritical ethylene decomposition are restrictive for compressor design temperature and pressure and have to be considered during design of the overall topology of the compression system and the safety system.

Relevant constraints for the overall compressor system topology need to be introduced with respect to piping arrangement and process cooler design. This is done with focus to the process temperature control in order to avoid liquid formation but at the same time to assure highest level of system safety towards ethylene decomposition. Since performance control for refrigeration service and product compression service may interfere due to compressibility effects, generic recommendation for the overall control synthesis are derived. Furthermore, recommendations are made for the system verification process for transient operation which is structured in those which are planned like start up, shut down and depressurization and those who are forced by failure modes.

The process design for compression of ethylene product is compared to the pumping of liquid ethylene to understand the main cost drivers for the two solutions with respect to invest and operating cost. Furthermore, the different process solutions bear different characteristics in control and operability particularly regarding transient operating conditions including turndown and start-up, which need to be taken into account.

2. Physical Properties of Ethylene

For applications with trans-critical product compression services the system must be carefully designed with regard to the physical properties of ethylene and their respective transient changes. Herein trans-critical denotes sub- to supercritical compression or reverse expansion.

Figure 2 shows typical change of state of ethylene for refrigeration service, C2 splitter operation and product compression in a Mollier (pressure, enthalpy) - diagram of pure ethylene [3]. For product compression inter-stage cooling was assumed with approximately 30 °C ambient temperature. Inter-stage cooling is required to avoid potential for polymerization or explosive ethylene decomposition. The limit for safe operation is shown as the red region indicated in Figure 2 [4].

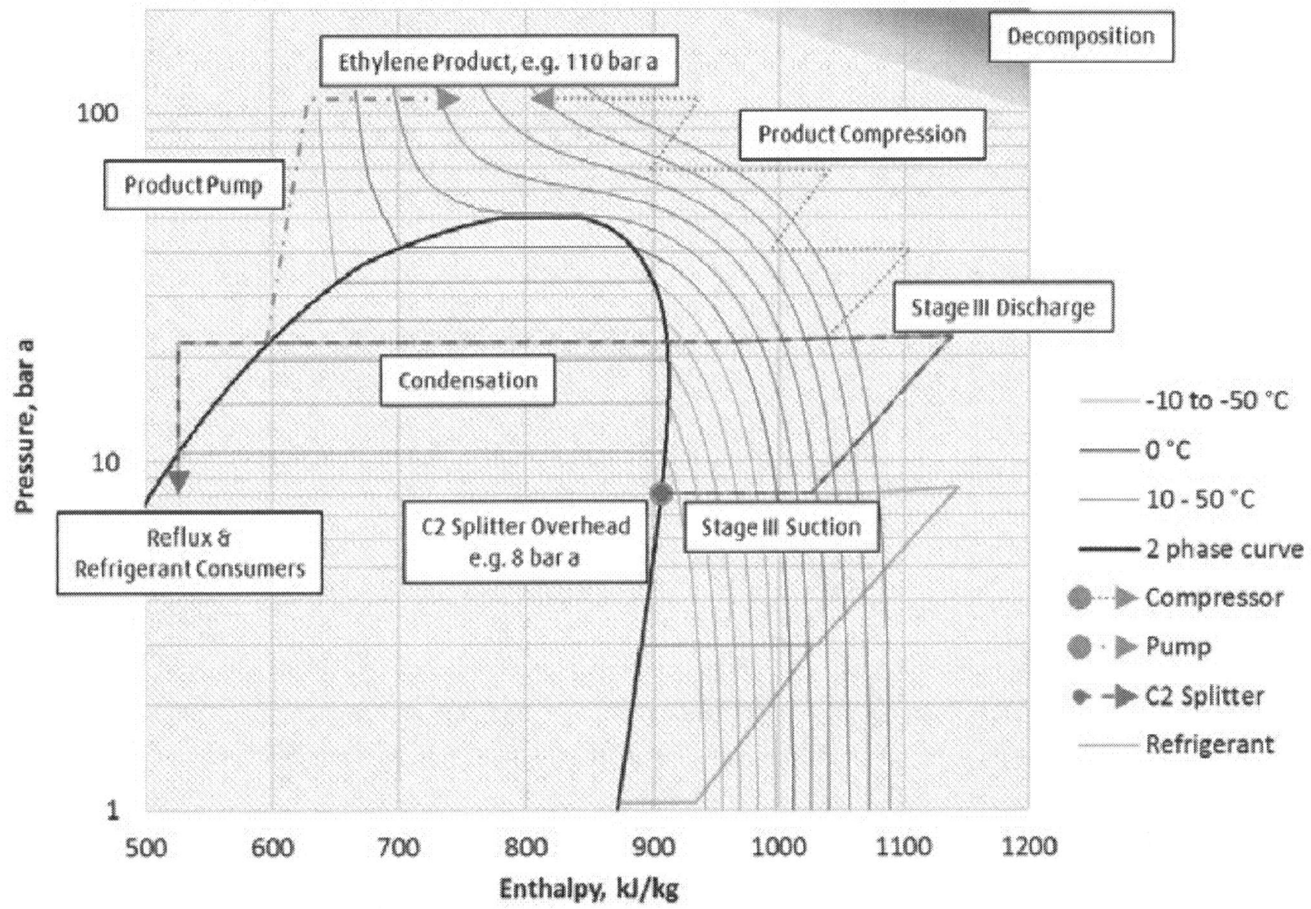

Figure 2: *Mollier- (pressure-enthalpy) - diagram for pure ethylene and ethylene process flows. The black solid line indicates the 2-phase region. The grey solid lines indicate the change of state for refrigeration services (compression and condensation). Compressor Stage I and II are used for refrigerant service only. The red dashed line indicates the change of state for ethylene used as reflux. The green dotted and dash-dotted lines show change of state for product pressurization, either by pump or by direct compression. Direct compression is carried out in multiple stages with limited discharge temperature and cooling to ambient condition. Ethylene decomposition limits are indicated in the upper right region.*

Figure 3 shows the volumetric properties of ethylene for trans-critical compression at approx. 30 °C. The change of the isentropic coefficient and the ratio of specific heat will affect the machine Mach number and specific volume ratio and therefore, will have impact to compressor aerodynamics. This is also shown by the thermal expansion factor plotted on the secondary axis of Figure 3 with significant pressure dependent variations [3].

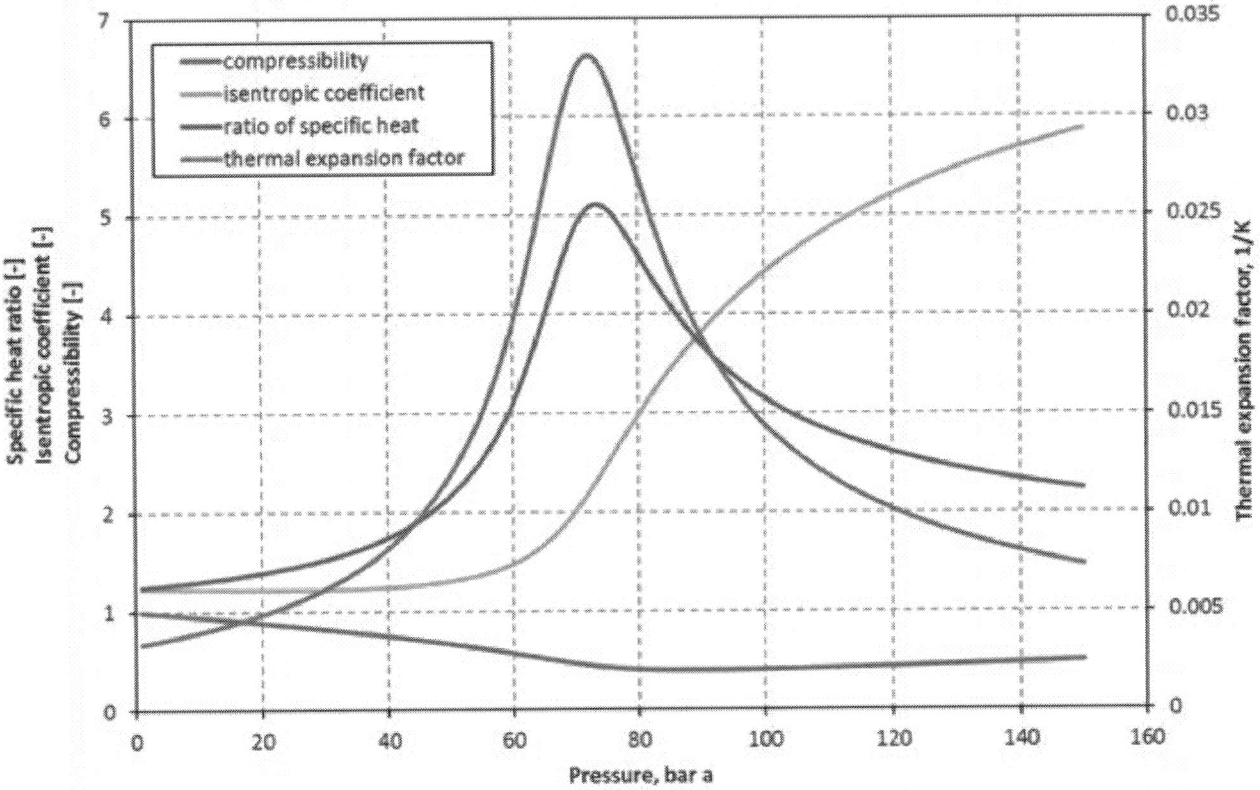

Figure 3: *Volumetric properties of trans-critical ethylene at ambient temperature (approx. 30 °C). On primary axis the non-dimensional coefficients namely compressibility, isentropic volume coefficient and specific heat ratio are shown vs. pressure range from sub to super-critical pressure. On secondary axis the thermal expansion factor vs. pressure is shown. Significant change of values can be observed for trans-critical compression (pressures higher than 50 bar a).*

The significant changes in properties are furthermore exemplarily displayed by change in densities as a function of pressure and temperature, as shown in Figure 4.

Besides that, thermal expansion and the change of compressibility will influence compressors performance control and surge protection. The latter itself has some potential to induce temperature/volume- fluctuations while acting on anti-surge control valves with significant Joule-Thompson effect up to 10 K per 10 bar pressure drop.

In case of trans-critical expansion there is also potential for liquid formation, if compressor effluent is sufficiently cooled down before isenthalpic expansion in bypass valves. This could already happen for example in the trans-critical compression stage shown in Figure 2 and a temperature decrease below approximately 30 °C before expansion. The resulting risk of liquid formation is to be mitigated by compressor design and/or control.

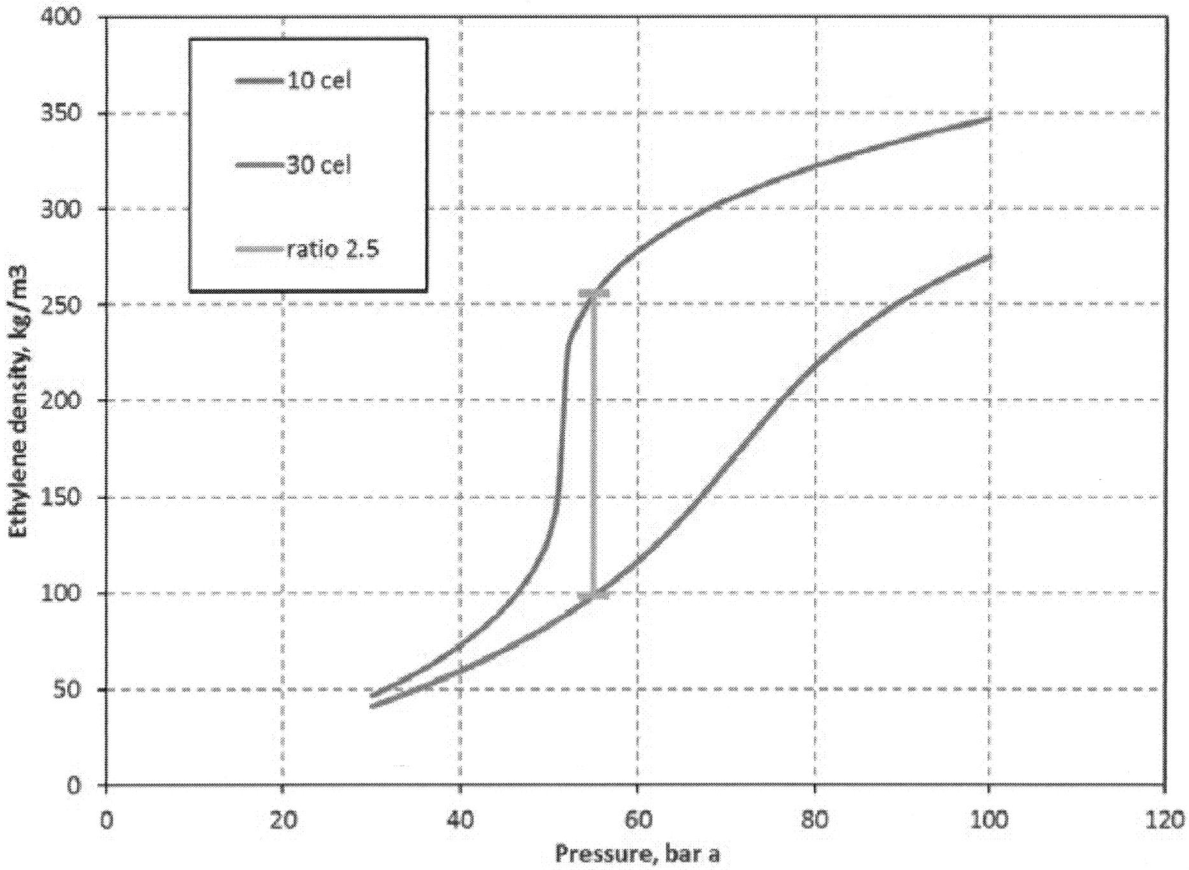

Figure 4: *Ethylene density vs. pressure for different temperatures (10 °C and 30 °C). A temperature change of 20 K might be induced by Joule Thompson effect on compressor bypass valves. Near critical pressure this turns out in density changes by factor of 2.5.*

3. Compressor Mechanical Design

The different process tasks to be executed and the different constraints induced by physical properties result in a radial turbo compressor design with multiple compression and process stages. One embodiment for a design is shown in Figure 5.

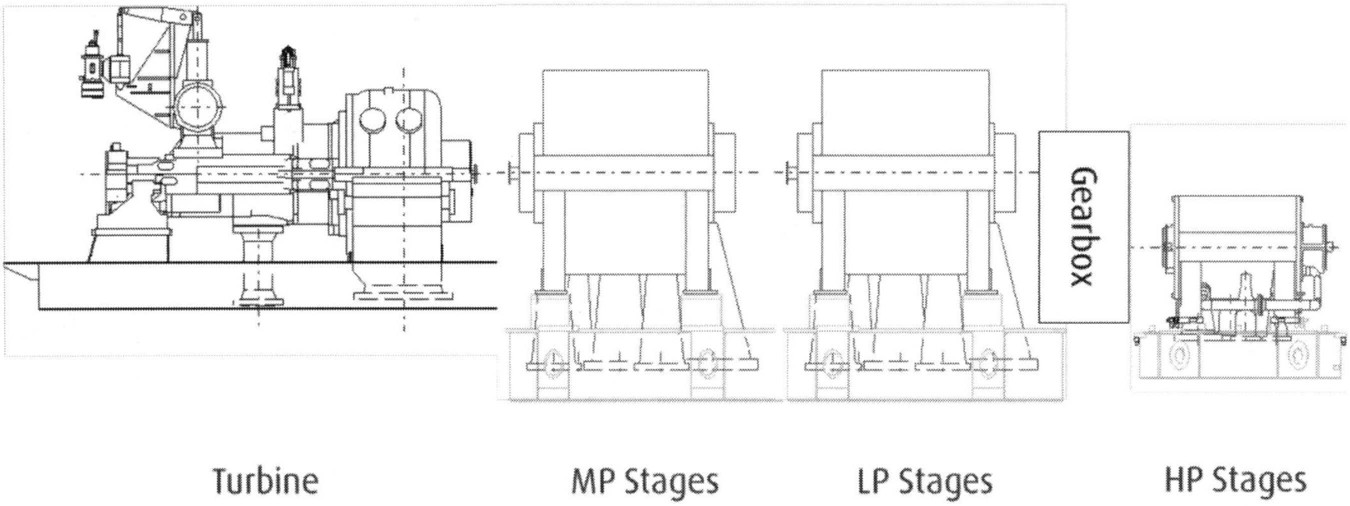

Figure 5: *Ethylene turbo compressor utilized for refrigeration, C2 splitter heat pump and product compression. Ethylene is compressed in 6 process stages, each consisting of several compression stages (impeller), which are arranged in three casings. All compression stages are driven by one steam turbine. The low pressure (LP) and medium pressure (MP) stages are mounted on the same shaft as the drive. The last two high pressure (HP) compression stages are in a separate casing and are mounted on a separate shaft, which is connected by a gear to the other shaft.*

A typical design for a turbo compressor consisting of four process stages and delivering ethylene to battery limit pressures between 30-50 bar a is carried out in two casings. The two low pressure (LP) stages utilized for refrigerant compression are arranged in one casing. The 3rd stage (utilized for refrigerant compression, C2 splitter heat pump and product compression) and the fourth stage (utilized for product compression) are arranged in a second medium pressure (MP) casing. Both casings are typically mounted on the same shaft as the steam turbine drive.

For supercritical product compression, more process stages are required. In the example shown in Figure 5, the according process stages are arranged in a third high pressure (HP) casing and are mounted on a separate shaft, which is connected to the LP/MP compressor shaft by a gear. Accordingly, the different compression stages can be operated at rotating speeds suitable for the respective aerodynamic conditions.

4. Compressor Control and Safety Functions

The integration of the compressor to the process system and relevant control loops are shown in Figure 6 at the example of a five-stage compressor system, with stage four being the trans-critical compression stage and stage 5 being a supercritical compression stage. The basic performance control of the compressor is done in a way that a critical pressure from process linked to Stage III suction pressure is kept constant via speed variation [5]. Stage I pressure is controlled via the anti-surge control system by help of suction pressure override functions. The pressure low override on Stage I avoids vacuum conditions while acting on the 1st stage bypass valve. The pressure high override on Stage I manipulates the 3rd stage bypass valve and is thus indirectly influencing the speed of the compressor.

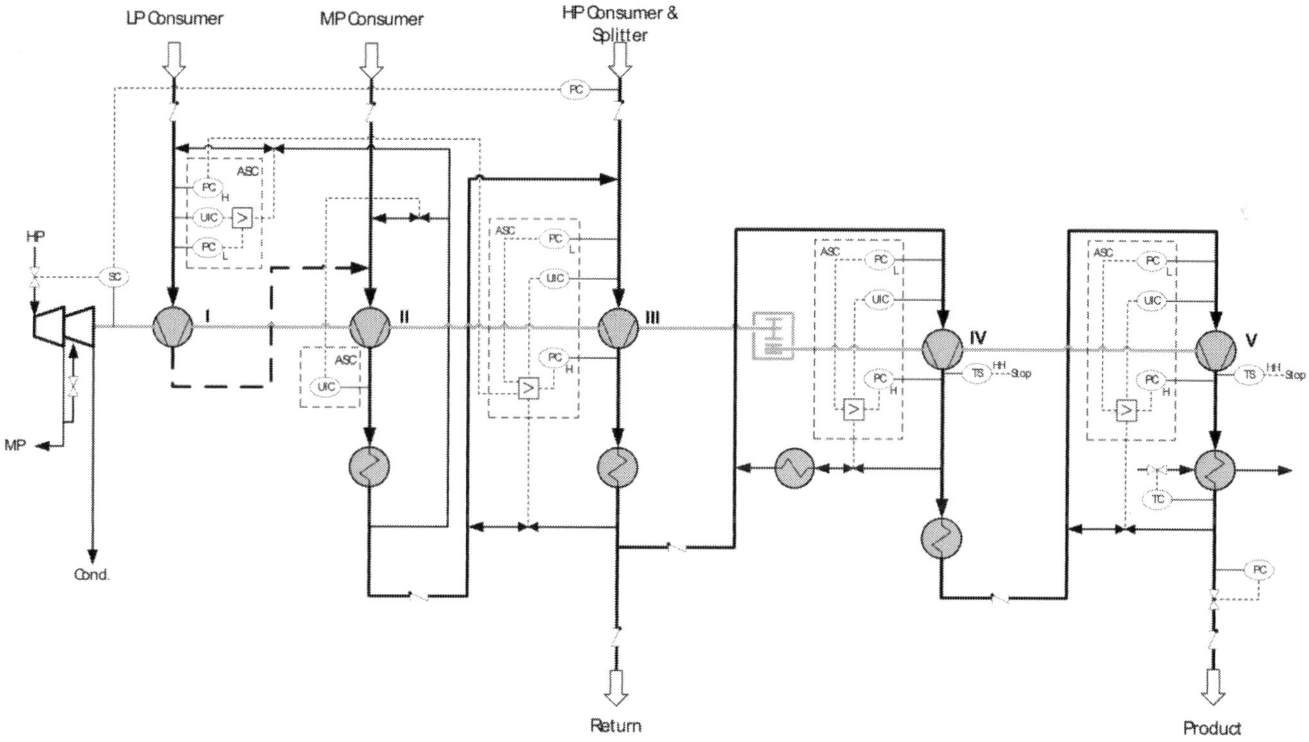

Figure 6: *Ethylene Compressor Control Topology. Pressure override functions of the anti-surge controllers are used to maintain the overall pressure profile of the machine with regard to required pressure for refrigerant consumers and safe operation of the supercritical product compression. Temperature controller and temperature switches on product compression are used to protect the compressor from high temperatures and liquid formation.*

Since required operating speed is defined by Stages I-III the remaining product compression service must be controlled by other means. Comparable to a four-stage machine [5], the low- and high-pressure overrides on all product compression stages can be used to adjust the inter-stage pressures. In this regard special care must be taken for proper design of the product compression bypass lines and valves with regard to high flow rates in recycle mode caused by density change as shown in Figure 4.

Due to very high overall pressure ratio of the machine there is potential to over-pressure the downstream system. As another layer of safety, discharge pressure override function can be used to keep pressures within the system design range. Further protection is needed for high temperatures at the discharge of each product compressor stage with regard to ethylene decomposition wherefore temperature high switches are applied.

In order to avoid any potential for liquid formation on the trans-critical stage (here: Stage IV) a different bypass topology is chosen, whereby the uncooled gas from the discharge first expands through the bypass valve before it is cooled. The cooling of the bypass can be done either with a dedicated bypass cooler (as shown in Figure 6) or via the foregoing inter-stage cooler.

The last compressor stage, when fully supercritical on suction and discharge, won't need additional measures to protect against liquids in normal operation. However, during start- up, the stage will undergo a transient from sub- to supercritical. This can be solved with temperature control on the downstream cooler e.g. by cooling water throttling and in addition, equipped with an additional temperature low switch in the bypass line to protect against a temperature control loop failure during start up. During normal operation the temperature low switch might be inactive, dependent on supercritical pressure, in order to improve reliability and avoid inadvertent alarms or shutdown signals.

5. Compressor Design Verification

For verification of the system design & layout and proper control synthesis it is recommendable to investigate transient changes e.g. by cause and effect study, HAZOP or via dynamic simulation of the system. The scope of verification should be at least the initial start-up, the restart from settle out condition, shut down scenarios including emergency shut down and controllability of the system in part-load operation. Furthermore, failure modes of the various control loops should be considered. Again, special focus should be set to the product compression system, which could cause significant pressure fluctuations by malfunction of the related anti-surge valves. The most critical event is the initial start-up of the compressor train, whereby most of the protective control loops will be in manual mode. It is therefore necessary to define a safe and reliable start up procedure. Main goals of the start procedure are:

- Avoid any liquid formation or liquid ingress to the machine and seal system caused by excessive cooling or trans-critical expansion.
- Avoid any trans-critical pressure changes for time restrictive sequences, e.g. during the acceleration from low idle speed to minimum operating speed with pre-defined ramp rate.
- Assure that the procedure of pressurization can be interrupted at any time without risk of damaging any equipment.

Figure 7 shows the simulation result for such a procedure for start-up and pressurization. The principle schedule of events is to run the turbine operation at low idle speed (heating up) with all anti-surge controllers in manual mode and valves in fully open position. At this time the system is initially pressurized with ethylene. If applicable, temperature controllers acting on cooling water flow should be in auto mode. Due to very low power submission the controlled cooling water will be almost closed. Next step is to accelerate the turbine in accordance to their start up procedure. Due to the specified initial pressure of the system it will neither undergo a trans-critical change nor will face vacuum conditions on suction side. However, system pressure on the low-pressure stages will decrease while gas is distributed to the high-pressure stages. Accordingly, the filling process with ethylene needs to re-start again. When the pressure set point of the PIC-SIC control cascade is reached by filling, then the anti-surge controller on Stages I-III are activated without risk of vacuum conditions, as the train is on minimum governor speed. Further filling of the system will allow activating the master controller of the PIC-SIC cascade. From thereon only minor fluctuations on the suction pressure 3rd stage can be expected. Continuous filling will cause slow acceleration of the compressor train and pressurization of the product compressor system. This process of filling can be interrupted at any time, e.g. when critical events occur like failure of temperature controller. When the set points of the suction pressure override functions of the HP system are reached, then the anti-surge-controller of the product compressor and their override functions can be set in auto mode. From thereon, the pressure profile of the overall machine is maintained automatically, independent on plant load wherefore the operators can concentrate on the plant cool down procedure.

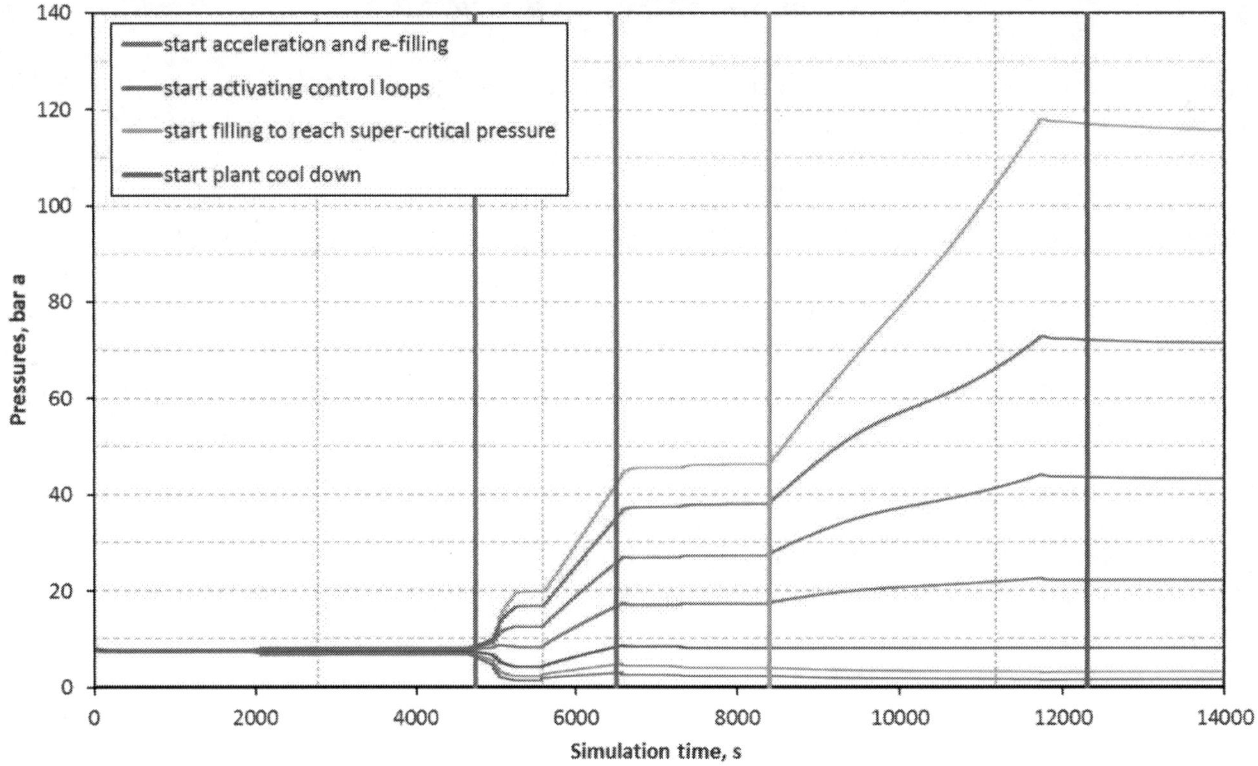

Figure 7: *Initial Start-up sequence: Suction and discharge pressure vs. time.*

6. Pumping of liquid Ethylene or Compression of vapor Ethylene

Considering the LP C2 Splitter process, supercritical product delivery can be carried out either by compression of vapor ethylene or pumping of liquefied ethylene product. Figure 8 shows an example of a simplified process flow of the whole ethylene system. For product compression, a more sophisticated ethylene compression system is required (indicated in green), as described in the previous chapters. Pumping to supercritical conditions requires some more heat exchanger area, particularly for condensation of the vapor ethylene product. Accordingly, a larger Propylene refrigerant machine is required (not shown in Figure 8). Finally, the liquids are pressurized by a dedicated pair of pumps and cooled. Resulting main cost drivers for the energy and utility balance are shown in Table 1.

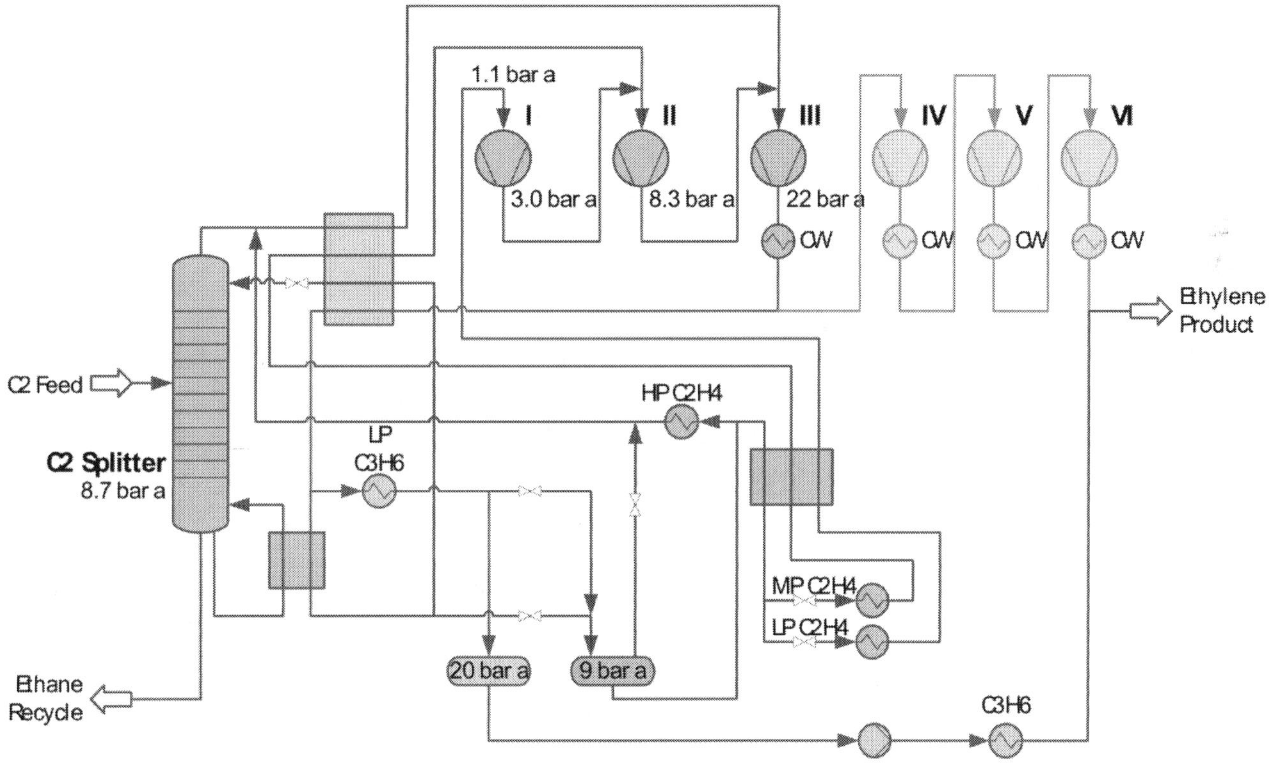

Figure 8: *Supercritical ethylene delivery either by direct compression from the gas phase (green), or by pumping of liquefied ethylene product (red).*

As indicated in Figure 8 and shown in Table 1, the ethylene compressor requires less shaft power, if the ethylene product is pressurized by dedicated pumps. However, the propylene compressor will become significantly larger due to increased liquefaction requirements. Additionally, shaft power is required by the pumps. Additional introduced heat and lower enthalpy of the battery limit product in case of pressurization by pumps needs to be compensated by additional cooling, as shown in Table 1.

In summary, additional operating costs in the order of magnitude around 1 MM€/1000 kta ethylene can be expected in case of pressurization by pumps.

Table 1: *Main cost drivers for energy and utilities comparing compression of vapor ethylene product and pumping of liquefied ethylene product.*

	Δ Pump - Compressor kW/tonne C_2H_4	Δ Pump – Compressor %
Ethylene Compressor Shaft Power	-50	-22
Propylene Compressor Shaft Power	+52	+75
Product Pump Shaft Power	+9	+100
Total	**+11**	**+4**
CW Ethylene Compressor Interstage Cooler	-115	-45
CW Propylene Condenser	+147	+75
CW other Heat Exchanger	+3	+1
Total	**+35**	**+5**

Resulting main cost drivers for invested costs are shown in Figure 9. Applying the pump solution, the main cost savings arise from a simpler ethylene compressor. However, these savings are compensated by more required heat exchanger area, additional product pumps and a larger propylene compressor. In summary, the invested costs for both solutions appear to be in a comparable order of magnitude and invested cost advantage of either solution may rather arise from project specific boundary conditions instead of inherent design characteristics.

Apart from investment and operating cost differences, there are also differences in operation and control comparing both solutions. A four-stage compression machine with three process tasks including refrigerant delivery, C2 splitter operation and product compression is state of the art. Two additional compression stages for product delivery to supercritical conditions yield further operational aspects as described in the previous chapters. Special attention must be paid to compressor kickback operation cases, which may not only be required for surge control reasons but also become necessary to accomplish other performance control issues, e.g. during start-up or cold cooling water conditions. This requires a more complex control concept as pressure ratios of compressor stages may vary. Moreover, the risk of condensation of supercritical ethylene after expansion in kickback line must be appropriately addressed in design and/or safety/operation concept.

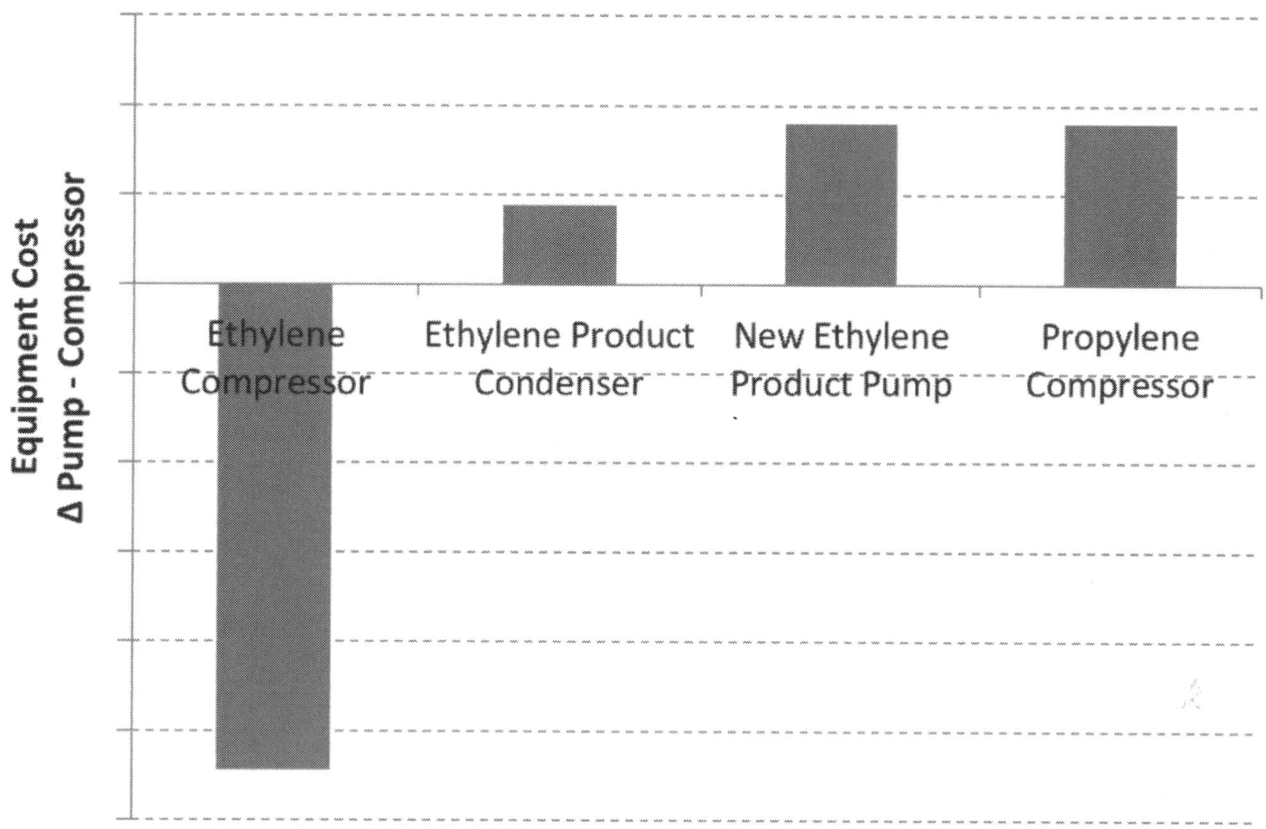

Figure 9: *Main cost drivers for investment costs comparing compression of vapor ethylene product and pumping of liquefied ethylene product.*

The product pump solution is assumed to be simpler in operation. The according three stage C2 compressor has only two process tasks (refrigerant delivery, C2 splitter operation) and the product is processed in dedicated pumps. Alternatively, a dedicated product compressor could be considered at the expense of an additional driver.

If variations of supercritical product rates are expected in a particular project, possibly due to more frequent turndowns or additional battery limits and frequent export elsewhere, e.g. by liquefaction, the compressor solution might be unsuitable, particularly when all tasks are carried out in a machine with a single drive. It is expected that different operational scenarios can be better handled, when the different tasks are carried out in dedicated equipment due to a higher number of degrees of freedom.

7. Conclusions

The system integration of open loop ethylene refrigeration and trans-critical product compression is feasible but the design of machine and control topology must account for requirements induced by physical behavior of trans-critical ethylene. In order to make the system reliable it is important to clearly define procedures for pressurization of the system during start-up of the machine.

Comparing pumping of liquid ethylene with compression of vapor ethylene, both solutions bear comparable characteristics with respect to investment costs. Operating costs appear to be somewhat better for the compression solution, provided that there is a single and defined operating case. As a consequence, for individual projects a tailored solution can be applied by choosing either of the two solutions and taking all important boundary conditions into account.

8. References

1. Zimmermann, Heinz and Roland Walzl (2012), „Ethylene" in "Ullmann's Encyclopedia of Industrial Chemistry", Wiley-VCH Verlag GmbH & Co. KGaA, Weinheim
2. Linde, Hermann (1985), „Gewinnung von Ethylen" in „Tieftemperaturtechnik", Springer
3. LINDE's general multiphase property system GMPS
4. Zimmerman, Thomas and Gerhard Luft (1994), „Untersuchungen zum explosiven Zerfall von verdichtetem Ethylen", Chem.-Ing.-Tech., 66, 1386-1389
5. Häfele, Martin et al. (2012), "Dynamic Simulation of an Ethylene Recycle Compressor System" in AIChE Annual Meeting, Pittsburgh, PA

AIChE Paper Number: 78a

Troubleshooting Vacuum Operation of an Inter-After Condenser Unit in an Ethylene Plant

Thomas Wieslaw
SASOL

Jeff Nichols
SASOL

Norman Lieberman
Process Improvement Engineering

Eric Michael Johnson
Graham Corp

Prepared for Presentation at the 2018 Spring National Meeting
Orlando, Florida, April, 24, 2018

Troubleshooting Vacuum Operation of an Inter-After Condenser Unit in an Ethylene Plant

Thomas Wieslaw
Sasol North American Operations
Jeff Nichols
Sasol North American Operations
Norman Lieberman
Process Improvement Engineering
Eric Michael Johnson
Graham Corp

Abstract:

An ethylene refrigeration compressor in ethane cracker unit is powered by surface condensing steam turbine that operates under vacuum. The vacuum is sustained by inter and after condenser. The condenser utilizes two steam ejectors a primary and secondary to pull a vacuum. Ejectors and inter-after condenser units constitute a critical part of stable turbine and surface condenser operation. Constant and sustainable vacuum is crucial for turbine and compressors they power. The 15 year old inter-after condenser had to be taken offline due to performance issues. The condenser suffered from two main problems: one it was unable to maintain design vacuum for the turbine's exhaust at no load condition; second the vacuum under load was not stable to the point that it would slowly deteriorate over time reaching turbine's trip point. An auxiliary "hogger" ejector had to be utilized to provide necessary vacuum by plant operation. The "hogger" ejector exhausts steam to atmosphere without condensing it. A rigorous troubleshooting study was initiated to bring online the condensing unit along with primary and secondary ejectors in order to save on steam usage and eliminate the noise level caused by the hogger jet. The paper focuses on design, principle of operation, troubleshooting and observations of single inter-after condenser unit along with its steam ejectors and surface condenser it services. Understanding of surface condensing units along with steam ejectors spans industries and lessons learned tend to be the same across. The study was formulated to systematically evaluate each piece of equipment in the condensing unit. Preliminary inspection has shown no tubes leaks between the shell and cooling water tubes or erosion in the ejectors. During tests further inspection revealed steam traps that were not properly specified, isolation bypass valve that had a leak and low compression ratio at the secondary ejector which was leading to higher back pressure causing further low compression ratio at the primary ejector. Secondary ejector had a degraded diffuser and loose steam nozzle where portion of the steam bypassed the nozzle and contributed further to overloading of the ejector. Another trial was conducted to test for internal condenser baffle leak. To test for suspected baffle leak between inter and after areas of the condenser, secondary ejector was unloaded and exhaust direct to atmosphere rather than to after condenser. Process data was collected along with each trail. After tests and data collection, unit was brought online with modified thermostatic vents on steam traps, bypass valves isolated and secondary ejector nozzle replaced. Turbine's surface condensing unit achieved stable vacuum operation range of 61-89 mmHgA vs. 125 mmHgA (original and unstable vacuum pressure with slow further deterioration prior to this troubleshooting study) under full plant load. The suction pressure to the secondary ejector decreased by a significant 114 mmHg. The troubleshooting study showed that inter after condenser was suffering from multiple bottlenecks that acted simultaneously to the detriment of the system performance.

This paper originally appeared in special focus section of Hydrocarbon Processing August 2017th issue and the paper is modified to fit EPC purposes.

Introduction

Inherent in the operation of a typical ethane cracker unit is a system of compressors powered by surface condensing steam turbines. These turbines run by extracting work from high pressure steam while a surface condenser condenses the turbine's exhaust to both maximize compressor horsepower and recover valuable condensate. In the surface condenser, a vacuum is created by the condensing steam. This vacuum is maintained by exhausting non-condensable load from the surface condenser via steam ejectors and inter-after condensing units. There are number of sources where non-condensable gases, which must be purged from the system, can originate: CO_2 entrained in the steam and air leak through shaft seals into the low pressure area of the surface condenser are two examples.

The technology to exhausting non-condensable gases to sustain vacuum has been around for more than a century. While the systems tend to have simple layout and are not overly complicated in terms of hardware, troubleshooting loss of vacuum or underperformance of these units is not straightforward. It requires a systematic approach to identify and rectify any issues that contribute to deteriorated performance. A process system operating with unstable and/or low vacuum directly affects turbine performance, turbine's steam consumption, and overall compressor efficiency.

In an ethane cracking unit, a troubleshooting study was undertaken to investigate an inter-after condensing unit and vacuum instability in the surface condenser. The troubleshooting study, summarized here in this paper, consisted of field observations, equipment review, trial runs, and data collection. The goal of the study was to:

- Evaluate unit's present operating conditions.
- Understand its operating constraints.
- Implement correction to sustain stable vacuum operation.

While ethylene plants are typically not known for wide utilization and operation of their vacuum systems, troubleshooting and lessons learned from steam ejectors and condensing units span many process industries.

System Overview

In the subject unit, a compressor turbine isentropically expands high pressure steam to a significantly lower pressure level. The expanded steam enters through the top of the condenser as shown on Figure 1. Once it enters the shell, steam spreads horizontally along the length of the shell while moving downward over the tube bundle. In this surface condenser, steam condenses on the shell side of the exchanger while heating the circulating cooling water inside the tubes. The condensed vapor falls to the bottom by gravity into the hotwell to be pumped out as valuable condensate. The surface condenser's vacuum created by the condensing steam is limited by the vapor pressure of water at the cooling water inlet temperature plus some approach temperature. Therefore, cooling water temperature determines the best possible operating pressure—that is, minimum achievable vacuum pressure—of the condenser.

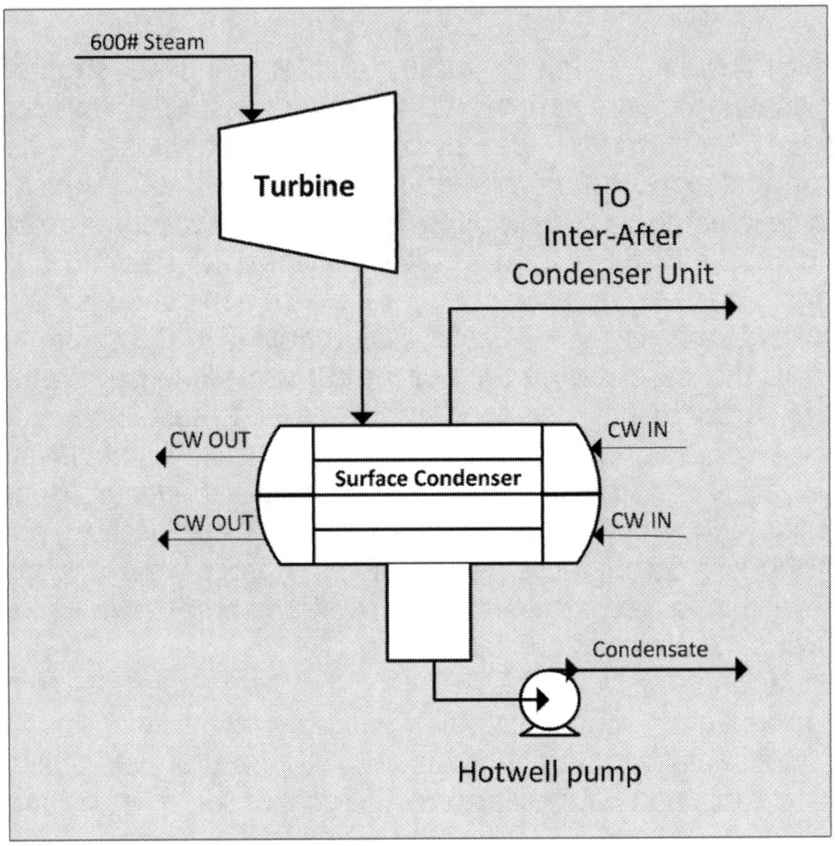

Figure 1. Surface Condenser Flow Diagram.

For example, with a cooling water inlet temperature of 90°F, a condensing temperature after approach of 100-110°F is possible in the unit. This would correspond to a minimum possible operating pressure of 66 mmHg (absolute) based on water vapor pressure. Cooling water flow also affects the LMTD of the surface condenser. Lower cooling water flowrate will result in decreased LMTD across the exchanger. The unit's surface condenser is a fixed tube sheet split exchanger with cooling water on the tube side as shown on Figure 1. The shell is split into two independent compartments which permit isolation of one side and periodic cleaning of cooling water tubes. The arrangement allows for continuous operation of the surface condenser, albeit at reduced capacity, during maintenance cleaning times.

Marginally lower pressure in the surface condenser reduces steam consumption and improves the turbine's efficiency. The condensing steam is creating of a vacuum inside the surface condenser. As the system operates under vacuum, any non-condensable gases from steam or an external leak source will accumulate in the low pressure area. These gases will quickly increase the operating pressure of the condenser because the non-condensable gases that cannot be evacuated blank the tubes and reduce the capacity of the condenser. They must be vented from the condenser in order to help maintain vacuum. In the plant, the main devices used to vent the non-condensable gases are two-stage steam jet ejectors coupled with horizontally installed inter-after condensing units as shown in Figure 2.

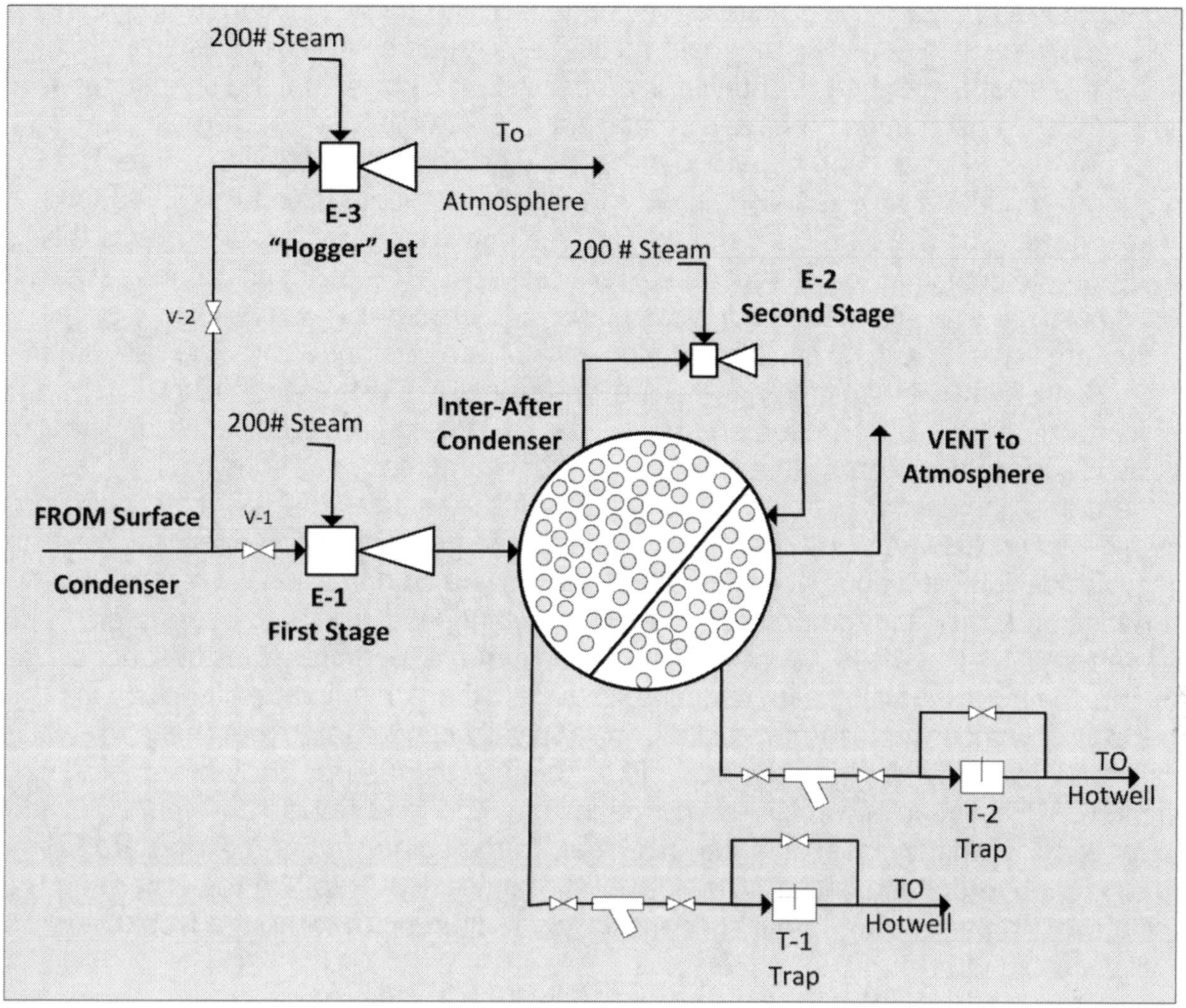

Figure 2. Inter-After condenser unit flow diagram.

The design of the ejectors E-1, E-2, and E-3 in Figure 2 includes five main parts each: a motive steam nozzle, suction chamber, inlet diffuser, throat section, and outlet diffuser.[3] Across the motive steam nozzle, 200 psig steam enthalpy is converted into kinetic velocity. Similarly to the steam enthalpy conversion to velocity across turbine's inlet nozzle this is also an isentropic process. At the nozzle's discharge the expanded steam creates a low pressure, which entrains process load into the high velocity steam. Steam and non-condensable gases mix entering the inlet diffuser. In the inlet diffuser the velocity of the process flow decreases as it enters the diffuser throat. The throat section is the transition piece between the converging supersonic inlet diffuser and diverging subsonic flow outlet diffuser. As the process flow moves through the throat it transitions to subsonic flow creating a supersonic shockwave. In the outlet diffuser the flow velocity is reduced further and essentially that velocity is converted back into pressure. In summary, the ejector uses kinetic energy in the form of a high velocity stream to push back on the pressure located at the outlet of the ejector. The outlet pressure that an ejector is subjected to is the result of the downstream equipment and is not set by the ejector. The ejector E-3 ("hogger jet") in Figure 2 is an auxiliary ejector used primarily to evacuate the condenser and turbine casing during start up. In case of a performance issue with the main holding ejectors it can be used as a temporary back up. Each part of the ejector is critical for proper vacuum operation and seemingly negligible changes in geometry can have a significant impact on system performance.

The compressed steam and non-condensable gases from the ejector E-1 enter the fixed tube sheet inter-condenser where a significant portion of that steam is condensed against cooling water flowing on the tube side. In addition, the non-condensable gases are further cooled, reducing volume load to the second stage ejector E-2. The condenser unit is separated by a welded partition that spans the entire condenser and creates two areas called, quite logically, the inter-condenser and the after-condenser. Ejector E-2 maintains the vacuum of the inter-condenser while exhausting non-condensables into the after-condenser. The ratio of absolute discharge pressure to suction pressure is the ejector compression ratio. Similar to vacuum in the surface condenser, an ejector's individual compression ratio is mostly load related. The first-stage ejector E-1, piped directly to the shell side of the surface condenser. Given the air load is below design, air removal system is in good condition, steam and cooling water meet the design conditions, the ejector E-1 will have a compression ratio set primarily by the surface condenser and the pressure the surface condenser wants to operate on.

The after-condenser does not contribute to vacuum and operates at atmospheric pressure. It recovers additional steam condensate while exhausting non-condensable via atmospheric vent without the need for a silencer. Typically condensate from the inter-condenser must be drained back to the hotwell, as the inter-condenser is always operating under vacuum. This may be accomplished by means of a condensate trap or, where elevation allows, by means of a highly reliable water loop seal. In the subject plant, condensate from the after-condenser is drained back to the hotwell through a float trap. On the other hand, after-condenser can be set up to drain condensate to an open drain without a trap.

Unit Operation and Problem Definition

Prior to the troubleshooting study, when ejector E-3 "hogger jet" was in operation, vacuum in the hotwell would operate at 125 mmHg (absolute). Switching the unit to inter-after condenser operation would cause slow deterioration in vacuum as shown on Figure 3.

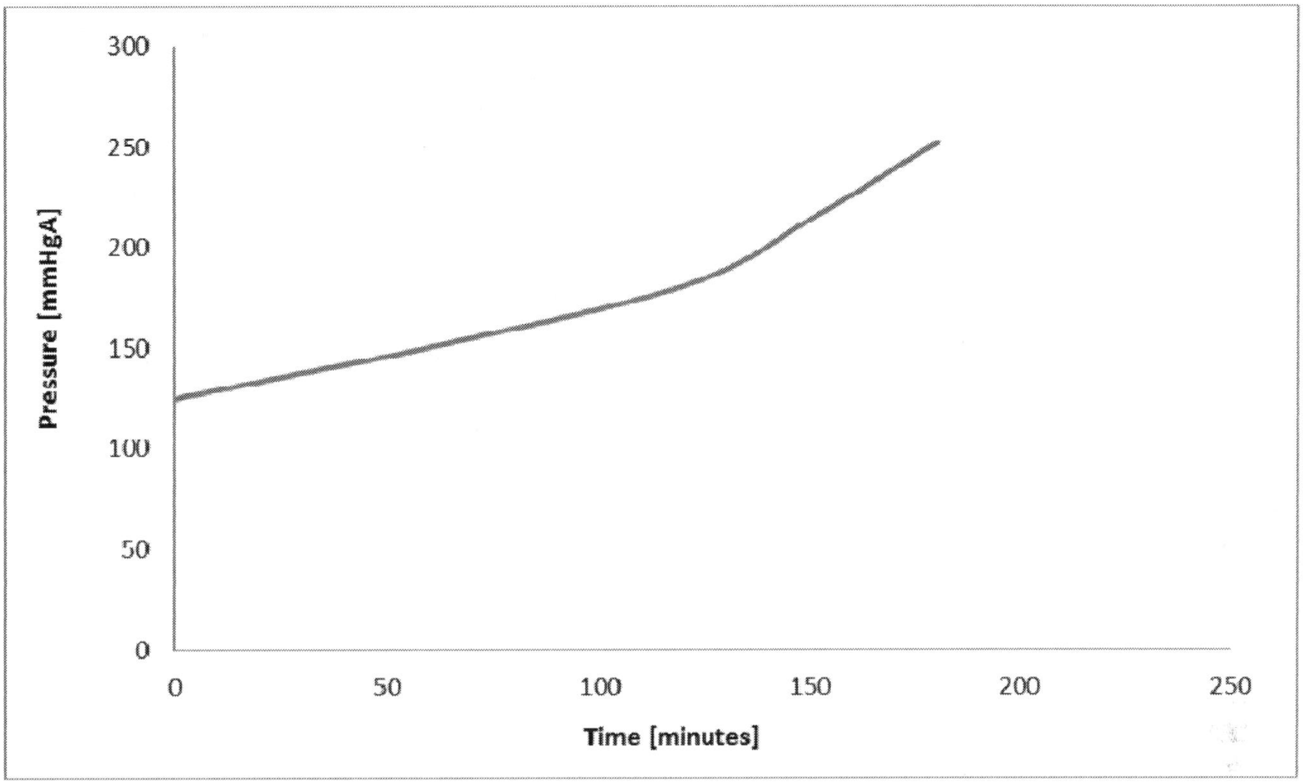

Figure 3. Vacuum measured above the water level in the surface condenser's hotwell.

There are two areas within this unit where system performance deterioration can occur. One is within the surface condenser and the other within the inter-after condenser unit, including ejectors. The surface condenser unit operates with cooling water inlet temperature range of 87°F to 91°F and cooling water temperature difference between inlet and outlet in the range of 17°F to 10°F vs. design at 17.5°F. The measured hotwell temperature averages 120°F. The corresponding water vapor pressure to this temperature is an indication of theoretical possible vacuum pressure inside the surface condenser given that non-condensable load is below design. These measured temperatures are well within design values of the surface condenser. However, if cooling water temperature were to rise, the surface condenser available LMTD would decrease. The system vacuum would be limited by the exhaust turbine temperature. The temperature of 120°F in the hotwell corresponds to water vapor pressure of 87.4 mmHgA . This temperature-pressure correspondence is close to the initial pressure of 90.9 mmHgA at the inlet to the inter-after condenser unit, measured with a pressure gauge. Rather than rely on installed DCS instrumentation (pressure transmitter), one important step was to perform a pressure survey between the hotwell and inlet to the inter-after condenser unit to ensure there was no unnecessary significant pressure drop between the two units. The pressure survey between the surface condenser and the inter-after condenser unit revealed only a small pressure drop of about 7.6 mmHg measured with the same pressure gauge.

An increased vapor flow and higher temperature into the inter-after condenser unit can indicate an air baffle leak in the surface condenser. In large surface condensers, in order to force steam and non-condensable vapors to go through the entire bundle, an air baffle is used. The air baffle runs the entire length of the surface condenser and is welded to the shell to avoid steam bypassing the surface condenser and flowing directly into the inter-after condenser unit. Throughout the data collection period, an average temperature range of 103°F to 115°F was recorded for vapor flowing into the air removal unit. A measured temperature of 115°F corresponds to a water vapor pressure of 76 mmHgA which is well below the measured and deteriorating system vacuum as shown on figure 3 when the surface condenser is operating on with a full load. With cooling water conditions operating below design, above suggests that load temperature from the surface condenser is not a limiting factor. Therefore the factors that limit the performance can either originate from high non-condensable loads (which can be checked at the inter-after condenser unit's vent discussed later in the paper) or there is a bottleneck within the inter-after condenser unit itself. When running steam ejectors it is important to look for surging or loud popping sounds coming from the steam ejectors. While there might be many causes for ejector surging, it is an indication of a vacuum break. Ejector surging can often lead to very hot vapor load temperatures flowing into the inter-after condenser as steam will want to flow backwards for example from the ejector E-1 towards the surface condenser while the ejector recovers its shockwave. Resulting hot load temperature can lead to misleading assessment as to where one's troubleshooting attention should be focused.

In order to troubleshoot the system performance, a rigorous raw data collection plan was developed. The following data collection points were identified and prepared prior to initiation of the trial:

- Suction (inlet pipe or ejector suction chamber) and discharge pressure on each of the two ejectors E-1 and E-2.
- Pipe surface temperature on cooling water inlet and outlet.
- Cooling water inlet and outlet pressure for the condenser.
- Exhaust & suction temperature on the ejector jet.
- Vapor flow from the after-condenser outlet atmospheric vent.
- Motive steam pressure and temperature.
- Pipe surface condensate drain temperatures.

Instead of using compound gauges, in order to ensure both precision and accuracy, self-compensating, absolute-vacuum pressure gauges with isolation root valves and pigtails were installed at the two ejectors and leak-tested. It is also recommended to use a single electronic absolute pressure gauge for consistent readings throughout the trials.

Inter-After Condenser No-Load Test

While the surface condenser was in operation running on the "hogger jet" ejector E-3, a "no-load" test was performed on the inter-after condenser unit. The inter-after condenser unit was isolated by closing a load valve at the suction inlet to first ejector followed by opening motive steam to both ejectors. Without any load present from the surface condenser, once steam was introduced, the ejector exhausted from the first stage suction chamber to the minimum absolute pressure it was capable of producing. Typically, if the test showed that the ejector was operating at close to its design shut-off pressure and before introduction of load from surface condenser, acceptable performance could be assumed along full ejector's performance curve. The system reached a vacuum of 99.5 mmHgA compared to the design expected no-load vacuum of ~10 mmHgA. While condenser small amount of leakage through the isolation valve might be expected, this dramatic performance deterioration was not rational. The compression ratio of the primary ejector E-1 was 1.9:1, and the ratio of the secondary ejector E-2 was 4.0:1. This indicates that, in addition to the inter-after condenser unit being unable to sustain stable vacuum, it was also unable to operate at no-load design vacuum. Based on this no-load test's compression ratio result, system troubleshooting could now be focused exclusively on the ejectors and inter-after condenser unit.

Steam Supply

Both higher and lower steam supply pressures (versus specified design) at -0%/+20% will have detrimental effects on the stability of any or all of the ejectors. For example, motive steam pressures in excess of design can create choked flow and decrease the ejector's capacity. On the other hand, with lower than design steam pressure supply, the ejector will not be provided with sufficient energy to compress vent load from the surface condenser. During the trials, steam pressure was measured at 210 psig, about 10% above design. The pressure was deemed as acceptable and not detrimental to the performance. Following the trial, the steam pressure to the inter-after condenser was brought down to design specification to save on steam consumption and condensate recovery.

Regarding steam quality, a source of dry steam close to saturation is preferable for the ejectors. Moisture in the steam will erode the internals as well as reduce suction load capacity for the ejector. To ensure dry steam, the insulated inlet line in the subject unit is routed from the top of steam supply header. If such a configuration is not available or if moisture is known to be present in the steam, a separator and floater trap can be used to improve steam's quality. One can tell if wet steam is present by measuring the temperature on the discharge of the ejector or inspecting the internals. The temperature of the vapor at the discharge of the ejector will be much cooler if a high degree of liquid is present.[4] The steam outlet temperature was measured on the "hogger jet" by inserting a thermometer into the exhaust stream. The steam temperature measured at the outlet of the "hogger jet" was in the range of 255-260°F, an expected temperature range for the exhaust steam to the atmosphere. Erosion of nozzle internals and diffuser, as mentioned, is evidence of wet steam presence; this requires internal inspection of the ejector, to be performed at a later date.

Condensate Recovery

It is important to ensure an unobstructed pathway for condensate drainage. If a strainer, separator, and/or trap are present, they should be inspected and cleaned after the lines are blown clear. It is very easy for a strainer to be plugged during initial start from deposits or rust contained in a stagnated unit. The subject unit contains two traps: one on the drain from the inter-condenser and one on the after-condenser. Low flow of condensate to the traps due to a plugged 1" strainer or undersized traps will cause eventual backup of water into the condenser.

A flooded condenser will lower available surface area for additional steam condensation from the ejector and cause unstable or deteriorated vacuum. Free water exhausting from the vent is an obvious sign that condensate has backed up and the after-condenser has flooded. A flooded condenser will make subcooled condensate. During the trials, the traps were opened and inspected for any mechanical issues which might inhibit function, as well. The traps on the unit were floater/thermostatic traps. A thermostatic part of a floater trap is a vent spring valve that, based on measured temperature, will vent air and non-condensable gases. The thermostatic vent in a normal floater trap application is designed to open at a few degrees below saturation. The thermostatic element contracts pulling the valve head off of the valve seat. The trap opens and discharges air and condensate, which is not desirable, especially if trap is connected to the hotwell and condensate is subcooled. The typical application of these traps is discharge of condensate in modulating conditions like traditional heat exchangers, air-handling coils and typical steam header stations. In a vacuum system, it is recommended to use a simple floater trap (no thermostatic element) or a loop seal. As shown on Figure 4, the thermostatic valve part of the condensate trap was removed and blocked with NPT plug during the troubleshooting trials.

Figure 4. Floater trap with thermostatic vent plugged.

It is also recommended that the condensate drain piping from the inter-condenser not tie in to the after-condenser piping at any point before hotwell. This will avoid any condensate backup problems in the drain lines, i.e. backing condensate up from one condenser into the other as they operate at different pressures. The difference in pressure may also allow air load from the after condenser to recycle back into the inter condenser. The subject unit both lines drain condensate independently.

Ejector Inspection

The primary and secondary ejector steam chests and outlet diffusers were disconnected and inspected. The secondary ejector diffuser was not smooth and showed signs of some erosion, corrosion and fouling. If the motive nozzle and diffuser throat have increased by about 7% by area or 3% by diameter versus initial installation, it is recommended to replace the ejector. Furthermore, a typical mechanical issue to look for when inspecting the steam chest is a loose motive steam nozzle. During inspection a slightly loose motive steam nozzle was found on second stage ejector E-2. A loose nozzle will easily leak steam across its threads and bypass discharge of the nozzle. The steam that bypasses the nozzles becomes a supplemental and unnecessary load in the air chamber of the ejector and decreases operational vacuum and stability. Inspect the nozzle and look for signs of wear and corrosion. To fix a loose nozzle one can temporarily wrap threads of the existing steam nozzle with a sealing device like Teflon tape for the duration of trials. A long-term fix is to seal weld the nozzle to the spacer of the motive chest or replace the motive chest entirely.

Unit Leaks

Vacuum systems are vulnerable to vacuum pressure deterioration when non-condensable load to a vacuum system increases above specified design. Besides expected non-condensable loads coming from condensing steam, or a condensate traps that load the surface condenser through thermostatic vents, there were also two possible areas identified where extraneous introduction of non-condensables might occur. The first was an ambient air, as both the exhaust of the steam turbine and the condenser operate under a significant vacuum; air is always likely to leak into the system. This leakage can occur through flanges, threaded piping, valve pickings, or from expanding seal, joint failure on the turbine itself. One of the ways to test for external leaks is with ordinary shaving cream. When applied to all potential leak piping and joints, shaving cream will be sucked into any unsealed opening. On the cooling water side, the unit was also checked for cooling water tube leaks. A cooling water tube leak into the shell of the inter- after condenser unit can overwhelm traps, back-flood the condenser and plug condensate strainers over time. Channel head covers at each side of the condenser unit were removed, and steam was introduced to the shell side in order to check for tube leaks. Secondary option can be also to perform a hydrotest.

Another leak source can occur from the process itself. When the inter-after condenser is operating, the ejector E-3 "hogger jet" is offline and it is isolated by valve V-2 on Figure 2. If the isolation valve leaks, it could easily suck in air from outside and overload the secondary ejector. The unit also had bypass valves around floater traps, which were identified as possible leak sources, as well. A vacuum deterioration occurred when block valves on the condensate drain were opened to the hotwell prior to introducing motive steam to the inter-after condenser. In response, the bypass valves around traps were taken out and blinded during the troubleshooting. Condensate trap bypass valves can only be opened temporarily to drain the condensate while the trap is being serviced for example. Without the trap in service the surface condenser pressure will be affected.

Ejectors can also get overloaded from an internal partition failure.[1,2] The inter-condenser is separated from after-condenser by a partition plate welded to the shell. The secondary ejector exhausts steam into the after-condenser which operates at higher pressure (atmospheric) than the inter-condenser, which operates under vacuum. A partition failure resulting in a leak would expose higher pressure after-condenser to the inter-condenser operating under vacuum. This would lead to operational upset on the first ejector. To offline test for internal partition leak, the system was set up in the following way:

- Outlet of the second stage ejector was routed to the atmosphere.
- Inlet to the after-condenser and its atmospheric vent were blinded and blocked off.
- Both shell sides of the condenser were drained and condensate lines were isolated.
- A vacuum pressure gauge was connected to the after-condenser.
- The unit was isolated from the surface condenser. The "hogger jet" E-3 remained online to sustain the vacuum in the surface condenser required for plant normal operations.

When motive steam was introduced to the second ejector only, the isolated and blinded after-condenser remained at atmospheric pressure while the inter-condenser held at a stable vacuum. Had the internal partition been compromised, the after-condenser pressure would fall to vacuum. Because the pressure in the after-condenser remained near-atmospheric, the test showed that the internal division plate had not failed.

To qualitatively determine excessive total leakage throughout the system, one should measure the non-condensable vapor flow. After modifications mentioned above were complete, a quantitative test was used to actually measure flow of saturated vapor and non-condensables exiting the unit at the after-condenser atmospheric exhaust vent. Shown in Figure 5, a small grocery 4-gallon trash bag was attached with duct tape and rubber bands to the exhaust vent. A stopwatch was utilized to measure the fill time of the bag, giving an estimate of total volumetric vapor flow through the 1.5" vent.

Figure 5. Trash bag test on After-condenser vent to measure total non-condensable load from the Surface Condenser and Inter-After Condenser units.

The flow was calculated to be about 15 SCFH at a temperature of 95°F. Based on the measured temperature, vapor is about 5.5% (vol) saturated water. The flow rate for the non-condensable gases then calculated to be about 1 lb/hr, well below the maximum load design value of 33 lb/hr for the inter-after condenser. For periodic performance checks down the road, it is recommended to have a permanent air leakage meter installed if one is already not part of the unit.

Unit Fouling

Within the subject plant, surface, inter- and after-condensers are subject to the same fouling mechanisms, as are all other cooling water exchangers. This fouling might occur at any portion of the condenser. Fouling on tube side (cooling water) or shell side - depending on unit's steam quality, or magnitude of a condition known as vapor binding of the tubes[5] - will lower heat transfer, condense steam at degraded pressure, and affect the vacuum performance. Severely fouled the condenser will not be able to condense motive steam at the design operating pressure. The surface condenser was taken offline (one cooling water side partition at a time) and tubes were cleaned. Once both shell sides were fully cleaned, the temperature inside the hotwell decreased by an average of 8°F. Given design flexibility in any plant, authors recommend for routine maintenance procedures to include periodic cleaning of condenser bundles. Prior to the initiation of the cleaning step, it is also worthwhile to inspect for tube leaks in the surface condenser. If one side of the condenser is isolated, monitor for water conductivity change with a meter in the condensate from the hotwell. If a conductivity meter is not available, test for the hardness with use of pronated dye solution such as Eriochrome Black T solution. The solution changes color in an alkaline environment from blue (no hardness indication) to purple (hardness present) when complexed with magnesium or calcium. A large conductivity or hardness color change would indicate one side of the condenser leaking cooling water more than the other into the steam condensate system. If a large tube leak is detected, caution needs to be exercised when opening the shell side to plug the tubes, as air will be sucked into the surface condenser.

Inter-After Condenser Follow-Up Analysis

The inter-after condenser unit was placed back into operation to record final data after the trials and modifications above. The unit achieved stable vacuum operation upstream of the primary ejector E-1 in the range of 61 to 89 mmHgA under full plant load operation vs. initial pressure of 125 mmHgA (prior to vacuum deterioration as per Figure 3) at the beginning of the study. The suction pressure to the secondary ejector E-2 decreased by a significant 114 mmHg. With thermostatic vent on floater traps plugged, bypass around traps to the hotwell isolated, and other leak checks completed, there was no evident high non-condensable vapor flow coming from the after-condenser vent.

Conclusion

Ejectors and inter-after condenser units constitute a critical part of stable turbine and surface condenser operation. Consistent and sustainable vacuum is crucial for turbines and the compressors they power. Without these seemingly simple ejector/condenser units, stability is jeopardized and turbine steam efficiency is compromised. The inherent simplicity of operation should not be taken for granted when multiple bottlenecks may be acting simultaneously to the detriment of any system performance. Answers to the operational problems exist in the field. They will always become more visible with systematic approach and good field testing plan. With frequent performance checks established, avoidance of future process problems, as well as minimization of operating costs, is an achievable goal.

Acknowledgments

Special thanks to Matthew Fullington (Sasol Operations Superintendent), Gary Guidry (Sasol Operations) and remaining Ethylene Unit Operations team for providing assistance, resources and their full support in these troubleshooting efforts.

References

1. Lieberman, N.P. *Troubleshooting Process Operations,* 4th Ed., PennWell Corp, 2009
2. Lieberman, N.P. *Troubleshooting Vacuum Systems,* 1st Ed., John Wiley & Sons, 2012.
3. Birgenheier, D.B., Butzbach, T.L., Bolt, D.E., Bhantnagar, R.K., Ojala, R.E., Aglitz J. "Designing Steam jet Vacuum Systems", *Chemical Engineering,* July 1994.
4. Lines J.R., Smith, R.T., "Ejector System Troubleshooting", *The International Journal of Hydrocarbon Engineering,* UK, 1997.
5. Lieberman, N.P. "Improving Heat transfer in Reboilers and Condensers", *Petroleum Technology Quarterly,* UK, 2016, Q1.

78b-Reduced Flaring Options during Ethylene Plant Start-up and Shutdown

Ramakrishnan Sankara Subramaniam
Process Start-up Manager
TechnipFMC Process Technology

Muhammad Imran
Lead Process Engineer
TechnipFMC Process Technology

Thomas Pickett
VP of Ethylene Technology
TechnipFMC Process Technology

Prepared for Presentation at the 2018 Spring National Meeting

Orlando, Florida, April 22 – 26, 2018

UNPUBLISHED

Reduced Flaring Options during Ethylene Plant Start-up and Shutdown

Ramakrishnan Sankara Subramaniam
Process Start-up Manager
TechnipFMC Process Technology

Muhammad Imran
Lead Process Engineer
TechnipFMC Process Technology

Thomas Pickett
VP of Ethylene Technology
TechnipFMC Process Technology

Keywords: Flaring, Start-up, Shutdown, Hydrocarbons, DCS, ESD

Abstract

Ethylene producers and licensors are facing the challenge posed by environmental regulations to minimize the emissions of Highly Reactive Volatile Organic Compounds (HRVOC) during plant start-up, shutdown and emergency conditions, as well as the economic loss from the flaring of valuable feedstock. For many years, various strategies have been employed to minimize flaring, with varying success.

TechnipFMC Process Technology has developed a scheme/ strategy to minimize the flaring, which would also help to reduce the steam import requirements at super high pressure (SHP) or high pressure (HP) level, required to start the main process compressors. The strategy is also aimed at reducing flaring during the preparation for a scheduled turnaround/ shutdown, purging the system of benzene while shutting down various units and reducing the days required for maintenance preparation.

Various recycle lines have been added to establish circulation through the plant, allow for chill down and stabilize various plant sections close to normal operating conditions, prior to the fresh feed introduction to the furnaces. This makes the unit ready to receive fresh feed and process the cracked gas, to minimize the time required to achieve on-spec production. The basic concept presented shows using cracked gas compressor (CGC) on nitrogen or methane rich natural gas, with provision to circulate through the feed system. The flow scheme provides the flexibility to circulate through the cracking furnaces, which allows for the generation of SHP steam, without fresh feed cracking.

The scheme presented here-in is for an ethane cracker, but the concepts can be customized for liquid cracker schemes as well, to meet the operator's specific requirements and applicable conditions. Existing plants can also be retrofitted to implement many of these concepts to reduce flaring and improve profitability.

1. Introduction

Flaring from ethylene plants is an unavoidable requirement, due to the nature of the plant flow scheme and process. Flaring is required for the safe disposable of hydrocarbons. The amount and duration of flaring varies, depending on the plant operating conditions, design features and flexibilities built into the design. Typically, maximum flaring happens during start-up, major plant upsets and shutdown. Flaring is a high temperature oxidation process, which is used for safe disposal of the hydrocarbons, with minimum environmental effects. However, sometimes flaring leads to excessive release of smoke, due to insufficient mixing of air for combustion. In addition, excessive flaring can also produce noise and brightness in the night which affects the neighborhoods and multimillion dollar waste of useful products. The environmental impact of flaring also depends on the type of hydrocarbons flared. Release of unsaturated hydrocarbons like ethylene, propylene, butadiene etc. are not desirable. These are called "Highly Reactive Volatile Organic Compounds (HRVOC)". There is much stringent restriction in the release of such compounds. The environmental regulations are getting increasingly stringent posing greater challenges to both licensors and operators.

Various operators and licensors have been continuously working on the strategies to reduce flaring from ethylene plants. It is getting increasingly important to have certain flexibilities built into the plant to minimize flaring. Similarly, operating practices should be developed to reduce flaring. The operations team must be trained in best practices and the maintenance team must ensure reliability of equipment and instruments. One of the most common causes of excessive flaring is the reduced reliability of the equipment and instruments. Very reliable supply of utilities like power, steam, water etc. is also essential to ensure stable operation. To achieve reduced flaring during start-up, major plant upsets and shutdown of grassroot plants, the licensor, engineering contractor and owner must work closely from the start of the project to develop the required flexibility in the process design and specify the reliable hardware. Licensor and owner can also work together to study and implement the required modifications for minimum flaring in existing plants. Each plant is unique and requires dedicated study for minimum flaring implementation.

Options to minimize flaring are discussed in the following sections. One important strategy to reduce flaring during start-up of a new ethylene plant or after a major shutdown or turnaround of an existing plant is to prove the reliability of critical equipment and most of the instruments by starting up various sections by closed loop methane or nitrogen recycling prior to the main feed into the cracking furnaces. The closed loop recycling helps in delaying the main feed introduction which otherwise leads to flaring until the plant is stabilized. It is not practically possible to achieve "Zero flaring or No flaring". It is only reduced flaring that is targeted. The main reason is that purging out of nitrogen with hydrocarbon will lead to minimum flaring of hydrocarbons.

The following discussion is for ethane crackers with non-heat pumped front-end deethanizer flow scheme:

2. Typical Locations of Flaring in Ethane Cracker Ethylene plants

Following are typical locations from where the furnace effluent is flared during start-up/shutdown of a front-end deethanizer type ethane cracker ethylene plant and the reason for flaring. As explained in Section 1, closed loop recycling of methane or nitrogen can reduce flaring from these locations. See block flow diagram in Annexure 1.

2.1. Quench water tower over head / cracked gas compressor 1st stage suction

Flaring at high furnace effluent load is required until the cracked gas compressor is started. This is also required if cracked gas compressor trips during normal operation. In some designs 50% of furnaces are automatically switched to hot steam stand-by (HSSB) mode by tripping the fresh hydrocarbon feed to the furnace. Typically, this flaring could be for two to four hours depending on the cracked gas compressor stabilization time after furnace fresh feed-in and the availability of SHP/HP steam for the cracked gas compressor turbine start-up at reduced furnace load.

2.2. Cracked gas compressor final stage discharge to drier feed knock out drum/ HP deethanizer inlet

Flaring is slowly shifted from quench water tower overhead to the cracked gas compressor final stage discharge. This is to stabilize the caustic wash tower to achieve specification for carbon dioxide and hydrogen sulfide (CO_2/H_2S). CO_2 removal is required to avoid any potential freeze-up issues in the chilling train and off-spec ethylene product. H_2S is removed to avoid potential poisoning of the C2 hydrogenation catalyst and off-spec ethylene product. Typically, flaring from the CGC final stage discharge could be up to four hours, depending on the caustic wash tower stabilization time.

2.3. C2 hydrogenation effluent/ demethanizer chill down train inlet

Flaring is required until the HP deethanizer overhead is on-spec for 1-3 Butadiene. C2 hydrogenation reactor is bypassed during this time. Flaring time depends on the stabilization time of the HP deethanizer. Two to four hours is a reasonable time to consider.

2.4. C2 splitter inlet/ demethanizer bottom

C2 hydrogenation reactor and chilling train is put into operation after HP deethanizer overhead is on-spec for 1-3 Butadiene. Flaring is now required at C2 splitter inlet/demethanizer bottom until the feed is on-spec for acetylene and methane to avoid off-spec ethylene product. Flaring time depends on the C2 Hydrogenation reactor stabilization time, based on the catalyst response and feed fluctuations. It could be anywhere between 8 hours to 24 hours.

Any contamination of the C2 Splitter/ C2 refrigeration compressor system, could require flaring for approximately 6 hours to achieve on-spec ethylene product, if there is no off-spec ethylene product storage, considering very short duration contamination. Otherwise, flaring duration could be much higher depending on the level of contamination since the entire system must be diluted.

2.5. Off-spec tail gas from regeneration gas knock out drum

Typically, the tail gas which is from the demethanizer and feed chilling train will be off-spec for ethylene till the required temperatures are achieved. The ultra-low NOx type furnace burner designs tend to get coked up, since the tips are very small diameter. Hence it is important to reduce the ethylene content of the tail gas to minimum possible prior to diverting to the cracking furnaces. In addition, in case of the regeneration requirement of the primary driers, it is desirable to have lowest possible ethylene in the tail gas. This might result in flaring of hydrogen and methane rich fuel stream. If the auxiliary boiler burners could handle this tail gas with higher ethylene content, then it could preferentially be routed to be consumed there and later switched to cracking furnaces.

3. Guidelines for Minimum Flaring in Ethane Cracker

Below discussion is specific to the start-up of a grassroots or after a turnaround, front-end deethanizer ethane cracker. Grassroot start-up is the first-time start-up for new projects after precommissioning and commissioning activities are completed. All the rotating equipment and instruments are new and the reliability is yet to be established. Most of the delays in start-ups are attributed to failures of the rotating equipment or instruments. Readiness of the operations team also plays an important role. Each day of delay costs millions of dollars to the operating company.

It is very essential to prove the reliability of the entire plant, by operating most of the sections as close as possible to the normal operating conditions. This is to ensure that there will be minimum interruption during start-up after fresh feed-in to the cracking furnaces. Once fresh feed is introduced, it is not possible to stop flaring from various reasons, until the main ethylene and/or propylene products are on-spec, suitable for consumption by user plants.

During start-up, the entire plant will be under nitrogen blanketing after the removal of oxygen. This nitrogen needs to be purged using compatible streams for various sections. For example, polymer grade ethylene to be used for C2 splitter and ethylene refrigeration systems, polymer grade propylene to be used for propylene refrigeration system, methane rich natural gas to be used for cracked gas and chilling train sections. Typically, major release of hydrocarbons happens during system purge to remove nitrogen and during ethylene, propylene and methane inventorying/ chilling process.

Depending on the contaminants, methane will require removal of CO_2 and H_2S to avoid contamination/ plugging of the downstream sections, which otherwise would result in excessive flaring during start-up. Typically, CGC is used for initial air blowing and dry-out of the entire plant and hence would have been tested to large extent. Furnaces would have been taken on-line partly to complete the refractory dry-out after the steam system chemical treatment and taken to the HSSB mode, generating steam as per design.

The balance sections of the plant should be checked by methane circulation through CGC for reliable operation as much as possible before feed-in. Closed loop methane circulation is established from chilling train, back to feed system, quench oil / quench water tower inlets or to cracking furnaces to generate SHP/HP steam to meet steam demand. However, the furnace coil outlet temperatures should be maintained much lower to avoid ethane (if present) cracking as much as possible. It is technically feasible to commission most of the sections by methane circulation depending on the resources available at each site. Any delay due to equipment/ instrument malfunction after feed-in could result in excessive flaring and cause significant loss of production. Methane circulation followed by equipment/instruments testing strategy also provides the plant personnel with the best chance to have hands-on experience of the design features and identify any design deficiency, distributed control system (DCS)/ emergency shutdown systems (ESD) configuration problems etc.

Availability of back-up steam generation system (auxiliary boilers) of adequate capacity for SHP/HP level steam generation is very important to achieve reduced flaring during start-ups and minimize flaring during major plant upset. Methane circulation from chilling train back to the furnaces, however, has an additional benefit of generating SHP steam from the furnaces. Depending on the number of furnaces in HSSB plus methane circulation, the required amount of SHP or HP steam import can be significantly reduced. Both licensor and plant owner need to work together to define the basis of steam import and number of furnaces in HSSB/methane circulation.

Table 1 shows the typical means of proving various sub-systems prior to furnace fresh feed-in.

Table 1

MAJOR SYSTEMS OF THE PLANT	METHOD OF PROVING
Furnace	During refractory dry out and steam drum safety valve hot setting at HSSB conditions
Quench water	Fill with water/ condensate/ boiler feed water (BFW) and establish quench water circulation, at close to normal rates
Quench oil	Fill with flux oil and establish circulation. Start heating using low pressure steam
Dilution steam	Process water stripper and dilution steam generation can be started using make-up BFW to either quench water tower or dilution steam drum, once the furnace is ready for refractory dry out. Routing of HSSB furnace effluent to quench water tower, will also allow start-up of dilution steam generation system
Cracked gas compressor	Typically, this compressor is used for air blowing and dry out. Most of the sub systems such as caustic wash tower, make-up caustic, spent caustic treatment etc. could be filled with water and gasoline-type hydrocarbon to establish circulation
HP deethanizer	Partly tested during nitrogen or methane run of the CGC and chill

	down. By adding some % of ethane during methane circulation, it will be possible to establish some traffic across the tower and in the downstream sections
C2 hydrogenation	The reactor could be pressurized and circulation established with nitrogen or methane and is free of impurities prior to cracked gas feed-in. This would help to reduce the chance of high bed temperatures/ potential run away conditions.
Demethanizer feed chill down train / demethanizer	Circulation can be established, if CGC is run on nitrogen or methane and chill down started once the propylene and ethylene refrigeration systems are stabilized
C2 splitter	C2 Splitter is started once the C3 refrigeration system is stabilized. This system is normally inventoried with ethylene and put on total reflux operation, which establishes the reflux and re-boiling loops.
Propylene compressor	This compressor is stand-alone system and hence can be started up once the other utilities systems such as steam, cooling water, etc. are ready.
Ethylene compressor	This can be started up once propylene compressor is stabilized
Fresh feed system	This is the last system to be commissioned once the plant is ready for feed-in. It is possible to establish nitrogen or methane circulation through the vapor feed (ethane, propane) systems, along with the demethanizer and downstream systems
C3 splitter	Inventory with propylene and establish total reflux operation. It is important to properly clean the system before start up since the system is very large and is of carbon steel, with potential for lot of rust particles being present in the system
Utilities (steam, air, water, nitrogen etc.)	All utility systems would be commissioned prior to starting up process systems. It is a critical prerequisite.

Licensor and owner can also work together to study and implement the required modifications for minimum flaring in the existing plants. Each plant is unique and requires dedicated study for minimum flaring implementation.

4. Sequence for Ethane Crackers Minimum Flaring Start-Up

Ethane feed crackers are relatively easy to start using the closed loop circulation concepts prior to fresh feed-in, compared to liquid crackers. It is technically possible to sustain CGC operation without any fresh feed to furnaces, by recycling streams from downstream chilling train to the quench water tower, to achieve close to design composition, provided steam is available for the turbines.

Annexure 1 shows a typical front end deethanizer ethane cracker scheme showing the proposed start-up methane or nitrogen recirculation loops back to quench water tower. This circulation will help to commission the entire downstream sections and establish close to normal operating temperatures before starting fresh feed to the furnaces. These start up loops can be implemented in older plants to speed up commissioning and start-up. However, a thorough design review will be required. Depending on the need for additional steam, furnaces could be commissioned with recycle methane or nitrogen stream. Coil outlet temperature should be monitored to avoid potential coke build-up due to ethane cracking (if present in methane stream).

The sequence of starting up the various sections could be changed, depending on available resources and flow scheme. In addition, methane or nitrogen can be used for establishing circulation. Ethane can also be added to help with condensing and establishing reflux in HP deethanizer and demethanizer. The choice between prechilling using methane or nitrogen depends on the operator decision, based on risk assessment of potential leak during restart/ commissioning and availability of methane/ nitrogen. Flange leak is potentially possible when liquid starts accumulating across the equipment flange, due to differential expansion, which can't be observed during nitrogen prechilling method.

The choice of using methane or nitrogen is explained below.

Option 1: Occasionally, operators prefer to start-up the propylene refrigeration (C3R) and ethylene refrigeration (C2R) compressors prior to starting up the CGC on methane. This is based on the consideration that the CGC would have been run on air already for air blowing the downstream sections and hence tested. Nitrogen purging would have been completed with a slightly positive nitrogen pressure maintained in the system. Once both refrigeration compressors are on-line, methane can be charged and CGC can be started up on methane before furnace feed-in to carry out the purging of the nitrogen to the flare and pre-chilling of the separation sections by closed loop recycling of the methane back to the quench system/furnaces. To save start up time, it is also possible to start CGC in parallel to the C3R/C2R compressors.

In addition, ethane can be introduced in the system in a sufficient quantity to start condensing in deethanizer and demethanizer towers. Purging the nitrogen to flare should be controlled to avoid extinguishing the flare system flame. Nitrogen can also be partially purged out to fuel gas system at a controlled rate so that nitrogen content is limited to the design calorific value of the fuel gas, which would eliminate/reduce release of hydrocarbons to flare.

Recirculation of the CGC final stage discharge back to the quench system ensures that the compressor discharge is free of CO_2 and H_2S by establishing the caustic tower operation. If the same is recycled back through the feed system and furnaces, it would help generate additional steam. Once feed is introduced in the furnaces, cracked furnace effluent could be routed directly downstream without any interruption in a seamless fashion, thus avoiding the flaring associated with stabilization of the CGC and downstream separation section. Even the C2 Hydrogenation reactor could be taken on-line with inert methane being circulated and thus would help to achieve on-spec earlier. This however needs to be discussed and agreed with selected catalyst vendor. Initial methane circulation helps to control the bed temperatures better.

There should, however, be enough availability of SHP import steam to support simultaneous start up requirements of CGC and C3R/C2R compressors. Both C3R and C2R compressors will require approximately 80% of the normal power demand, whereas the CGC can be started up on methane circulation at 40% of the normal power demand. The reason that the CGC requires only 40% of the power is due to the lower discharge pressure required during initial methane circulation run. Later, once additional SHP steam is available thorough methane run from the furnaces, the CGC speed can be increased to achieve the full discharge pressure for methane condensing and running machine at approximately 80% of the normal power.

Option 2: In case of availability issues of the SHP steam import, modify Option 1 by starting up the CGC first on methane before starting the C3R/C2R compressors. Depending on import SHP/HP steam availability, C3R compressor can also be started in parallel to the CGC since it is a small compressor and steam demand will be lower. Before starting the CGC, nitrogen purging would have been completed with a slightly positive nitrogen pressure maintained in the system. Pressure purging of the nitrogen can be carried out to the flare during this time. Purging the nitrogen to flare should be controlled to avoid extinguishing the flare system flame. Nitrogen can also be partially purged to the fuel gas system at a controlled rate so that nitrogen content is limited to the design calorific value of the fuel gas, which would eliminate/ reduce release of hydrocarbons to flare. Closed loop recycling of the methane from final stage discharge back to the quench system/furnaces is started. Chilling of HP depropanizer section can also be started once propylene refrigeration is available.

Recirculation of the CGC final stage discharge back to the quench system shall ensure that the compressor discharge is free of CO_2 and H_2S by establishing the caustic tower operation. If the same is recycled back through the feed system and furnaces, it would help to generate additional steam. The C3R/C2R compressors can be started once enough SHP steam is available from the furnaces on methane circulation. In addition, ethane can be introduced in the system in a sufficient quantity to start condensing in deethanizer and demethanizer towers. After stabilizing C3R/C2R compressors, chilling of demethanizer section can be started to allow recycling of separation section streams to the CGC.

Once feed is introduced in the furnaces, cracked furnace effluent could be routed directly to the downstream without any interruption in a seamless fashion, thus avoiding the flaring associated with stabilization of the CGC and downstream separation section. Even the C2 hydrogenation reactor could be taken on-line with inert methane being circulated and this would help to achieve on-spec ethylene product earlier. However, this needs to be discussed and agreed with selected catalyst vendor. Initial methane circulation helps to better control the bed temperatures.

Option 3: Further improvement of Option 2 could be done by routing HP deethanizer reflux (C2 rich) stream to the CGC through the methane recycle loop (tail gas expander outlet liquid recycle stream) to allow further chilling below C3R compressor levels prior to starting C2R compressor.

During initial start-up, ethylene refrigeration system is filled up with nitrogen. Purging of nitrogen from the system is normally done by introducing ethylene vapors into the system which gradually displace nitrogen to the flare or fuel gas system. However, this results in loss of valuable product and adds to the HRVOC to the flare system. To avoid flaring the HRVOC, an alternate approach is to produce a methane-free and contaminant-free ethane stream by demethanizer re-boiling and use it for inventorying the C2R compressor system.

Later, the C2 splitter system can be inventoried with ethylene product. The ethane inventoried earlier shall come down to the C2 Splitter bottom once the reflux is established and then could be recycled to feed/furnaces, to establish the recycle loop. This method can also be useful if ethylene is not available for start-up of C2R compressor. This option of using HP deethanizer reflux drum liquid would depend on the C3R compressor design basis to meet start-up scenario load.

In addition, availability of intermediate storage for off-spec ethylene/C2s will substantially reduce the flaring and can be used effectively to recover product every time there is an upset during normal operation. Also, if the flaring of off-spec C2 Hydrogenation effluent is done from the C2 Splitter inlet, it would allow the recovery of hydrogen and methane streams, thus reducing the net import of external fuel gas which could be a limitation during start-up.

Option 4: Another variation is to start the CGC on nitrogen and carry out the prechilling of the downstream systems up to the demethanizer. This helps in pressure testing at higher pressures without hydrocarbon and checking for any leaks during chill down.

However, this option will require a large quantity of nitrogen which can be very expensive in some plants. Nitrogen venting to fuel gas and flare will also be an issue and will take extended amount of time before fresh feed in introduced to the furnace. The CGC operation thus will be interrupted, resulting in the chilling train warming up depending on the time required to restore forward flow.

4.1. Liquid feed crackers grassroot start-up (naphtha, atmospheric gas oil etc. feed crackers)

Liquid feed crackers are more difficult to start compared to gas crackers, as there are more systems to be proven and the cracked gas effluent molecular weight is much different from methane rich natural gas. The CGC must be checked for the operating conditions when operated on methane rich natural gas. The concepts mentioned above for gas crackers are also valid for liquid crackers.

The difference is the additional recycle requirements from downstream systems. Some systems such as the depropanizer and debutanizer are typically started once the furnace feed is started, as these are relatively smaller systems and can be stabilized with minimum flaring. If the feed is not off-spec for lighter components, there should not be any major venting to flare except for purging residual nitrogen from the condenser along with some hydrocarbons.

5. Turnaround Shutdown

During turnarounds, the system must be de-inventoried prior to handing over the system for maintenance. During the process of shutting down the plant. there would be a substantial amount of flaring which cannot be fully eliminated.

However, flaring could be substantially reduced by taking the following actions:

a) Gradually stop the furnace fresh feed-in, while keeping the CGC running on recycles from different sections of the plant and the minimum flow recycle, provided the steam balance allows this mode of operation.

b) Reduce the liquid inventory of LPG streams in the entire plant to the minimum required for sustained operation by recovering them as products/by-products. This is done over one or two days as the plant load is reduced.

c) Minimize the release of HRVOC by replacing them with saturated components such as methane and ethane and sending HRVOC compounds to the respective storage tanks or neighboring plants.

d) Maximizing the recovery of the light hydrocarbons through the fuel gas system, if the furnace burner designs allows, or alternately send the fuel gas to the HP steam boilers.

e) Integrate with adjacent ethylene unit, if available, to recover the inventory through the nearby CGC system.

f) Add provisions to the design for maximizing the recovery of hydrocarbons during de-inventorying by providing off-spec storage. For example, C3R refrigeration system de-inventorying would require separate pump. If this is not provided then the inventory must be flared, releasing large amount of HRVOC components.

g) Utilization of residual clean hydrocarbons or natural gas or ethane to dilute the CGC system to purge benzene, after all furnace feed is stopped, since it is required to meet safety requirements for releasing the system for maintenance. This would help to reduce the number of days required for preparation.

h) Designing the system for higher settle out pressure, so that boil-off due to ambient heat leak does not cause over pressurization and flaring. For example, the propylene system can be designed to match the saturation pressure at maximum ambient temperature since this will not increase the cost of the equipment to a large extent.

Some of the systems handling C2s or lighters, will require flaring to maintain the system pressure due to ambient heat leak (weathering/boil-off). Provision of a vent control valve could reduce the chance of large release of hydrocarbons when the relief valve/safety valve lifts. If the inventory in the C2 Splitter/ethylene refrigeration system could be routed to off-spec storage, this will reduce flaring substantially.

6. Plant Upset / Emergency

During major plant upset/ emergency there will be substantial flaring, depending on the type of upset. One way of reducing the flaring is to cut back on the fresh furnace feed and utilize more of the internal recycles to meet the turndown requirements of the separation section.

Typical major emergencies applicable for ethylene plants are detailed below.

1) **Loss of CGC:** This leads to immediate flaring from the quench water tower overhead and cannot be stopped. However, the amount can be reduced by switching some of the furnaces to HSSB mode automatically and therefore taking out the hydrocarbon feed from a few furnaces. It is important that plant develops steam balances for all major plant emergencies and a procedure for steam load shedding to minimize impact and reduce flaring. In addition, there will be some amount of venting of boil-off gases from the demethanizer feed chill down, demethanizer, deethanizer and C2 Splitter/C2 refrigeration systems, which cannot be prevented.

2) **Loss of refrigeration compressors**: This would require stoppage of forward flow to the Demethanizer chill down train and hence the net furnace effluent must be flared either from the CGC suction or discharge. As in the case of loss of CGC, the fresh feed to furnaces must be reduced. Venting of boil-off gases from the demethanizer feed chill down, demethanizer, deethanizer and C2 Splitter/ C2 refrigeration systems, cannot be prevented.

3) **Loss of cooling water:** This would mean that most of the cooling capacity will be lost. The extent of the impact and flaring would depend on the availability of other cooling systems using air fin fan coolers. In the case of designs with substantial amount of air fin fan cooler utilization, the impact will be much less. The flaring would be basically from the quench water tower overhead due to the loss of the CGC. In addition, there will be flaring from the C3R compressor discharge.

4) **C2 Hydrogenation reactor trip:** In case of tripping of the C2 Hydrogenation system, the system would get isolated and depressurized rapidly, thus releasing the entire system inventory in a very short period. Since this is a safety requirement, it cannot be eliminated. There will be additional flaring during start up since the C2 Splitter feed will be off-spec and hence, it will be required to reduce fresh furnace feed.

If the reduction of furnace feed is automated as part of the shutdown systems, then operator intervention and delays could be avoided and flaring minimized. Also, if the ethylene product with higher acetylene can be routed to non-polymer plants which can handle higher impurity levels, it would save flaring during off-spec operation. Additional high integrity protection systems (HIPS) can be implemented in new or existing plants to reduce the flaring loads. For example, the C3R compressor can be tripped on two out of three voting on low pressure of cooling water header.

7. Conclusion

Reduced flaring concepts can be applied in varying degrees on new projects and existing plants depending on the flow sheet flexibility, availabilities of utilities, flexibility to integrate with adjacent units, etc. It is lower cost and much simpler to build maximum flexibility during design stage. However minimum flaring techniques can be implemented in older plants as well. A detailed review of the existing facility is recommended to implement such improvements. Future generations will benefit as sustainability targets can be achieved and resources conserved.

With the global focus on reducing environmental impact, it is in the best interest of every licensor and operator to consider implementing minimum flaring approaches from the design stage, or alternatively modify existing plants to incorporate these features. Since plant owners need to fulfill the legal requirements by regulatory authorities and satisfy the interest of neighboring communities, consideration for such features needs to be discussed early in design for new facilities, even at the expense of minor additional investment.

Annexure 1

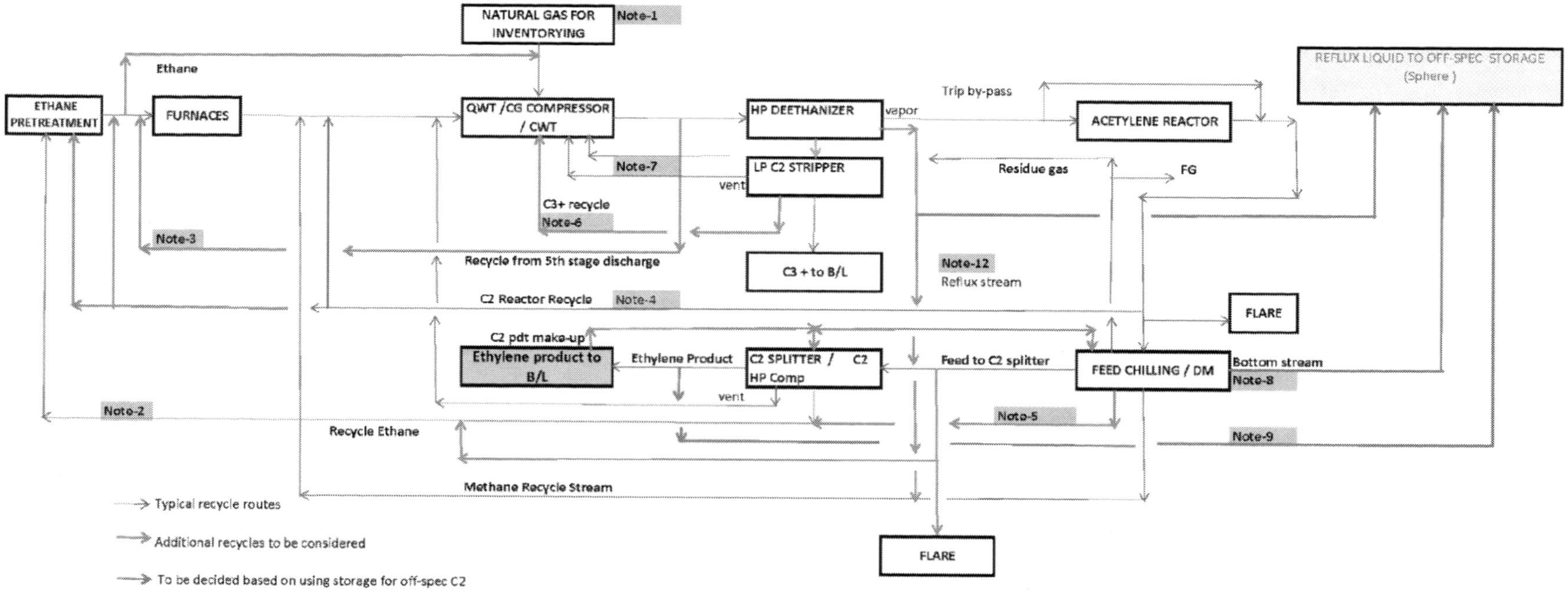

Notes

1) Natural gas inventorying into CG compressor 1st stage. Additionl to 4th stages to increase system inventory prior to starting compressor.
2) Ethane recycle to CG compressor / feed treatment to inventory C2 stream for recycling and prechilling.
3) Recycling from 5th stage discharge to have complete forward flow from upstream of CG driers and clean up the system
4) Recycling from HP Deethanizer overhead (By-pass C2 reactor) to CG compressor / furnace feed
5) Recycling from Demethanizer bottom / C2 splitter feed to CG compressor / furnace feed
6) C3+ stream recycle to CG compressor
7) Use of existing Residue gas/ methane recycle / LP C2 stripper overhead recycle to CG compressor
8) Liquid C2 stream from Demethanizer bottom to off-spec C2 storage sphere, if available
9) Off-spec C2 product to Off-spec storage sphere / salt dome
10) Main advantage is to establish hydrocarbon flows to remove any foreign particles / objects from the system and allow clean-up of core exchanger filters prior to feed-in and flare only natural gas / Ethane.
11) All utilities shall be on-line and available. C3R compressor would be started either before or later starting up of CG compressor.
12) Routing of HP Deethanizer reflux to CGC through Methane recycle route

510219 Olefin Plant Turnaround Optimization

Abstract:

Abstract

Maintenance of olefin plant equipment causes a significant contribution to the total cost of plant ownership; however, the money invested in maintenance activity also delivers value by ensuring the plant remains available for operation in accordance with business needs. Some plant maintenance is possible with the plant in normal operation, but the plant has to be shut down periodically in order to allow maintenance of equipment items that are essential to plant operation and not spared. These periodic shutdowns are commonly known as plant Turnarounds (T/As), and these T/As are of major significance in the determination of olefin cracker performance.

The HSB Solomon Associates LLC (Solomon) *Worldwide Olefin Plant Performance Analysis* (*Olefin Study*) and Solomon *International Study of Plant Reliability and Maintenance Effectiveness* (*RAM Study*) provide information that can be used to help companies optimize T/A performance. Using information from Solomonâ€™s extensive *Olefin Study* and *RAM Study* performance databases, this paper summarizes:

- Industry trends in T/A cost, interval, and duration.
- Strategies for minimizing annualized T/A production loss.
- The relationship between effective T/A scoping and execution and plant reliability between T/As.

Insights into the characteristics and practices that lead to improved T/A performance will also be covered.

AIChE Paper Number: 78d

Impact of Cracked Gas Drier Adsorbent Pressure Drop Increase on Plant Profitability

Recai Artan
Lead Process Engineer
Sabic Geleen, the Netherlands

Wim Reel
Senior Marketing Manager
Honeywell UOP, Antwerp, Belgium

Bruno A. De Jonckheere
Adsorption Technology Knowledge Manager
Honeywell UOP, Antwerp, Belgium

Prepared for Presentation at the 2018 Spring National Meeting

Orlando, Florida, April, 24, 2018

Impact of Cracked Gas Drier Adsorbent Pressure Drop Increase on Plant Profitability

Recai Artan
Lead Process Engineer
Sabic Geleen, the Netherlands

Wim Reel
Senior Marketing Manager
Honeywell UOP, Antwerp, Belgium

Bruno A. De Jonckheere
Adsorption Technology Knowledge Manager
Honeywell UOP, Antwerp, Belgium

The impact of pressure losses on the profitability of a steam cracker is important and can be quantified. This is illustrated with the changes SABIC Geleen has implemented on Olefins IV during its most recent turnaround. One of the modifications has been to add a third drier to the existing two-bed cracked gas dehydration system.

In all steam cracker configurations, cracked gas dehydration plays a critical role to enable reliable operation of the downstream cryogenic distillation train. The dehydration unit is located upstream "cold end" of the plant. As cracking occurs at near atmospheric pressure and separation is done at elevated pressure, the furnace effluent stream must be compressed. The cracked gas dehydration unit is at an intermediate compression station or downstream of the compression train, and removes moisture to less than one ppmv to avoid hydrate / ice formation in the downstream cold fractionation train. Pressure drop over these, often large, dehydrator vessels must be kept as low as possible, because pressure drop comes at a cost.

This paper summarizes what different mechanisms can contribute to pressure drop build-up and how to estimate the cost associated with these. It also provides insights into how to detect these mechanisms early on and proposes actions to minimize the pressure drop build-up.

Introduction

Refining or natural gas products such as naphtha, NGL, LPG and ethane are being used as a feedstock to the steam cracker in order to produce the primary petrochemical building blocks: ethylene, propylene, C4 olefins and some aromatic byproducts.

Paraffinic feedstock is heated in the furnaces to elevated temperatures where olefinic molecules are thermodynamically more stable than paraffinic molecules. To maximize olefin yield, steam is introduced as a diluent to keep the molecules separated. For the same reason the furnace is operated at near atmospheric pressure.

The effluent of the furnace thus contains a mixture of olefins, paraffins, hydrogen, aromatics and heavy cracked gas components, which needs to be separated into mainly single component streams. This is achieved in the various fractionation trains (cryogenic and hot). The cryogenic fractionation requires deep cooling (down to -160°C, -260°F). The shift in temperature from the furnaces to atmospheric temperatures needs to happen very quickly to minimize back conversion of the alkenes into alkanes, and this happens in the primary fractionator. The cryogenic section needs to operate at higher pressure (typically >30 bar g, 435 psi g) in order to enable condensation and separation of the light hydrocarbon components, hydrogen and methane. The cold for condensation is introduced by various refrigerant systems and by cold recovery in the various product streams. The energy to produce the cold is introduced in the compression step.

In short, low pressure gas from the furnaces is compressed, expanded and separated into its individual component streams.

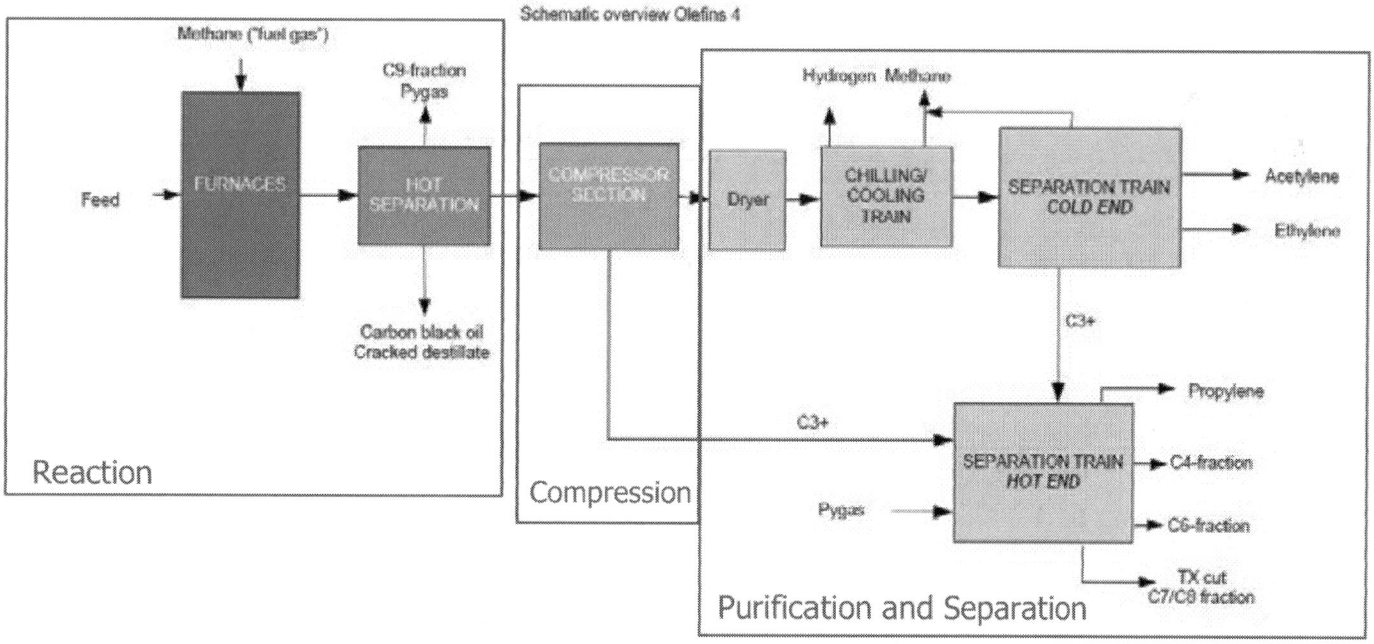

Figure 1: Block flow diagram of the Sabic Geleen naphtha cracker IV highlighting Reaction, Compression, Purification & Separation.

Value of pressure drop

Energy is introduced to the cracked gas through compression. Therefore, one must be careful about any pressure losses introduced downstream of the compressor as these impact on operating cost through energy efficiency. In every system, there is a threshold pressure that must be achieved at a certain point to have the plant running properly. In the case of Sabic Geleen cracker IV, the limiting factor is the pressure of the hydrogen produced and sent to the grid (grid operates at about 28 bar g / 406 psi g). The compression train is limited to the design pressure of the equipment downstream of the compressor. In practice, there is a maximum pressure loss that can be tolerated before the plant throughput must be reduced. A continuous capacity reduction is highly undesirable and when this occurs, a shutdown of the plant is typically scheduled to correct the problem.

The hydrogen pressure is of course only relevant in the case at hand here and the value equation of pressure drop is very specific to every plant, based on its geographical location, plant architecture, integration with other processing complexes, local energy mix, etc.

During normal operation, an 8 - 10 bar (116 - 145 psi) pressure loss between the compressor discharge and the hydrogen discharge valve is observed. The major contributors to this pressure loss are the thin plate aluminum heat exchangers in the "cold box", various KO - drums and S&T heat exchangers. In the methane wash column, remaining ethylene is recovered down to several ppmv level from the cracked gas stream in the cold end of the "cold box" (-125°C or -190°F range). This stream needs to be routed to the top of the demethanizer about 50 m (165 ft) above ground level. The cryogenic temperature is unfavorable for the use of a pump and therefore the designs foresee to transport the liquid stream using sufficient pressure drop between the methane wash and the demethanizer column.

As explained above, there is a limited margin on rise of 4 bar (60 psi) pressure drop on this system. If the pressure drop in the driers or in the cold box would increase, this would have to be compensated by a higher discharge pressure of the last stage of the compressor. The cost of compression has been calculated to be approximately 3 – 4 MW per bar. This is a relatively high energy cost and is in the range of 10% of the total energy consumption of the compressor.

Pressure losses can thus be compensated by additional increase of discharge pressure as long as the main gas compressor is not operated at the maximum discharge pressure of max load on its turbine. The main gas compressor in a steam cracker is usually steam driven with steam produced in the furnaces. The steam produced in the furnaces is typically 2/3 of the total steam requirements of a cracker. Note that other steam consumers are: reboilers of columns, turbines of various compressors and pumps. Some electrically driven pumps are also spared with a steam driven system. As 2/3 of the steam is produced from the furnaces, 1/3 must be purchased or produced by dedicated steam boilers on site. Depending what source of fuel is available, this can have considerable cost impact. Clearly this fuel cost differs significantly from region to region.

Lowering the pressure in the demethanizer column results in a decrease in ethylene recovery with losses of ethylene to the fuel system. Depending on the configuration of the demethanizer column (with / without pre-cut demethanizer column), a decrease in pressure of about 1 bar results in doubling the ethylene losses to the fuel system. This loss can increase up to more than 1 ton/h. At 3 USD/kg ethylene and a loss of 300 kg/h this corresponds to 900 USD /h or 792000 USD /year.

In summary, the value of pressure drop can be captured in an equation, but that value equation is highly plant specific. The value equation of pressure drop calculates both additional cost as well as yield losses as a consequence of increasing pressure drop. Elements of that value equation can be compression cost, energy costs (turbine compressor), but also increased ethylene losses to the fuel system, if the pressure drop losses are not fully compensated. Understanding this value equation is critical as it allows us to focus on this hidden cost called pressure drop.

Causes of pressure drop increase

After having given the overall plant perspective on pressure drop, we will now focus more on pressure drop associated with the cracked gas dehydration unit. It is however understood that there are many other sources of pressure drop which are not related to the adsorption system, such as fouling of heat exchangers, etc.

The first and foremost pressure drop issue associated with the dehydration unit is inadequate drying performance. Moisture getting into the cold section will inevitably cause hydrate/ice formation which limits or blocks flow across equipment like strainers, the thin plate heat exchanges in the cold box or heat exchangers in general. It is important to keep the pressure drop over the cold box as low as possible as pressure drop increase eventually impacts on overall plant capacity as discussed earlier. Hydrates are formed when low levels of moisture are slipping from the cracked gas or cracked liquids dehydrators. It should be noted that low level moisture slippage is not the same as the typical moisture breakthrough which occurs when the adsorbent capacity is exhausted (see Figure 2).

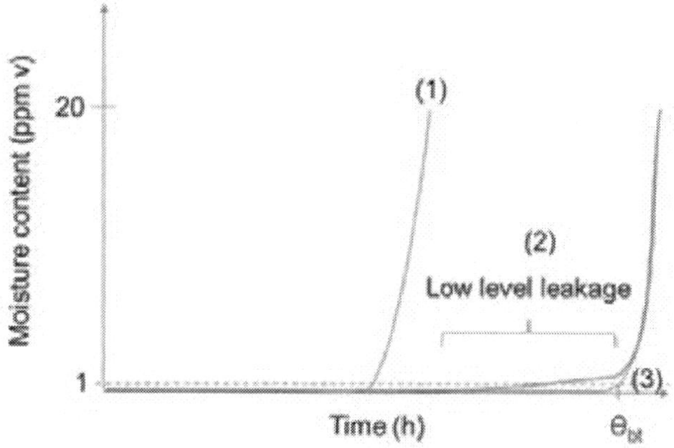

Figure 2: Moisture breakthrough profile: (1) early breakthrough caused by adsorbent deactivation due to high uncontrolled adsorbent reactivity, (2) moisture slippage, (3) normal dehydration performance.

Dehydrator early moisture breakthrough can be caused by many factors. The top 3 ones are:

- Adsorbent reactivity issue causing adsorption capacity reduction. Once the beds cannot be regenerated properly during the adsorption step time, the adsorbent is at its end of life and needs to be changed out. Minimizing reactivity is therefore critical (Kanazirev and Smith).
- Secondly, liquid carry-over can cause moisture to effectively bypass the bed as the liquids act as a carrier and basically bypass the adsorbent.
- Thirdly, just to give the top 3, bed configuration disturbances due to, for example a bed support failure.

Pressure drop build-up across fixed bed adsorbent dehydrators is always experienced to some degree. However, the extent to which the pressure drop increases over the life of the adsorbent is highly dependent on how the adsorption unit has been operated. Pressure drop increase over the life of the adsorbent can evolve in different ways. Figure 3 and Figure 4 represent some typical patterns of how pressure drop has evolved over time.

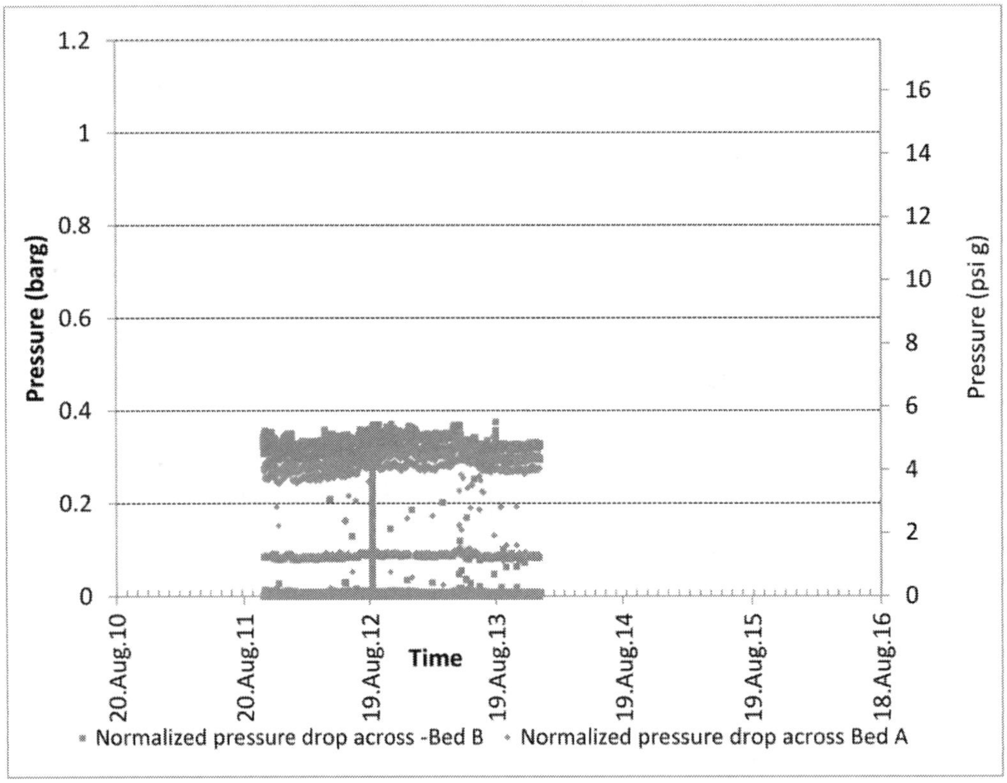

Figure 3: Stable pressure drop evolution in a properly designed and operated cracked gas drier. 2-bed dehydration unit.

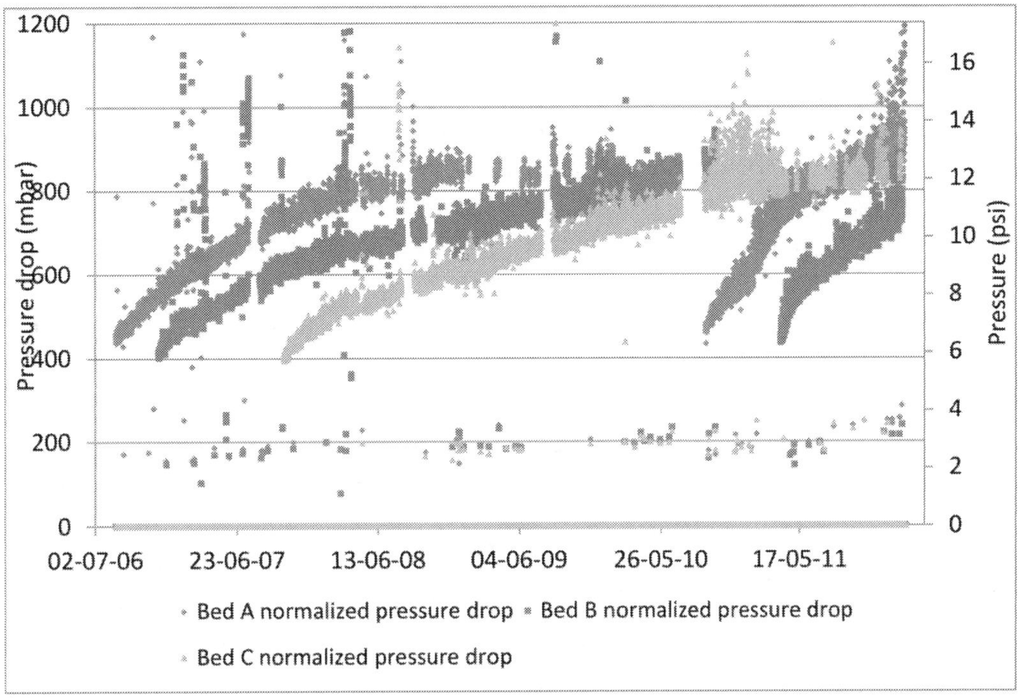

Figure 4: Pressure drop evolution over the cracked gas driers (3-bed system) in a non-optimized designed unit. Note the shape, slope and width of the normalized pressure drop curve.

What are the reasons for pressure drop build-up during the adsorbent service life?

Different elements contribute to this pressure drop build-up, but they can be classified in two major reasons: chemical reasons and mechanical reasons.

The chemical reasons for pressure drop build-up are related to the presence of liquid water which originates from liquid carry-over (adsorption phase) or results from condensation of desorbed water (regeneration phase), which is commonly referred to as refluxing. Carry-over liquids are typically a mix of hydrocarbon liquids and water but can also contain traces of caustic. Free water tends to be trapped in the macropores of the adsorbent as typically the micropores are already saturated. (De Jonckheere; Hatami)

Pores in material are divided in three groups: micropores, mesopores and macropores.

Micropores are pores that are very small and have about the same size as our most common simple molecules. Micropores are defined by the IUPAC as pores not exceeding about 2 nm (20 Å) size.

Mesopores are pores of intermediate size. What is special about mesopores is that capillary condensation can occur in there. Mesopores are found in desiccant grade aluminas and range between 2 nm and 50 nm or 20 Å and 500 Å.

Pores exceeding 0.05 µm or 50 nm (500 Å) are called macropores. Macropores are not able to induce capillary condensation, but they can retain sponged liquids. At UOP, sponged liquids are defined for zeolite based adsorbents as the liquids one cannot drain and that are not adsorbed.

A dehydration grade adsorbent for cracked gas drying is essentially a zeolite – binder matrix whereby the zeolite crystals are combined with a binder to form the actual adsorbent particles. The binder is a combination of clays with specific additives based on actual application. Binder selection requires extensive testing as it directly impacts on reactivity, mechanical strength, adsorption characteristics etc. of the final adsorbent. The void space which is created between the zeolite crystals and the binder is referred to as the macropores. Macropores will not get filled if contacted with a water-saturated vapor stream but will only get filled when contacted with an over-saturated stream containing liquids.

These liquids that are held in the macropores and in part interparticle void space, are called the sponged liquids. These liquids are not drainable and represent between 0.15 to 0.25 volume fraction of the bed dependent on the adsorbent. Liquid carry-over is difficult to identify as it cannot be directly measured; it can sometimes, but not always, be identified through a saw-tooth profile (see Figure 4). The pressure drop increases over individual adsorption cycle as void space gets filled with liquids which cause increased flow resistance. During regeneration, these sponged liquids are removed through vaporization, and the subsequent adsorption step starts again with a lower pressure drop compared to the end of the previous adsorption step. It is during this initial vaporization that moisture can break the oxygen bridges between the zeolite and the binder structure. This causes the particle (bead or pellet) to segment with loss of particle integrity, and repeated over many cycles this causes pressure drop to build-up. A similar mechanism occurs during regeneration. At the start of the regeneration, moisture is gradually desorbed from the bottom section of the bed and flows with the gas towards the top of the bed (case of downflow adsorption, upflow regeneration). The upper sections of the bed are still at or close to the lower adsorption temperature, so heat transfer will cool the gas and heat up the solid adsorbent until they are in equilibrium; then the gas continues its path to the exit of the bed, again equilibrating out with the solid phase. As the regeneration gas contains desorbed water from the lower (hot) section of the bed, and as it cools down in the upper section, eventually the moisture partial pressure will be above the vapor pressure and condensation will occur. The adsorbent bed is acting as a coalescer and the droplets are sponged into the adsorbent. This phenomenon is happening in every TSA (Thermal Swing Adsorption) system, but the degree at which it will happen is highly dependent on the ability of the gas to carry away the moisture. In this context, high pressure regenerations are much more demanding than low pressure systems. The temperature of the bed at the start of the regeneration is the second important parameter. Lower temperatures will cause larger amounts of condensation and sponging to happen. The delta between the top temperature and the bottom temperature is therefore important and so is the absolute temperature of the top section. A top section at 2°C (36°F) will condense substantially more water than a top section at 16°C (601°F) for example.

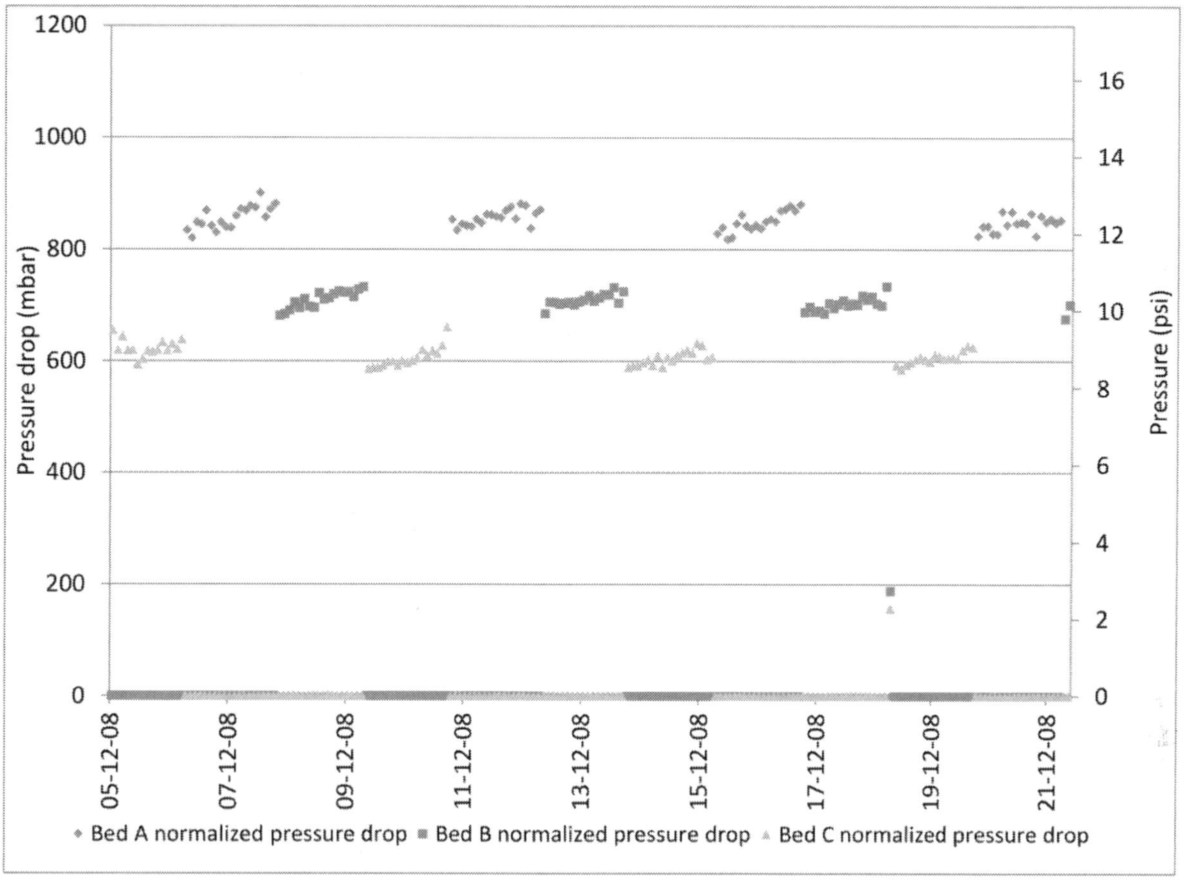

Figure 5: Saw-tooth profile which is indicative for liquid carry-over. Three-bed unit operated in one bed in adsorption, one bed in regeneration and one bed in rotating stand-by mode.

The degree of acidity or basicity of the liquid carry-over is also important in terms of potential impact on pressure drop behavior of the adsorbent bed. Acid components will cause cation exchange resulting in salt deposition and a protonated zeolite. Protonated zeolites are not all stable at elevated temperature, for instance it is well documented that A-type zeolites (LTA topology, Si/Al = 1) will collapse and lose micropore volume if the temperature is elevated above 120°C (250°F). The net result is a loss of adsorption capacity. Amine carry-over will cause a similar issue as acid carry-over but here ammonia is formed from the breakdown of organic amines. The ammonia can be exchanged against sodium and when the ammonium exchanged zeolite is heated up during the heating step, it will decompose causing a loss in zeolite crystal structure as well as loss in mechanical strength.

Besides the chemical reasons for pressure drop build-up, there are 3 main mechanical reasons for pressure drop build-up: bed support failures, fluidization and freeze-thaw effects.

Bed support failures have different root causes like design flaws, inadequate material selection and installation errors. So, we will only discuss here a more common example of bed support failure with a mechanical beam/screens support system. With this system, typically a bed support grating is installed on top of a support ring and a set of I-beams. The support grating leaves an annular gap close to the vessel wall which is filled with ceramic rope packing (see Figure 6) in order to allow thermal expansion of the support grating during the initial phase of the regeneration, when there is a large temperature delta between vessel wall and support grid. On top of the grating a rigid 4 US mesh is installed with a diameter equal to the vessel ID. On top of that mesh a final 20 US mesh screen is installed. The 20 US mesh screen should be cut at the exact same diameter as the vessel. However, it is often observed to have an overlap with the vessel wall of 20 cm (8 inches), i.e. its diameter is larger than the vessel ID. This is not a good practice since the overlap with the wall on paper is perfect, but in reality, it leaves openings of 10 – 20 mm (1/2" – 1") at a location where the ceramic ball support material or the adsorbent is at best 3,2 mm (1/8") or more commonly 1,6 mm (1/16"). The buckling 20 US mesh screen basically creates a by-pass around the support balls, which are the layers of defense to avoid having the adsorbent leaking in the bottom head. Ceramic balls are laid in successive layers to provide a gradual particle size change. These layers are bypassed by the buckling mesh and thus the adsorbent can find its way to the gap with rope packing. These adsorbent particles can get crushed between wall rope packing and support grating resulting in fines which eventually fill up the annular gap. If that happens there is a risk for the support gratings to deform as well. Especially sensitive to this are the stainless-steel support gratings with a high outer band height (see Figure 6). Once material starts flowing from the bed into the bottom head it will cause all the adsorbent above to settle and eventually the top bed hold-down ceramic balls can find their way into the bed causing a channel to be created. This channel can cause moisture slippage during adsorption and localized fluidization during the up-flow regeneration. This causes dust formation and when particles are reduced in size then pressure drop increases.

Figure 6: Left, V-wire profile bed support with rope packing in the annular gap of the cracked gas drier; Right, V-wire profile bed support which failed due to ingress of adsorbent in the annular gap.

Fluidization can occur without bed support failure: pressure surges in up-flow directions, deviations from design conditions (higher flow and / or lower pressure of upflow steps), and applying a top layer of ceramic hold-down balls without intermediate mesh can all cause fluidization or at least fines creation. Once fines are created they tend to create more fines which is a self-feeding effect or snowball effect.

Freeze-thaw break-up is a third cause of mechanical pressure drop build-up. Freeze-thaw is a phenomenon that is experienced when moisture is present in the macropores of the adsorbent and the adsorbent is exposed to temperatures below the freezing point of water. When water freezes, it will reduce in density and hence expand; this expansion force is strong as it is capable of breaking rocks (and roads), but also adsorbent particles. So, we require two phenomena to be present at the same time; temperatures below the freezing point of water and water in the macropores. Temperature below the freezing point can occur because of depressurization. Temperatures can also drop due to vaporization of liquid hydrocarbons either after changing the pressure (= depressurization) or the partial pressure by purging the vessel with a stream containing little or no of the liquid phase molecules. For cracked gas driers, the regeneration typically starts with a depressurization step to the quench tower, followed by initial purging or immediate heating with Demethanizer Overhead gas, a very lean gas. So, the depressurisation Joule Thompson effect and hydrocarbon flashing in case of carry-over can cause temperatures low enough to cause ice formation.

The second element that is required to break up the adsorbent with ice formation in the macropores (freeze-thaw break-up) is water in the macropores.

A zeolite based adsorbent used in cracked gas dehydration mainly contains macropores and micropores. Moisture adsorbed in micropores will not freeze like water in the macropores or in a lake. At temperatures below 200K (-73 °C, -100°F) the mobility of the water molecules in the micropores is reduced and at 77K (-196°C, -321°F) the water molecules form an ice-like structure. Explanation of these phenomena can be found in Breck.

Moisture in the macropores can originate from liquid carry-over from the upstream separator. During depressurization, freezing will occur and fines will be formed. This phenomenon can be more pronounced by specific unit lay-out; long poorly insulated or non-insulated lines between the separator and the dehydration unit itself.

How can we detect signs of a future increase in pressure drop over the cracked gas driers?

Here is a check list:

- Liquid carry-over: possible indication saw-tooth profile of the pressure drop over the vessels. No pressure drop increase over the adsorption cycle doesn't mean there is no liquid carry-over occurrence, but there is no indication of it, at least not in the measured pressure drop.
- Subzero temperature after depressurization: subzero temperature must be avoided in the bed.
- Relative pressure drop evolution

What can be done to avoid pressure drop build-up?

As there are many causes to pressure drop build-up, it is important to identify the root cause. General good practices include:

- Select a strong adsorbent to start with a low reactivity. UOP has especially developed its Ethylene Plant Grade (EPG) adsorbents (EPG product line) to meet the challenging operating conditions in most cracked gas driers. Members of this product line are 3A-EPG, EPG-2, EPG-N and HgSIV™ 3.
- Adsorbent particle size or combination thereof to maximize adsorption capacity whilst minimizing pressure drop. (Sicard and Richman) In certain cases the TRISIV® shape can provide an optimal solution (Ausikaitis).
- Ensure adequate design and operation of the separation upstream the dehydration vessel as this is critical to allow a moderate pressure drop increase over the adsorbent life.
- Critically review plant configuration and thermal insulation needs.
- Ensure proper design and installation of the bed support system. Work with your adsorbent vendor to capture experience in this field.
- Change the process architecture: Sabic Geleen investigated all the contributing factors very thoroughly and made two major modifications to the dehydration system and surrounding piping: Sabic added a 3rd vessel to the 2-bed system. So, the unit was a 1oo2 and became a 2oo3 dehydration system. Sabic also replaced the inlet and outlet piping which represents 100 m of piping and was able to reduce the pressure drop over the dehydration section by 29 psi (2 bar) with these two measures.
- EOR Pressure drop is proportional to the initial pressure drop (SOR). Units starting with a high initial pressure drop, pressure drop will increase relatively toward the initial pressure drop as would a "low initial pressure drop unit". However, the absolute increase will be much larger for high initial pressure drop units. (see below table)

	High initial pressure drop	Low initial pressure drop
Fresh DP	5 psi / 0.35 bar	2 psi / 0.14 bar
Cycled DP	7 psi / 0.5 bar	2.8 psi / 0.2 bar
Pressure drop increase	40% increase	40% increase

Conclusion

Several factors are contributing to pressure drop build-up over a steam cracker plant and specifically over the cracked gas driers. Given the multitude of phenomena associated with pressure drop increase over the cracked gas driers, close cooperation between plant engineers and adsorbent vendor specialists is recommended to identify root cause and corrective actions. Starting off with a strong adsorbent and with a low pressure drop adsorbent configuration is critical to minimize the risk of unplanned shutdown due to unacceptable pressure drop increase.

References

1. Vladislav Kanazirev and David L. Smith (2004) "Adsorbent Reactivity: An Important Factor in Contaminant Removal", AIChE Ethylene Producer's Conference, Paper Number 55d
2. Bruno A. De Jonckheere (2010), "Charge Gas Dryer Tutorial" AIChE Ethylene Producer's Conference
3. Allan Hatami (2017), "Charge Gas Dryer Tutorial" AIChE Ethylene Producer's Conference
4. Donald W. Breck (1973). Zeolite Molecular Sieves: Structure, Chemistry Wiley, New York and London
5. Jean-Paul Sicard and Philip Richman (1988) "Product and Process Improvements in Cracked Gas Drying"
6. Joseph P. Ausikaitis (1983) "TRISIV® Adsorbents - The Optimization of Momentum and Mass Transport Via Adsorbent Particle Shape Modification", International Conference on Fundamentals of Adsorption.

Disclaimer

AIChE PAPER NUMBER: 78e

CRACKED GAS DRYER MOLECULAR SIEVE STEAM DEACTIVATION

Megan Finstad
Production Engineer
Dow Chemical Canada

Prepared for Presentation at the 2018 Spring Meeting
Orlando, FL, April 22-26, 2018

CRACKED GAS DRYER MOLECULAR SIEVE STEAM DEACTIVATION

Megan Finstad
Production Engineer
Dow Chemical Canada

Abstract:

At the Dow Fort Saskatchewan hydrocarbons plant, the cracked gas dryers are critical to remove moisture from the cracked gas stream in order to meet product specification and prevent hydrate formation in the back end of the plant. The molecular sieve in the FS Cracked Gas Dryers needs replacement approximately every 3 years in service. Bed life is limited by water adsorption capacity and run length. In order to safely remove the mole sieve from the vessel, it must first be deactivated to prevent undesired pyrophoric reaction. Water flooding is considered best practice to deactivate dryer mole sieves, however, the Fort Saskatchewan dryers were not designed to be water flooded. There have been two previous process safety near misses at the Fort Saskatchewan plant due to unsuccessful mole sieve removal causing smouldering. In 2017, a new procedure was trialed using steam to deactivate the mole sieve in order to deactivate the bed without exceeding the weight capacity of the vessel. This method was documented in the mole sieve vendor literature but very little practical experience or data existed on this method. Using this new procedure, the plant has demonstrated the ability to safely deactivate the mole sieve and prevent additional process safety near misses. Additionally, the process has resulted in decreased bed removal time, reducing downtime and cost associated with the outage. This paper and presentation will focus on the process safety incidents encountered during mole sieve change-outs, the success of the new procedure, and improvement plans which will be further trialed during the next mole sieve change out.

Introduction

The cracked gas dryers are a critical part of the ethylene plant. The purpose of the dryer system is to reduce the water content of the cracked gas stream to prevent hydrate formation in downstream equipment and meet product specifications. This is accomplished by adsorption of water vapor from the cracked gas in a packed molecular sieve (mole sieve) bed. Mole sieves are crystalline, hydrated metal aluminosilicates with the ability to adsorb molecules in the many pores present in their crystalline structure. The mole sieves have a pore size large enough to allow water molecules to enter but not large enough to absorb the larger hydrocarbon material present in the cracked gas. Over time the mole sieve beds must be changed out to maintain water specification. Typically this is a large maintenance operation and requires a method to safely remove the spent mole sieve from the dryer vessel. The Dow ethylene plant in Fort Saskatchewan, AB, Canada has faced many challenges in previous mole sieve changeouts to safely and efficiently remove the mole sieve from the primary cracked gas dryers.

After two challenging bed removals in 2014 and 2015, it was decided that an alternate method must be developed and trialed for the 2017 mole sieve removal to prevent future risk and extended outages.

Dryer Mole Sieve Life

Over time in operation, the adsorption of hydrocarbons reduce the sieve capacity and causes coking during regeneration to the point when the entire mole sieve bed must be replaced. At the Fort Saskatchewan site, dryer bed lifetimes are typically 3-4 years. The pressure differential across the bed is monitored but in recent years this has not been the limitation for dryer life. The current trigger for bed replacement is when the run time of the online dryer is less than or equal to the regeneration time. Long runtimes and known water carryover issues have led to difficult mole sieve changeouts due to the requirement to break apart agglomerated mole sieve in the vessel.

Bed Replacement Strategy

It is necessary to establish safe confined space entry (CSE) conditions for the catalyst removal contractor. The contractor must enter the vessel to aid in the mole sieve removal. Since heat is generated when the mole sieve absorbs water, if the mole sieve is exposed to atmospheric air and absorbs moisture from the air, it may lead to localized hot spots. Because of this potential for heat up, it is necessary to remove all hydrocarbons from the vessel to prevent a fire. If the mole sieve can be completely deactivated prior to entry, this greatly aids in the ease of establishing safe entry conditions and reducing the safety risk associated with entering the vessel.

Industry best practice and vendor recommendation (1) is to use a water flood of the dryer vessel to deactivate the mole sieve in place. Unfortunately, the dryer vessels at the Fort Saskatchewan plant were not designed to be liquid full. Historically the plant was able to successfully remove the mole sieve by allowing CSE under a nitrogen purge (typically deemed an "inert entry"). Based on unfortunate industrial learnings, the Dow Chemical Company globally no longer allows immediately dangerous to life and health entry conditions due to the risk to the personnel involved. Without flooding or inert entry as an option, in 2014 and 2015, the team attempted to remove the mole sieve from the dryers using other methods.

Plant History & Challenges with Bed Replacements

Dryer Changeout 2014

To safely remove the mole sieve, the team attempted to establish safe entry conditions by purging the bed since deactivation was not possible. The first step was to purge vessel with nitrogen until a safe atmosphere was established for line and equipment opening. The vessel was then blinded and prepared for contractor entry. At this time the entire bed was fresh air purged to establish an appropriate atmosphere for contractor entry. While the contractor was preparing for entry, their detectors measured carbon monoxide and hydrogen, leading to concerns over the safety of the atmosphere for entry. These concerns resulted in the decision to purge more air through the system in an attempt to remedy the issue. Hindsight would later reveal this error in judgement. Additionally, this caused a long delay between the time the vessel was open to air until the contractor was ready to enter. To avoid significant delays to the overall outage schedule, a decision was made to proceed to vacuum the top layer of inert supports and begin side dumping mole sieve through the side man-way without contractor entry. Several hours later the contractor reported seeing the mole sieve "glowing red" through the man-way and smoke exiting the vessel. Fortunately, operations was able to respond swiftly to close the man-way and then began to pad/depad the vessel with nitrogen. It was later determined that a nitrogen purge rather than pad/de-pad is more effective at removing the heat due to the heat capacity of nitrogen. In order to prevent further smouldering, several days were spent dumping and vacuuming mole sieve with a continuous nitrogen purge. Safe CSE conditions were only able to be established once ~95% of spent mole sieve had been removed. At this point, the contractor entered and removed the small amount of mole sieve still present.

In the end it took 22 twelve hour shifts to remove the mole sieve but even more importantly, the event was considered unsuccessful due to the risk posed to personnel by the smouldering mole sieve and labelled a process safety near miss.

Investigation determined the root cause of the events to be the extended air purge through the bed; oxygen reacted with the pyrophoric material resulting in oxidation and incomplete combustion. Additionally, exposing the dryer to moist air for an extended period of time caused the adsorption of water on the still partially active mole sieve, causing the heat up to occur.

Dryer Changeout 2015

In 2015, another dryer bed was due to be changed. With still no identified method of deactivation based on the limitations discussed, the plant was forced to proceed with a similar method to 2014. Based on the findings from 2014, the plant modified the procedure used to only purge the vessel vapour space above the mole sieve with fresh air rather than the entire bed. Additionally, the bed was nitrogen purged until the contractor was ready to begin the CSE. This was to avoid the delay that happened in 2014 and exposed the mole sieve bed to moist air for an extended period of time. The entire job was carefully planned to avoid any logistic delays which would lead to increased exposure time of the bed. The method of air purging the vapour space allowed the contractor to enter the bed to begin vacuuming the top layer. After a few days of vacuuming and dumping, the contractor detected rising temperature on their CSE monitoring equipment and evacuated the vessel. Operations arrived and observed smoke and heat and responded as they had in 2014 to safely close the vessel and purge. The rest of the job continued similar to 2014 which again led to significant delays in removing the mole sieve.

Overall, the removal of the mole sieve took 15 twelve hour shifts, an improvement from 2014 but because of the risk involved with mole sieve smouldering it was clear that a better method needed to be developed for future mole sieve changeouts.

Deactivation by Steaming Trial

For the mole sieve changeout in 2017, the plant elected to use a steaming method to deactivate the mole sieve prior to dumping the bed. Vendor documents provided a skeleton procedure (1) but had little practical experience. Additionally, there was minimal experience within the Dow global ethylene business for steaming mole sieve dryers. The procedure used a similar methodology to water soaking but was advantageous for the Fort Saskatchewan plant by reducing the weight load put on the vessel. Civil engineering analysis confirmed that although the dryer vessels were not designed to be liquid full, they were rated for the theoretical mass of the steam required to deactivate the mole sieve, within a conservative safety factor. The procedure for steaming addressed the risk of condensate accumulation in the bed leading to weight concerns.

The basic procedure for deactivation was to first complete a final regeneration and keep the bed hot. Next an inert gas (nitrogen) was used to purge the vessel and then steam was connected to the vessel. At Fort Saskatchewan, careful care was taken to insure the bed stayed warm and the pressure was controlled to avoid any condensate build up which could exceed the vessel weight rating. The plant steamed the bed in multiple directions for approximately 60 hours to insure distribution throughout the bed. A main disadvantage of steaming, as noted by the mole sieve vendor, UOP, is that poor distribution can lead to incomplete deactivation (1). For this reason, the plant elected to reverse flow several times during the steaming process to try and improve distribution. Unfortunately, as the source of steam was hoses connected to utility stations, the flow rate of steam available was limited.

Cooling the Bed

The bed was then cooled with an inert gas. This was another disadvantage noted by UOP; the need for a large volume of inert cooling gas (1) and indeed the plant faced significant challenges trying to cool the bed. It is necessary for the bed to be lowered to a maximum of 35°C to allow safe contractor entry. Originally, the bed was cooled with plant nitrogen. Initially, the bed did not seem to be excessively warm, however this was based on false information due to the location of temperature indication around the bed. Unfortunately, the vessel does not have any bed temperature indication and all indication is on the inlet or outlet piping. When nitrogen flow was started, the temperature began to rise indicating the bulk volume of the mole sieve was still hot. After more than 24 hours of cooling with plant nitrogen, the bed failed to show any significant signs of cooling. A high pressure nitrogen truck was brought in to try to increase the cooling rate. Again, as the flow of nitrogen was increased, the temperature of the outlet probes increased as the bulk heat from the bed was being removed and flowing past the probe location. The process was excessively slow and after a day with still no reduction in temperature it was decided that a change must be made.

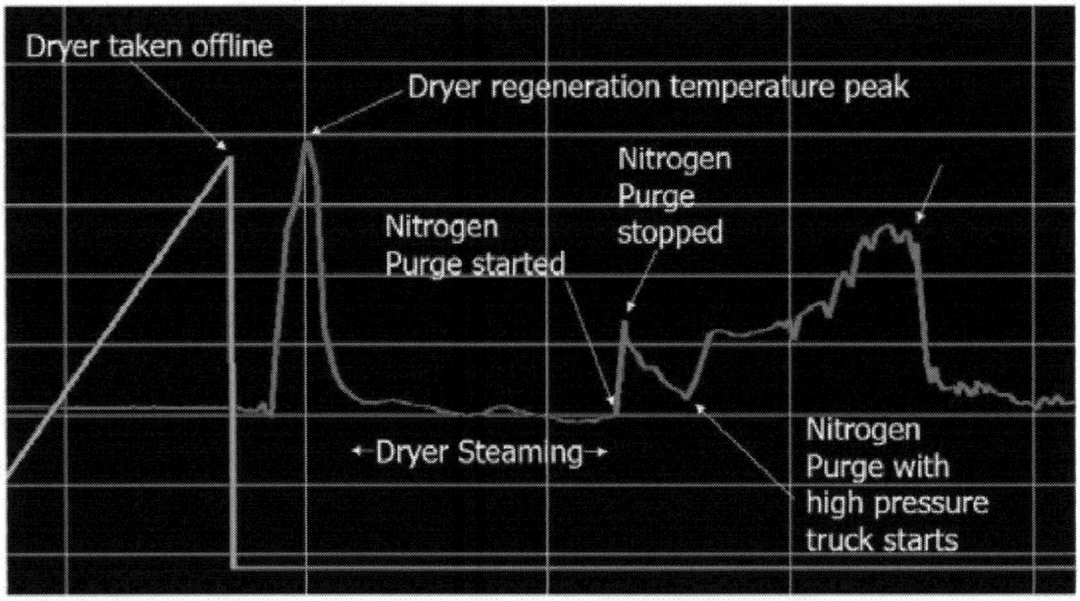

Dryer Bed Runtime and Temperature Profile

Regeneration Gas as a Means to Cool the Bed

In hindsight the use of nitrogen as a cooling medium was an obviously poor choice. The dryer vessels are large and heavily insulated so a high heat transfer medium should be used to cool the bed. The plant had originally avoided a suggested option of using regeneration gas (the hydrogen rich off-gas from the turbo expanders) due to the risk of reintroducing hydrocarbons onto the already purged and deactivated mole sieve. Additionally, using the plant regeneration gas meant careful coordination to not interrupt the regeneration system and put the online dryers at risk. However, the advantage of hydrogen as a cooling medium over nitrogen was obvious due to the high heat capacity and high flow available. After careful consultation, it was decided to utilize the plant off-gas to cool the bed. With hydrogen as a cooling medium, the plant was able to cool the bed in less than 10 hours. Upon completion of cooling, the bed was purged multiple times with nitrogen to insure any hydrocarbons reintroduced on the bed were removed.

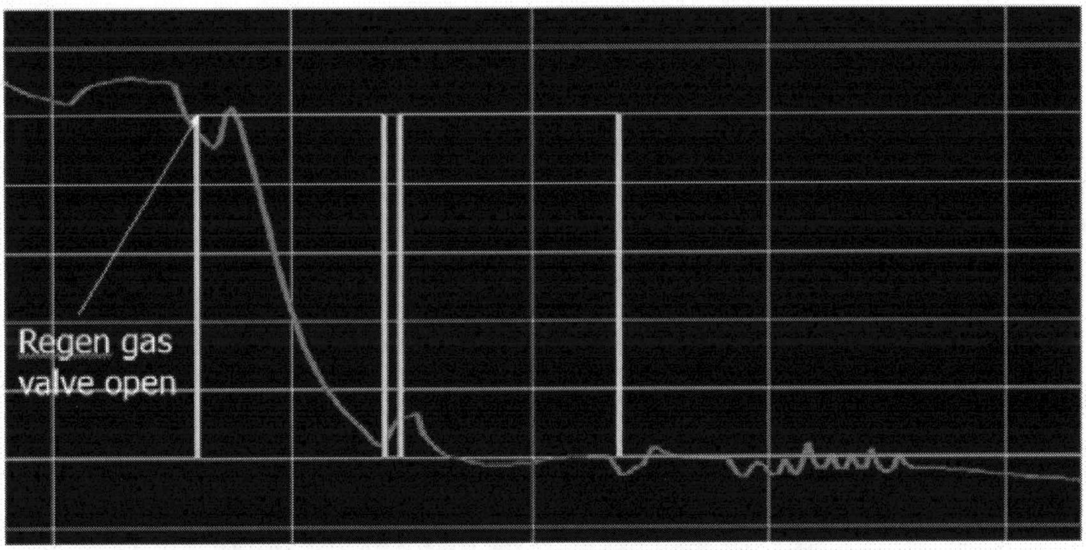

Temperature Profile and Regen Gas Valve

Mole Sieve Removal after Steaming

Once the bed was cooled and purged, it was possible to establish safe entry conditions for the catalyst removal contractor almost immediately. The contractor crew was able to remove the mole sieve in approximately 8 twelve hour shifts; a significant improvement over the previous mole sieve changes in 2014 and 2015. However, most important was that no incidents of smouldering occurred. No personnel were put at risk and no process safety near miss occurred. The mole sieve changeout was considered a success.

Analytical Comparison of Spent Mole Sieve after Steaming

It is typical practice for the Fort Saskatchewan plant to have spent mole sieve analyzed by UOP to determine remaining levels of activity. A %H2O capacity test is utilized to determine the residual H2O capacity of the spent mole sieve compared to a fresh mole sieve. This can be used to quantity the amount of deactivation that occurred in the steaming process compared to the previous methods. When the mole sieve from 2017 was compared to the mole sieve removed in 2015, it showed an average 14% decrease in %H20 capacity. This indicates that at least some level of deactivation occurred as a result of steaming. This degree of deactivation seemed to be enough to safely remove the bed however in order to achieve a more complete deactivation, future uses of the steaming procedure will make use of a temporary steam header, rather than hose connections, to increase the volume of steam available.

Summary & Path Forward

The initial trial of steam as a means to deactivate spent mole sieve was extremely successful for the Dow plant in Fort Saskatchewan, AB, Canada. Although improvements to the procedure were noted for future use, the procedure was successful to allow the plant to safely and efficiently remove the mole sieve. Improvements noted include the need to use plant regeneration gas as a cooling medium and to use additional steam sources to increase volume to more fully deactivate the mole sieve. The steam procedure will be refined with these modifications for additional use in upcoming mole sieve removals.

References

1. Handling of zeolite molecular sieve adsorbents in process units, UOP: A Honeywell Company, April 2009, www.uop.com

AIChE Paper Number: 78f

ETHYLENE FURNACE PRE-COMMISSIONING AND COMMISSIONING – NEW FURNCACE OR AFTER MAJOR FURNACE REVAMP

Kari Enqvist
CEO/Chief Consultant
ENQ-Consultancy Oy

Prepared for Presentation at the 2018 Spring National Meeting
Orlando, FL., April 22-26, 2018

ETHYLENE FURNACE PRE-COMMISSIONING AND COMMISSIONING – NEW FURNCACE OR AFTER MAJOR FURNACE REVAMP

Kari Enqvist
CEO, Chief Consultant
ENQ-Consulting Oy

Abstract:

This paper will give the overview of decisions and considerations for the furnace project schedule planning, design phase, inspections during equipment manufacturing, construction and hydrotesting to guarantee good pre-commissioning and commissioning, focus being in the pre-commissioning and commissioning itself, to finally achieve successful start-up and feed-in. All these activities have significant importance for attaining long, trouble free operating life for critical furnace components. The learnings and experience, summarized in this paper, are collected from working in a numerous grass root and revamp projects in different operating companies and with different EPC contractors.

When starting the project to build grass root ethylene unit, new furnace in the existing plant or just doing a major revamp and technology upgrade to old furnace in older ethylene unit, to get smooth first start-up and feed-in as well as to achieve the design performance, if not with the first run but maybe the second run, the planning for the pre-commissioning and commissioning must to be kept in mind throughout all the phases of the project.

In early planning and design, the first considerations are where to locate the tie-ins, when to install those, how to isolate the construction site from the rest of the plant or how to minimize the disturbance to the rest of the plant ongoing construction, how to allow for safe and complete line cleaning, steam blowing, chemical cleaning etc. and to do all that while in the middle of the operating plant or in the middle of the project site where the construction work is still going on?

In project scheduling, the priorities for sequence of construction has to be set to allow for early start of pre-commissioning, but same time to allow for remaining construction to continue.

During the design, serious consideration has to be given about the prerequisites for hydro testing, chemical cleaning and piping blowing/flushing methods in order to minimize e.g. the eventual need for golden welds or to minimize the waiting time for already cleaned systems before the commissioning and burner ignition is possible.

Inspections in vendor shops are important e.g. for before fabrication checking critical piping/tubing process side condition for scale and coating. To speed up pre-commissioning, the mechanical cleaning might be needed before pipe bending and spools production. All large bore process gas piping is better to check carefully before spools production and to do the mechanical rust and scale removal, before lifting pipe sections on pipe racks and structures.

Automation loop checking, functional checking etc. before burner ignition testing is time consuming, which is often underestimated. BMS and controls test activities might cause limitations also for finalization of other still bending construction work.

Furnace start-up to Hot Steam Standby soonest possible after Calcium Silica refractory installation and chemical cleaning completion is mandatory to prevent any Alkaline Hydrolysis damage in the refractory and to get the initial Magnetite layer to protect the freshly chemically cleaned, very reactive BFW and Steam systems metal surfaces.

If all well planned and diligently done, operator is expected so see smooth start-up and feed-in with less or not at all furnace cold shut downs for venture, FG & Pilot gas filter and nozzle plugging, less coking due to corrosion products coming with HC Feed or Process Steam etc.

Ethylene Furnace Pre-commissioning and Comissioning – New Furnace or After Major Furnace Revamp

1 Introduction - Summary

This paper will give the overview of decisions and considerations for the furnace project schedule planning, design phase, inspections during equipment manufacturing and construction including pressure testing followed by thorough pre-commissioning and commissioning, focus being in the pre-commissioning and commissioning itself, to finally achieve successful start-up and feed-in. All these activities have significant importance for meeting the expected run length, if not with the first run, but maybe second or third, as well as attaining long, trouble free operating life for critical furnace components. The learnings and experience, summarized in this paper, are collected from working in a numerous grass root and revamp projects in different operating companies and working with different EPC contractors.

The considerations presented in the paper cover grass root project, new furnace in existing ethylene unit for capacity expansion and the extensive revamp of the furnace that is located in between the operating furnaces inside the existing ethylene unit, which is in production. The new furnace or the furnace refurbishment inside the operating plant, has less "freedom" as the live hydrocarbon containing piping and equipment is in vicinity, which is causing, besides all inconvenience to construction and potential disturbance to production, is serious safety issue. The paper covers also a brief description of the pre-commissioning, commissioning and the start-up, just to give a feeling of acceptance criteria and the start-up sequence of the furnace, including lessons learned from the previous projects.

Fig. 1; Furnace Section with Revamped and New Furnaces

2 Project Planning – Study Phase

This paper will concentrate on those considerations, decisions and actions, which are important for smooth and efficient pre-commissioning and which are contributing to trouble free operation after the start/up.

In project planning and study phase such matters are:

- Agree about the procurement/purchasing specifications and requirements for piping material which covers piping ID surface quality requirement of corrosion and scaling resulting from long storage in high humidity, maybe also in saline atmosphere. ID surface coatings are not accepted.
- Allow time and funds for expediting and inspection of the material, to see what has been bought, before the delivery to site or to spool production sub-vendor and/or allocate time for corrosion and scale mechanical removal at spool production sub-vendor shop, before the spool production starts
- Where to locate the tie-in's valving and the design of the tie-in point, i.e.:
    - Locate as close to existing headers as possible to minimize the "dead ends", of which cleaning during pre-comm. is difficult and which might act as a trap for rust and particles, when doing line cleaning
    - Provide the tie-in's with construction side vents and drains
    - Provide the construction side of the valve with such spool piece, which can be replaced with blow-out connection e.g. for steam blowing
- Consider having break flange/flanged connection with spacer/blind for all process piping (process steam and hydrocarbon feed lines) connecting to the convection section to ease line blowing, without sending the dirt to convection modules, which have been delivered to site in clean condition
- Deciside which process, fuel gas etc. lines are to be cleaned in pre-comm. by steam blowing to make sure that those piping and piping components are designed to take the steam blowing temperature
- For BFW and VHP Steam piping, consider where to locate the Chemical Cleaning connections to get the efficient chemical cleaning circulations, for full coverage of all piping and to minimize the number of golden welds
- When considering Hydro Testing and Chemical cleaning, it has to be decided where to accept the golden welds. In this case, it is also important to figure out how to do the final ASME Stamp Hydro Testing for Furnace BFW and all VHP Steam piping and systems, including convection modules, Steam Drum, Quench Exchanger's and Down-comer and Riser piping. If the pressure testing requires pipe cutting, the pipe cutting locations, i.e. the golden weld locations should be chosen so that same cuttings can be used for both Chemical Cleaning and Hydro Testing

3 Schedule Planning

The most important thing in schedule planning is to achieve the shortest possible project schedule till the Feed-in. The schedule planning should also consider the order of priority of different piping, equipment and systems completion to allow sufficient time for the time-consuming pre-commissioning activities and to minimize the disturbance to construction, when the construction and the pre-commissioning are going on simultaneously.

Following activities are always problematic:

- Chemical cleaning of the BFW and VHP Steam systems, which, especially in case of Acid Pickling, stops pretty much all the construction work and other pre-commissioning activities around the furnace
- Timing of Burner Management System (BMS) loop and functional checking, which need to be completed before Burner ignition testing. However, the ignition testing, requires the FG system pressurizing with FG and therefore these functional checks are the last pre-commissioning/commissioning activities before firing burners for refractory dry-out, Magnetite Layering and VHP Steam lines STM Blowing
- In case the furnace refractory/insulation is Calcium Aluminate concrete materials, after the completion of refractory installation, the furnace should be ready for burner ignition and steam generation due to the urgency to get the refractory dry-out done soonest possible to avoid Alkaline Hydrolysis damage. Dry-out is important also for Ceramic Fiber insulation as the chemical bonding of some of those materials, takes place only during the dry-out temperatures
- If the furnace start-up is delayed or scheduled to take place several weeks/months after the refractory is installed, especially in case if the Calcium Aluminate refractory material is used, then the refractory pre-dry-out soonest possible after installation is mandatory
- Note that in case of Convection Section module fabrication, the refractory insulation is always installed before the intermittent tube plate and tube bundles assembly. In most cases the refractory is Calcium Aluminate based, which requires dry-out soonest possible after installation in-order-to prevent Alkaline Hydrolysis damage. Due to manufacturing sequence and the fact that the convection modules fabrication is completed weeks or months before the lifting of modules in furnace structure is possible. Even after all modules are in place, the furnace readiness for burner ignition is still months later. So, the convection module dry-out (pre-dry-out) is to be done in vendor shop, as soon as possible after refractory installation. It should be noted that the convection module refractory/insulation inspection for any damage is very difficult after the tubes are installed. This refractory/insulation is expected to last for the entire furnace life. Any repair without removing complete tube bundles is not possible
- After the BFW and VHP Steam System Chemical Cleaning is completed, including the Passivation, the reactive, clean metal surface is protected only for 1 to 2 weeks, therefore it is urgent to proceed to Refractory Dry-out to get to Magnetite Layering for producing the initial protective Magnetite on clean metal surfaces
- Both, the refractory dry-out soonest possible after refractory installation and getting the Furnace to Steam production soonest possible after completion of the chemical

cleaning are very important targets to create bases for the >15 years trouble free refractory, Quench System and steam generation equipment life.

4 Design Phase

The design specifications will be chosen when selecting the Furnace Technology Licensor. Normally licensor specifications shall be used, completed with client company own or preference specifications and/or amendments to include client specific requirements.

Tie-in locations and design:

- In furnace revamp cases, locate the tie-in's as close to the headers as possible
- All tie-in's should be provided with vent and drain connection on furnace side of the tie-in valve
- In case of the project to build a new ethylene unit, the tie-ins are not needed except in cases when e.g. one or two furnaces are to be started-up to steam production for steam blowing purposes, well before the start-up of rest of the furnaces. In such case the need of tie-ins and tie-in design needs to be considered case by case
- In case of both BFW and VHP Steam line tie-ins, the went and drain sizes should be such that these connections can be used also for the Chemical Cleaning purposes

Early involvement of permitting authorities for details and acceptance of BMS:

- In case of Olefins Cracking Furnaces, the BMS Standard requirements allow some interpretation and the spirit of the standard is often to allow the use of alternative solutions and ways of implementation "provided the same level of safety is reached". To reach the common understanding about "alternative solutions" with the permitting authority and finally getting approval from them, could require several meetings with the authority. As the furnace automation design, implementation and getting the functional testing completed and to get to the burner ignition testing, is on the critical path, it is very important to get the discussions with authority to be completed as early as possible

Hydrotesting/Pneumatic testing, ASME Stamp requirement:

- ASME spec breaks and ASME test pressure issues has to be clarified in early design phase:
    - Furnace system related ASME classifications are:
        - Furnace boiler system is ASME I, because of furnace coils
        - Quench exchangers are ASME III
        - Based on rules of ASME I, furnace piping is ASME B31.1 up to the spec break. Beyond the speck break piping could be B31.3. Spec break locations are to be decided in design phase
        - The client or permitting Authority is to make a decision about ASME Stamp requirements

- Quench Exchangers BFW side, Steam Drum and all Economizer, BFW Preheat and VHP Steam Superheat convection modules shall be Hydrotested in the vendor shop
  - After all jump-overs connecting convection modules, all BFW piping, Riser and Downcomer piping, all Saturated Steam piping and VHP Steam piping has been installed at field, the BFW and VHP Steam side of the equipment and piping are to be hydrotested for the proof of all welds done to complete the system
- Hydrocarbon/Process side pressure test issues require the decision about which way to go:
  - All HC Feed Vaporizer and Preheat convection modules shall be hydrotested in the vendor shop
  - After all Hydrocarbon feed piping, Process Steam piping, feed preheat convection modules connective jump-overs, Radiant Coils and Process Gas piping downstream the furnace itself are installed, the whole furnace process side and Hydrocarbon/Process Gas side has to be pressure tested for the proof of all field welds. This pressure test (from feed control valves to Transferline and Decoke valves) can be either Hydrotested or Pneumatic tested. The decision between Hydro and Pneumatic test causes some thinking because both of them have pros and cons:
    - Advantage of Pneumatic test is that then all furnace process side remains dry till the furnace start-up. However, in case of furnace revamp the huge safety area around the furnace, required by the test, is not possible due to operating plant around the construction side
    - Hydrotesting is more complicated to arrange due to big water volumes and need to design and built the process gas header supports downstream the furnace to take the weight of complete water filling. Another disadvantage is that the furnace Radiant Coils, and in some cases also High Temperature Feed Preheat convection modules and X-over headers form a pocket(s), which are very difficult, if not impossible to empty from water. If the water can not be removed, it will sit there till the furnace is ready for Refractory Dry-out
    - Based on above considerations, the Pneumatic pressure test can be used only for crass root plants and the Hydrotest is for the furnace revamps and for the new furnace to be built inside the operating plant

Chemical Cleaning – Connection Locations for Chemical Circulations:

- Alkaline Boiling do not require any additional connections for circulation. The treatment is based on using just furnace burners for heating purposes and the Process Steam flow for starting the thermosiphons. However, it must be made sure that the chemicals addition to steam drum is possible during the treatment as the chemical consumption might be higher than expected when doing chemical loading before the start of the treatment
- Acid Pickling method needs external/portable boiler for heating the chemical circulation, as well as external filtering system to remove any debris from chemical circulation during treatment. These are normally to be provided by the contractor doing the work. The need for additional connections or checking if the existing vent

and drain connections are sufficient, is to be checked with contractor as early phase of the project as possible, preferably already in design phase. There might be opportunity to use e.g. same on-purpose connections for acid pickling which are needed for hydrotesting of BFW/Steam systems

Fuel Gas and Pilot Gas Distribution Piping:

- Fuel Gas and Pilot Gas distribution piping to burners, downstream the pressure control valves, under the furnace, are always quite complicated from cleaning point of view due to numerous branches, connections and low points
- Each system (Pilot Gas, FG to Start-up Burners and FG to Main Burners) headers under the furnace, should have end flanges to allow efficient Steam/Air Blowing and debris removal. The normal vent and drain connections at the end of the header are not sufficient for that purpose

5 Procurement and Expediting

When purchasing equipment, construction materials, piping materials and piping components, the owner or, in case of lumpsum-turnkey project, the contractor tend to try to find the most economical options. Normally the Project Specifications define material qualities and manufacturing details and the Inspection and Test Plan gives the acceptance criteria for inspections during manufacturing. However, less attention is paid and less money spend on the following considerations.

Although the material composition spec requirements are met, the piping materials, especially for the Boiler Feed Water and VHP Steam system piping, including particularly Ricer and Down-comer piping material, might have been laying in storage area in high humidity, high ambient temperature, maybe also in saline atmosphere close to harbor for a long time. The inspection of these critical piping materials needs to be carried out, preferably before purchasing or latest in the sub-vendor shop where the pipe bending and spools pre-fabrication is taking place. In these inspections the status of inside/process side surfaces is to be evaluated for corrosion, scaling and coatings. If found, then e.g. mechanical cleaning by sand or shot blasting save a lot of time and effort in pre-commissioning.

Pre-cleaning, normally mechanical, can be done to spool and piping material before or after bending. However, it is better to be done before welding bent spool components together.

Note that not any type of pipe inside surface coating is allowed. If found, the coating must be removed before spool production and before installing any coated straight pipe section in furnace piping.

If coating is found, the best practice would be to give a sample to Chemical Cleaning contractor for testing, if it can be removed in Chemical Cleaning. If not, like e.g. Japanese Black, then it must be possible to be removed mechanically or such coated pipe material needs be rejected.

To have the reference for piping/tubing material ID surface minimum surface quality requirements for BFW and VHP Steam piping, it could be e.g. based on NACE surface quality criteria, e.g. as WJ4 of NACE No. 5/SSPC-SP 12, Table 1: Visual Surface Preparation Definitions, with exception that there may not be any coating, which can't be removed by chemical treatment. Verify and confirm that all BFW and Steam piping material meets this standard before assembly starts. This should be fine for the metal surfaces, which will still get Chemical Cleaning treatment.

6 Piping Spools Pre-Fabrication

In this category, both process piping and any BFW and VHP Steam system related piping spools and components are considered. The pipe bending and pre-fabrication, including hydrotesting for shop welds normally takes place in sub-vendor shop. Bending has to be done in special bending shop. The piping sections delivered to bending shop, have to be free from any excessive Steam/BFW side corrosion, scale and coating.

The BFW and VHP Steam piping spools and components are:

- BFW piping
- Riser and Down-comer piping
- Condensate blow-down headers and piping
- Jump-over's connecting/related to BFW Preheat (Economizer) and VHP Steam Superheat Modules, including all jump-over and header spools, which are prefabricated in vendor shop
- Saturated VHP STM piping from STM Drum to Superheater
- VHP Superheated STM Piping from Superheater to VHP Steam Header, including start-up vent connections

BFW piping from main header to Economizer Convection Module inlet is in some cases just having hydrojetting for final cleaning, but very often it is included also in the Chemical Cleaning.

VHP Steam system related piping up to VHP Steam Header, will see the Chemical Cleaning and Steam Blowing, but for both BFW and VHP Steam piping, to do the preliminary mechanical cleaning at vendor shop, will speed up the pre-commissioning and by removal of major scale and corrosion product, help Chemical Cleaning to give uniformly cleaned pipe surface quality. For prefabricated spools and piping sections, this is the last chance to inspect thoroughly the piping water/STM side condition of the whole piping length, as spool sections.

When these piping are erected in the furnace structure, internal inspection is only possible in very limited areas like in areas close to vent and drain connections or at the control valves and check valves, of which bonnet and internals can be removed.

The Process Gas piping spools, most of them quite large diameter piping, are prefabricated either at construction site or in the sub-vendor/piping erection contractor's off-site shop. In either case the removal of rust and scale is better to do mechanically with steel brush, HP hydrojetting or by sand/shot blasting, before spool production and lifting the spools to furnace structure or piperack. During Pre-commissioning, Process Gas piping cleaning is by Air Blowing and by Steam Blowing.

Process piping and spools ID surface quality should meet e.g. close to NACE WJ-1 before final piping installation.

For process/HC feed piping, upstream the convections section the Air Blowing is only good for removing the loose items, sand and construction debris. Steam Blowing, would be much more efficient, but it requires that the piping and pipe components are designed for minimum 160 C, to allow Low Pressure Steam blowing.

For furnace HC feed piping Steam Blowing is often not done for safety reasons and/or the Steam is not available.

For good rust and scale removal by Steam Blowing the use of Medium Pressure Steam, in combination with heating-cooling cycles is needed.

During pre-comm./comm., all the Feed Preheat Convection Modules, Radiant Coils and all Process Gas piping downstream the furnace will get Steam Blowing with Process Steam to Decoke system. However, that will be done only after the burner ignition testing is complete, parallel with the Refractory Dry-out.

7 Equipment Fabrication

After the equipment vendor has been selected, the first thing is to have the equipment manufacturing kick-off meeting. The equipment manufacturing ITP is to be approved in this meeting.

On top of all normal manufacturing ITP items, the cleanliness requirements of the equipment BFW, Steam or HC/Process Gas containing parts, when it is leaving the vendor shop, is to be defined very carefully. The inspections during manufacturing has to be timed and classified as Hold points for owner or client to make sure they are offered the opportunity to see critical parts before the further manufacturing steps make it impossible or difficult to review the later date. Each inspection should have the acceptance criteria listed in the ITP.

For any equipment item, which will be hydrotested, drained, dried to -10 C - -20 C due point, inertized with N_2 to <0.5 % mol. O_2 and "bottled" at 0.3 .... 0.4 barg pressure, in the vendor shop for temporary storage either in vendor shop or at construction site. These preservation conditions, maintained and regularly checked and replenished if needed, is good up to some months storage of the steam generation equipment and piping.

The corresponding conservation conditions and relevant acceptance criteria is to be agreed for:

- All Feed Preheat Convection Modules

- BFW Preheat and VHP Steam Superheat Convection Modules tubing
- Quench Exchangers
- Steam Drum
- Radiant Coils

Piping spools needed for connecting Convection Section Modules, piping spools and straight sections for BFW piping, Riser & Downcomer piping, Saturated and Superheated VHP Steam piping cannot be kept in inertized conservation conditions due to huge number of different components, but these piping items have be kept in covered, dry storage till the erection starts and each item ID condition needs to be checked before erection and welding starts on site.

7.1 BFW Preheat and VHP Superheat Convection Modules

BFW Preheat/Economizer and VHP Steam Superheat 1st and 2nd Convection tube banks shall be included in the Chemical Cleaning. However, although the Chemical Cleaning/Acid Pickling removes the corrosion products and scale, it would be good to have in ITP the hold-point for the client to inspect the tubing material ID surface status before the finning and manufacturing of the tube bundles starts.

Inspections and acceptance criteria:

- All tube material, all 90° and 180° bends, header and jump-over material inside surface should be at or close to status WJ4 of NACE No. 5/SSPC-SP 12, Table 1: Visual Surface Preparation Definitions conditions, with exception that there is no any coating, which can not be removed by Chemical cleaning
- No any coatings allowed. If some coating found and if it is not known what it is, then first give a sample to Chemical Cleaning Contractor to test, if it is removed by chemical cleaning/Acid Pickling. If the coating is not dissolved, then it has to be removed by mechanical cleaning, like e.g. high pressure hydrojetting or by sand/shot blasting, before the welding of the tube banks starts. Note that the BFW and Steam tube/piping/bends material procurement specification should not allow purchase of tube/piping material with inside surface coating

7.2 Feed Preheat and Process Steam Superheat Convection Modules

All Hydro Carbon Low Temperature Preheating bundles, Process Steam Superheating and Mixed Feed Preheating/high temperature preheating tube bundles welding and the module assembly as well as the modules tubing/piping cleaning and hydrotesting will be done in sub-vendor shop.

Not any of these Convection Modules is included in the Chemical Cleaning. During the upstream piping air or steam blowing and/or flushing, the feed piping shall be cleaned only up to the first flange i.e. break flange of the related Convection Module.

Before starting the welding and assembly of the tube bundles in Hydrocarbon and Process Steam Superheat tube bundles, it is very important to check the process side surface condition of these tubes, bends, headers etc., for scale and rust. If any scale or heavy rust is found, then it is better to do the mechanical cleaning either by sand blowing or by high pressure hydro jetting. Cleaned tubing and pipe components need to be dried to minimize the further rusting/corrosion before assembly/welding of the module tube bundle.

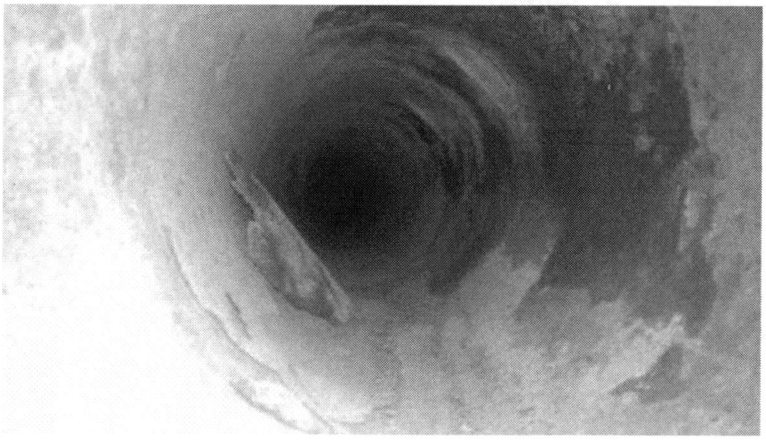

Fig. 2: The Inlet Header of Feed Preheat Convection Module before Sending to Site – the Module had been Cleaned, Hydrotested, Drained, Dried, Inertized and Capped, Waiting the delivery to Site for about Two Months on Sub-vendor's Lay-down Area

During Pre-commissioning, a good care has to be taken, to avoid moving any dirt from upstream piping to Convection Modules. For removing any dirt coming from piping to Convection HC Feed Pre-heat or Process Steam Superheat Modules next opportunity for cleaning is only after the Furnace is ready for firing. i.e. during Commissioning when furnace is being heated up and the process steam flow is established through the convection banks to provide cooling for convection modules, radiant tubes and quench exchangers is started to allow Flue Gas temperature increase to refractory dry-out temperatures.

7.3 The Refractory of All Convection Modules

Convection Module Refractory is to be installed on casing walls of all Modules before installation of intermittent tube supports and before the tube bundle welding starts. The interesting thing from Pre-commissioning/Commissioning point of view is that, the refractory material is in most cases Calcium Aluminate based, which means that the refractory dry-out needs to be done soonest possible after the refractory installation in-order-to prevent the Alkaline Hydrolysis damage. Normally, after the manufacturing is completed, these Convection Modules have to wait in vendor storage area or at the site several months before the lifting and assembly of Modules in Furnace structure is possible and after that is done, it is still a long time before the Refractory Dry-out is possible.

Pre-dry-out of Convection Modules in vendor shop is much better option for there the Dry-out can be better controlled, and each Module gets the same treatment. During actual

Refractory Dry-out on site, the Flue Gas temperature, especially in topmost Convection Modules will be very low for dry-out purposes anyway?

7.4 Riser and Down-comer Piping, Quench Exchangers and Steam Drum,

7.4.1 Riser and Down-comer Piping

Piping inside surface quality requirement before tube assembly and welding starts:

- All Riser and Down-comer bends and piping material inside surface should be at or close to status Heavy (H) of NACE No. 5/SSPC-SP 12, Table 2: Flash Rusted Surface Definitions conditions
- No any coatings allowed. If some coating is found and if it is not known what it is, then the sample must be given to Chemical Cleaning Contractor for the test, to see if it is removed by Chemical Cleaning/Acid Pickling. If the coating is not dissolved, then it is to be removed by mechanical cleaning in spool manufacturing vendor shop, like e.g. high pressure hydrojetting or by sand/shot blasting, before the welding of the piping spools starts. Note that the BFW and Steam tube/piping/bends material procurement specification should not allow purchase of tube/piping material with inside surface coating

7.4.2 Quench Exchangers

- Quench exchangers vendor normally takes a good care of cleaning the water/steam side of the tubes during all manufacturing phases. However, all tubing inside/waterside surfaces, including in double tube designs the inner tube outside surface, have to be in accordance with agreed ITP requirements and to be verified by inspection. Exchangers hydrotesting both water/steam side and process side is done in the vendor shop and this test is witnessed by client. The inspection of the water/steam side tubes condition must be done before the vendor welds the pressure test caps on riser and down-comer piping connections and any vent and drain connections. That is also the last moment to remove any possible manufacturing debris from any part of the exchanger
- After hydrotest, the draining of test water is important, but vendors lifting and turning capability of these heavy equipment is good, so there is no danger of leaving any test waters in any pockets of the equipment. After water removal, proper drying is necessary, followed by N2 inertizing and leaving the equipment in low N2 pressure, which is to be monitored regularly and replenished if needed till the exchanger is lifted to furnace steel structure and piping is aligned and ready for welding
- After the delivery from vendor shop the Quench exchangers are normally in very good shape from pre-commissioning point of view.

7.4.3 Steam Drum

- The Steam Drum is manufactured in vendor shop and goes through similar inspections and hydrotest as the Quench Exchangers. Steam Drums are easy to clean after construction and easy to inspect. Also Steam Drums are delivered to the site all hydrotest caps in place and in positive N2 blanketing, so no concerns till the piping welding start and the connections are opened for welding allowing the ambient air in causing rust formation?

8 Construction Erection

Erection of Radiant Chamber, steel structure, platforms, stair cases etc. structural items do not require Pre-comm. and Comm. and therefore those are excluded in the paper.

8.1 Quench Exchangers

On construction site, after exchangers are lifted to furnace steel structure and piping welding starts, the pressure test caps and other connection caps should not be removed before all other piping is completed, except the Riser/Down-comer welding to Quench Exchanger. The whole Water/Steam system could get quite significant rust layer before whole system is completed and all connections closed. This is experienced especially in case if the Water/Steam system welding and installation is done in autumn rainy season. Very often these plants are located not far from seaside of high humidity and saline atmosphere which is speeding up the rust formation.

8.2 Steam Drum

With Steam Drum it is also important to minimize the ambient air entering the Steam Drum and causing rust formation. The connection caps should not be opened before the Riser/Down-comer piping welding is completed, except the connections to Steam Drum, still pending welding. Only then open the connection caps and complete the Riser/Down-comer welding to Steam Drum.

8.3 Piping

Piping installation important thing is the prevention of sand and dirt being left in the piping (pipe spools, components and straight sections) when moving the material from storage to construction site and lifting to structure. When lifting pipe straight or bend sections to structure, every pipe component is to be checked that there is not anything inside that does not belong there.

8.4 Refractory Installation

Without going to any details about the radiant box and bridge wall insulation or refractory installation, it should be noted that if the refractory is based on castable, Calcium Aluminate concrete material, the refractory installation should be left to as late as the schedule allows. After refractory installation is completed, to avoid Alkaline Hydrolysis damage, the waiting for ignition of burners and readiness for Refractory Dry-out, should be as short as possible. If, for any reason the refractory installation has be done early, then the refractory Pre-dry-out is to be done.

Note that, in case of Ceramic Fiber insulation, besides the normal dry-out, the insulation material needs Dry-out temperatures for setting the chemical bods. As-long-as the Dry-out is not done, the Ceramic Fiber is very sensitive for water. In case material is made wet, e.g. during line flushing on top of the furnace, the Ceramic Fiber will not set any more even if the Dry-out is done. The wet Ceramic material gets damaged and it have to be replaced.

9 Pressure Testing

When the piping gets completed, the hydro pressure testing is to be done section by section or if convenient, several sections combined. Without going in into details of the pressure test procedure itself, I would like to highlight the importance of the removal of the test waters from the piping and the equipment:

- From the all tested piping and equipment included in the test, the test waters must be drained off carefully and the piping and the equipment needs to be dried by air blowing to minimize any further rusting and corrosion. At the time of pressure testing, the Furnace construction and piping assembly is still ongoing and therefore the inertization with Nitrogen is causing safety risk and therefore not allowed
- Most likely the Furnace Hydrocarbon piping, i.e. piping from feed control valve to convection Feed Preheat modules, Radiant Coils, Quench Exchanger and process Gas Header all the way to the first blind, will be hydrotested. In that case it is very important that the all Feed Preheat convection modules and the Radiant Coils are drained carefully. The water removal from the Radiant Coils is very tricky as there is no drain connections. How to do it, is to be figured out case by case
- If the water is left sitting there:
    - Depending on test water purity it might cause some corrosion damage
    - In furnace commissioning/start-up the water hammering will be inevitable and might cause mechanical damage
    - In northern countries, if the test water is left in radiant coils before or during the cold, below freezing point temperatures, the test waters freezing will cause mechanical damage of the Radiant coils

- In both cases, the water hammering and/or freezing, the damage might also be seen in lowest convection modules and the lower part of the Quench Exchangers

10 Choosing the Pre-commissioning Methods

Based on my experience, the following pre-commissioning methods for different piping and equipment has given the best results. However, in some cases the status of construction, site specific design details, piping/equipment material selection, utilities availability etc. might cause the need to choose some other method. Those cases will be briefly commented at relevant cases.

10.1 Fuel Gas Lines

Fuel Gas lines, including Pilot Gas supply lines, from tie-in up to the burner valve of each burner and pilot:

- Steam Blowing, provided that the piping design and piping components are designed for it. Note that the timing of the cleaning of Fuel Gas lines is often such that the Steam is not available or the Steam blowing of Fuel Gas line would be safety issue and/or would stop too many construction activities. So, contractors in many cases end up doing just Air Blowing
- Alternatives;
  1. Air Blowing or rupture disk pressurization/depressurization. Not very effective for removing scale and corrosion products, only good for loose dirt removal
  2. Water Flushing. Also not effective – worse than Air Blowing
- FG and Pilot Gas piping has high cleanliness requirement due to small drillings in burner/pilot nozzles. FG/Pilot gas piping can be very long. There are filters for both systems, located at furnace main block valves. Downstream the FG/Pilot gas distribution piping to burners, under the furnace is very complicated with many branches, low points and pockets it is difficult and time consuming to clean. There are normally not any filters protecting individual burners or burner tips

10.2 Hydrocarbon Feed Lines

Hydrocarbon feed lines from tie-in to the break flange at inlet to convection section:

- Steam blowing, provided that the piping design and piping components are designed for it. However, the availability of LP steam, the safety issues and the location of these piping on furnace structure and platforms, makes often steam blowing not possible. Alternatives;

1. Air blowing, not very effective, but other methods are complicated or just not possible to implement
2. Water flushing, not effective and the handling of water above radiant box not recommended due to danger of vetting the refractory

10.3 Process steam and all purge steam lines

Process steam lines cleaning will be done in two steps. The first step is a kind of preliminary cleaning by air blowing from tie-in to break flange on furnace convection wall:

- Preliminary cleaning by air blowing from tie-in point to break flange
- Final cleaning by steam blowing
- Alternatives;
  1. Do the first step – from tie-in to convection wall – by steam blowing. Note that this gives better results than air blowing, but is more complicated from safety point of view
  2. No other alternatives needed

10.4 BFW lines

Preheating exchangers outside the furnace, convection economizer/preheat modules and the line from convection to Steam Drum:

- Include in the chemical cleaning
- Alternatives;
  - Just water flushing. No good as the only cleaning method. Water flushing is normally done as a first step of Chemical Cleaning
  - Hydrojetting - good and efficient method, provided special equipment and/or specialized contractor is available

10.5 Saturated and Superheated VHP Steam line

Saturated VHP Steam line from Steam Drum to VHP Steam Superheat convection modules and Superheated VHP Steam line from superheater to the VHP Steam Header tie-in, including all branches:

- Include in the chemical cleaning and
- Steam blowing from Steam Drum up to VHP Steam header tie-in during Refractory Dry-out or as a one step of the furnace commissioning after the Refractory Dry-out and magnetite layering
- Note that there are cases when the Steam Superheat convection modules are made from stainless steel and the Chemical Cleaning Contractor is concerned about corrosion if those modules are included in the chemical circulation? That

complicates the situation somewhat as the VHP Steam superheat convection modules must then to be isolated/bypassed from chemical cleaning circulations – this means also minimum two more golden welds. When chemical cleaning is completed, the VHP Steam superheat modules are then to be reconnected to Saturated VHP Steam line and Superheated VHP Steam lines and be included later in the steam blowing, which is not a problem and the desired cleanliness will be achieved

- Alternatives;
  - No need for alternatives

10.6 Hydrocarbon Feed Preheat Modules, Cross-Over Line, Radiant Coils, Quench Exchanger Process Gas Tubes, Cracked Gas Piping and Headers, Decoke Piping and Headers

All convection modules, radiant coils and quench exchangers has been pre-cleaned in the vendor shop. Connective piping, like jump-overs, cross-over line, process gas/cracked gas lines and headers, should have been checked and cleaned for heavy corrosion and/or scale before erection. However, any debris from construction and remaining corrosion products and scale should be removed.

This cleaning is to be by steam blowing with Process Steam after burner ignition testing is complete, during refractory dry-out and steam system magnetite layering. The steam is to be by normal route from Process Steam supply via Process Steam-HC mixing point. The Feed Preheat modules upstream the mixing point should possible to blow with Purge Steam, via connections down-stream of feed control valves.

Note that alternative, or preliminary cleaning can be achieved, by:

- Keep Feed and Process Steam inlet connections of these convection modules blinded at convection module inlet break flange
- Install plastic folio rupture disk, thick enough let 3...max 4 bar air pressure, on e.g. furnace Process Gas line Manway, upstream the TLV and Decoking line valve
- Pressurize the whole Process Gas system form convection HC and Process Steam inlets all the way to TLV with air, and do the normal Rupture Dis Air Blowing
- Note that this Air Blowing is not very efficient, as e.g. the dirt in convection modules has very a long way to move before it exits at the Rupture disc opening
- Note also that any dirt or big particles removing from convection modules, could plug the Critical Flow Venturies in Radiant Coil inlets. During Air Blowing it is impossible to see if the venturies are plugged or no? It can only be seen when the furnace is getting to HSSB during Refractory Dry-out by ventury pressure drop or by Radiant Tube color, when in high temperature. The same problem is in case, if the Air Blowing was not done, but the cleaning is by Process Steam "Blowing", when going to Refractory Dry-out

11 Pre-commissioning – How to do it?

11.1 Manual/Mechanical Cleaning

- Can be applied only on large diameter piping (e.g. ID >30") to allow safe entry, like in some sections of furnace outlet process gas line and transfer line headers. However, this is better to do for large bore piping and header sections, before lifting them on pipe rack for welding
- By using el. driven steel brush machine or handheld steel brush. In case of heavy scaling, also light hammering could help to get the scaling loose
- Acceptance criteria:
  - Manual cleaning is always or at least in most of cases just a pre-cleaning, which require e.g. steam blowing for completion. As in case of furnaces, manual cleaning can be used only in some sections of Process Gas line or in Transfer Line headers and decoke lines
  - Acceptance criteria for "manual" part of cleaning is to remove all scale and rust layers as close as possible to get to metal surface. After that brushing and vacuum cleaning to remove loose material – final cleaning is by Steam Blowing during pre-commissioning

11.2 Air Blowing

- Alternative ways to do it:
  - Just blowing through the piping (minimum velocity?)
  - To do the blowing using rupture disk or pressurizing and depressurizing by (sacrificial) valve at the end of the line
- Acceptance criteria:
  - Exiting air must be free from particles observed by using cardboard target plate

11.3 Steam Blowing

- Blowing with steam, using preferably same pressure steam that the line is designed for:
  - Just blowing, which includes piping heating slowly to full steam temperature watching piping movement on pipe supports, blowing and cooling down – repeat this few times this heating-blowing-cooling down to get the scale and heavy rust loose (e.g. 3...4 blows of 10 minutes in full temperature)
- Acceptance criteria:
  - Using target plate, the acceptance criteria i.e. number of particle hit signs depending on use of the line. If the line is Steam line to turbine, the turbine vendor steam quality targets sets the acceptance criteria

11.4 *Water Flushing*

- How to do it:
  - Use potable water. In case of furnace BFW and steam generation system flushing use demin water
  - Just by water flow through the piping/equipment, min velocity 5 m/s or
  - By filling the system and then draining as fast as possible – repeat several times
- Acceptance criteria:
  - When draining water collect, sample in bottle. Sample must be clear and free of particles

11.5 *Hydrojetting*

- Alternative ways to do it:
  - Using "inhouse hydrojetting equipment" – e.g. by using same equipment as for TLE hydrojetting. The problem is normally that good cleaning results requires specially designed nozzle for each different pipe diameter. When used for piping, the hose length limits the length of line sections to be cleaned
  - Contracting the work to the specialized hydrojetting contractor – the advantage is that they have optimum size nozzles available for wide range of pipe diameters, hose length and "hose feeding" system can cover up to 200...300 m pipe lengths even if there are some 90° bends. Those companies can also do the "passivation" and drying of the treated pipe sections, to delay corrosion/rusting so long that the reinstatement and inertization of the piping can be done
- Acceptance criteria:
  - Use NACE No. 5/SSPC-SP12 standard surface cleanliness definitions as acceptance criteria. E.g. in case of BFW lines from Deaerator all the way to furnace battery limit, from where the downstream piping/equipment is in chemical cleaning. The standard criteria WJ-2 gives sufficient surface quality almost comparable to what can be achieved with chemical cleaning/acid pickling
  - In case of comparing different hydrojetting methods and setting the acceptance criteria for them, good practice would be to go to project pipe material storage and choose relevant pipe material, diameter and rust/scaling etc. ID surface quality, pick some sample sections and ask the contractor to demonstrate the effectiveness of hydrojetting for cleaning. This demonstration, is good opportunity for client and contractor to reach common understanding and agreement of acceptance criteria and reference for later date inspection after piping cleaning

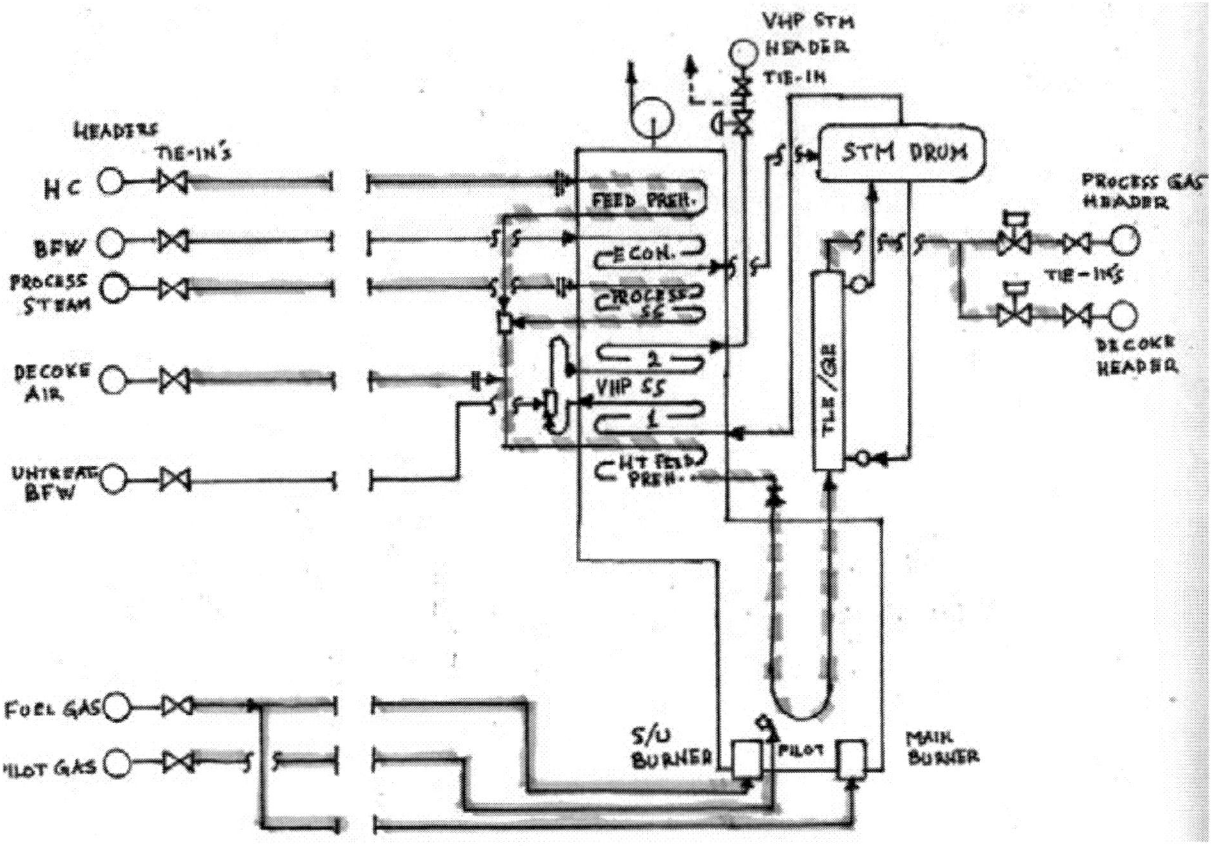

Fig. 3, The Flow Diagram of Furnace Showing Different Pre-comm Method Areas

11.6 Chemical Cleaning

- Alternative methods:
  - Acid Pickling/Chemical Cleaning, removes quite well all the grease, loose and attached rust and scaling. However, the Chemical Cleaning is not removing some of the coating materials, like e.g. Japanese Black. Chemical cleaning contractor is not giving warranty for the quality of the Acid Pickling if Japanese Black is found. If the piping is heavily corroded and having scaling already when purchased, the mechanical pre-cleaning is recommended to speed-up chemical cleaning, the piping inspection after Chemical Cleaning is very superficial, as most of the piping and piping connections are by welding which causes limited access for inspection. Only locations where the lines are cut for installation of connection for chemical cleaning, some valves where the bonnet can be removed or some vent and drain connections from where the borescope can be inserted, provided the connection is big enough for the purpose
  - Alkaline Boiling-out/chemical cleaning, removes grease and all loose rust and dirt. Alkaline Boiling-out does not remove attached rust, scaling and coatings,

like Japanese Black. Alkaline Boiling-out is only done/effecting to water containing parts of the Steam Boiler/Furnace.

Alkaline Boiling-out is good only for the piping, steam generators etc. equipment which are in good shape and clean already when transferred from vendor shop to the site. So, all the scale, attached rust/corrosion products and coatings must have been removed by vendor, prior to manufacturing/assembly. Alkaline chemical solution is not allowed to enter steam piping, superheaters and desuperheater water injection piping, i.e. those parts of the furnace steam generation system are to be cleaned/treated by Flushing, Hydrojetting and Steam Blowing.

Due to mentioned limitations and shortcomings of the Alkaline Boiling-out, compared to Acid Pickling, in case of new furnace or major revamp of the old furnace, the Acid Pickling is recommended for the chemical cleaning method. The details, like e.g. selection of acid, acid combinations and concentrations etc. details, should be chosen together with the chemical cleaning contractor based on careful evaluation of status of the piping and equipment water/steam side surfaces, and preferably, based on tube section sample testing in contractor laboratory.

The following Chemical Cleaning acceptance criteria is listed only for Acid Pickling, done based on Hydrofluoric Acid treatment. The Acceptance criteria are only indicative. Acceptance criteria is to be always discussed with Chemical Cleaning contractor after all details of the chemical cleaning has been agreed and the chemicals/acids has been chosen:

- Acceptance Criteria:
  - Pre-flushing:
    - Flow rate >0.5 m/s
    - Acceptance; turbidity <0.2
  - Soaking and flushing with "wetting agent":
    - Flow rate >0.5 m/s
    - Acceptance; turbidity <0.2
  - Acid treatment:
    - Heating by external boiler 50.....60 C
    - Flow rate about 0.5 m/s
    - Acid and inhibitor (chemical that controls the acid to react rather with corrosion products and scale than the base metal) concentrations as per Contractor, based on his evaluation and testing done during site visit inspection of different parts of the treated system condition and samples taken
    - Acceptance; Acid concentration is stabilized and Iron content stabilizes (note: normally after 2 ....3 hours circulation, but could be significantly longer than that)
  - Flushing after Acid solution has been drained off:
    - Flow rate about 5 m/s
    - Acceptance; after 3...4 fillings conductivity <20 μS/cm

- Passivation:
  - Filling with Demin Water, pH >9.5
  - Temperature >40 C, using hot water or heating with external boiler
  - Acceptance; proof of positive Redox Potential
  - Steam Drum final inspection after the passivation. During this inspection all steam drum connections (instrumentation, water and steam connections) have to be cleaned from inside the drum as very often some debris is collected there blocking the connection. Check also riser piping inlet boxes for any dirt. Check the tightness of all bolted connections inside the drum

- Final acceptance by Contractor and operator joint evaluation by visually and borescope inspection from as many locations as possible

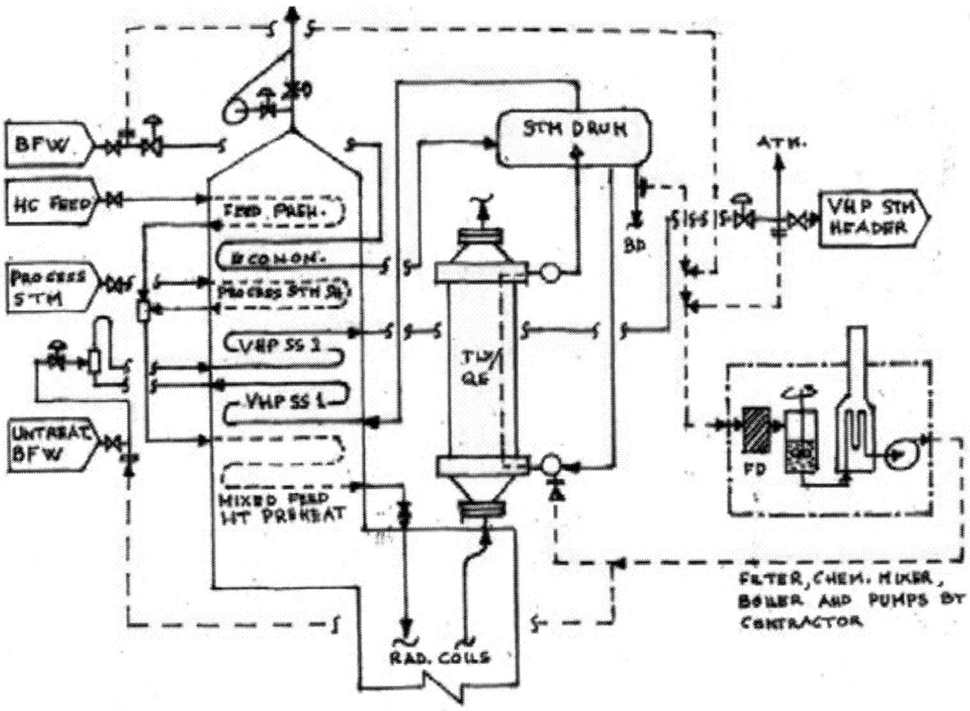

Fig. 4: Chemical Circulations for Acid Pickling Chemical Cleaning. The Yellow Colored Piping and Equipment will be Treated

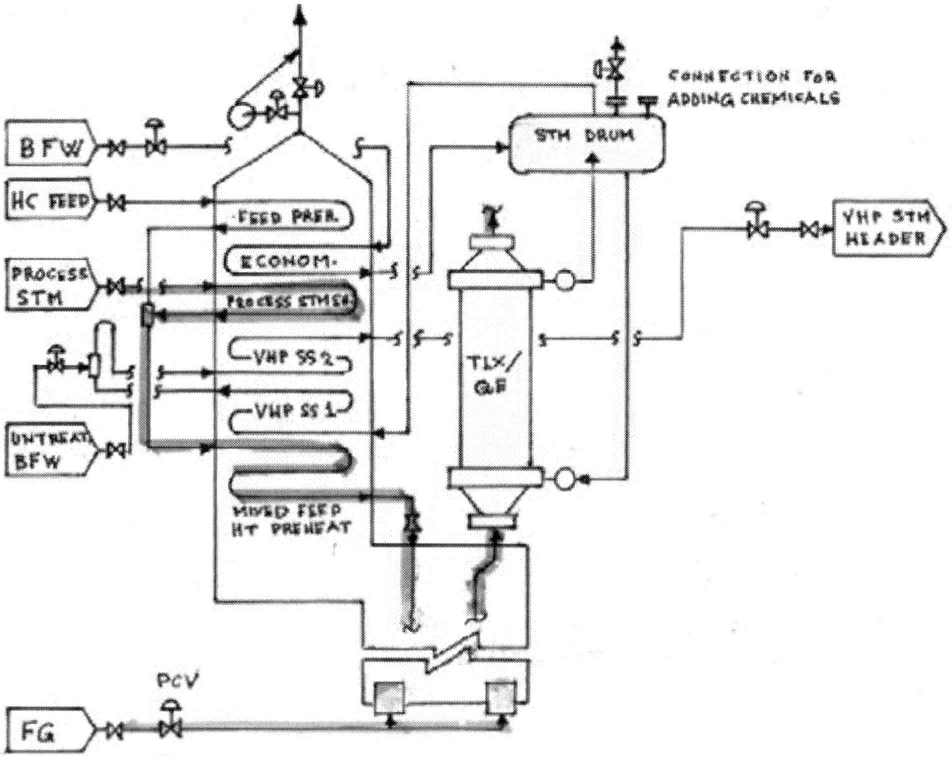

Fig. 5: Alkaline Boiling-out Treatment. The Yellow Colored Piping and Equipment will be Treated

12 Preservation

Needed after chemical cleaning and reinstatement is completed, but there is up to some days or weeks waiting for the furnace burner ignition and start-up are after chemical cleaning has been completed. Alternatives are to do either Wet or Dry Preservation:

- Wet Preservation:
  - Complete filling with Demin Water or BFW pH 9.5 – 10.5
  - Hydradzin (or corresponding other Oxygen Scavenger content:
    - 100 ppm wt, for short, just few days preservation
    - 350 ppm wt, less than 1 month preservation
    - 500 ppm wt, for longer than 1 month preservation
    - Keep in pressure of 5 barg, re-pressurizing by manual pump when needed
- Dry Preservation:
    - Remove/drain water off completely, dry the system to –10 C - -20C dew point
    - Nitrogen purging and filling, to O2 <0.1 % vol.
    - Keep in Nitrone pressure of 0.3 – 0.4 barg, check and refill daily

13 Commissioning

Furnace Start-up activities when moving from Pre-commissioning to Commissioning is somewhat gray area about which activity belongs to where. In following "classification", only Refractory Dry-out is a kind of Pre-commissioning activity, but it can only be done after Fuel Gas system is already pressurized, burners ignition testing completed and the whole furnace is in practice ready for feed-in.

Without going too much into details of the Commissioning, following are few notes about the most critical activities:

- ID Fan mechanical/performance test run must be completed and fan suction and by-pass dampers functional checking must be completed before BMS functional checking
- Loop and Functional Checking of the furnace instrumentation
  - These tests are very time consuming, should be started as early as possible, after the reinstatement of piping components and instrumentation is completed after line cleaning and chemical cleaning is completed
- Functional checking of BMS and Furnace Safety Interlock systems
  - Here there is need to simulate FG lines pressurization/depressurization when running:
    - The sequence controls of FG system leak/tightness testing
    - Sequence controls and/or permissive related to cracked gas and decoking gas valve operation
    - Testing the function of different Furnace Shut Down levels
- Burner ignition testing
  - Fuel Gas system inertization and pressurization with FG/Pilot Gas means that the work permit system comes mandatory, most of the construction is completed or have to be stopped to guarantee plant safety when the first HC system is pressurized and being leak tested
    - Note that parallel to FG System pressurization, the BFW system needs to be commissioned, including Steam Drum level control and the BFW level has to be established in the Steam Drum
  - Burner ignition testing starts with Pilot ignition and flame detection testing continued with automated pilot ignition sequence testing – when all pilots are ignited, then
  - Start-up Burner ignition testing, each burner individually, all S/U Burners need to be functional before moving to next step, but only few are needed to reach self-ignition temperature, e.g. >750 C or as required in Radiant Chamber by BMS Standard EN 746-2
    - Note that during Start-up burner ignition testing, the Process Steam flow need to be commissioned to provide cooling for the radiant coils and feed preheat Convection Modules
  - Main Burner ignition testing for each burner individually can be done only after all S/U Burners have been tested and the radiant box temperature is >750 C, as per EN 746-2, i.e. well above FG autoignition temperature. The

other BMS standard's temperature requirement might be somewhat different

- Refractory Dry-out
    - Refractory Dry-out is to be carried out, based on refractory vendor instructions for flue gas temperature ramping-up, hold time periods etc. and in the end temperature ramping down. However, during Dry-out, the VHP Steam production starts, which give the nice opportunity for parallel Magnetite Layering.
- Magnetite Layering
    - Is a simple procedure to operate the furnace at 30...40 bar VHP Steam pressure, with Oxygen free BFW for 10...20 hours, depending on expected Boiler Water increased pH value, to speed-up the formation initial magnetite layer in Steam generation equipment and piping surfaces to provide solid and tight (non-porous) magnetite protection for metal surface fresh from chemical cleaning, just before the furnace cracking operation starts

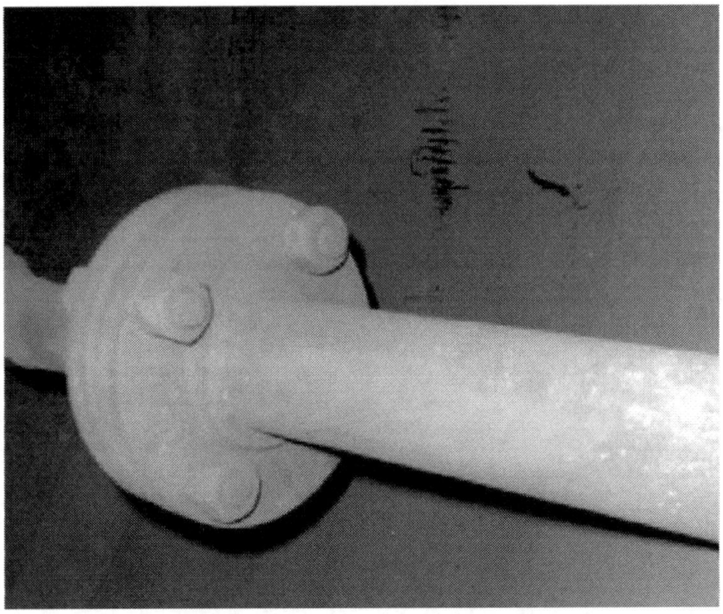

Fig. 6: Poor Quality Magnetite Layer – Solid, Tight Magnetite is missing, Wiping of Porous, Loose Surface Magnetite off Discloses Reddish Metal Surface

- Last inspections and checks before final closing-up
    - At this point in most cases, the furnace is still cooled down to low enough temperature to enter the radiant box and Bridge Wall for checking refractory for any potential damage during the Dry-out. The acceptance for the refractory can be signed if everything is OK
    - This is also the last moment to inspect the burner tiles and burner tips for any damage or fine toning e.g. to adjust pilot tip distance to burner tip for

safe ignition. Also, in case if during burner ignition testing there was any problem with burner tip plugging, now is the good opportunity to clean them

14 Furnace Start-up, Feed-in

Following is just a very brief overview of the furnace start-up, mainly to highlight the sequence of different start-up actions, for the purpose to help making sure that the timing of the completion of different pre-commissioning activities is right, in-order-to have everything ready when needed, but not too early to avoid waiting times and unnecessary conservation.

Fuel Gas headers and piping inertization – fuel gas system is the first system to be pressurized with Hydrocarbons to allow the burner ignition testing to start.

Furnace HC feed system, furnace itself and the cracked gas lines and header inertization – all remaining furnace Hydrocarbon piping inertization should be done same time with Fuel Gas system. When the hydrocarbon piping of the furnace is being inertized, that is stopping most of the construction work for safety reasons.

Start-up sequence:

- Filling Boiler Feed Water system and Steam Drum
- Fuel Gas and Pilot Gas system pressurization with fuel gas – Burner ignition testing
- Refractory Dry-out, followed by Magnetite layering of the furnace VHP Steam generation system
- Magnetite Layering
- VHP Steam line from furnace Superheater exit to Tie-in at the VHP Steam Header Steam Blowing
- Furnace cooling down for final inspection of burner tips/burner tiles and the furnace refractory of radiant chamber and bridge wall
- Feed system pressurization with HC
- Furnace restart and ramping up to Hot Steam Stand-by
- Feed-in
- Ramping-up the Feed and stabilizing the process conditions to design values

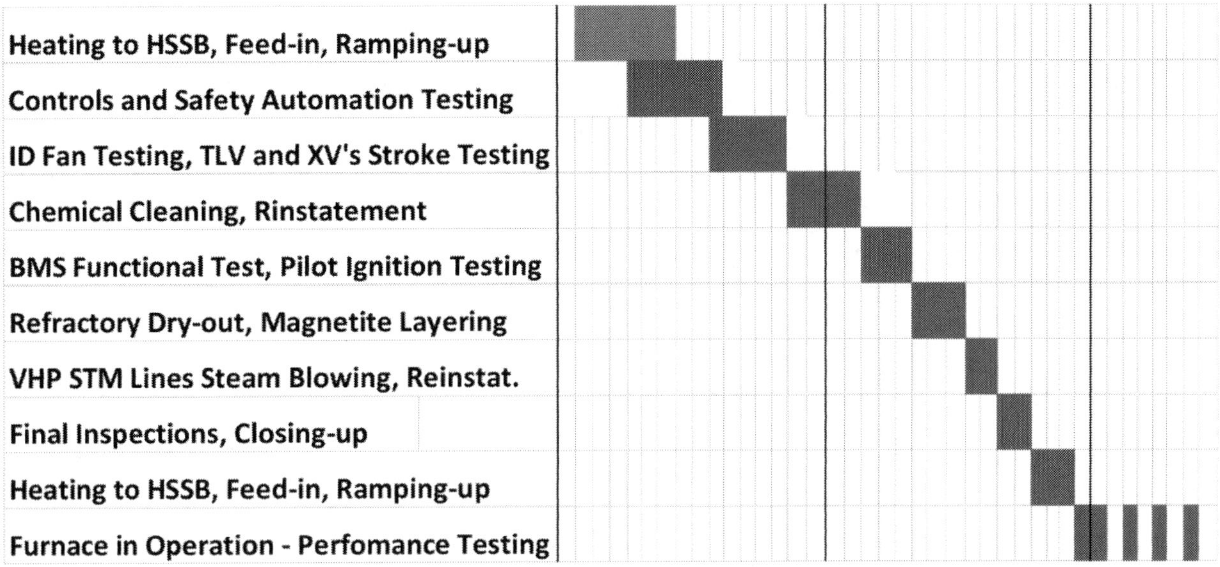

Fig. 7; Schematic Optimum Pre-commissioning/Commissioning/Start-up Schedule

15 Summary

When starting the project that include building one or more new furnaces or major revamp of the existing furnaces, there are numerous actions and decisions which are to be considered already when kicking-off the project, like setting priorities and sequencing of actions in scheduling. The project specifications should cover well the minimum quality requirements for e.g. piping and tubing materials, including the process side surface conditions. In manufacturing the equipment and piping spools, ITP's should have hold points to allow inspections in vendor shop are important as there is still possibility to do some preliminary cleaning, before final assembly and delivery to site. Construction at site should be scheduled to allow frequent inspections and to optimize the pre-commissioning and start-up sequence and timing in-order-to avoid after construction completion the long waiting times before the pre-commissioning, inertizing and/or conservation is possible. When approaching start-up, the last few remaining ones has to be in certain order like chemical cleaning, burner ignition testing, refractory dry-out, Steam blowing of VHP Steam lines, final inspections. When final closing and boxing-up of furnace equipment, radiant box and flue gas channels are done, the furnace is ready for start-up and feed-in.

If everything is properly checked before this point, the furnace start-up and feed-in should not cause any unexpected problems. The furnace should reach the design conditions already during the first run. However, the first run might not be as long as expected, as there could still be e.g. some minor corrosion particles coming from upstream HC feed or process steam lines, causing catalytic coking. After the first Decoke, the furnace should be ready for the performance test run.

References:

1. NACE No. 5/SSPC-SP12, Surface Preparation and Cleaning of Metal by Waterjetting Prior to Recoating
2. VGB PowerTech, Guideline / Internal Cleaning of Water/Tube Steam Generating Plants and Associated Pipework, VGB/R 513 e, First Edition 2002
3. European Standard EN 746-2; Industrial Thermoprocessing Equipment – Safety Requirements for Combustion and Fuel Handling Systems

AICHE PAPER NUMBER 128a

Exploring the use of Fitness For Service (API 579) in Ethylene Production Facilities

Scott Bouse,
Senior Associate
Stress Engineering Services, Inc

2018 Spring National Meeting, Orlando, Florida, April 22-26, 2018

Exploring the use of Fitness For Service (API 579) in Ethylene Production Facilities

Scott Bouse,
Senior Associate
Stress Engineering Services, Inc

Abstract:
The API 579 / ASME FFS-1 Fitness For Service standard is a remarkably useful tool for assessing damage that is routinely encountered during plant operations, but we find that many Ethylene plants have only limited experience with the document. This paper will provide a general outline of the FFS document as a whole, with special attention paid to the damage most-commonly encountered in the production of Ethylene. Specifically, generalized case studies will be presented covering the following topics:

1. Corrosion Under Insulation (CUI) and other wall loss damage
2. Furnace Tube Creep
3. Brittle Fracture (including Auto-Refrigeration risks)
4. High-Energy Piping (Superheated Steam and others)
5. Reactor Runaway Over-heating (Shell Distortions and material degradation)

Introduction

The API 579 / ASME FFS-1 joint Fitness For Service standard serves a role to standardize the engineering approaches to a wide variety of process-related damage mechanisms. As an industry-led initiative, representatives from refiners, gas plants, chemical companies work together with technical experts from engineering companies to establish common rules based on good engineering practice.

Despite the functional and stated goals of bringing 'best practices' to a wide variety of industries, we (as consultants) sometimes encounter groups or industries with little exposure to these tools. In this paper, we hope to provide a basic explanation of the methods and limits included in API 579, along with a few short case studies on various topics common to Ethylene Production Facilities. The most important sections for Ethylene Producers cover topics on Corrosion Under Insulation, Furnace Tube Creep, Brittle Fracture, High-Energy Piping, Distortions due to hot spots or other loading, and Crack-Like Defects.

Structure of API 579

General Philosophy

API 579 is based on two concepts working in tandem: (1) the design philosophies for most major codes are intended to reduce workload of practicing engineers by making assumptions about loading conditions, and covering potential gaps in those assumptions with relatively generous margins on failure (as high as 5x for old equipment in the USA), and (2) detailed analysis can sometimes show that a feature with sinister appearance actually has little or no effect on the final performance of the structure.

To these ends, API 579 formalized a concept known as the Remaining Strength Factor (RSF), whereby the ultimate load capacity of a damaged component is compared against the load capacity of an undamaged component (or sometimes more simply against the hand-calculated expected load capacity). Considering the assumptions included in item (1), above, it is relatively common that when an engineer considers all the possible and likely loads on a damaged component that it does not approach failure at the Code-required margin. For items such as these, the Remaining Strength Factor approach permits continued use of many components that are technically in violation of their Code of Construction. Even when the margin is shown by advanced methods to be reduced, the RSF can permit continued operation by virtue of the improved detail of engineering performed in the area.

To that end, almost all major states and jurisdictions in the United States of America have enshrined some acceptance of the methods in API 579 for the ongoing assessment of damaged components in the materials processing industries. In addition, the methods within

API 579 are considered to be Recognized and Generally Accepted Good Engineering Practice (RAGAGEP).

Explanation of Assessment Levels

All parts of API 579 are broken into three degrees of analytical intensity, meant to be added upon one another. For example, if the easiest and most-conservative assessment technique (Level 1) is deemed to provide an unacceptable outcome, an organization or responsible engineer may elect to perform an intermediate assessment (Level 2). Level 2 assessments generally require more information about the vessel and have certain limits as to their applicability, but will generally allow a Remaining Strength Factor of less than 1.0 to be permitted by the responsible parties. In practice, this means relatively uncomplicated conditions can benefit by about 10% more permissible corrosion than Construction Code equations would otherwise allow, exclusively because a more-detailed engineering analysis has been performed.

If the rules within a given Part do not allow Level 2 analyses to be performed (or if the results are unsatisfactory), advanced analyses are permitted in all cases, at Level 3. Level 3 analyses frequently, but do not always, leverage Finite Element Analysis (FEA) software to compute the predicted structural response of a damaged component. For instance: instead of assuming a locally thin region can rely on its surrounding material for 10% of the imposed loading without detriment, the FEA analysis would directly compute the degree of load sharing around the defect and determine the actual margin against structural failure (Protection Against Plastic Collapse).

Level 3 assessments do not require the use of FEA, and may be used for any defect type under any loading, so long as a responsible expert evaluates the results carefully. This author has, personally, performed a Level 3 assessment using only Roark's Formulas and the Machinery's Handbook.

Responsibilities for Each Section

Level 1 analyses are generally equal to Code calculations for the concerned component. In practice, these assessments are generally performed by API inspectors such as 510/570/583 inspectors at the time of condition assessments. Plant engineers are likely to receive an inspection report identifying a thickness reading near or below the minimum thickness necessary to satisfy the code criteria or 'tmin', and the inspectors' justification for documenting the deficiency will be replica Code calculations.

Level 2 analyses are generally performed by plant Fixed or Stationary Equipment Engineers, and require that the responsible engineer perform some additional calculations or checks against loading conditions that could invalidate the assumptions either buried within the Code of Construction or that are used in the methodology of the Level 2 approach. The benefits are usually provided in the form of justification for continued operation, until such time that a more-formal repair can be performed. Mostly due to the early reporting of problematic components by inspections services providers, a large majority of API 579 assessments are found to be acceptable via Level 2 procedures. Examples of Level 2 analyses might be Brittle Fracture assessments, where the pressure in a vessel is compared to the Minimum Acceptable Temperature for all possible pressures, instead of simply at the Code-Defined coincident Maximum Allowable Working Pressure (MAWP)/Minimum Design Metal Temperature (MDMT). The resulting calculation could show that a vessel carrying ethylene is safe to operate, even if ethylene can flash-boil to a lower temperature than the MDMT since the coincident pressure during this event is most likely less than the MAWP.

Level 3 analyses generally rely on specialized software or deep technical expertise in a field of study, and so are generally performed by Subject Matter Experts (SME's). Not all companies have access to internal SME support, so a large amount of Level 3 work nationally is performed by advanced engineering contractors. These experts will consult with unit operators, inspections personnel, and maintenance engineering staff to ensure that all relevant concerns are incorporated into a detailed engineering study. Due to the cost of such engineering work, most Level 3 assessments are limited to long-lead items or large, complex components with no Code calculation coverage or those requiring sophistication beyond the traditional design methods. Examples of Level 3 analyses might be CUI around a conical transition on a tower subject to wind loading, or detailed analysis of a dented storage vessel.

Arrangement of Sections

API 579 is generally arranged by problem type. For example: assessment of pitting corrosion can be performed largely by opening to the Part 6 and following the instructions contained therein. In some cases, the parts are inter-dependent – gouge assessments frequently must also be considered as crack-like defect assessments to ensure fracture-type failure does not govern the component's safe use. Part 2, however, is global and applicable to all assessment types.

- Part 1: Introduction and Identifying Terms
- Part 2: General Assessment Procedures and Equations
- Part 3: Assessment of Existing Equipment for Brittle Fracture
- Part 4: Assessment of General Metal Loss
- Part 5: Assessment of Local Metal Loss
- Part 6: Assessment of Pitting Corrosion
- Part 7: Assessment of Hydrogen Blisters and Hydrogen associated with HIC and SOHIC
- Part 8: Assessment of Shell Misalignment and Shell Distortions

Part 9: Assessment of Crack-Like Flaws
Part 10: Assessment of Components Operating in the Creep Regime
Part 11: Assessment of Fire Damage
Part 12: Assessment of Dents, Gouges, and Dent-Gouge Combinations
Part 13: Assessment of Laminations
Part 14: Assessment of Fatigue Damage

Example of a Basic Corrosion (Part 4) Defect

A quick example of general thinning might go as follows: a section of pipe has a Code-required minimum thickness of 0.210 inches, but has been found with a spot of CUI at 0.196. By assessing that the pipe is not severely loaded and following the procedures within Part 4, the responsible engineer could conclude that the minimum required thickness for a Level 2 analysis is 0.9*0.210 = 0.189 inches. The calculation establishes that the CUI spot is acceptable as-found, with a Future Corrosion Allowance (FCA) of 0.007 inches.

Such small reductions in required thickness are not intended for new-construction designs, but are generally accepted as responsible proof of safety while a repair can be planned.

Simplified Case Studies

Corrosion near Saddle Supports

In this first example, one of our clients has encountered some Corrosion Under Insulation (CUI) near the bottom of a horizontal vessel. The Figure below shows a segment of this horizontal vessel, and the saddle support around an area of thinning:

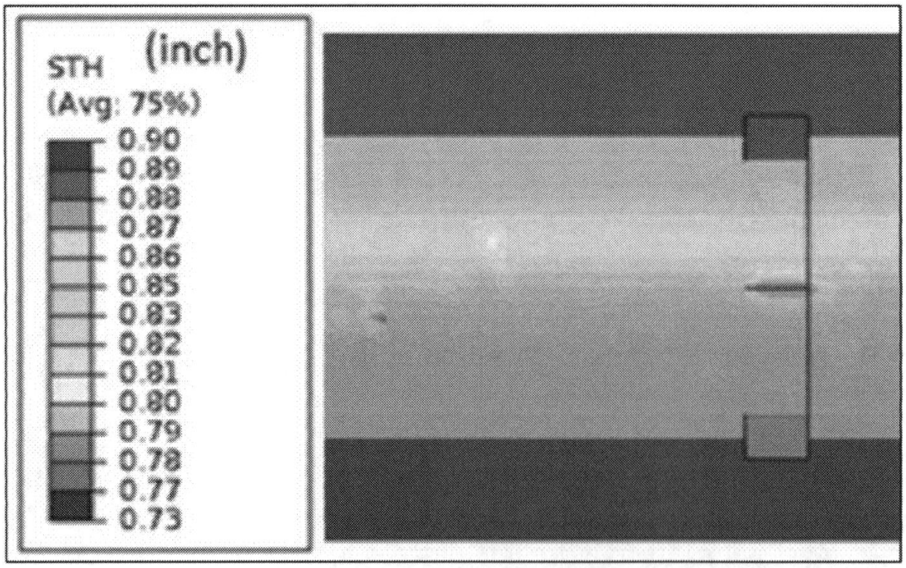

Figure 1 – Thickness Survey of CUI Damage (t-min = 0.825 inches)

In Figure 1, note the area of thinning marginally below the minimum required thickness at the saddle support. Due to the complex interactions between the shell and saddle support, API 579 prohibits the use of simplified calculations to establish Fitness For Service. Instead, a Level 3 analysis is performed using the ABAQUS FEA software. Thicknesses are mapped into a model representing the vessel, and the future corrosion state is increased until the margins fall below Code required minimums.

This is an ASME Code-Stamped Section VIII, Division 1 pressure vessel (VIII-1), constructed before 1999 (older allowable stress limits). This sets the required design margin equal to 3.6 {4.0 x (RSF_a = 0.9)}. The figure below shows the vessel's response to increasing load, with the Code-required and API-579 limits shown:

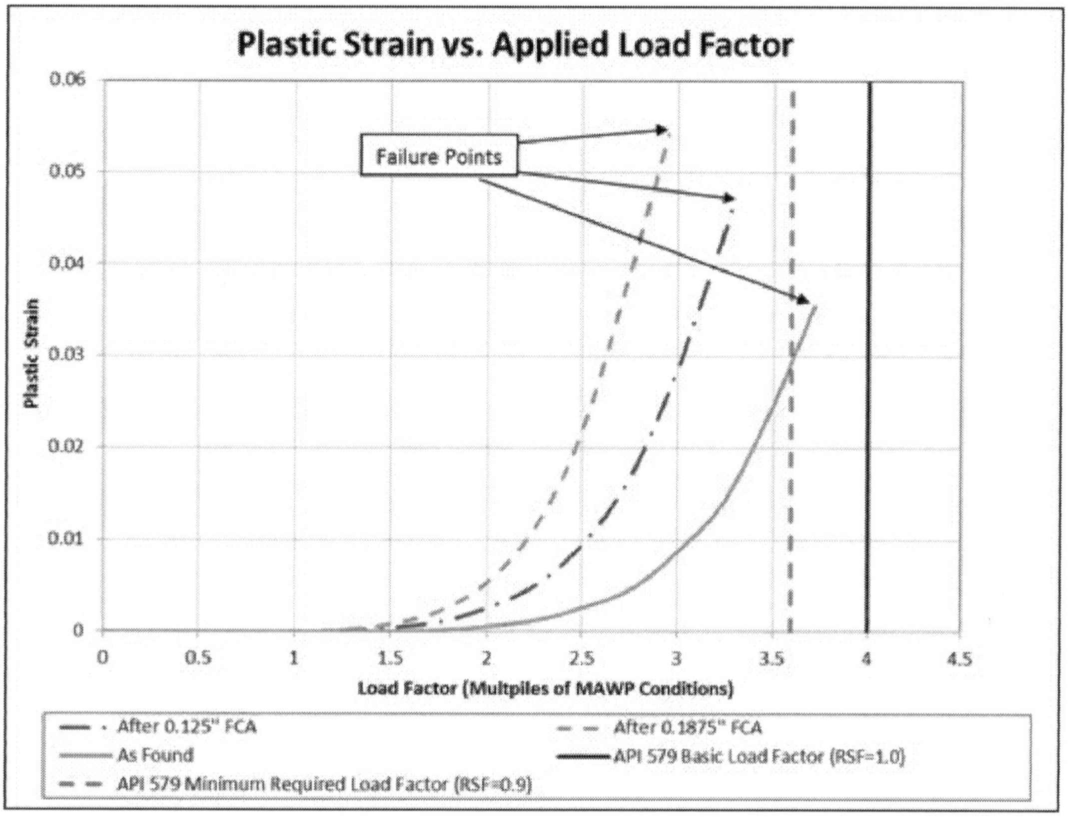

Figure 2 – Vessel Response to Increasing Load

In this case, the detailed analysis shows that the vessel as-found can be operated under an API 579 Fitness For Service assessment, but that increasing Future Corrosion Allowance (FCA) will result in the vessel falling short of the required margin. The margin could be satisfied if the FCA is reduced to a maximum value of 0.040 inches. With this adjustment, the client was able to continue operation of the vessel under an enhanced CUI-protection program until the next major turnaround.

Furnace Tube Creep

While not strictly an Ethylene Plant concern, the example below shows an outlet pigtail for a Steam-Methane Reformer in a refinery. The flowing fluids are typically very hot, but the tube that makes up the pressure boundary (T11 in this case) is protected by a sleeve. As the intermediate refractory begins to break down, the pressure boundary (shown below) begins to get warmer, and approaches design limits. The client needs to understand how warm the pipe may become before significant damage is accumulated over the estimated 10,000 hours until a repair can be performed.

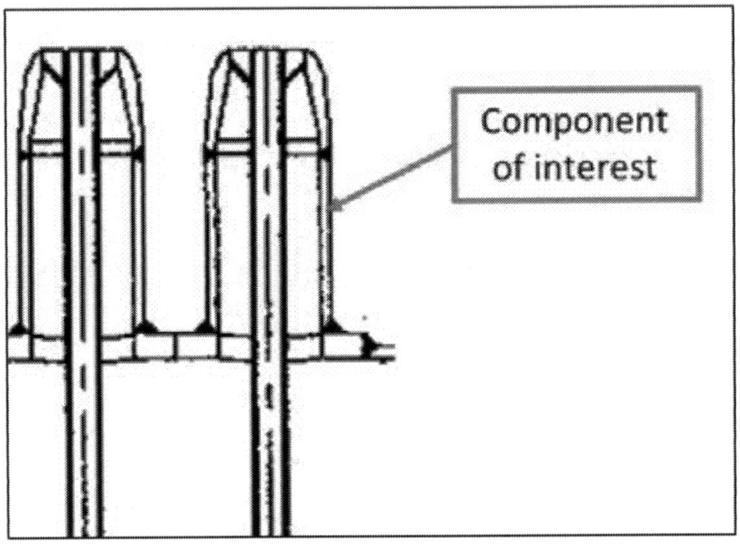

Figure 3 – Sketch of Tube Subject to Overheating

Per the Level 1 Creep procedures in API 579, this component may be subject to assessment via the screening curves. After determining that the critical state of stress in the tube is 3 ksi, the screening curves are entered at 3.0 ksi and a permissible damage rate of 1/10,000 hours (1E-4):

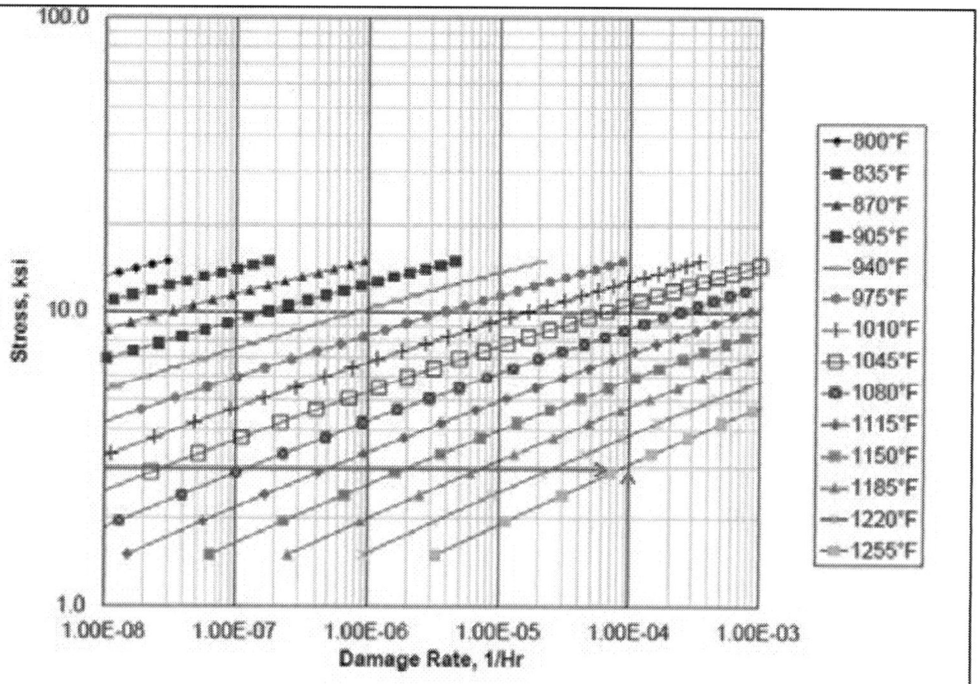

b) Damage Curve

Figure 10.7 – Level 1 Screening Criteria for 1.25Cr-0.5Mo – Annealed

Figure 4 – API 579 Part 10 Creep Screening Curve for T11

Following this simplified method, the engineer can establish that the T11 tubing may approach 1255°F before damage rate exceeds the established maximum permitted rate. In reverse, these curves may also be used to estimate heater tube creep per the same method. Basic creep damage of an HK-40 tube can be estimated by the figure below:

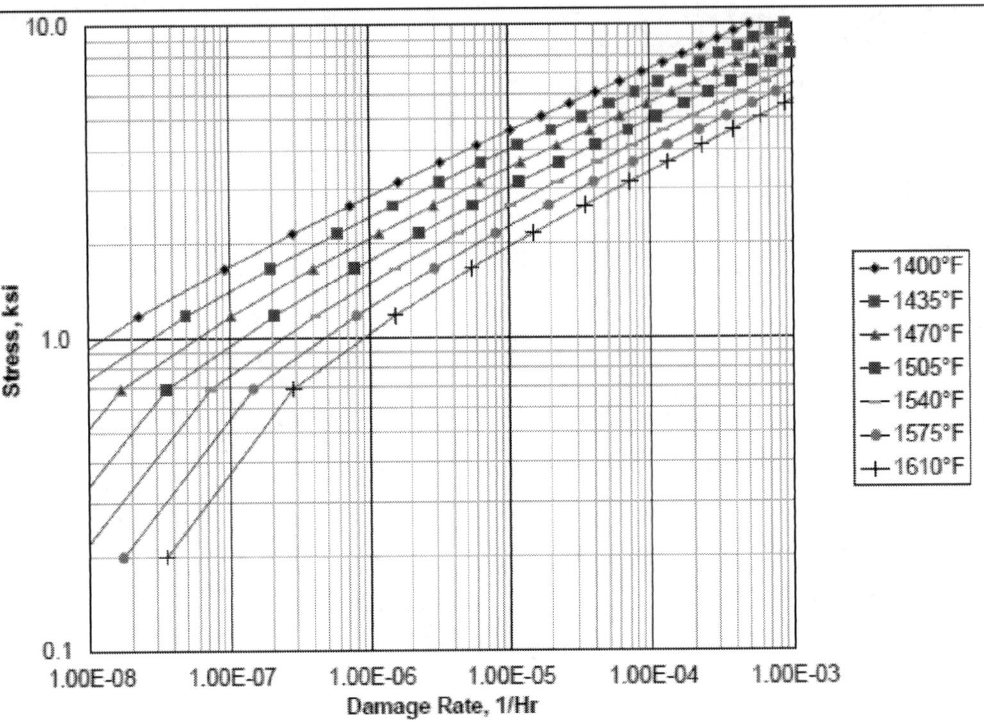

b) Damage Curve

Figure 10.26 – Level 1 Screening Criteria for HK-40

Figure 5 – API 579 Part 10 Creep Screening Curve for HK-40

In cases where a material is not generally included in the tables (such as HP-modified or HP-micro), vendor data may be used in conjunction with the methods described in Part 10, paragraph 10.3.5. In this method, Larson-Miller or Omega creep parameters are developed from supplied data, and can then be used in a forward-looking life predictions. This approach may require the involvement of an expert in high-temperature materials degradation to ensure conservative results.

Brittle Fracture

In this example, a client has a need to place a carbon steel pressure vessel of unknown material quality into temporary propane storage service. The vessel is 1.25 inches thick, and presumed to have been spot radiographed. The concern is that while the Minimum Design Metal Temperature (MDMT) for the vessel at 250 psi is 80°F, propane is capable of boiling at much lower temperatures. In this case, an API 579 assessment can be helpful, comparing the propane boiling point at various pressures versus the allowed reduction in the Minimum Allowable Temperature (MAT). The figure below shows the resulting calculation:

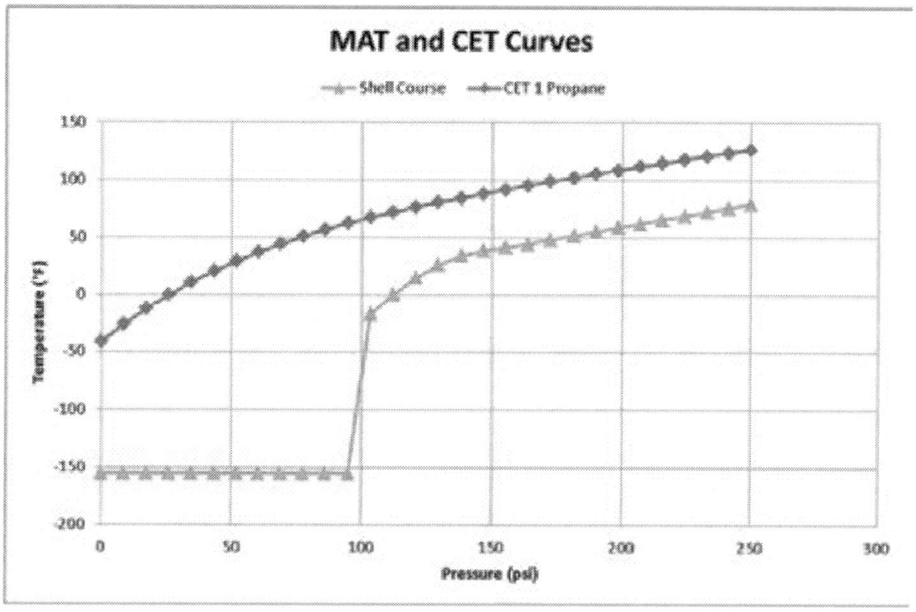

Figure 6 – Results of a Brittle Fracture Assessment on a Propane Vessel

This figure shows the required metal temperature versus applied pressure for all points between MAWP and zero. The Critical Exposure Temperature (CET) curve for propane is also plotted; showing that at all points propane is unlikely to flash-cool and place the vessel into immediate jeopardy. The conclusion from this basic study would confirm that temporary use of this vessel for propane service is permissible. A more-thorough Process Hazard Analysis might identify a two-step brittle fracture risk (flash-cool followed by repressurization), this scenario must be considered but was not applicable for this situation.

Reactor Runaway (Overheating) Leading to Distortion

On occasion, in process industries, runaway reactions occur within reactors, leading to shell distortions and (possibly) compromised material properties. In this example, we provided a Level 3 Fitness For Service assessment on a reactor shell overheated by just such an event. The figure below shows the distorted geometry as captured by laser scanning:

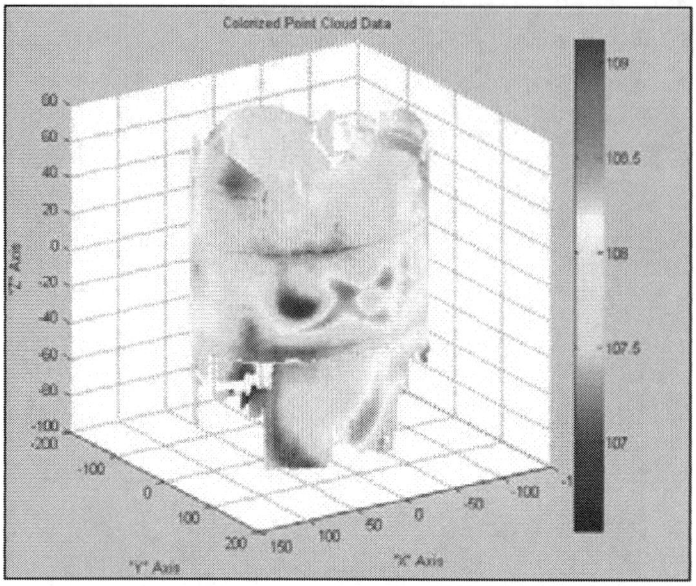

Figure 7 – Raw Laser Scan Mapping of a Distorted Vessel

In the figure above, the distortion is mapped as a function of deviation from nominal radius, and the information used to generate a Finite Element mesh suitable for assessment. The completed FEA mesh is shown in the figure below:

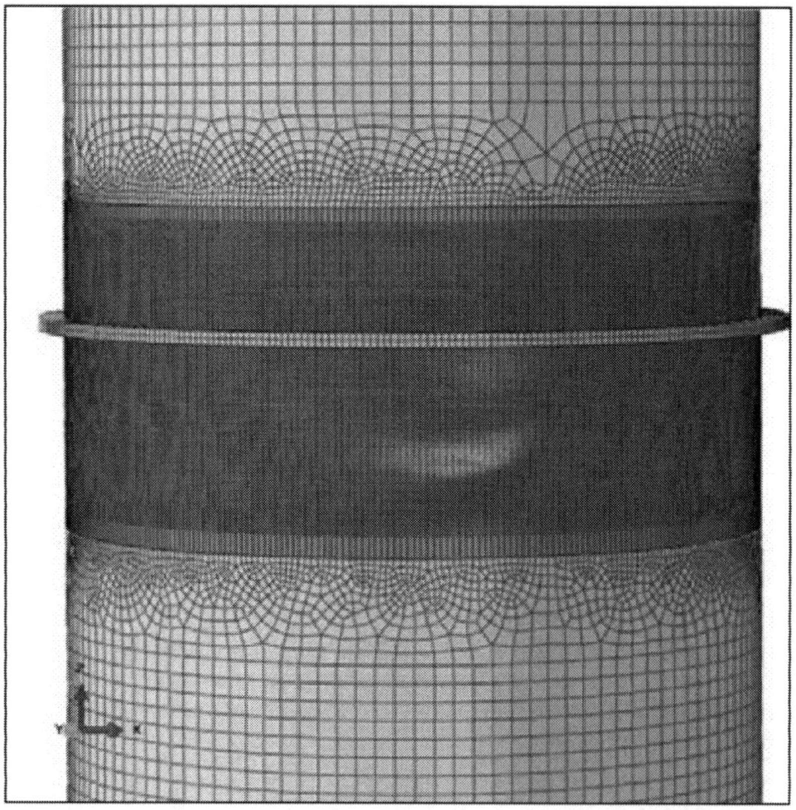

Figure 8 – FEA Mesh of the Mapped Distortion Field

After the model was prepared, we applied loads (in this case, Internal and External pressure are both possible, and the vessel exists in a seismically active region. After running a series of analyses to check for all relevant combinations of loads, the following Load Factor vs. Deflection chart was derived. The results indicated that the distorted region is stable against all expected load combinations up to and beyond the required 4.0x design margin required by the Code of Construction.

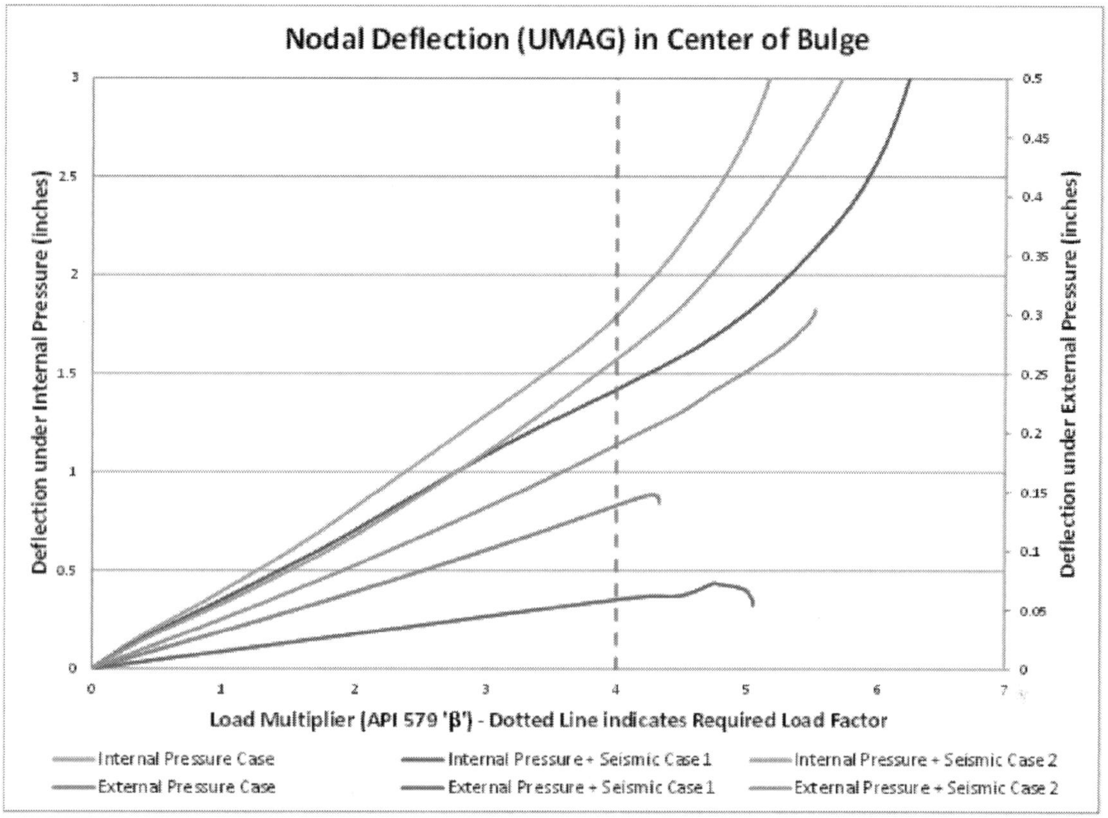

Figure 9 – Final Results Showing Exceedance of Code-Required Margins

Conclusions

The API 579 standard can be a very useful tool in the hands of both plant maintenance engineers and engineering Subject Matter Experts. While the rules cannot cover all possible types of damage, the guidance included provides an excellent jump-off point for detailed engineering assessments.

References

1. API 579 / ASME FFS-1 (2016)

AIChE Paper Number 128b

TLE Fitness for Service Analysis Case Study: Assessing Erosion for Remaining Life Prediction

Chevron Phillips Chemical Company. LP:
Jimmy Cleavinger
Sr. Asset Integrity Engineer
Bala Devakottai
Sr. Olefins Technical Support Engineer
Tom Garza
Principal Engineer, Fixed Equipment & Inspection
Bill Valerioti
Principal Engineer, Materials, Corrosion & Mechanical Integrity

ARVOS GmbH, SCHMIDTSCHE SCHACK Division:
Dr. Joerg Weidenfeller
Director RnD / Technology
Andreas Linnenkohl
Head of Structure Analysis and Simulation

Knighthawk Engineering:
Cliff Knight,
President and Chief Engineer
Carlos Corleto
Materials Lab Director and Senior Mechanical Consultant

Prepared for Presentation at the 2018 Spring National Meeting
Orlando, Florida, April 22-26 2018

TLE Fitness for Service Analysis Case Study: Assessing Erosion for Remaining Life Prediction

Chevron Phillips Chemical Company. LP:
Jimmy Cleavinger
Sr. Asset Integrity Engineer
Bala Devakottai
Sr. Olefins Technical Support Engineer
Tom Garza
Principal Engineer, Fixed Equipment & Inspection
Bill Valerioti
Principal Engineer, Materials, Corrosion & Mechanical Integrity

ARVOS GmbH, SCHMIDTSCHE SCHACK Division:
Dr. Joerg Weidenfeller
Director RnD / Technology
Andreas Linnenkohl
Head of Structure Analysis and Simulation

Knighthawk Engineering:
Cliff Knight,
President and Chief Engineer
Carlos Corleto
Materials Lab Director and Senior Mechanical Consultant

Abstract:

Ethylene furnace transfer line exchangers (TLE's) are often designed with minimal erosion or corrosion allowance on the tubes in order to keep them as thin as possible. This is done to keep the high tube side gas temperatures from overheating the tubes which are cooled by the boiler feed water on the shell side. Typically, when tube wall loss from inner diameter erosion (or other damage mechanisms) or outer diameter corrosion causes the tube thickness to fall below design thickness, the tube must be replaced. However, if the thinning is confined to a small, local area, a fitness-for-service (FFS) analysis may allow for continued, safe operation with tubes that have thinned below design thickness, provided that the FFS established local area minimum thickness is not exceeded. This paper provides a case study of a FFS analysis that allowed for the continued, safe operation of TLE tubes that were thinned beyond original design thickness. The owner/user, TLE original equipment manufacturer (OEM), and a third party engineering contractor collaborated on this analysis. This paper will discuss how the results of this analysis can be used to replace tubes selectively and continue safe operation.

Introduction

The Chevron Phillips Chemical Cedar Bayou plant produces Ethylene, High Density Polyethylene, Low Density Polyethylene, Linear Low Density Polyethylene, Normal Alpha Olefins, Poly Alpha Olefins, and on-purpose 1-Hexene. Cedar Bayou's 1592 ethylene plant (a Lummus design) began production in 1977 with one SRT-1 and twelve SRT-3 furnaces. An additional SRT-3 furnace was added later. In 2004, the TLE's were replaced in order to increase steam production and reduce pressure drop. Nominal TLE operating pressure is 1,500 psig.

Around 2010, routine inspection revealed problems with moderate to severe erosion of the TLE tubes. This erosion was typically located within the first two inches of the tube just past the inlet tubesheet (see Figure 1 and 2). Figure 3 shows a cross section of a tube at the location of maximum erosion damage, 1.5 inches above the bottom tubesheet. Initially, Remote Field Testing (RFT) was used for inspection of the tubes, however, Internal Rotary Inspection System (IRIS) inspection of the tubes was later used to better quantify the damage. The corrosion/erosion allowance on the tubes is minimal. At first, the eroded tubes were simply plugged; later, they were replaced. Figure 4 shows a graphical representation of the nominal tube wall thickness versus the minimum original design thickness. Since the corrosion allowance was so small and the damage limited to such a small area Chevron Phillips Chemical determined that a Fitness for Service (FFS) be performed to confirm if we could reduce the minimum required thickness (t-min) so that tubes did not have to be replaced as often.

Figure 1: Typical erosion damage on TLE tubes – Example 1

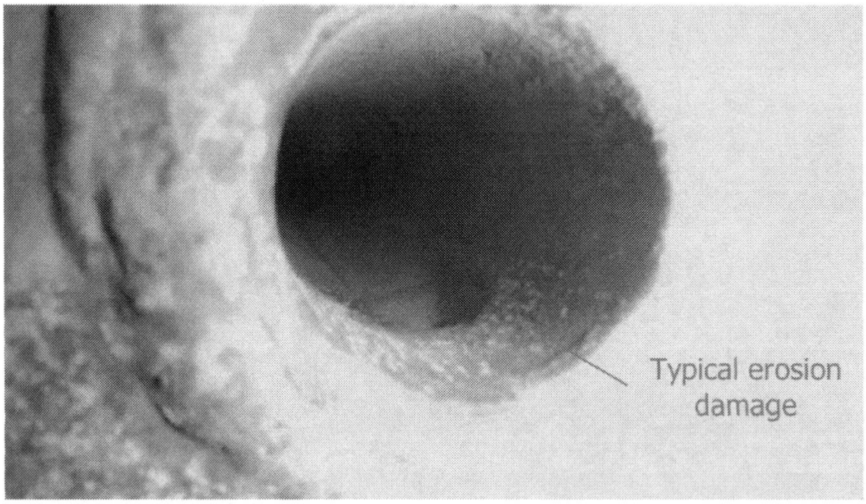

Figure 2: Typical erosion damage on TLE tubes – Example 2

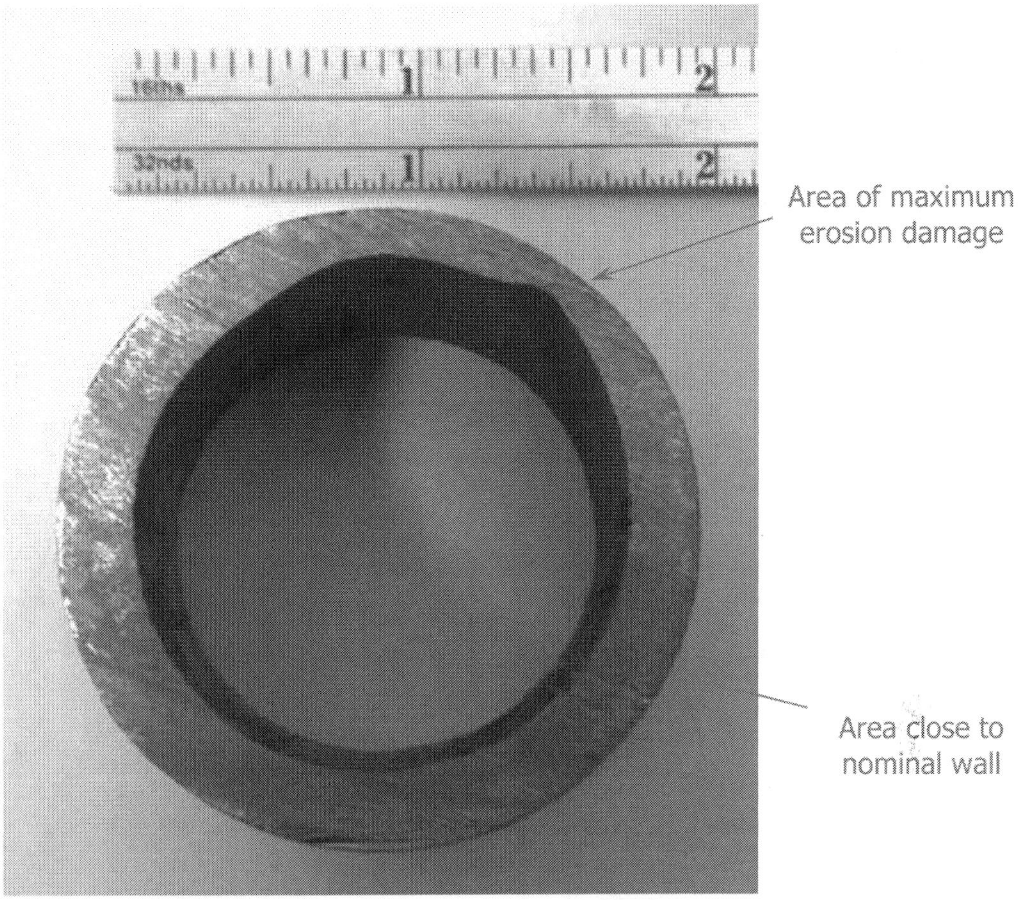

Figure 3: Cross-sectional view of maximum erosion damage

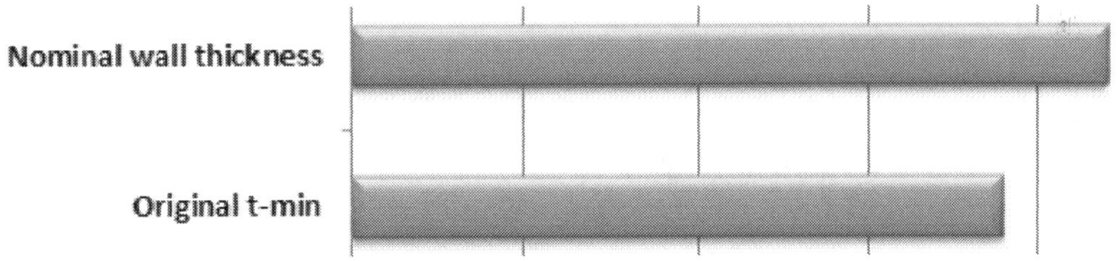

Figure 4: Graphical comparison of nominal and original t-min tube thickness

Chevron Phillips Chemical contacted the TLE's OEM, SCHMIDTSCHE SCHACK Division of ARVOS GmbH for assistance.

First assessment of Erosion Damage

SCHMIDTSCHE SCHACK, formerly known as ALSTOM Power Energy Recovery as OEM for transfer line exchanger (TLE) designed and manufactured the bathtub TLE's which are characterized by short process gas residence time and therefore higher Ethylene yield. The design principle consists of double tubes with oval headers, which is a typical feature for all SCHMIDTSCHE TLE's. The mechanical design for the new TLE's was performed according ASME Section VIII division 1, Edition 2001 where minimal corrosion/erosion allowance was considered for the inner tube, which carries the process gas. This philosophy enables keeping the wall thickness of the inner tube as thin as required by the ASME code which improves the heat transfer from hot process gas to the boiler feed water flowing with high velocity in the annulus between outer and inner tube. This results in low material temperatures for the inner tube as well as low temperature amplitudes at water side during steam bubble growth and separation which is beneficial for the integrity of the magnetite layer as corrosion protection.

It is important to mention that in general the results of accurately conducted tube wall thickness measurements do not allow deriving erosion or wear rate which is important information for a condition based maintenance program. The reason for that is that the real wall thickness deviates from the nominal wall thickness, typical by ±12.5% based on the tube manufacturing process. This spread complicates the determination of a wear rate since the installed new and clean wall thickness is not known which is qualitatively shown in Figure 5. Depending on the real wall thickness the remaining operation time until reaching the minimum allowable wall thickness significantly varies.

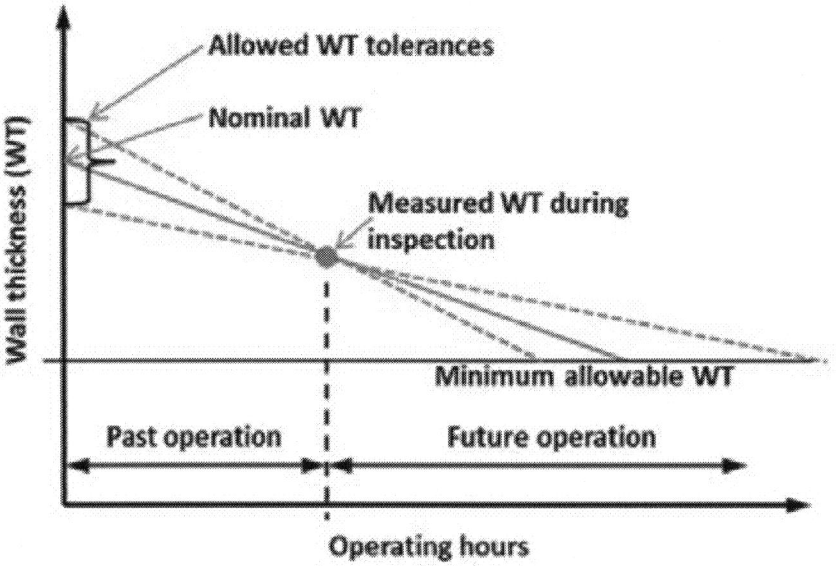

Figure 5: Qualitative wear rate depending on tube wall thickness tolerances

For that reason, complete information is necessary for the condition of the installed tube. This information comprises of regular wall thickness measurements as well as mechanical assessments based on these measurements. However, the baseline measurement inspection of the new and clean tubes, which can be easily and reliably performed at OEM's workshop, is usually not considered by customers due to additional cost.

The data of the ultrasonic tube wall measurement of different TLE tubes provided by Chevron Phillips Chemical were assessed by SCHMIDTSCHE SCHACK and the results are shown in Figure 6. The inspection report provided a single value of the remaining wall thickness for each tube which is the lowest wall thickness reading obtained on that tube. The frequency distribution of measured values was approximated with a Weibull distribution which is common for technical applications. In addition, the distribution of the wall thickness for the new tubes is shown assuming a normal distribution with a standard deviation considering the tube manufacturing tolerance. From this figure a tube wall thickness was chosen for the further mechanical assessment. This wall thickness is the minimum value of all approximated wall thickness distributions.

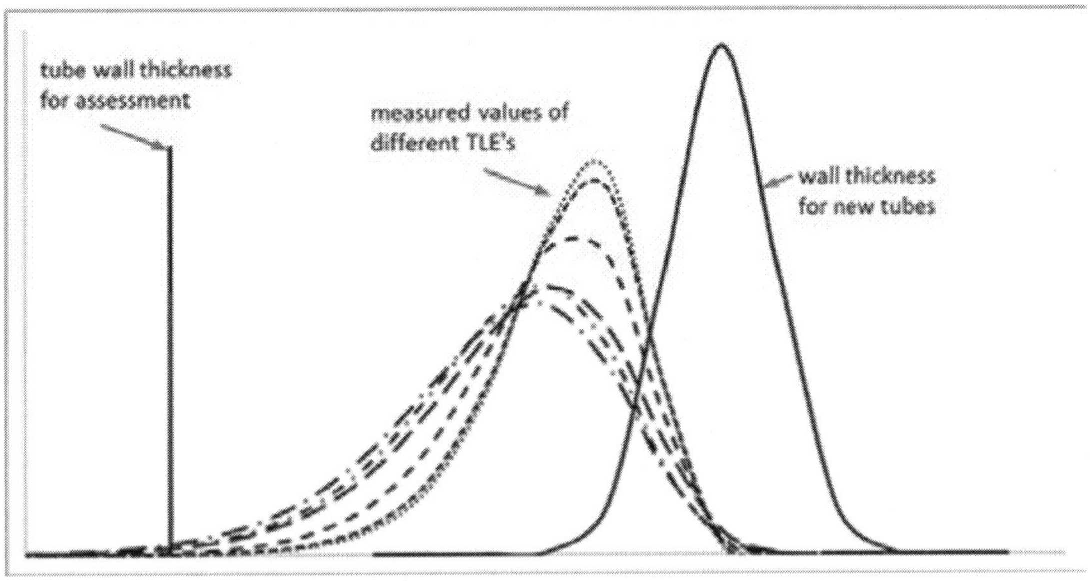

Figure 6 Frequency distribution of measured tube wall thickness and comparison with new values

In addition, the inspection report identified the axial location and the length of the of tube wall thinning area.

With this information a 3D-model Finite Element Model (FEM) for the TLE was set-up using latest ANSYS software. The wall thickness defect based on the available information was considered as uniform over the axial and circumference location within the tube inlet area. This approach is in the style of the fitness for service ASME FFS-1, Part 4, which describes the procedure for general metal loss. Modelling an area with a minimum uniform wall thickness applied to all tubes is for the current case a conservative approach. Based on the detailed stress analysis with FEM, this procedure in principle corresponds to a level 3 assessment of ASME FFS-1, part 4. The load case which was agreed with Chevron Phillips Chemical was chosen and applied to the model. As a result the maximum intensity of the linearized stress values were below the allowable material values in the area of the thinned tube wall.

Although this approach is conservative, it has the advantage of guaranteeing tube integrity case of:

- multiple flaws in circumference which have a local reduced wall thickness which is greater than the value assessed above,
- flaws which increase over operating time in length and width

For that reason this conservative approach is justified. If another and local wall thickness is below the assessed value an assessment according to ASME FFS-1 Part 5 has to be performed. This can result in local lower required minimum wall thickness.

The assessment result with the thinner t-min is represented graphically in Figure 7

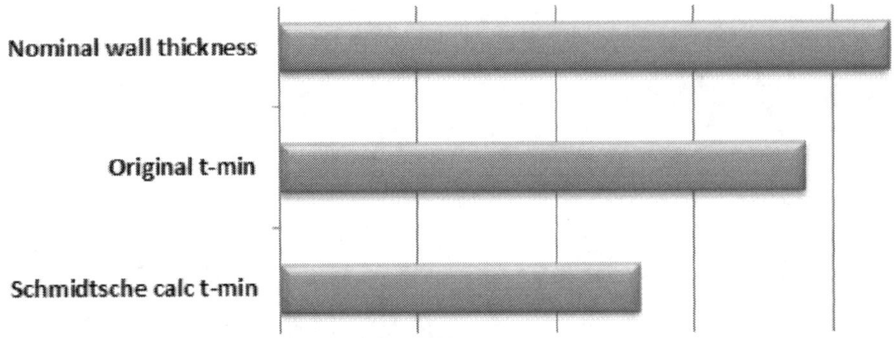

Figure 7: Graphical comparison of nominal, original and Schmidtsche t-min tube thickness

Since the erosion damage was not a uniform metal loss, but was very localized and linear in shape, Chevron Phillips Chemical contracted with KnightHawk Engineering, Inc. (KHE) to conduct further analysis using API 579-1/ASME FFS-1 2007 Fitness for Service (FFS) code.

Erosion Damage Characterization

KHE used 3D laser scans to determine the minimum wall thickness of two TLE tubes showing the most localized erosion damage provided by Chevron Phillips Chemical. These tubes and their 3D laser scans are shown in Figure 8. This erosion damage is consistent with a groove-like flaw as defined in Part 5 of the FFS Code. Tube B was thinner than Tube A at the deepest point of the groove region. Away from the groove, both tubes showed a uniform wall thickness. These 3D laser scan measurements were verified using a micrometer and were lower than those obtained by IRIS inspection.

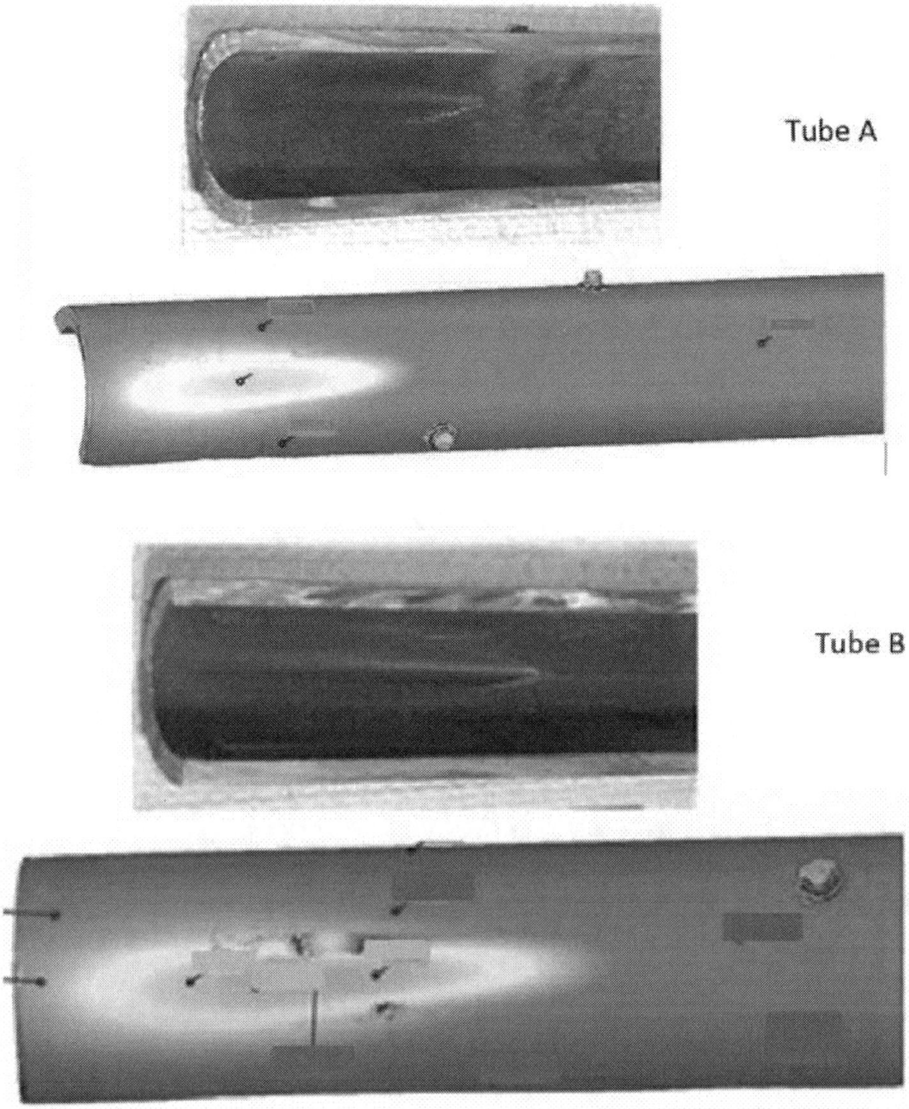

Figure 8: Erosion Damage on TLE Inner Tubes

As shown in Figure 9, wall loss inside the tubes resembles an ellipsoidal geometry nearly parallel with the longitudinal axis of the tube. The damage can be characterized by three parameters: length (L), depth (d) and width (w), with t-min defined as the nominal thickness (t) of the tubes minus d. This localized erosion damage was found only at the inlet tubes from the gas inlet side of the TLEs.

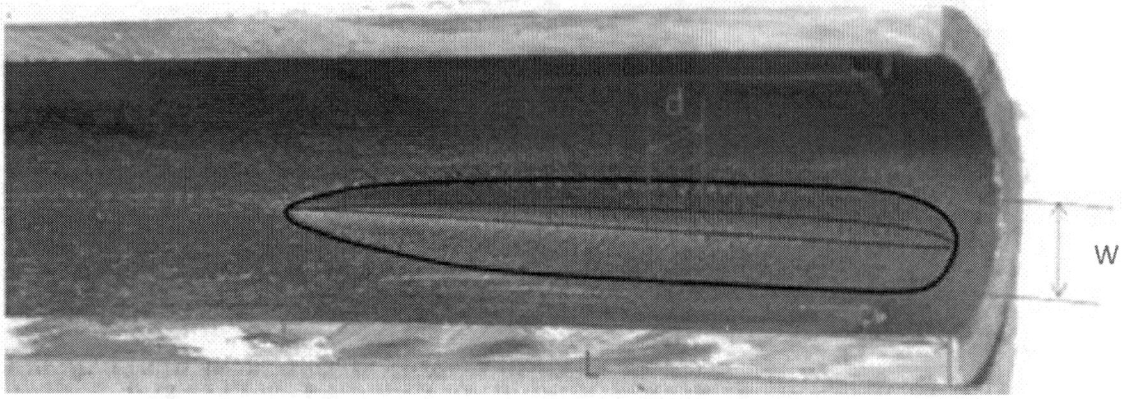

Figure 9: Characterization of Groove-Like Eroded Tube

Fitness for Service Assessment

This assessment was conducted to determine the t-min at which the TLEs could be safely operated based on FFS Code guidelines. The technique and acceptance criteria used followed API 579-1/ASME FFS-1 2007 Part 5, Level 3 assessment guidelines (Section 5.4.4). The assessment included elastic-plastic stress analysis techniques using Finite Element (FE) analysis methods provided in Annex B1 of the FFS Code, suitable for evaluating regions of general or local metal loss in pressure vessels, piping and tanks with complex geometry and loading like these TLEs that have experienced groove-like eroded inner tube damage.

Groove-Like Eroded Tube Models:

KHE developed five solid models of the groove-like eroded tubes using 3D laser scan cross-sectional profiles of the tube with the lowest t-min shown in Figure 8 (Tube B). The models had the same nominal thickness and overall length, the same w/L and d/L ratios of tube B, but different t-min.

Figure 10 illustrates one of the models developed. The models were combined with TLE solid models and used to determine and validate the t-min at which the TLEs could be operated based on this FFS assessment.

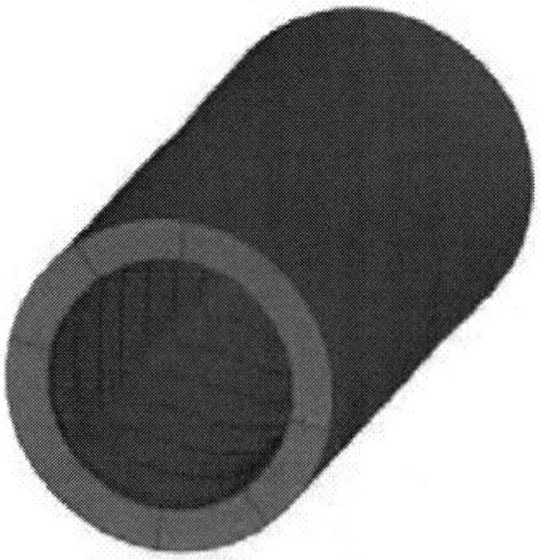

Figure 10: 3D Solid Model of Groove-Like Eroded Tube

TLE/Eroded Tube FE Models

KHE developed five 3D full tube length quarter tubesheet FE models of the TLEs, one for each of the groove-like eroded tube models described in the previous section using the ANSYS FE software. The 3-D FE models were developed based on TLE drawings provided by SCHMIDTSCHE SCHACK ANSYS Solid 185 and 186 elements were used. These elements have eight and twenty nodes, respectively. The Solid 186 models were used exclusively to mesh the eroded tube models. Figure 11 shows one of the TLE FE models with an eroded tube placed at the gas inlet side. The models had approximately 470,000 to 630,000 elements each, depending on the number of eroded tube models placed at the gas inlet side. Symmetric thermal and mechanical boundary conditions were applied but the tubesheet was restricted vertically at its longitudinal ends.

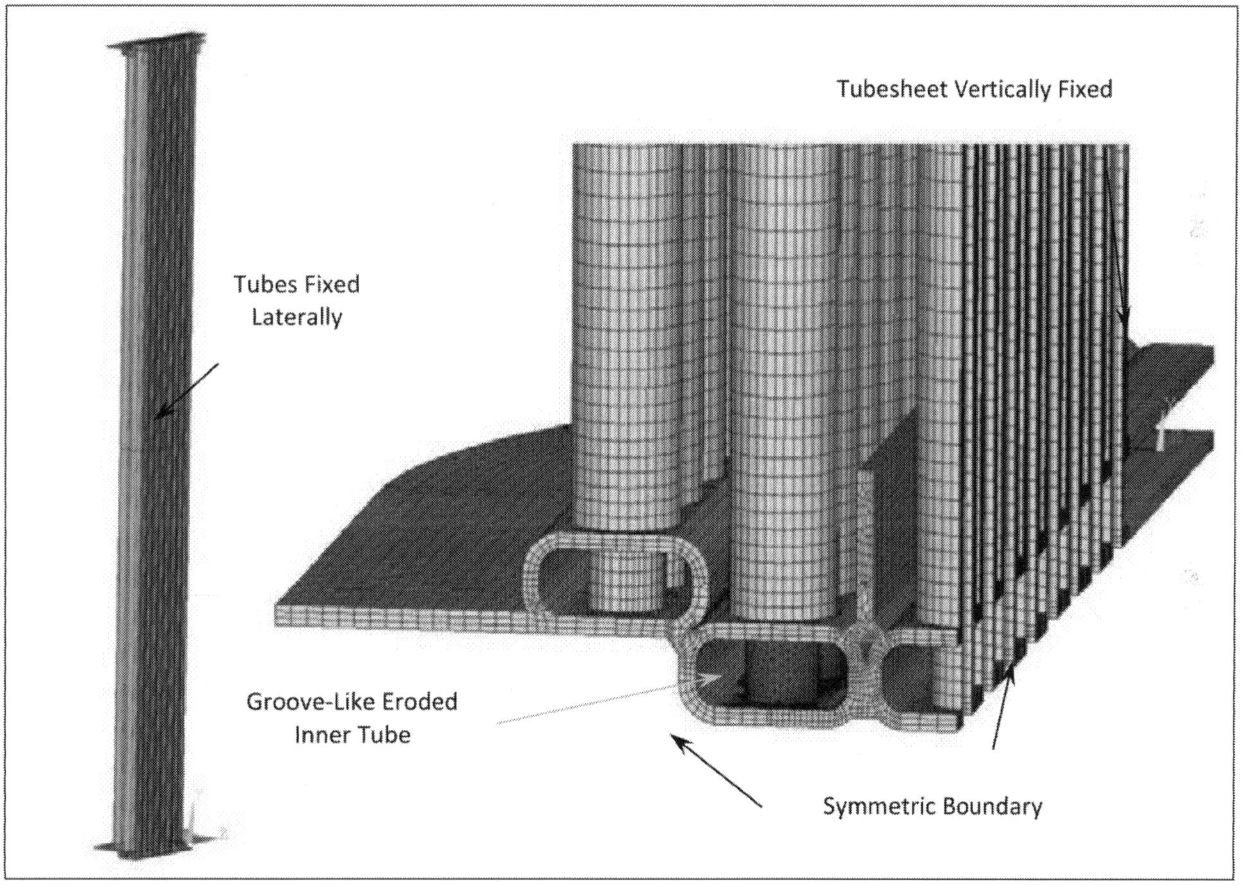

Figure 11: FE Model of SCHMIDTSCHE SCHACK Bathtub Vertical Flow TLEs with Eroded Tube

Elastic properties, thermal properties, yield strengths, and stress-strain data for the materials of construction of the TLE shell, tubesheet ant tubes were obtained from the 2010 ASME Boiler and Pressure Vessel Code, Section II, Part D, as well as Annex F of the FFS Code.

Thermal Analysis

The five TLE/Eroded tube models with different t-min were thermally analyzed to determine temperature profiles. Different film coefficients were used for the shell and tube side. Gas side temperatures were applied at the inlet and at the outlet of the tubes and steam temperature on the shell side. The tubes were assumed insulated.

The temperature distributions for all TLE FE models were very similar. Figure 12 illustrates the temperature distribution at the inlet tubesheet and inner tubes for a TLE model with one groove-like eroded tube. The maximum temperature calculated was slightly lower at the gas inlet of the inner tube for the groove-like eroded tubes. The maximum temperature experienced by the TLE was well below the creep temperature range of the tubesheet and tube materials. Therefore, a creep analysis to calculate creep lifetimes was not necessary. These temperature distributions were used to run the FE stress analysis presented in the next section.

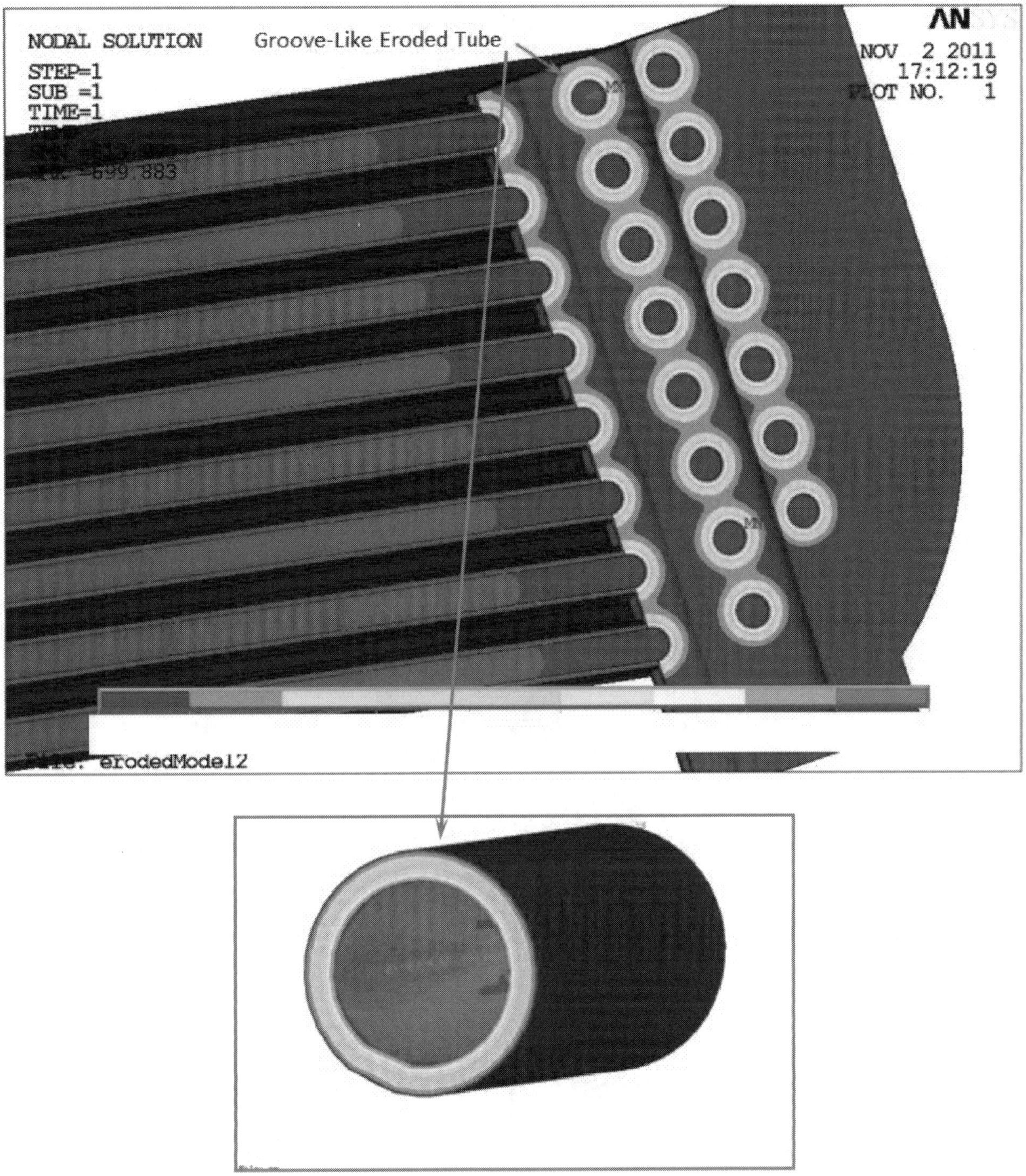

Figure 12: Temperature Distribution with One Groove-Like Eroded Tube

Minimum Groove-Like Eroded Inner Tube Thickness

Determination of t-min for a groove-like eroded inner tube to safely operate the TLEs was first conducted by performing elastic-plastic TLE/Eroded tube FE analyses using three different decreasing thicknesses below the measured t-min. The analyses were performed for pressure, weight and thermal loading equivalent to a load factor of 2.71. This load factor was established based on Table B1.4 of the FFS Code. Operating pressure loads were applied for the shell/steam and tube sides. The weight of the tubes was included as well as the weight on top of the TLE.

As expected the analyses, shown in Figure 13, indicate the maximum Von-Mises stresses and strains are located at the thinnest section of the groove-like eroded region of the inner tube. However, this location is in compression.

Figure 13: Stresses and Strains at Groove Region at a load Factor of 1.51

Results for protection against global and local plastic collapse are given in Figure 14. Global and local FFS Code criteria were satisfied only for t-min_1 and t-min_2. The tube with the smallest eroded tube thickness failed the global criteria (t-min_3). Based on these results, the next smallest eroded tube thickness, t-min_2, was selected as a candidate t-min for complete FFS Code validation.

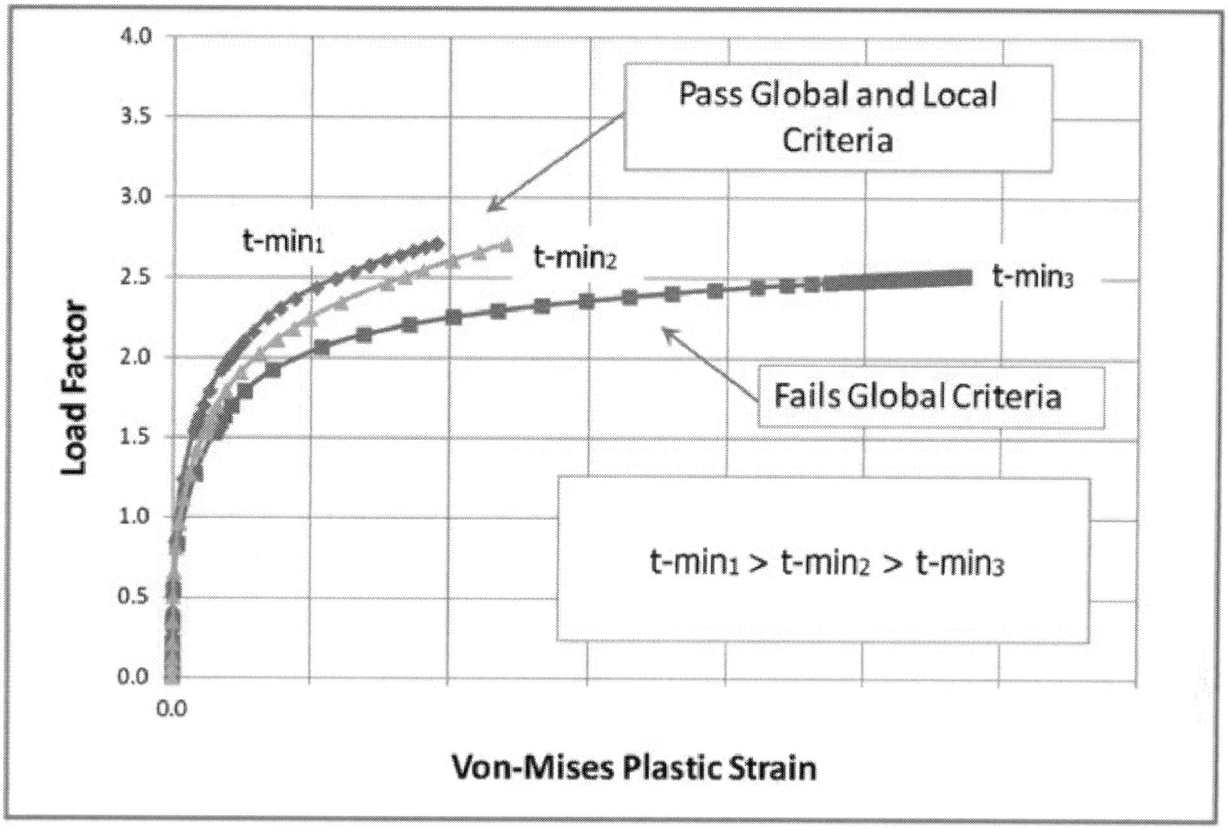

Figure 14: Results of Global and Local Plastic Collapse Analyses for Candidate t-min

Validation of the candidate t-min_2 required elastic-plastic analyses to assess protection against buckling, plastic collapse, ratcheting, and determination of a fatigue lifetime. Two elastic-plastic analyses were conducted to evaluate protection against buckling and global plastic collapse per paragraph B1.4.1 and B1.2.4.3. The first analysis was for a load factor of 3.15 (P + D), while the second was for a load factor of 2.71 (P + D + T). These load factors were determined using Table B1.4. of the FFS Code. Paragraph B1.3.3 was followed for evaluation against local plastic collapse at a load factor of 1.51 (P + D). The nominal thickness was also reduced as the request of CP Chem to evaluate the buckling stability of the tubes from generalized thickness loss. A TLE with three groove-like eroded tubes instead of only one was used for this t-min validation.

Plastic assessment results can be seen in Figure 15. The TLE with three groove-like eroded tubes with t-min_2 failed to converge both global load factor cases. The first case only reached 1.8 (P + D) while the second 2 (P + D + T). These results suggest the additional groove-like eroded tubes and reduction in inner tube thickness led to a significant increase in plastic deformation at the groove region and non-convergence of the buckling and global plastic collapse analyses. However, using a t-min_4 = 2 $tmin_2$ all criteria for protection against global, local and buckling plastic collapse were satisfied.

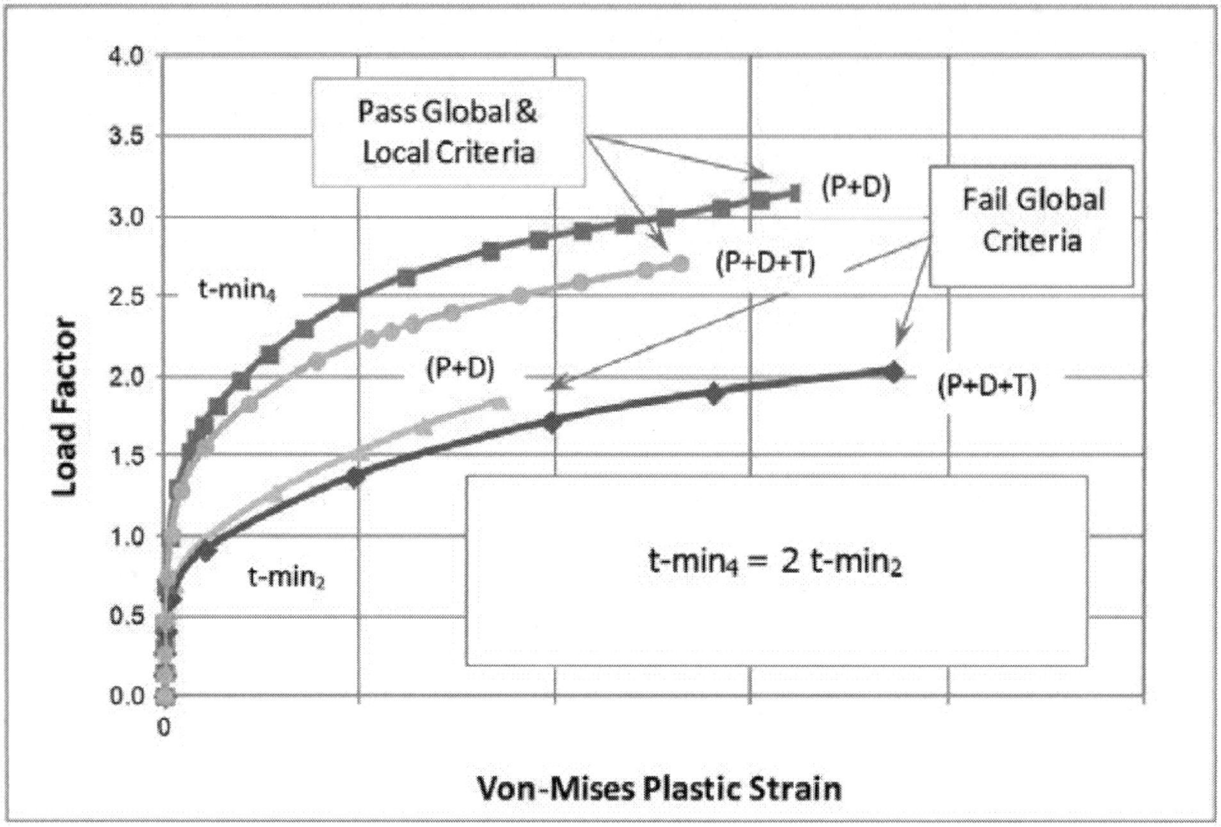

Figure 15: Plastic Collapse Assessment Results with a TLE with Three Eroded Tubes

Ratcheting Assessment

Protection against ratcheting of the TLE with three groove-like eroded tubes with a thickness of t-min4 was conducted per paragraph B1.5.7 of the Code using (P + D + T) loading. The analysis used Chaboche's multi-component nonlinear elastic-plastic, kinematic hardening material model that allows simulation of ratcheting and shakedown effects of the material. Results of this analysis, given in Figure 16 , show the TLE with the eroded tubes will eventually shakedown because the slope of the plastic strain vs. number of cycles is decreasing, indicating it will eventually converged/shakedown. Note that the results shown are for a TLE with a single groove-like eroded tube instead of 3. The reason for this was to simplify the analysis and based on the fact that, as shown in Figure 17, the highest plastic strain of the single groove-like eroded model, is similar in magnitude and trend with number of cycles of the TLE with three groove-like eroded tubes for 6 cycles.

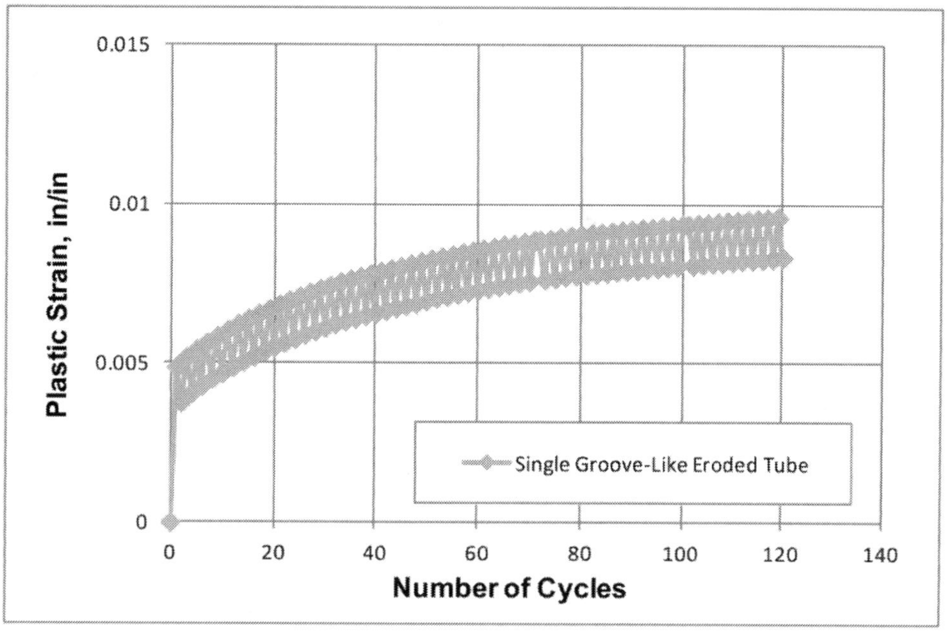

Figure 16: Single Tube Plastic Strain Vs Load Step at Thinnest Location

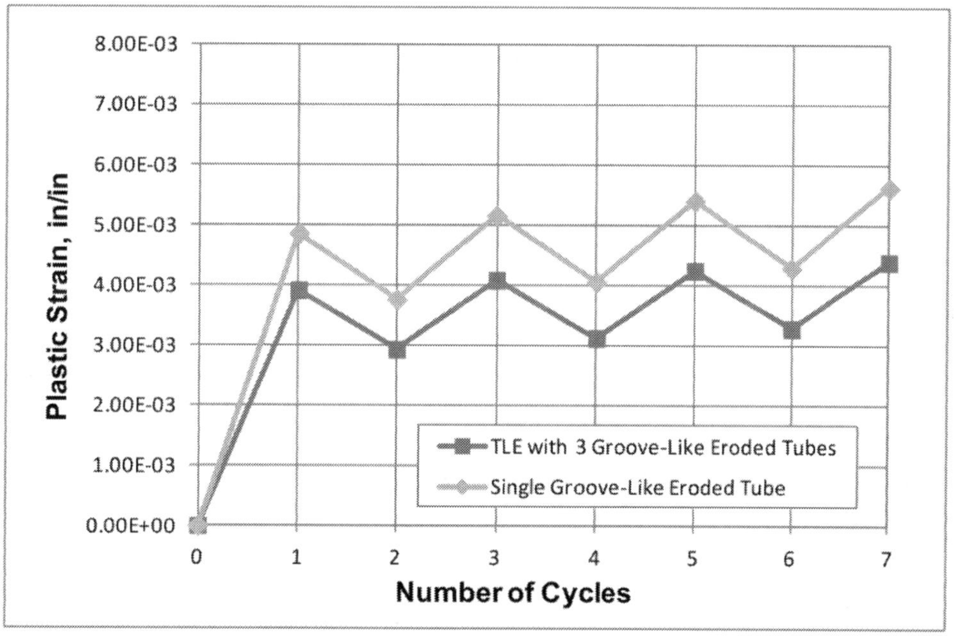

Figure 17: Plastic Strain Vs Load Step at Thinnest Region in Groove Region

Fatigue Lifetime

A fatigue lifetime analysis was conducted per paragraph B1.5.4 of the Code involving elastic-plastic material properties with stabilized cyclic stress-strain curve properties. The limiting fatigue lifetime of the TLE with 3 groove-like eroded tubes was calculated to be 4380 cycles.

Determination of Operating t-min by Chevron Phillips Chemical

It has to be emphasized that the validity of both assessments are restricted to the available information of the tube wall thinning, geometry and boundary conditions. However, the previous statement is in general valid for each simulation result.

Even though the local t-min$_4$ determined by the KHE FFS assessment was significantly thinner than the SCHMIDTSCHE calculated global t-min, the inherent uncertainties in any inspection wall thickness measurement and the geometric restrictions in the FFS assessment, prompted the Chevron Phillips Chemical decision to operate with a t-min that was between the KHE FFS assessment and that calculated by SCHMIDTSCHE. This operating t-min is shown graphically in Figure 18. Chevron Phillips Chemical has successfully used this t-min to increase the service life of, selectively replace tubes with localized groove-like erosion damage, and continue safe TLE operation for more than six years since the FFS assessment was conducted. Without this FFS assessment, the tubes would have been replaced at higher t-mins. This in turn would have resulted in a higher number of tube replacements.

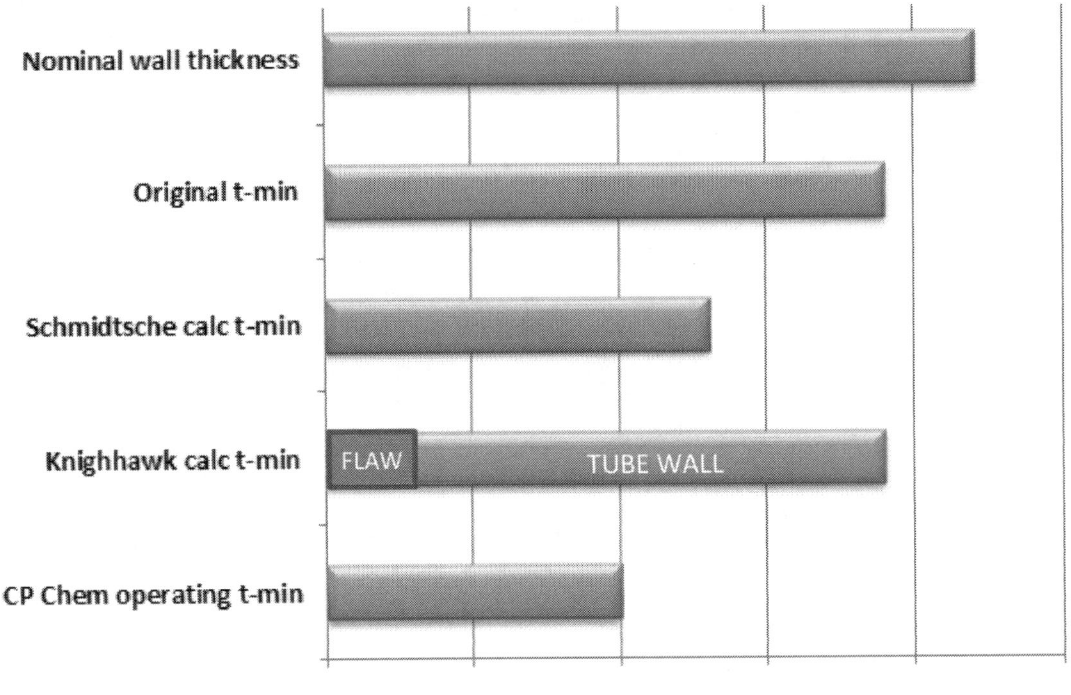

Figure 18: Comparison of Operating and Calculated t-min.

Conclusion

This paper illustrates the successful collaboration between TLE owner/user, OEM, and a third party engineering contractor to increase the service life and selectively replace TLE tubes with localized groove-like erosion damage using a Level 3 FFS Code assessment. Because circumstances involving TLE's are unique, each TLE owner/user would need to conduct its own appropriate testing and analysis (FFS assessment) before drawing conclusions about t-min adjustments for particular applications.

AIChE PAPER NUMBER 128c

PREDICTIVE ANALYTICS USING PROCESS & OPERATING DATA

Fréderick Motte
Customer Success Manager
TrendMiner Inc.

Sandy Peterson
Account Executive
TrendMiner, Inc.

Prepared for Presentation at the 2018 Spring National Meeting Orlando, Florida April 22-26, 2018

PREDICTIVE ANALYTICS USING PROCESS & OPERATING DATA

Fréderick Motte
Customer Success Manager, TrendMiner Inc.
Sandy Peterson
Account Executive, TrendMiner, Inc.

ABSTRACT

Asset performance in the petrochemical industry is an increasing area of focus for operating companies. Firms are working to control short-term maintenance and operating costs, as well as reach longer-term goals of operational excellence. The majority of traditional data modeling exercises, however, only provide a positive return on investment (ROI) for the 1% critical assets because they are typically very engineering-intensive projects which require a skilled data scientist who is not necessarily familiar with the process.

Self-service analytics provides engineers and operators with new ways to monitor assets in order to improve availability and overall effectiveness. Here, case studies are presented that reflect successes derived from applying an analytics platform to processes, and the benefits received.

Three areas related to asset reliability of particular interest for ethylene producers are discussed in this paper: valve performance, heat exchanger efficiency, and compressor reliability. For valve performance, the issue of drifting valve characteristics, and valve wear and tear are tackled. For heat exchanger efficiency, this paper demonstrates how prediction of fouling can improve overall production. Finally, in the compressor case, the focus is on understanding, and predicting oil temperature alarms.

These examples demonstrate how deploying self-service analytics helps engineers increase the reliability of operating assets in ethylene production by establishing baseline performances, identifying anomalies, and creating background monitors that act as an early warning system. For the company's bottom line, predictive, self-service analytics reduces stress on assets, avoids unplanned downtime, and helps establish timely and properly planned and prioritized maintenance timetables.

INTRODUCTION

Reliability improvements are an on-going directive for process and maintenance personnel. If all assets, not just the critical ones, can be operated more reliably and predictably, the twin goals of safety and profitability can be achieved.

One feature of daily engineering life is the increasing availability of data - sensor data, actuator data, pressures, temperatures, flows, rpms, viscosity, etc. The list is almost endless and it often requires data scientists to make sense of it all. But what if the people closest to the data - the process and reliability engineers – were enabled to leverage analytics in their day-to-day tasks? They would uncover a wealth of actionable information that can be applied to their daily operations.

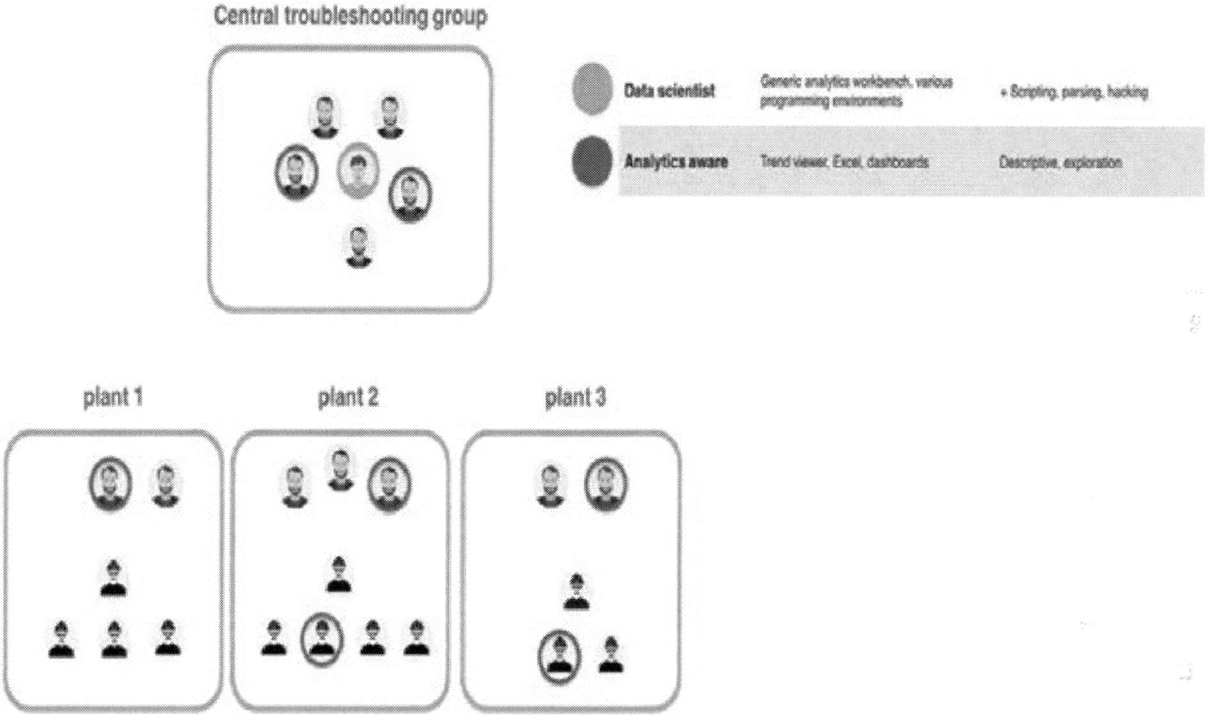

Figure 1.0—Generalized Engineering Organization

Figure 1.0 shows what a general engineering organizations may look like. Many individuals at the plant level use trend viewers, spreadsheets, and a variety of other dashboards. These tools help describe plant and equipment behavior, however when it comes to rapid searching of data, diagnosing root causes, predicting future responses, or recommending remediation actions, they are limited in their functionality.

Engineers at the corporate or divisional level, on the other hand, may have more analytics experts and actual data scientists at their disposal. Their tools such as workbenches and programming environments allow them to search, hack, and parse large datasets, as well as do a deep dive into certain situations. These tools can tease out unique insights and are extremely valuable, but without a tie back to the process, they can often end up being merely an intellectual exercise.

Self-service analytics can help bridge the gap between the centrally located data scientists who know how to analyze data, and the process and reliability engineers who bring deep subject matter expertise about how their plants run.

I. OPERATING USE CASE #1 - Deethanizer Reflux Valve Performance Degradation

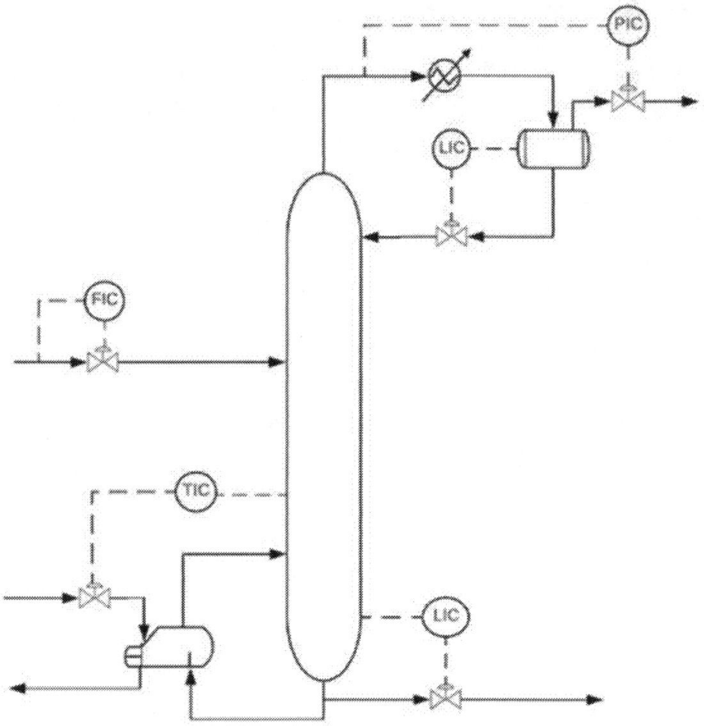

Figure 2.0—De-ethanizer control scheme. Valve performance of top LIC control scheme under review.

Background - The reflux valve in the de-ethanizer section would foul, upsetting the cold section of the plant. Valve performance typically degrades over time because internals corrode, repetitive movement causes stress on the equipment, the actuator loses potency, etc. Independent of the underlying reason, the issue will manifest itself as a changing valve characteristic.

For example, if a valve is opened at 75% at the beginning of service, one might expect that the same volume of flow runs through that same valve at that same opening sometime later in its life. However, if the valve begins to foul, the flow passing through will change. In this case, the de-ethanizer reflux valve began to foul and caused an upset that rippled through the entire cold section of the ethylene plant.

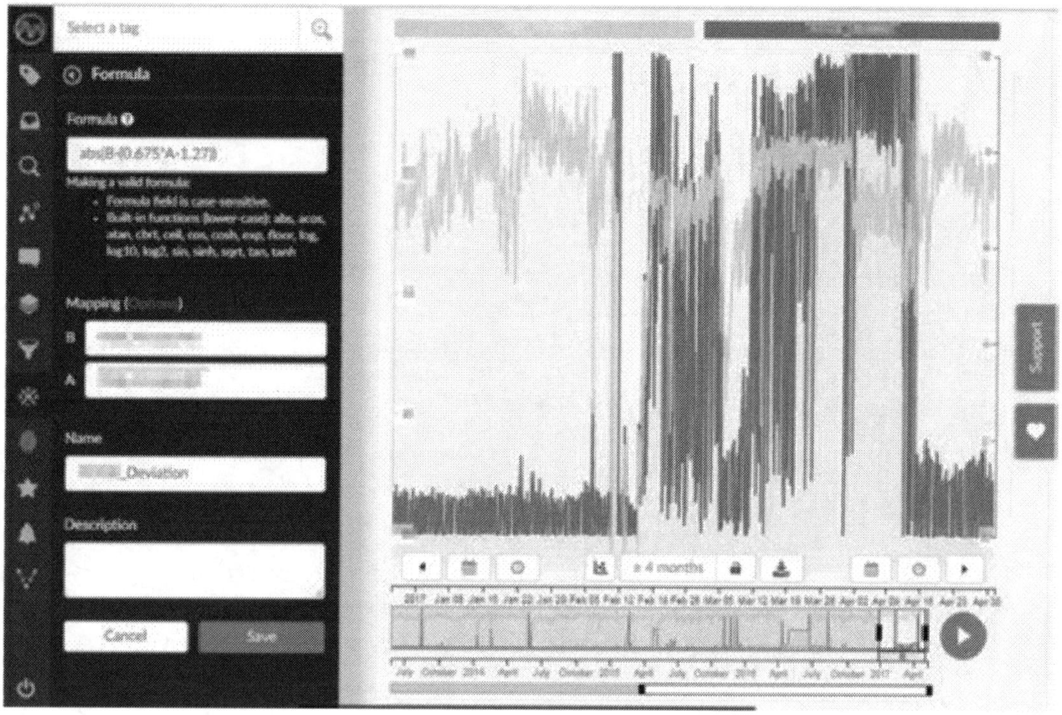

Figure 3.0 Calculated Deviation and Valve Output. Green signal represents deviation from normal expected valve position and should not exceed natural noise.

The simplest approach was to set up a univariate alarm when the valve opening nears saturation (e.g. >80% depending on valve characteristic). However, this simple approach has severe limitations. In one example during periods of reduced throughput, the valve could already be severely fouling but the absolute opening might still be relatively low. Serious problems could then arise when the throughput is increased again.

A multivariate regression is another option, but set-up is often time consuming and hard to generalize on many valves. Some of the challenges faced:

- One would like to capture as much variation as possible, but periods with disturbances make analysis harder.
- Comparison with vendor-supplied valve curves is not straightforward
- Periods of fouling are not excluded from the analysis

In general, the fouling is recognized too late and actions are reactive rather than proactive.

Solution & Results: Self-service analytics tools making use of real-time and historian data made it possible to rapidly search through operational history and identify all previous periods of fouling.

First, the data was quickly segmented and the current performance was compared to both the baseline behavior and the vendor-supplied valve curves. Second, these patterns were overlaid on periods of no fouling (just after maintenance, for example), and a suggested regression analysis was performed. As soon as performance began to drift, an alert was sent to the process and reliability engineers, and maintenance was scheduled before the process drifted into instability. Figure 3.0 above shows the calculated valve deviation (green) plotted with actual flow (yellow) to highlight areas of instability.

The outcome was a repeatable workflow that improved the overall stability and reliability of the entire cold section without replacing any valves or actuators.

II. OPERATING USE CASE #2 – REFRIGERATION COMPRESSOR TRIPS

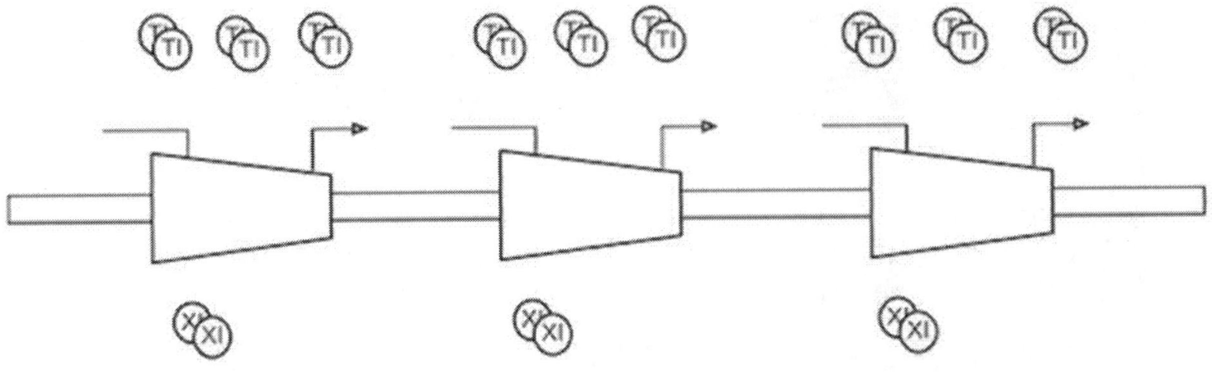

Figure 4.0—Simplified Compressor Train. Redundant temperature and axial displacement measurements indicated as Ti and Xi respectively.

Compressors are, for the most part, critical assets that are closely monitored, with built-in, high-speed monitoring and control systems for inlet/discharge pressure, rotation speed and protections against surge. Other constraints or alarms, however, may appear that are of great interest to engineers, especially when compressor behavior changes over time.

Background: One compressor began experiencing unplanned and unpredictable trips. Some hypotheses existed, but due to the sheer amount of data and variation, these could neither be validated nor rejected using traditional methods. Cause and effect were particularly hard to separate.

Solution & Results: Self-service analytics were used to pay particular attention to axial displacement, in particular when the axial displacement of the compressor began to cause high oil temperatures, rather than other causes.

Historical data was searched with self-service tools and all compressor trip events were examined. Once the compressor trips were all retrieved they were then overlaid with normal compressor behavior. A statistical comparison of those layers revealed an inconsistency in the suction valve patterns prior to the trips compared to normal behavior.

The hypothesis that those abnormal valve patterns were predictors of a trip was tested through additional searching. The number of compressor trips that occurred after the abnormal patterns was significant. Another mechanism through which trips could occur was elevated oil temperatures. Singling out these events and identifying which ones were related to axial displacement have enabled the engineers to take action at the right times. Other causes, like temporary cooling water changes can be differentiated.

Additionally, engineers were able to use the patterns they found to provide a "trip fingerprint". This fingerprint was overlaid on current operations and alerts were configured based on the identified patterns so that now staff can then be warned about an upcoming compressor trip and maintenance can be prepared. This leads to 1-2 hours less downtime of the compressors.

We will leave the savings to the reader.

III. OPERATING USE CASE #3 - HEAT EXCHANGER FOULING

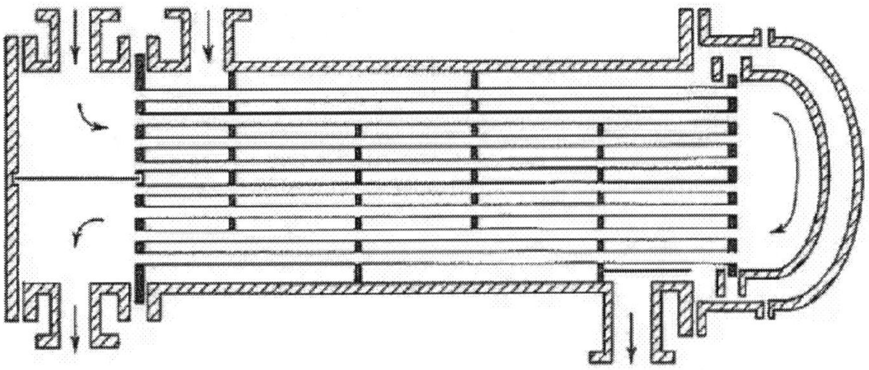

Figure 5.0 Simple Shell and Tube Heat Exchanger

Heat exchanger fouling is one of the most common problems faced by process and reliability engineers. It would be an ideal situation if different types of fouling could be predicted, and maintenance be proactive, rather than reactive.

Under normal process conditions, fouling behavior is typically noted by a rise in ΔP or a change in calculated heat transfer coefficient, *U*. Self-service analytics can be used to identify a general trend for predictive maintenance planning, alert and support a maintenance decision, and help find the right process parameters to tweak to lengthen the time between maintenance cycles.

However, fouling is also caused by special factors—the viscosity of the process fluid changed, the temperature of the heat exchanger fluid changed, etc. This type of situation is where self-service analytics can help diagnose and prevent these special cause fouling cases by detecting the root cause (i.e. process anomalies upstream), and providing engineers an alert based on prediction.

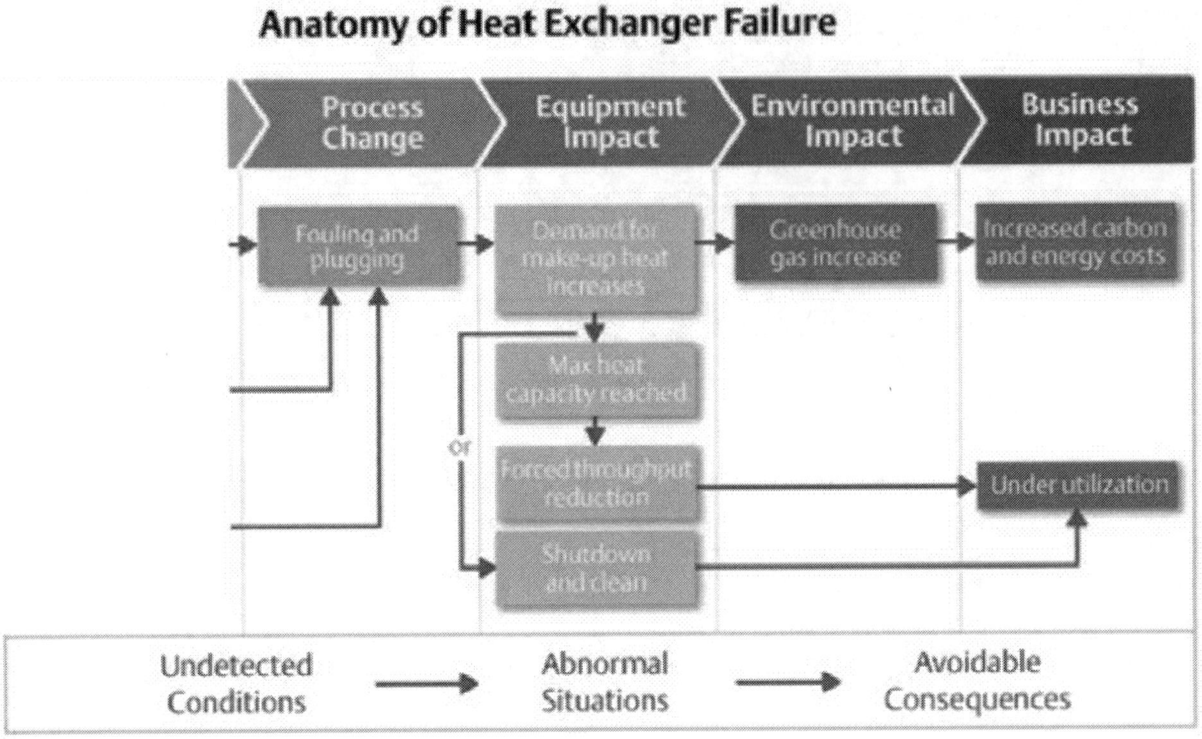

Figure 6.0 Anatomy of Heat Exchanger Fouling

Background: Periodic heat exchanger fouling leads to process upsets and unacceptable product quality. Prior to using analytics, plant personnel would switch to a spare heat exchanger after 2-3 days; however, this spare exchanger would foul as well. The cycle time to recover from fouled heat exchangers was generally two or more weeks. As noted, the "normal" solution would be to bypass the fouled exchanger and route to the spare.

Solutions & Results: Applying self-service analytics, engineers were able to identify a pattern of changes in the heat exchanger co-efficient. Once the pattern was identified, an alert was configured to notify the engineers that maintenance should be performed as soon as possible.

This approach was applied to many different heat exchangers, including the steam cracker transfer line heat exchanger, de-ethanizer and de-methanizer columns, heat exchangers on the refining side (including FCC preheat and hydrotreater units) and finally, the polymerization side.

Even though the residence time of the fluids in the heat exhangers might be very short, over longer periods, patterns of degraded performance start to emerge.

This identification of fouling patterns and the configured alert gave plant personnel a two to three week lead before fouling began to affect the stability of the plant and product quality.

CONCLUSION

All of the cases described had been ongoing operational and maintenance problems; and traditional methods of remediation failed to satisfactorily solve them. Rather than resorting to a major revamp or process redesign, the customer turned to the data they had at-hand, the process data stored in their historian, and applied self-service analytics techniques. The solutions that were discovered enabled them to improve process stability, avoid unplanned shutdowns, and improve the overall reliability of the equipment.

These engineers were not data scientists, but process and equipment experts. Once enabled with the proper tools and an increase in analytics awareness, these SME's were able to achieve significant results at a modest investment in time and technology.

AIChE PAPER NUMBER 128d

CONDITION BASED RECOILING

Stef Luijten
Senior Global Technology Manager, Ethylene Cracking Furnaces
The Dow Chemical Company

George Bellos
Senior Research Scientist
The Dow Chemical Company

Prepared for Presentation at the 2018 Spring National Meeting
Orlando, Florida, April 22 – 26, 2018

CONDITION BASED RECOILING

Stef Luijten
Senior Global Technology Manager, Ethylene Cracking Furnaces
The Dow Chemical Company

George Bellos
Senior Research Scientist
The Dow Chemical Company

Abstract:

Radiant coils in ethylene cracking furnaces are operated at high tube wall temperatures. Remaining coil lifetime is determined by metallurgical properties, including carburization and creep resistance, furnace design, and operating conditions. The cost of a furnace shutdown for recoiling can be significant and may include lost ethylene production.

Optimizing furnace recoil intervals, while maintaining furnace availability and reliability, is a key furnace ownership task. Recoiling a furnace with coils that still have significant remaining life is a suboptimal use of resources. On the other hand, if a recoil is scheduled with coils at end of life, then furnace reliability will suffer and there will be considerable risk of coil failure and unscheduled maintenance work.

Carburization is an important radiant coil damage mechanism. Today there are robotic techniques available to measure coil carburization in the firebox (using a carburization inspection robot). These measurements are non-destructive and can be done while a furnace is out of operation for other maintenance reasons.

Dow is using carburization data as part of a condition based recoiling program, to optimize furnace recoil intervals along with modeling to assess coil coking tendency and thermal exposure across the entire coil life.

This paper will cover some of the important coil failure mechanisms, radiant coil inspections with the carburization inspection robot, focusing on the associated modelling efforts and how inspection data is used and interpreted. Finally, we will explain how this affects our furnace recoiling strategy.

Introduction

Ethylene cracking furnaces are at the heart of an ethylene plant. Furnaces are the chemical reactors that convert feed into products like ethylene and propylene. When designing and building a new ethylene plant, cracking furnaces are a significant part of total invested capital. Once built, they essentially define plant economics, since plant yield and conversion energy is mostly determined by the furnace design and operation. So, furnaces are important in many ways, and good furnace maintenance practices are vital to ensure furnace availability and plant profitability.

Ethylene plants typically have 5 to 15 cracking furnaces in operation, allowing them to decoke or maintain one furnace while the other furnaces remain in operation, without significant reduction of total plant throughput, see Figure 1.

Figure 1: Cracking furnaces

Radiant coils are subject to temperature cycles. The pyrolysis reactions take place at high temperature. An undesired byproduct of the cracking reactions is coke, which deposits at the inside wall of the radiant tubes. Coke formation will deteriorate heat transfer, increase radiant coil pressure drop, and increase tube metal temperatures. At some point, either the coil pressure drop or tube metal temperature reaches a maximum allowable value, and the furnace needs to be decoked. Decoking is done by a controlled combustion of coke, sending a steam/air mixture at controlled temperatures through the radiant tubes. After all coke has been removed, the furnace is made ready for the next run.

Combustion of natural gas and/or plant offgas is the most effective method of realizing the high process temperatures that are required for the pyrolysis reactions (820-860°C). A number of parallel radiant coils are mounted vertically in an insulated firebox. Heat transfer from the burner flames to the coils is largely by radiation, see Figure 2. Due to the high outer tube wall temperatures (up to 1100°C) the radiant coils are manufactured from high temperature alloys. Most commonly used are centrifugally cast Cr-Ni-Fe alloys with good high temperature corrosion resistance, high temperature strength, and good machinability and weldability.

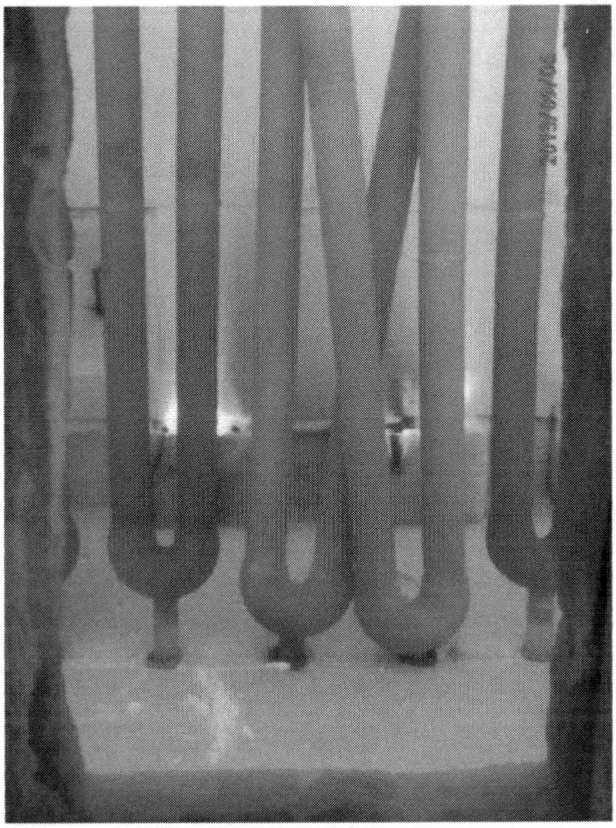

Figure 2: Radiant coils

As a result of the severe temperature conditions, radiant coils have a limited life. The coils are subject to ageing, creep and carburization. Process performance slowly deteriorates, at the same time the risk of coil failure increases. Radiant coils need to be replaced after 2 to 8 years of operation (see Figure 3), depending on coil metallurgy, cracking furnace design and furnace operations.

Figure 3: A recoil in progress

Recoiling Strategies

The cost of a furnace recoil can be significant and, depending on furnace availability, may also involve lost ethylene production. The scope of work sometimes includes additional maintenance or inspection activities, in some cases the duration of a furnace outage can extend to several months.

Historically, Dow ethylene plants have been recoiling their furnaces based on a fixed time schedule, where every furnace has a target coil life, and the next recoil date is set many years before it actually takes place. Target coil life is based on factors like furnace design and historical experience. This essentially results in an *educated guess* of the optimal interval, determined by furnace SME's in cooperation with plant engineers, and chosen such that the risk of premature coil failure due to ageing effects is thought to be minimized.

The downside of using fixed recoiling time intervals is that a particular recoil will always be either too early or too late, in terms of coil carburization and mechanical reliability. Recoiling a furnace with coils that still have significant remaining life is a suboptimal use of resources. On the other hand, if a recoil is scheduled with coils near or at end of life, or if the coils are actually being run to failure, then furnace reliability will suffer, leading to unscheduled maintenance work, higher maintenance cost, lower furnace availability, and longer furnace outages.

Scheduling recoils *just-in-time* implies that the plan is based on *coil remaining life*, and the planned date of a recoil is set, say, within a year before the recoil, and essentially becomes a tradeoff between furnace performance, reliability, and recoil cost. This strategy needs strong metrics for remaining coil life, together with a sound assessment of the risk of mechanical failure. It does however ensure better use of resources, as the recoils are scheduled at a carefully selected and acceptable level of coil performance and mechanical reliability.

Dow is introducing a condition based recoiling program, with the intent of recoiling *just-in-time*. We are beginning to optimize our recoil intervals, along with modeling efforts to assess coil coking tendency and thermal exposure across the entire coil life. Measuring coil carburization is an important part of the program, helping us to determine coil remaining life, as will be clarified in the next sections.

Failure Modes and Radiant Coil Carburization

The importance of having coil carburization available becomes clear when realizing that carburization is an critical factor in many radiant coil failure modes. To illustrate this, a few important coil failure modes are described below (see Ref [1]):

Brittle fracture occurs when a coil is carburized and exposed to fast cooling with a significant layer of coke present on the coil inside wall (see Figure 4). A carburized coil is brittle and will crack at low strains. The coke layer has a much lower thermal expansion coefficient than the tube metal. So if a furnace trips and cools down, the coke layer will resist the shrinkage of the metal tube, leading to a high stress in the tube.

Brittle fracture is a classic and frequently occurring failure mechanism in ethylene furnaces. It can result in many failed tubes after furnace trips. Tube failure after a furnace trip is more likely if the coil is carburized and if a thick coke layer is present.

Figure 4: Brittle coil fracture

Bulging and ovalization is a result of carburization. Carbon uptake leads to a higher specific volume of the formed carbides, leading to local coil deformations. The internal stresses can eventually build up to coil failure.

Creep elongation occurs because of creep by the weight of the tubes, at elevated temperatures. Carburization is a major contributor to lowering of the creep resistance of used alloys, and shortening remaining coil life. Coil failures can occur if coils reach the furnace floor. The coils can become bowed, resulting in even higher tube stresses and creep rates.

Condition Based Recoiling

Dow's condition based recoiling program is aimed at optimizing furnace recoil intervals, based on coil inspections plus process data. We intend to move away from fixed time base intervals towards a data-driven approach, without giving up on furnace reliability. It should be noted that our primary intention is not to move all recoils backward and save money. Condition based recoiling can work two ways: we might decide to postpone specific recoils, based on inspection and process data, but may also decide to move other recoils forward, if there is sufficient evidence that coils are approaching their end of life. The program is designed to ensure that recoils are scheduled at a quantifiable and acceptable coil remaining life.

We have put several tools, practices and metrics in place to support our condition based recoiling program:

1) Tracking of process data that are relevant to coil load
2) Collecting routine tube metal temperature data
3) Performing metallographic investigations, if and where needed
4) Performing routine full coil carburization measurements
5) Modeling and prediction of coking rates
6) Offer a web-based data selection, analytics and visualization tool

Each element will be described in more detail in the next paragraphs.

Process data and Pyrometer TMT

Furnace process data are stored in Process Information history servers for the purpose of e.g. performance evaluation and troubleshooting. A lot of measurements are also good indicators for the assessment of remaining furnace coil lifetime, including hydrocarbon flow rate, steam dilution ratio, cracking severity and coil outlet temperature. All are key factors in determining coil thermal exposure (*time-temperature history*) and thus remaining coil life. Additionally, furnace run length, the number of decokes and furnace trips need to be included in the assessment of remaining coil life.

Outside tube metal temperatures (TMT's) are measured manually (see Figure 5), using a pyrometer that measures radiation at one or more specific wavelengths, aiming at the outer coil surface. Outside tube metal temperatures are also an important factor for coil life. This is why, as part of the program, we have defined specific instructions and guidelines for taking TMT measurements, and store the results together with process data in our PI history servers.

Figure 5: Taking a pyrometer TMT measurement

We are keeping all relevant process data for all furnaces with a sufficiently long retention time (longer than 1 recoil interval), and make them available for easy visualization, review and analysis, modeling and use for recoil planning purposes. TMT data, together with coil outlet temperatures, are prime indicators of a coil's time-temperature history.

Metallographics

Metallographics is used in the program to study the physical structure of coil metals, after a recoil or after coil failure. Metallographic techniques include Scanning Electron Microscopy (SEM) and Energy-Dispersive X-ray Spectroscopy (EDS), see Figure 6. The investigations are used to measure Chromium depletion and the degree of carburization, and to identify creep voids. Metallographics is a valuable add-on to using ferromagnetic devices, as it will distinguish between OD oxidation and/or Chromium depletion, or midwall / ID carburization.

We also measure mechanical properties at room temperature and tensile strength at elevated temperatures. Post-mortem metallographics has an important role in root cause finding after a coil has failed, and to verify carburization depths that were reported by coil carburization inspections.

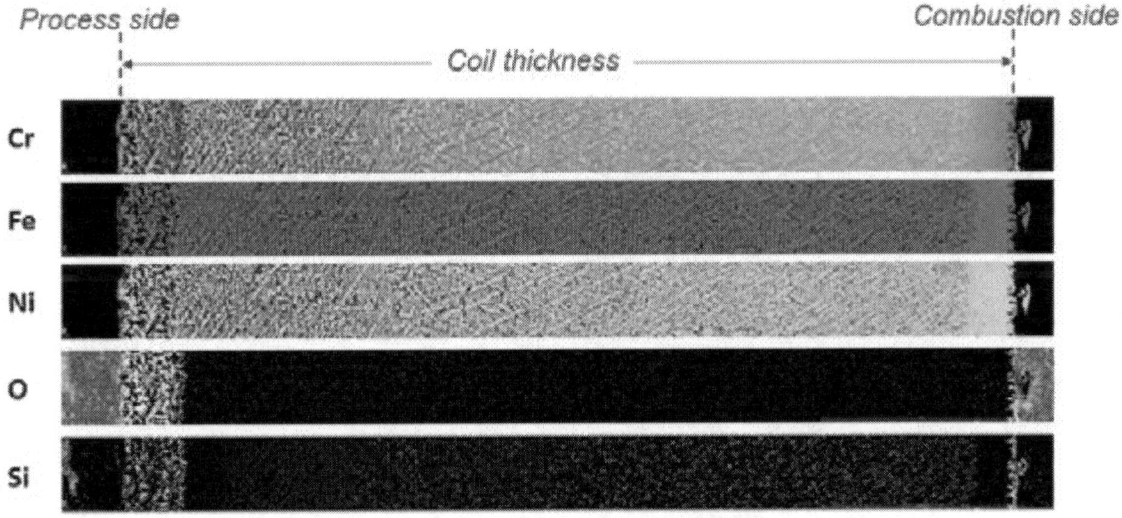

Figure 6: SEM-EDS analysis of an aged radiant coil

Carburization Inspections

Key element in our condition based recoiling program, and new to Dow, is the measurement of coil carburization. Today there are robotic techniques available to measure coil carburization in the firebox. The measurements are non-destructive and can be done while a furnace is out of operation for other maintenance reasons. So the furnace needs to cool down with the firebox made ready for entry.

Typically, we combine this inspection with other planned furnace maintenance activities, like hydrojetting of primary quench exchangers. The robotic device that is employed (see Lit. [5] for more details) is a significant improvement compared to handheld devices, as the measurements can be done without scaffolding, and in a safe, fast and cost-effective manner.

The carburization robot, see Figure 7, is equipped with a number of circumferential Hall-effect sensors, measuring local magnetic responses that result from Chromium depletion.

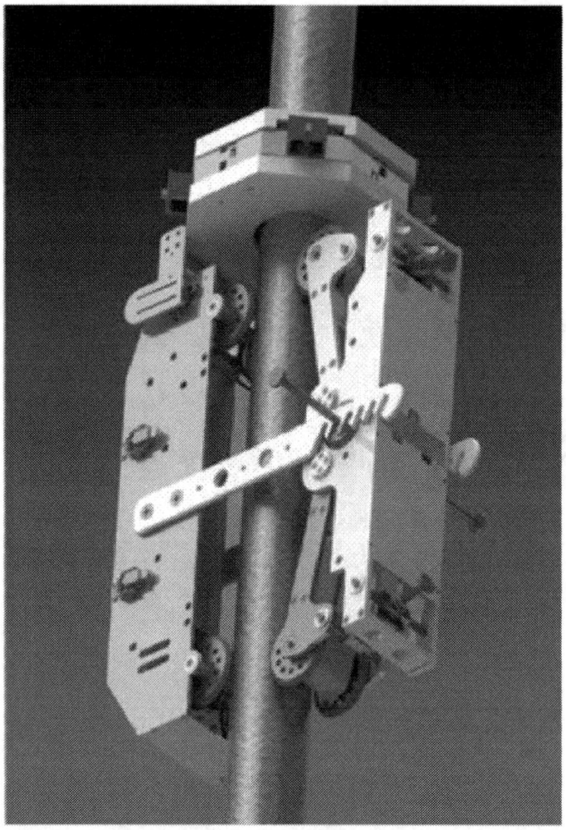

Figure 7: Carburization inspection robot

The translation from raw probe data into a % carburization depth is done using carburization curves. A carburization curve is specific for one combination of alloy composition and tube wall thickness.

Carburization inspections do not significantly affect the total downtime of a furnace. They can be scheduled during planned or unplanned furnace repair activities, including a furnace recoil.

The carburization inspection reports are quite detailed and contain all carburization measurement results, for every measured coil, see Figure 8. This information is analyzed and summarized for later use, and greatly helps us determine a target date for this furnace's next recoil.

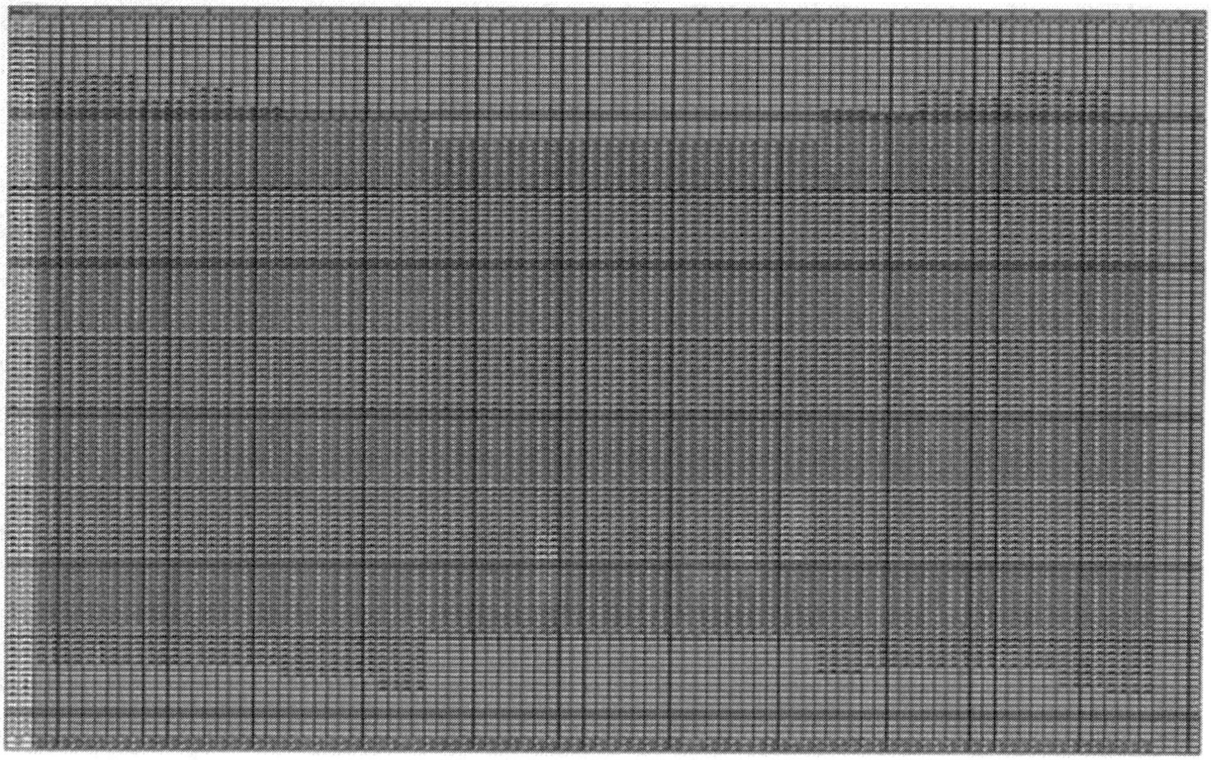

Figure 8: Typical carburization measurement result

Coking Modeling

Coking rate is recognized as being a key indicator for coil ageing. A coking rate model is developed to assist in evaluating observed coking tendencies and predict remaining coil lifetime. The model essentially forecasts the change of coking rates over time. A confidence interval is defined through statistical analysis, and together with an interpretation of actual ethylene furnace process data, this allows for an assessment of acceptable values for coking rate.

The model reads furnace process data and estimates a coke layer thickness that matches actual coil pressure drop, at actual operating conditions. Similar modeling efforts have been presented earlier in EEPC [3]. The calculations in the current model are done for each coil separately. The change of coke thickness over time is the quantification of coking tendency. Data filtering is crucial in order to get meaningful results. The data management is carried out step-wise, as presented in Figure 9. The following data reduction techniques are implemented:

- Eliminate data which do not correspond to cracking operations, such as decoke, heat up and cool down stages.
- Select one hourly data point per 24 hours to reduce data density and accelerate calculations.

- When a furnace trips in a run, significant fluctuations of coking tendency are expected which do not correspond to usual coke growth. An algorithm is developed which identifies furnaces trips and eliminates run data from the time the trip occurred, until time zero of the next run.
- Eliminate data when process upsets cause pressure drop swings until the next run. To detect the moment when a process upset starts, it takes noise reduction techniques so that the first derivative can be estimated with relatively low error.
- Remove outliers when process conditions are outside the lower and upper control bounds

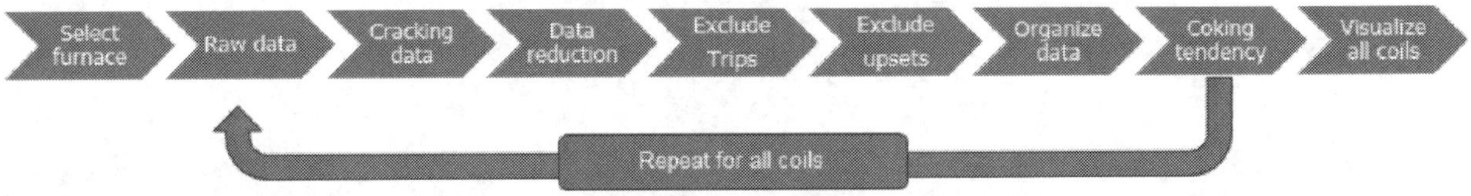

Figure 9: Calculation sequence for the coking rates

As mentioned earlier, the model calculates the coke thickness rate for each coil and run separately. A computing code is developed to store all data in a common database. Figure 10 shows the results of the model. Each blue dot represents the coking tendency for a single coil during a run. The same chart displays the respective whisker plots. The lower and upper bounds of the gray boxes correspond to the first and third quartile of coking tendency data, while the interface between dark and light gray indicates the median coking tendency. The database has the capability to blend the coking tendency data with information about trips. This is displayed in Figure 10 also. This is important as the user can evaluate the effect of substantial process upsets on coil performance.

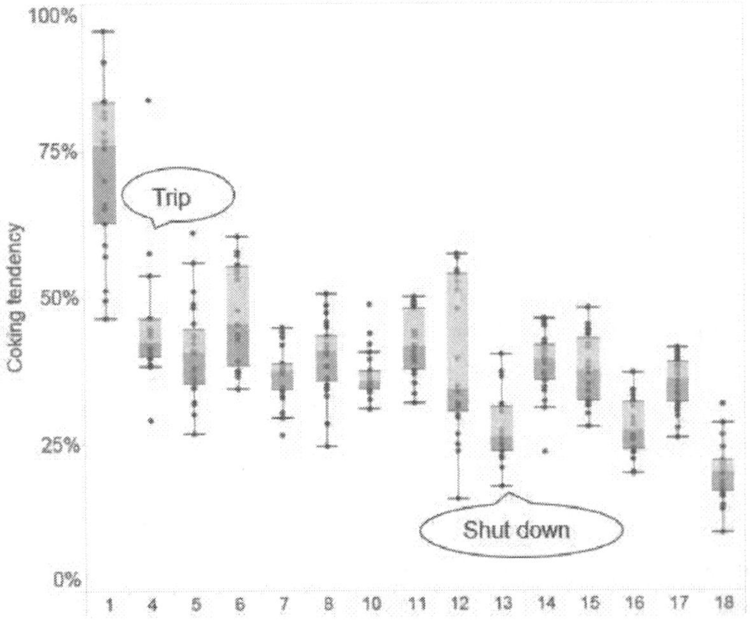

Figure 10: Coking rates per coil and run for one furnace

Data Visualization

Beyond the modeling of coking tendency, streamlining of process data is important to make decisions for recoils. A significant effort is made to collect data from all ethylene plants and report data in a common way, to allow fast analysis and comparisons. The result of this work can lead to several views, which are presented in this section. The first example is presented in Figure 11 with statistics for all furnaces in one plant. In the upper zone, the user can read the time fraction corresponding to steam cracking operations along with the typical run and decoke lengths. In some cases, the cracking time fraction can be lower than usual, therefore it is important to display if recoils occurred in the selected timeframe. The user gets an overview of the chronological distribution of decokes, process trips, and maintenance stops in the lower zone of Figure 11.

A closer look at coil performance is given in Figure 12. The user selects a furnace and a single coil cycle. Then the development of run length is displayed for the entire coil lifetime. The user can select to read only the runs that ended where specific end-of-run conditions were reached. The chart shows the change in run lengths over time. This method is not as accurate as the coking tendency model, from a coil ageing perspective, but it is a rapid and practical way to get an answer. The red line in Figure 12 represents the expected coil life for the selected furnace.

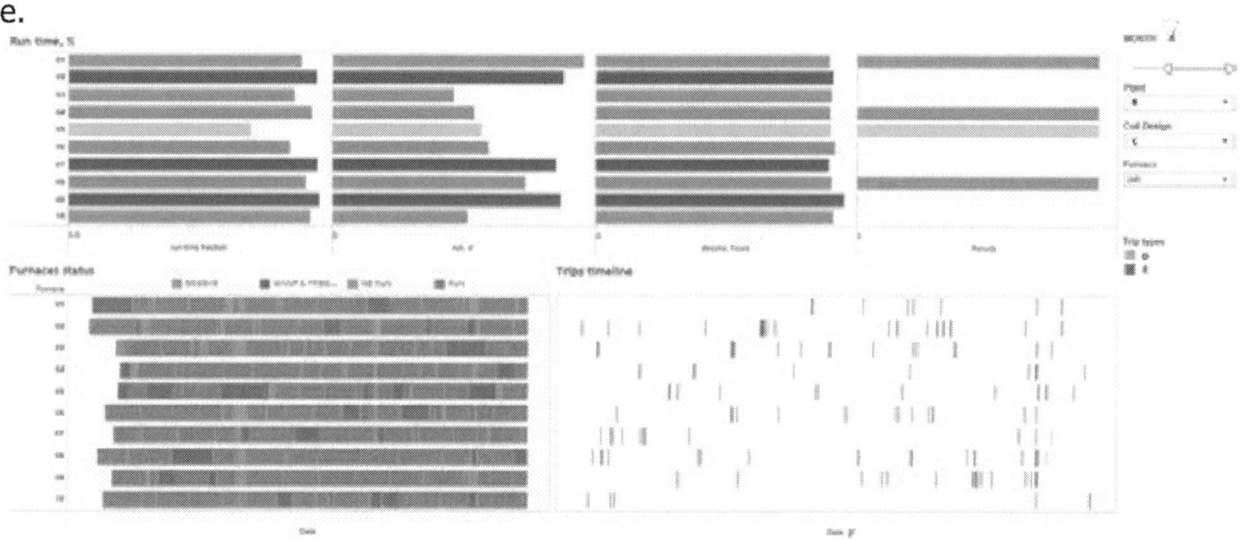

Figure 11: Distribution of runs, decokes, and furnaces trips over time for a selected ethylene plant

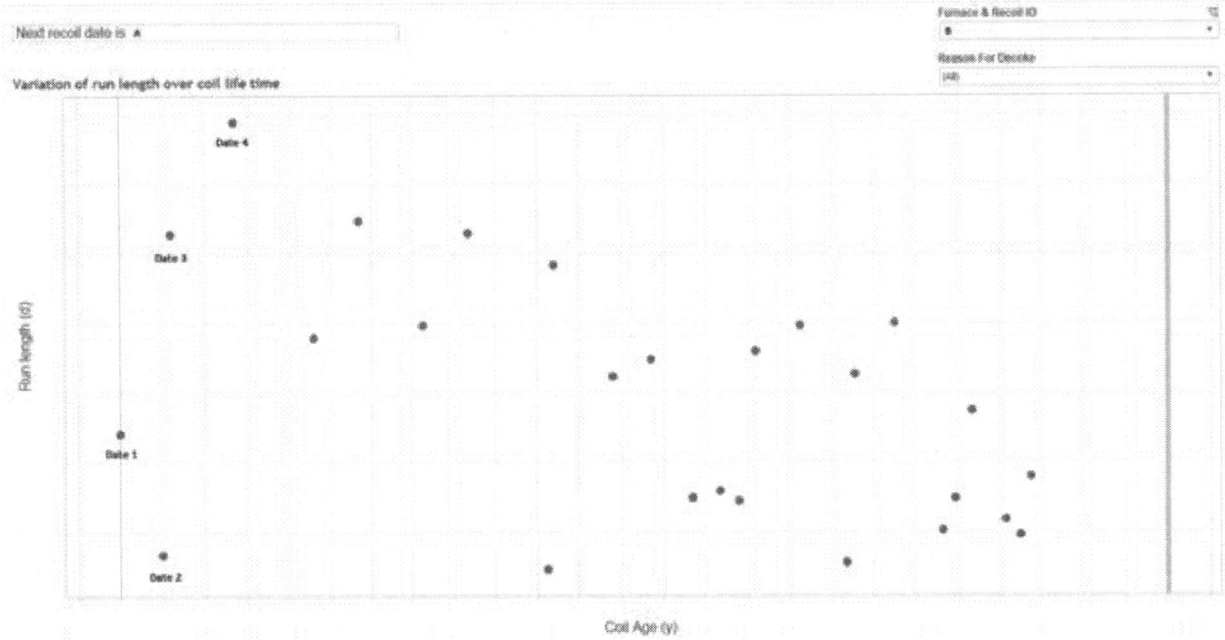

Figure 12: Actual run lengths throughout the entire coil life time for a single furnace

Recoil Planning

A team with representation of furnace SME's, R&D, maintenance, and purchasing, is overlooking our global furnace recoil schedule. This team is working with operations to determine the optimum timing for individual furnace recoils, and all decisions made are supported by process and inspection data, brought together in the visualization tool. Besides optimizing our recoiling intervals, the dataset is also supporting evaluations of any potential changes in coil design and/or metallurgy.

References

[1] D. Jakobi, R. Gommans: "Typical pyrolysis coil failures and how to combat them", AIChE EPC 2002 Spring National Meeting

[2] H. Chasselin: "Carburization in ethylene tubes", AIChE EPC 2013 Spring National Meeting

[3] A. IIiyas et al: "New approach for Anticoking Technologies Qualification for Ethylene Furnaces", EEPC 2015

[4] B. Braun, G. Bellos, S. Luijten, L. Li, M. McAdon, J. Hamilton: 'Connecting the Dots – Gaining Insight from Big Data in Hydrocarbons Manufacturing', AIChE EPC 2016 Spring National Meeting

[5] T. Tiyawongsakul: "Carburization inspection robots for radiant tubes inside ethylene cracking furnaces", Hydrocarbon Processing, August 2017

AIChE Paper Number 128e

High Temperature Hydrogen Attack in Ethylene Producing Plants

Flint Hills Resources

Tatiana Randall

Fixed Equipment Reliability Leader

Michael S. Cayard

Corporate Corrosion and Materials Leader

Prepared for Presentation at the 2018 Spring National Meeting

Orlando, Florida, April 22-26, 2018

High Temperature Hydrogen Attack in Ethylene Producing Plants

Tatiana Randall
Fixed Equipment Reliability Lead
Flint Hills Resources
Michael S. Cayard
Corporate Corrosion and Materials Leader
Flint Hills Resources

Abstract: The purpose of this paper is to highlight and discuss one user's methodology for the assessment of High Temperature Hydrogen Attack (HTHA) in equipment associated with the ethylene industry. This paper will specifically review the equipment susceptible to HTHA in production of "on-specification" ethylene product that uses high purity hydrogen to prevent catalyst poisoning in C2, C3 and cyclohexane hydrogenation reactors. This paper will concentrate on methods to quantify HTHA risk and associated methods to reduce the risk as outlined in API RP 941 "Steels for Hydrogen Service at Elevated Temperatures and Pressures in Petroleum Refineries and Petrochemical Plants". The methodology also will include a description of the Layers of Protection Analysis (LOPA) to be completed on any HTHA action items identified.

Introduction

Areas evaluated during HTHA assessment study included Furnace Section, Dilution Steam Generator Section, Primary Fractionator/Fuel Oil Stripper Section, Cracked Gas Compressor Section, Caustic Scrubbing Section, Cracked Gas Dehydrators Section, Demethanizer/Deethanizer Section, Methanators Section (Cold Train Overhead Gases), Depropanizers/Propylene Tower/Debutanizer Section and Acetylene Reactors/Ethylene Tower Section. Only the Methanators Section environment was identified as having the potential for HTHA and will constitute the balance of the discussion.

Acetylene (C_2H_2) is produced in the cracking operation as a byproduct of high temperature pyrolysis reactions that also produce ethylene and propylene. Production of "on-specification" ethylene product requires removal of almost all the acetylene from the deethanizer overhead stream. The deethanizer overhead stream is hydrogenated using catalyst to convert the undesirable acetylene to ethylene and ethane. However, this catalyst is subject to poisoning from both carbon monoxide (CO) and sulfur compounds. The methanator reactors are used to remove the CO from the high purity hydrogen fed to the acetylene converter. The methanator reactors contain enough hydrogen and operate at sufficient temperatures to warrant a review for HTHA. The sulfur poisons are removed from the cracked gases earlier in the process with the use of caustic treating, however this caustic treating process is outside the scope of this paper.

High purity hydrogen from the demethanizer cold train core exchangers is routed as methanator feed through preheaters where the temperature is increased from 60°F to 430°F. This combination of hydrogen content and temperature lends potential susceptibility to HTHA, thus requiring more detailed analyses. The basic process schematic for the methanator is shown in Figure 1.

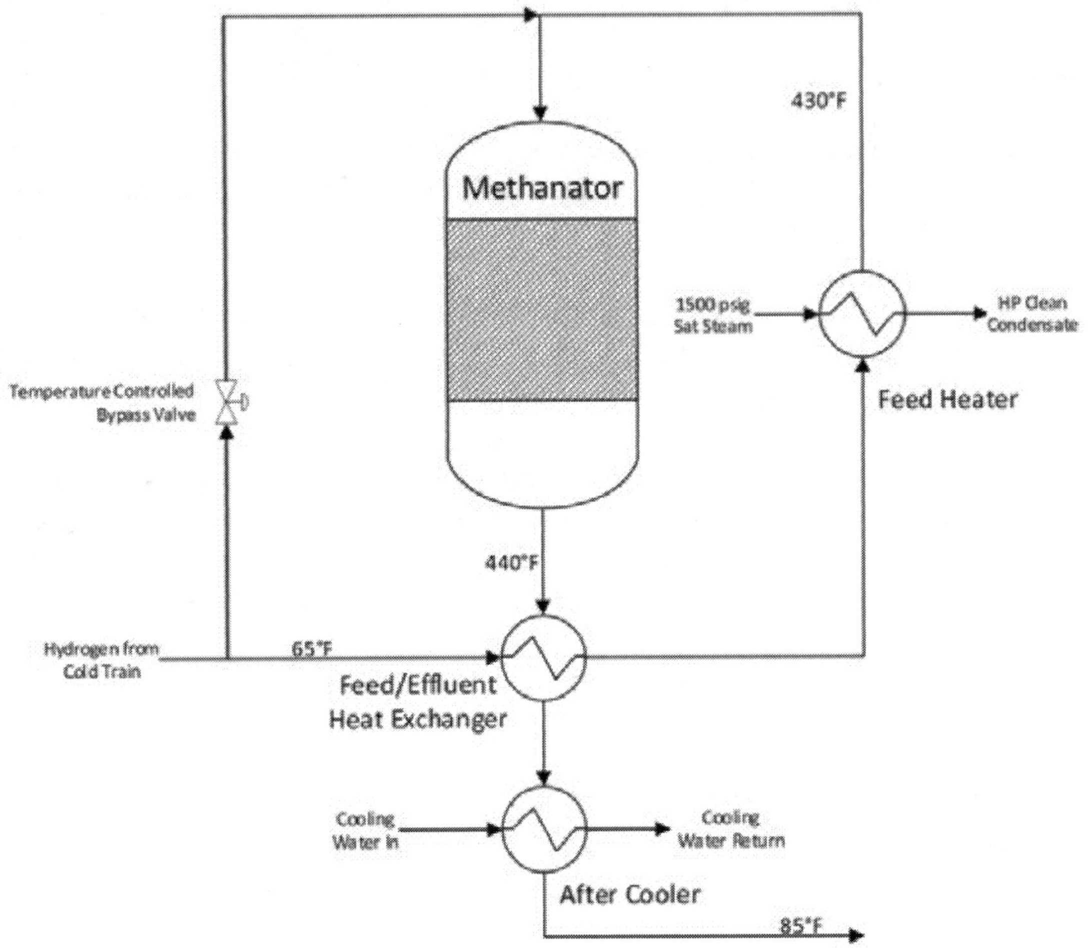

Figure 1 -Schematic representation of typical methanator reactor process.

HTHA Mechanism Description

While the refining industry has recognized HTHA as a concern for many years, awareness of its existence and potential to occur in the ethylene industry has not been well understood.

HTHA falls into damage mechanism categories that include environmentally assisted cracking, hydrogen assisted cracking, and high temperature degradation. HTHA only occurs with exposure to hydrogen at elevated temperatures greater than 375°F. The hydrogen reacts with carbon in steel to form methane (CH_4), which cannot diffuse through the steel and generates small pockets of methane pressure at the grain boundaries. The HTHA mechanism is shown schematically in Figure 2.

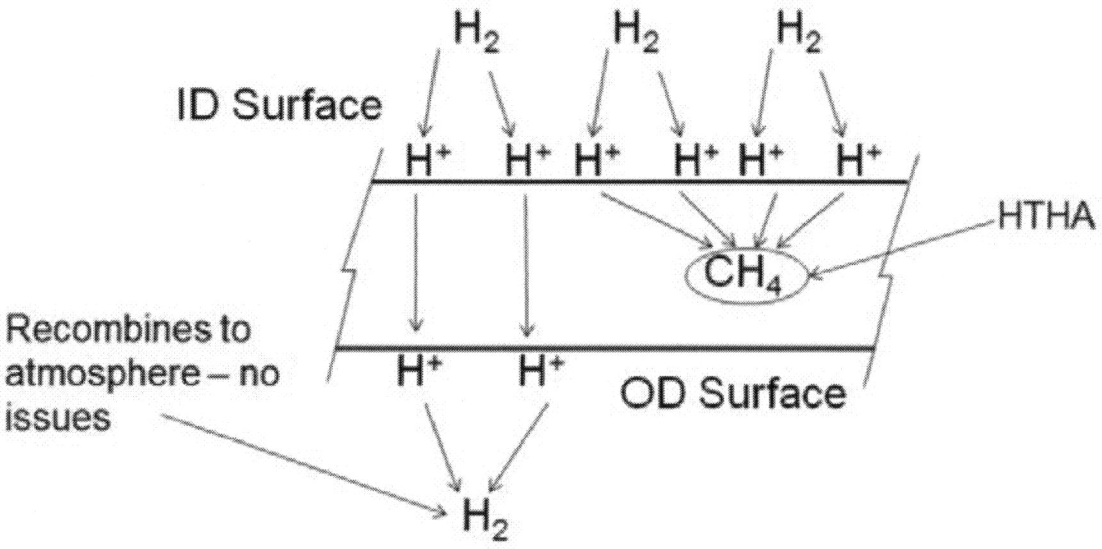

Figure 2 – Schematic representation of the mechanism of HTHA.

Consumption of carbon to form methane results in decarburization and associated loss in strength. Methane pressure builds up, forming bubbles or cavities, microfissures and fissures that may combine to form cracks as shown in Figure 3.

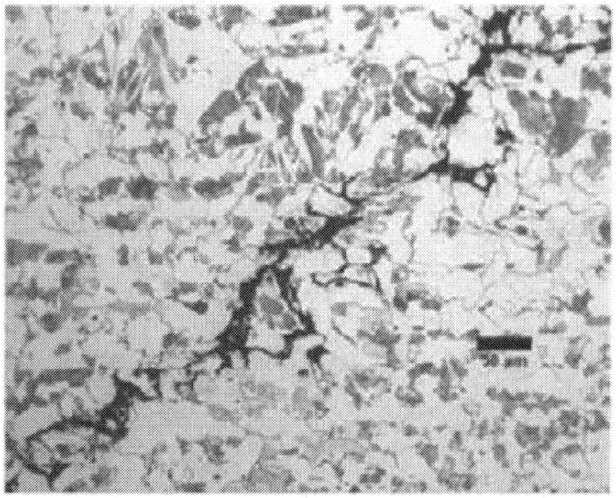

Figure 3- Typical macrocracking from HTHA observed in carbon steel. (Source: Anacortes NHT Investigation Report, Anacortes Refinery, Washington, July 21, 2011)

HTHA affects carbon and low alloy steels, but is most commonly found in carbon steel and C–0.5Mo steel that are operating above their corresponding Nelson curve limits represented in Figure 4.

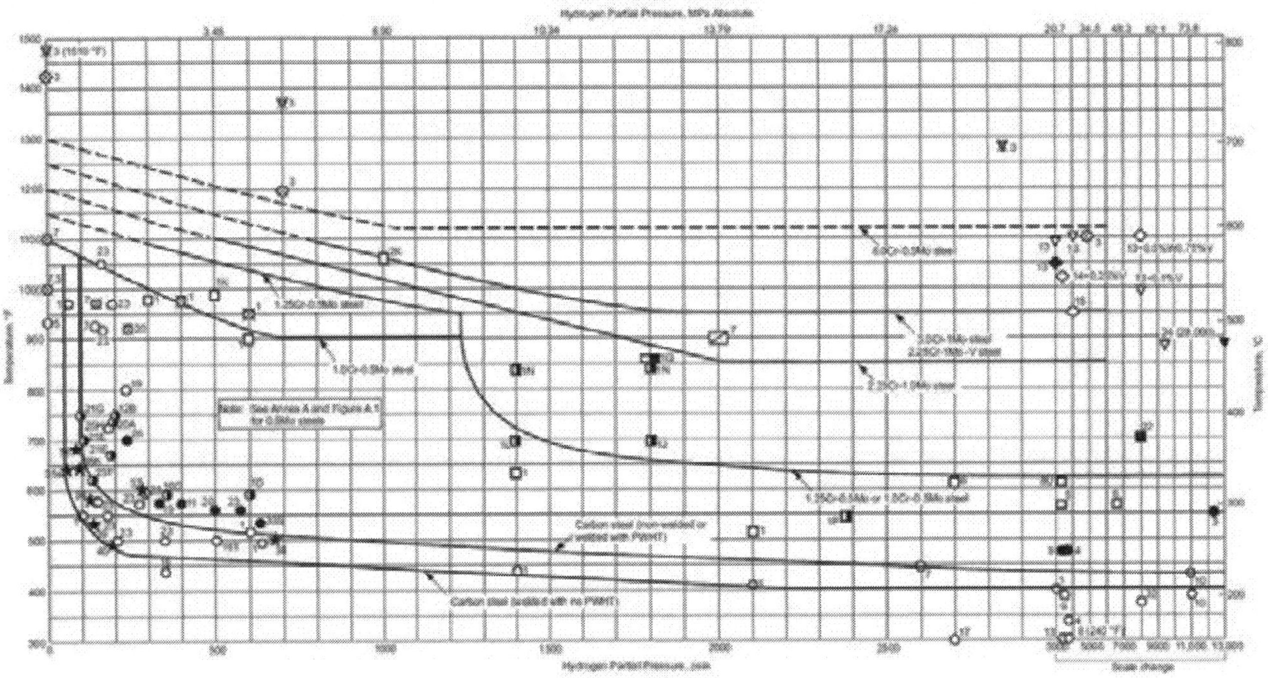

Figure 4- Schematic of the API 941, 8th Edition Nelson Curve.

The Nelson curves were developed back in the 1950's and were drawn based on industry experiences with HTHA. Points on or above the Nelson curve are expected to suffer HTHA, those operating below the curves are expected to be resistant to HTHA. Nelson curves have been developed for:

- Carbon Steel (welded with no Post Weld Heat Treatment (PWHT))
- Carbon Steel (non-welded or welded with PWHT)
- C–0.5Mo (since removed from the base curves due to failures below the curve)
- 1Cr–0.5Mo
- 1.25Cr–1Mo
- 2.25Cr–1Mo
- 2.25Cr–1Mo–V / 3Cr–1Mo
- 6Cr–0.5Mo

Methanator Section HTHA Equipment Assessment

The first step of the assessment involved identification of the boundaries for equipment operating in hydrogen service. The methanator section piping and equipment operating at more than 50 psia H_2 at temperatures greater than or equal to 375°F represent the boundaries for assessment as shown in Figure 5.

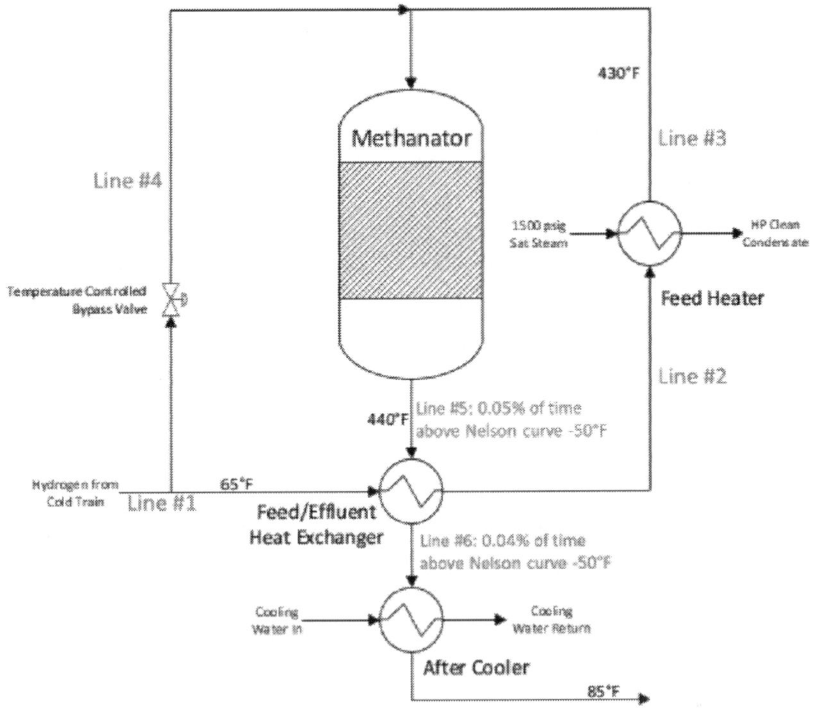

Figure 5 – Boundaries of the equipment in hydrogen service

More detailed data were collected by the Fixed Equipment Engineer, Process Engineer and Corporate Corrosion and Material Engineer for further analysis. The data collected included:

- Equipment description: Section Description, P&ID, Equipment/Line#(Figure 6)
- Metallurgy of the components: Pipe or Nozzle, Shell/Channel/Header Box, Cladding, Tubes, Tubesheet, PWHT Condition (Figure 6).

Methanators (Cold Train OVHD Gases)							
Description			METALLURGY				
Item #	Section Description	Equipment / Line#	Pipe or Nozzle	Shell/Channel/Header Box	Tubes	Tubesheet (Shell or Channel side)	CS PWHT (Yes, No, N/A)
1	Raw H2 to Feed/Effluent Heat Exchanger	line #1	Carbon Steel	N/A	N/A	N/A	No
2	Feed/Effluent Heat Exchanger Tubeside Inlet	Tubeside Inlet Channel	Carbon Steel	Carbon Steel	C-0.5Mo (A209 T1)	Carbon Steel	Yes
3	Feed/Effluent Heat Exchanger Tubeside Outlet	Tubeside Outlet Channel	C-0.5Mo	1.25Cr-0.5Mo & C-0.5Mo	C-0.5Mo (A209 T1)	C-0.5Mo	N/A
4	Feed/Effluent Heat Exchanger Tubside to Feed	Line #2	C-0.5Mo	N/A	N/A	N/A	N/A
5	Feed Heater Shellside	Shellside	1.25Cr-0.5Mo	1.25Cr-0.5Mo	C-0.5Mo (A209 T1A)	C-0.5Mo	N/A
6	Feed Heater Shellside to Methanator	Line #3	C-0.5Mo	N/A	N/A	N/A	N/A
7	Exchanger Bypass from TV to Methanator Inlet	Line # 4	C-0.5Mo	N/A	N/A	N/A	N/A
8	Methanator		C-0.5Mo	C-0.5Mo	N/A	N/A	N/A
9	Methanator to Feed/ Effluent Heat Exchnager	Line #5	C-0.5Mo	N/A	N/A	N/A	N/A
10	Feed/Effluent Shellside		C-0.5Mo	1.25Cr-0.5Mo	C-0.5Mo(A209 T1)	Carbon Steel Inlet and C-0.5Mo Outlet	N/A
11	Feed/Effluent Heat Exchanger Shellside to After Cooler Shellside	Line #6	Carbon Steel	N/A	N/A	N/A	No
12	After Cooler Shellside	Shellside	Carbon Steel	Carbon Steel	Carbon Steel(A214)	Carbon Steel	No

Figure 6 – Data collection on Equipment Description and Metallurgy

- Equipment Design Data: Design Pressure and Temperature, Nelson Curve Temperature, Nelson Curve-50F (Figure 7). Note the Nelson curve temperature was based on the assumption that 100% of the design pressure was hydrogen. This assumption is conservative at lower hydrogen partial pressures but provided a good starting point for the assessment.

Methanators (Cold Train OVHD Gases)						
Description			DESIGN			
Item #	Section Description	Equipment / Line#	Design Pressure	Design Temperature	Nelson Curve Temperature	Nelson Curve - 50 F
1	Raw H2 to Feed/Effluent Heat Exchanger	line #1	524	120	519	469
2	Feed/Effluent Heat Exchanger Tubeside Inlet	Tubeside Inlet Channel	600	500	516	466
3	Feed/Effluent Heat Exchanger Tubeside Outlet	Tubeside Outlet Channe	600	750	516	466
4	Feed/Effluent Heat Exchanger Tubside to Feed	Line #2	545	500	518	468
5	Feed Heater Shellside	Shellside	600	525	516	466
6	Feed Heater Shellside to Methanator	Line #3	545	500	518	468
7	Exchanger Bypass from TV to Methanator Inlet	Line # 4	545	500	518	468
8	Methanator		634	800	514	464
9	Methanator to Feed/ Effluent Heat Exchnager	Line #5	545	584	518	468
10	Feed/Effluent Shellside		600	800	516	466
11	Feed/Effluent Heat Exchanger Shellside to After Cooler Shellside	Line #6	545	300	518	468
12	After Cooler Shellside	Shellside	600	500	516	466

Figure 7- Equipment design data.

- Equipment operating data: Process data tag, Average and Maximum Temperature, Nelson Curve -50F Temperature, % of Time Above Nelson Curve -50F, % of Time Above Nelson Curve, Design Temperature, % of Time Above Design Temperature, and Maximum Temperature Dictated by HTHA or Design (Figure 8). The percentage of time was based on 5 years of data. Average and maximum temperatures were collected for information only. A reduction of 50°F off the Nelson curve was the limit chosen and represents industry RAGAGEP.

Methanators (Cold Train OVHD Gases)												
Description			OPERATING									
Item #	Section Description	Equipment / Line#	Temperature PI Tag	Average Temperature	Maximum Temperature	Nelson Curve - 50 F	% of Time > Nelson T-50F	% of Time > Nelson Curve Temperature	Design Max	% of Time > Design T	Max T Dictated by HTHA or Design?
1	Raw H2 to Feed/Effluent Heat Exchanger	line #1	L_TI 1	79	98	469	0.0%	0.0%	120	0.0%	Design
2	Feed/Effluent Heat Exchanger Tubeside Inlet	Tubeside Inlet Channel	L_TI 1	79	98	466	0.0%	0.0%	500	0.0%	HTHA
3	Feed/Effluent Heat Exchanger Tubeside Outlet	Tubeside Outlet Channel	L_TI 2	397	455	466	0.0%	0.0%	750	0.0%	HTHA
4	Feed/Effluent Heat Exchanger Tubside to Feed	Line #2	L_TI 2	397	455	468	0.0%	0.0%	500	0.0%	HTHA
5	Feed Heater Shellside	Shellside	L_TI 2	397	455	466	0.0%	0.0%	525	0.0%	HTHA
6	Feed Heater Shellside to Methanator	Line #3	L_TI 2	397	455	468	0.0%	0.0%	500	0.0%	HTHA
7	Exchanger Bypass from TV to Methanator Inlet	Line # 4	L_TI 2	397	455	468	0.0%	0.0%	500	0.0%	HTHA
8	Methanator		L_TI3	398	510	464	0.05%	0.0%	800	0.0%	HTHA
9	Methanator to Feed/ Effluent Heat Exchnager	Line #5	L_TI3	398	510	468	0.04%	0.0%	584	0.0%	HTHA
10	Feed/Effluent Shellside		L_TI3	398	510	466	0.05%	0.0%	800	0.0%	HTHA
11	Feed/Effluent Heat Exchanger Shellside to After Cooler Shellside	Line #6	L_TI3	398	510	468	0.04%	0.0%	300	0.0%	Design
12	After Cooler Shellside	Shellside	L_TI3	398	510	466	0.05%	0.0%	500	0.0%	HTHA

Figure 8- Equipment operating data

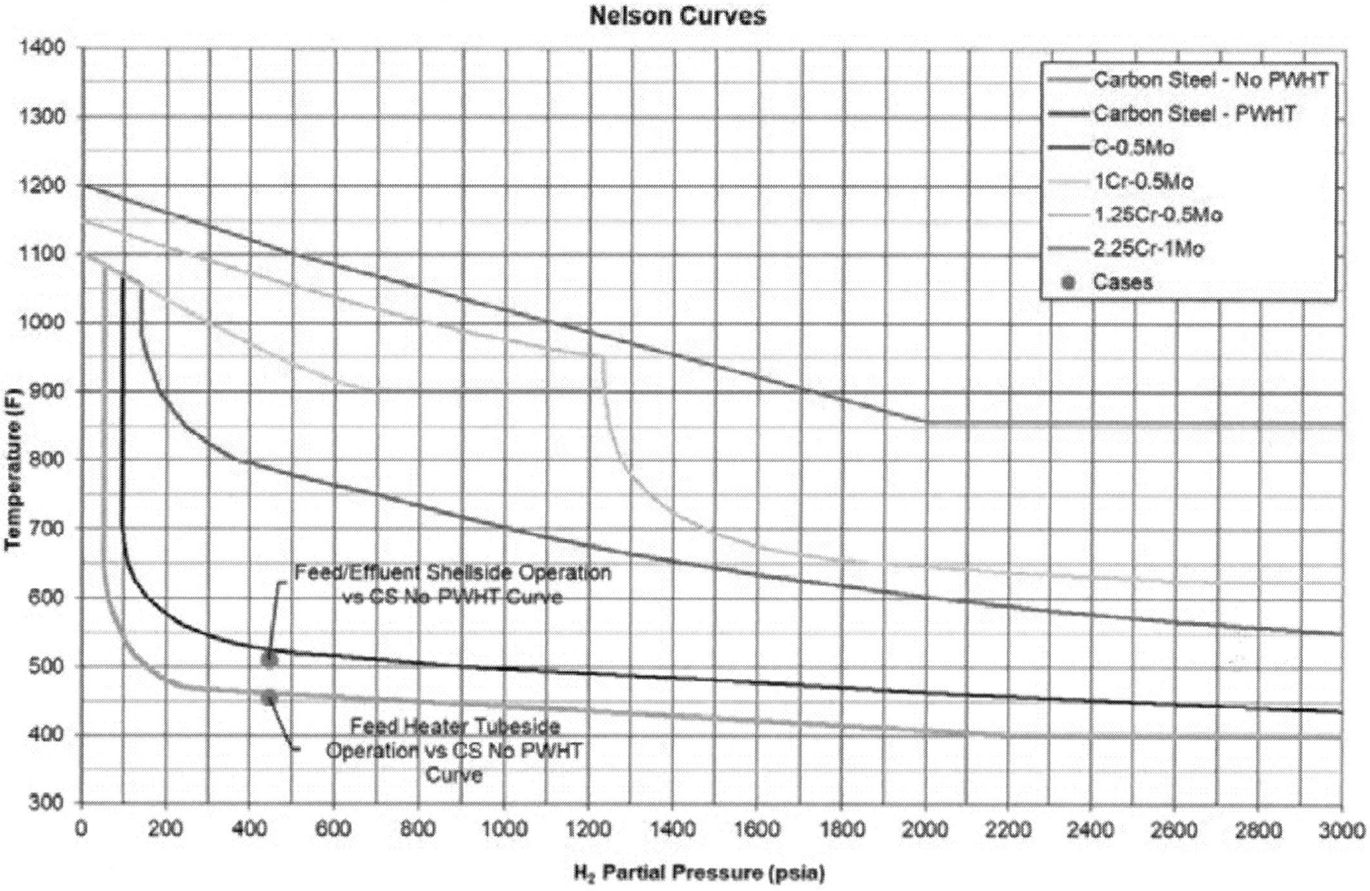

Figure 9-Feed Heater and Feed/Effluent Heat Exchanger operating conditions vs Nelson Curve.

General guidelines for mitigating HTHA should be evaluated per individual company risk profiles and could contain the following steps:

- Establishing an operating limit (Integrity Operating Windows, or IOW's) to protect equipment from process creep.
- Establish additional layers of protection to reach an acceptable risk per the company's risk profile (e.g., LOPA).
- Perform Fitness-for-Service (FFS) on the equipment operating in excess of the Nelson curve to ensure continued integrity until the next replacement opportunity.
- Equipment replacement/upgrades.

Layer of Protection Analysis (LOPA) of HTHA Action Items

Action items used to mitigate HTHA risk may not provide sufficient risk reduction. A "PHA" style approach was used to validate the ability of the identified controls to successfully mitigate the risk to an acceptable level. The analysis considers initiating events, failure severity, the frequency of initiating events and likelihood the initiating event leads to HTHA. Analysis defines the number of controls required to meet or exceed the company's risk profile.

The case study showed that several pieces of equipment run close to or at Nelson -50°F curve, but below Nelson curve temperature itself. Mitigation steps included establishment of operating limits and replacement of C–0.5Mo Feed Heater bundle and Feed/Effluent Heater with 1.25Cr–0.5Mo shell material.

Conclusions

Most inspection techniques for HTHA are not reliable. Advanced stages of attack are easier to detect, but at that point equipment should be taken out-of-service. Inspection techniques for finding advanced stages of HTHA at the surface include WFMT (Wet Fluorescent Magnetic Test), MT (Magnetic Test) and AUBT (Advanced Ultrasonic Backscatter Testing). More detail on these techniques can be found in the 8th edition of API RP 941. Several Joint Industry projects are working to develop more sensitive techniques, but at the time of this paper, none have been identified as offering much promise.

Consideration should be given to evaluating your processes and equipment for potential damage mechanisms. HTHA is simply one example which can lead to failure with little warning and requires proactive assessments such as those discussed in this paper.

Additional emphasis should be placed on establishing IOW limits for HTHA services to mitigate the potential increased process severity which many accompany future plant expansions and debottlenecks.

References:

1. API RP 571- Damage Mechanisms Affecting Fixed Equipment in the Refining Industry.
2. API 941 8th Edition - Steels for Hydrogen Service at Elevated Temperatures and Pressures in Petroleum Refineries and Petrochemical Plants.
3. Anacortes NHT Investigation Report, Anacortes Refinery, Washington, July 21, 2011.

AICHE PAPER NUMBER 128f

COLD SERVICE BOLTED JOINT RELIABILITY

Kathryn Worden
Applications Engineer
Flexitallic US, LLC

Prepared for Presentation at the 2018 Spring National Meeting
Orlando, Florida, April, 2018

COLD SERVICE BOLTED JOINT RELIABILITY

Kathryn Worden
Applications Engineer
Flexitallic US, LLC

Abstract:

Throughout the ethylene manufacturing process, the range of cold section thermal changes may be secondary to that of the hot section, but they are not without challenges. Contraction upon cooling can cause leakage if bolted joint conditions are not adequate or simply if the gasket cannot recover towards the flanges. Sealing two flanges of dissimilar metallurgy, for example an aluminum cold box flange mating to a steel process flange, presents additional challenges with differential thermal contraction upon cooling. This paper discusses options for maintaining sealability as well as their relative level of effectiveness.

Introduction

As with most matter, flanges of a bolted connection contract with decreasing temperature. Bolted connections are assembled at room temperature. As the process cools to the desired level, bolted joint flanges contract and move away from each other. A multitude of factors affect whether a gasket can maintain a seal throughout cooling. Some connections also experience frequent thermal cycling. The gasket must maintain adequate sealing through multiple cooling/warming cycles. Factors that affect sealing include rate of cooling, frequency and magnitude of thermal cycling, desired length of service, gasket type, bolt material, initial preload, flange condition and alignment.

Leakage

Also note that there is no industry accepted definition of a leak. In the purest sense, all joints have some measurable leakage unless they are welded. The question is how much is acceptable and is it being monitored. Other than valve bonnet gasket connections, most gasketed joints are not monitored for emissions via leak detection programs. Therefore, most leaks are not identified until there is visible fluid seepage or worse – gross leakage leading to shutdown, loss of containment or temporary clamps. Acceptability must be defined by the individual plant or company. For the sake of plant efficiency, reliability and safety, let's proceed targeting low emissions and zero visible leakage throughout plant operation till joint disassembly.

Regarding the concept of leakage, generally speaking the more load you apply to a gasket, the better it seals and assures an adequate seal will be maintained long term, especially after relaxation occurs. However, the maximum allowable gasket stress, flange stress, and bolt yield strength cannot be exceeded. This value is different for each type of gasket (including thickness of sheet gasket), type and size of flange, and type and size of bolt. And of course, not all gaskets handle high pressures or extreme loads, some gaskets relax more than others, and some gaskets recover from fluctuating load and/or loss of load better than others. The best representation of the leakage performance concept is the output of an EN 13555 leak test.

EN 13555 Leakage Curve

EN 13555 (1) is a testing standard used to develop gasket parameters for EN 1591 (2) flange design calculation method. An EN 13555 Leakage Curve can be generated for both sheet and semi-metallic gaskets and shows the leak rate performance as gasket stress increases. Standard test parameters are ambient temperature with internal Helium pressure of 40 bar (580 psi). Helium is a safe test media and provides a buffer to actual in-service leak rates. Extra data can be obtained at other gas pressures, for example 10 bar (145 psi) and 80 bar (1160 psi), if desired. The gasket is loaded and unloaded in stages and at a fixed rate. Leak rate is recorded after stabilization at each load. Refer to Chart 1 below for a graphical representation of the leakage curve. Note the leakage scale is logarithmic.

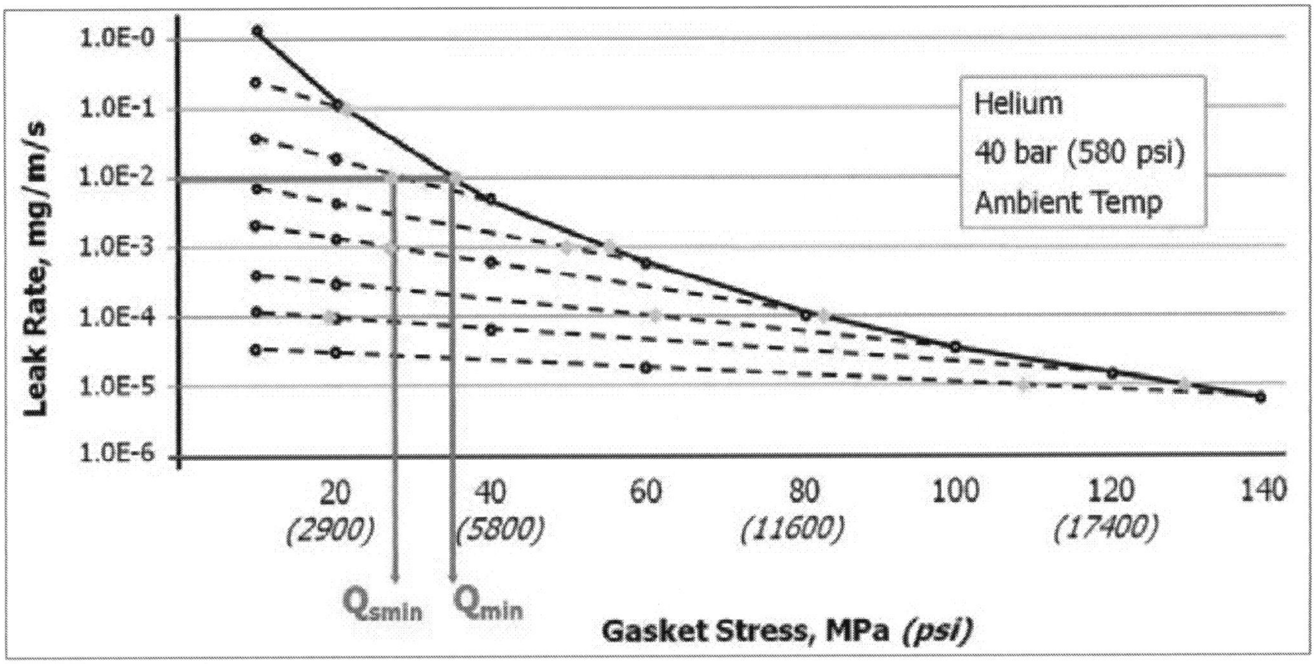

Chart 1. EN 13555 Leakage Curve Example

One decides what leak rate is acceptable. The curve identifies the gasket stress required to achieve that rate on loading and unloading (Qmin and Qsmin respectively). It can be useful to overlay leakage plots of different gaskets for performance comparison. Leakage curves will be shown for some gaskets further in the paper. Note that it is standard to collect low load data even if the gasket's minimum seating stress is higher.

Gasket Type & Performance

Spiral Wound Gasket

Because of their low standard cost and general benefits, spiral wound gaskets (SWGs) with inner rings are the first choice for the refining and petrochemical industry. The outer retaining ring ensures blowout prevention. The thin uniquely shaped winding wire provides resiliency (recovery). The graphite filler material conforms to the mating surface for good sealing while providing fire safe protection in hydrocarbon service. Finally the inner ring prevents buckling, widening and breakage.

One well acknowledged disadvantage is handling. Spiral wound gaskets have the tendency to spring apart when handled, an inconvenience that increases in frequency with increasing diameter. Second disadvantage is the seating stress required to create a minimum level of acceptable sealing. It is higher than gasket types to be discussed further in the paper. For those familiar with ASME Boiler & Pressure Vessel Code (BPVC) Y values (3), a spiral wound has a Y, minimum seating stress, of 10,000 psi.

Another shortcoming is plastic deformation upon compression. The thin metal winding wire undergoes plastic, or permanent, deformation when compressed. This limits its usable recovery, or live springback, while in a bolted joint. Figures 1 and 2 below depict SW winding wire shape before and while under compression. The red color through the center of the wire on the right depicts plastic deformation (not to scale).

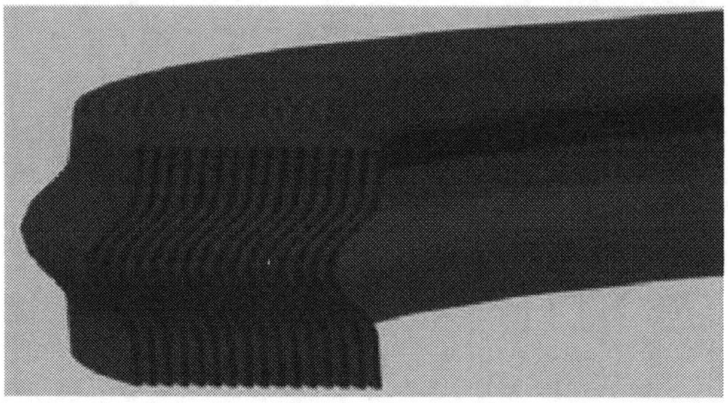

Figure 1. Spiral wound wire before compression

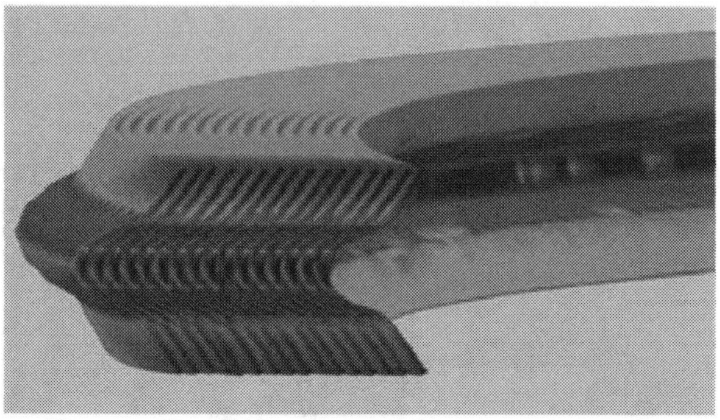

Figure 2. SW wire under compression

Kammprofile/GMCL

Commonly referred to as a kammprofile, this gasket type was dubbed a grooved metal gasket with covering layers, or GMCL, when added to ASME B16.20 **(4)**. This type made significant sealing improvements over double jacketed gaskets in heat exchangers and considerably improved handleability over spiral wounds. ASME BPVC does not offer a standard Y value for this type. Major United States manufacturers' published values average approximately 3100 psi.

With increased use over the last couple decades however, shortcomings have presented themselves. Aside from the serrations in each face, all kammprofiles (kamm) have a solid metal core. The facing material provides little towards flexibility. A kamm's lack of flexibility has been known to cause leaks in piping systems as they are subjected to bending loads and strain. The refining, petrochemical and chemical industries have or are in the process of eliminating kammprofiles from piping unless joints are well supported or not subject to appreciable external loads.

Another drawback of the GMCL's solid core is having very little recovery like all solid metal gaskets. Basic compression/recovery data illustrates this clearly. See Table 1 below. A kammprofile only recovers 6% of the distance compressed under 18ksi gasket stress. That is only 1.5% recovery of the initial thickness.

Table 1

Gasket Type	**% Compression** (18,000 psi gasket stress)	**% Recovery** (of Compression)	**% Recovery** (of Initial Thickness)
SWG w/inner	30	26	7.8
Kammprofile	25	6	1.5

Kammprofiles are known to leak after thermal cycling (depends on T range and frequency) unless they are retightened. However, retorquing adds time and cost and is not always feasible. Kamms have difficulty sealing large duty heat exchangers, specifically the large thermal variation across the face of a flange. This includes extreme situations like condensate settling at the bottom of a vertical flange. GMCL gaskets cannot compensate for the differential thermal contraction of mated flanges with significantly dissimilar metallurgy (sealing aluminum to steel for example). And finally, kamms cannot maintain a seal during rapid thermal changes like flash cooling and will struggle with assisted cooling. Note that standard spiral wounds would be incapable of sealing these situations as well.

A kammprofile's EN leakage curve results, or lack thereof, demonstrate their inadequate recovery. Three (3) of five (5) kammprofiles failed during the unloading phase of this test. They were unable to adequately recover towards the test plates to maintain a seal during unloading. Gasket stress unloading occurs at 0.5 MPa/s (72.5 psi/s), and failure is when internal pressure drops faster than 14.5 psi/sec.

One of the gaskets that did not fail is plotted against a spiral wound in Chart 2 below. Both are NPS 4" Class 300, 300 series stainless with graphite. Notice the slope of the unloading lines. Flatter unloading lines are preferred as it generally indicates better recovery. Regardless of initial stress levels, GMCL unloading line leak rates rapidly increase signifying its poor recovery. It's important to note that a kammprofile's ability to achieve very low leak rates counteracts its poor resiliency in standard situations. Also imperative is knowing that a SWG's leak rate is highly dependent on the density of construction and the filler material.

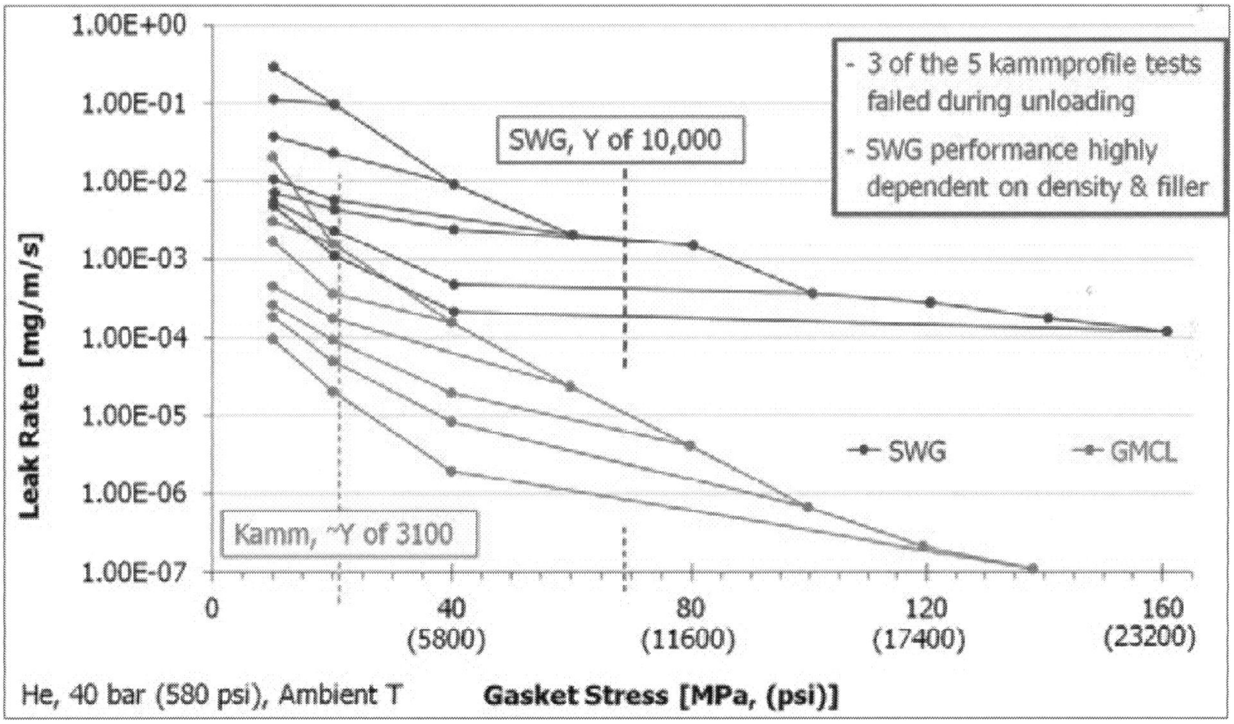

Chart 2. EN 13555 Leakage Curve for Stainless/Graphite GMCL & SWG

Dynamic Recovery Gasket

The latest generation of gasket is one with dynamic recovery designed to overcome the limitations of spiral wounds, kammprofiles and beyond. A dynamic recovery gasket, or DRG, combines aspects of both a traditional spiral wound and a kammprofile, improving upon both using reinvented winding wire approximately five times thicker than traditional spiral wound wire. The new heavy gauge metal wire is formed with a functional edge that simulates the serration profile of a grooved metal gasket. The heavy wire is wound like a SW, incorporating filler material, and is held together via a unique and optimized laser welding process. Stronger, thicker wire combined with laser welding offers significantly improved handling over a SWG. The DRG is finished with layers of facing similar to a kammprofile. Visual representations are shown in Figures 3 thru 6 below.

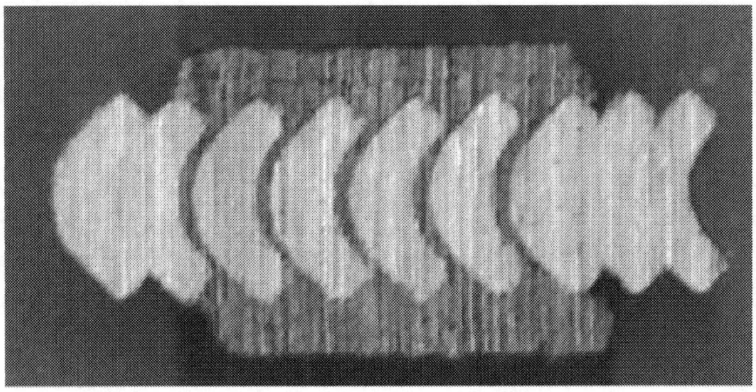

Figure 3. DRG cross sectional cutaway

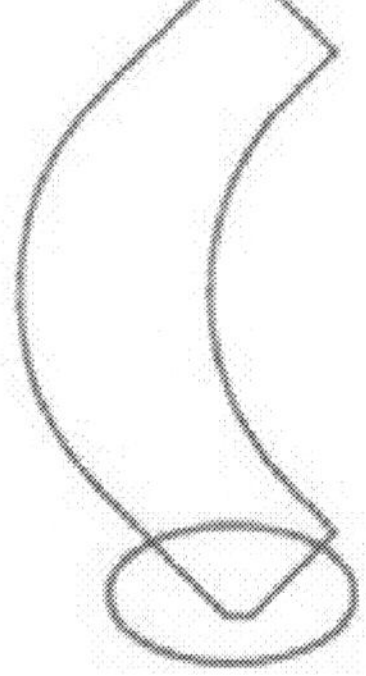

Figure 4. DRG heavy gauge winding wire and formed edge

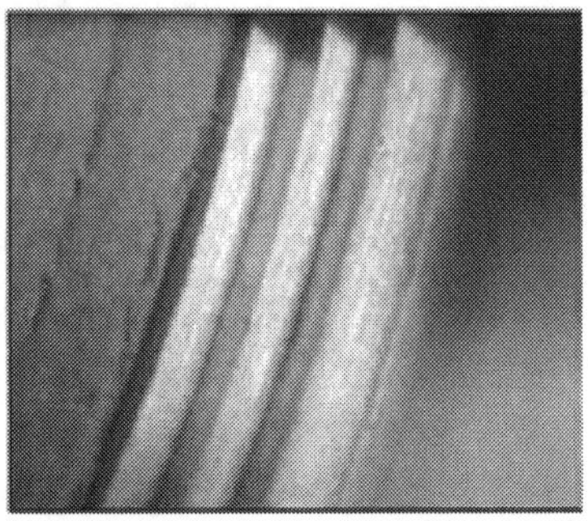

Figure 5. Wire's top edge beneath facing

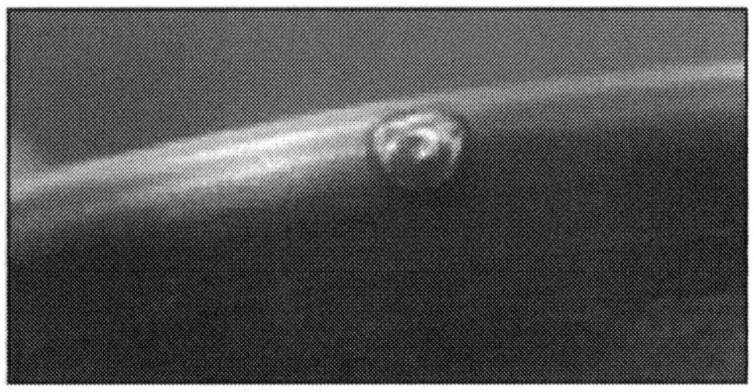

Figure 6. Laser weld at outer diameter

DRG recovery surpasses any gasket in the industry. Table 2 includes compression and recovery values for a DRG compared to the previous SWG and GMCL.

Table 2

Gasket Type	**% Compression** (18,000 psi gasket stress)	**% Recovery** (of Compression)	**% Recovery** (of Initial Thickness)
SWG w/inner	30	26	7.8
Kammprofile	25	6	1.5
DRG	**30**	**34**	**10.2**

Even more significant, because the DRG wire is thicker through the center, it does not undergo plastic deformation upon compression but stays elastic. Compression energizes the seal maximizing its usable recovery in a live joint. Figures 7 and 8 below depict DRG winding wire shape before and while under compression. No red color indicates no plastic deformation (not to scale).

Figure 7. DRG wire before compression

Figure 8. DRG wire under compression

As indicated by the high level of recovery and elasticity, the DRG has no issues maintaining sealability during the unloading phases of an EN leakage test shown in Chart 3 below. As the DRG is a hybrid of a spiral and kamm, its leak rate performance is between the two. All three gasket types have a leak rate less than 1E-2 mg/m/s prior to reaching their ASME Y value. The DRG has flat unloading lines until stresses are well below its Y value of 6,400 psi. This minimum seating stress was estimated to ensure sufficient load to activate the DRG's spring-like wire. Additional load energizes the wire further.

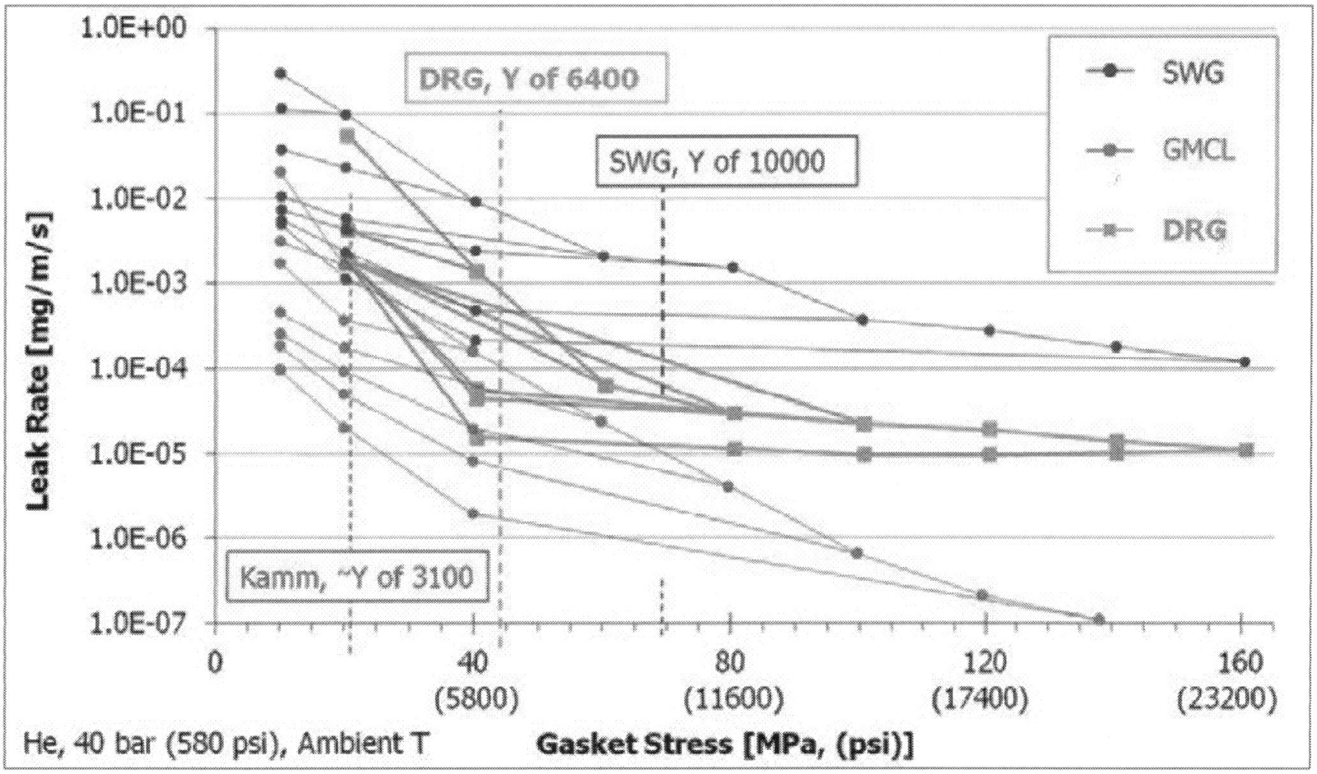

Chart 3. EN 13555 Leakage Curve for DRG, GMCL, and SWG

Successful DRG Applications in Cold Service

Since the DRG was introduced in 2012, its ability to maintain sealing through changing loads has surpassed all others tried in service.

Successful DRG Application – Liq O_2, Liq N_2

An OEM manufacturing cryogenic storage and transfer equipment had persistent leaks sealing liquid oxygen and liquid nitrogen using PTFE filled spiral wound gaskets. The design temperature is -450°F with operating temp as low as -300°F. Flash cooling is not frequent, and some services operate continuously. However, some flanges thermal cycle as frequently as once per day. Flanges are stainless steel ASME B16.5, 1" to 4" Class 150. In 2013, the OEM began using the DRG and now uses approximately 100 per year stating that "Unlike the previous Teflon gaskets we used, the (DRG) gaskets have proven remarkably reliable in our cryogenic applications. I cannot find one field complaint."

Successful DRG Application – Ethylene

This olefins and polypropylene manufacturer needed a problem-solving gasket after seeing failure of spiral wounds and kammprofiles in a challenging service. This application sealing ethylene normally operates at ambient and 1000 psi cycling to 400°F once a week for 10 hours. Pressure loss during any activity to take down a piece of equipment causes the ethylene to flash cool. Resultant temperatures can potentially reach as low as -150°F. To alleviate sealing issues associated with weekly thermal cycling and occasional flash cooling, this site converted to the DRG for this ASME NPS 8" Class 900 flanged connection back in May 2015.

Successful DRG Application – Cold Box

This gas processing and natural gas liquids (NGL) extraction and fractionation facility had difficulty sealing stainless steel flanges mated to tempered aluminum alloy flanges of their cold box. A cold box is a rectangular structure holding one or more braised aluminum heat exchangers (BAHX) along with its associated cryogenic equipment like drums, columns, piping, valves and instrumentation.

The OEM specified B8 Class 1 bolts loaded to 25,000 psi bolt stress along with a low stress (LS) spiral wound gasket for sealing. B8 Class 1 bolts have a minimum yield strength of only 30,000 psi. Class 1 bolts and low target bolt stress were specified here to limit the stress applied to the RFWN aluminum flanges. Aluminum flanges are weaker and raised face type can suffer permanent bending, or rotation, if over stressed. Having said that, this level of bolt stress is considered low. 30ksi bolt stress is considered a very general minimum to maintain a seal through service, after relaxation, without vibrating loose or self-loosening due to lack of bolt stretch.

A low stress spiral wound has a Y value of 5,000 psi compared to the standard 10,000. This gasket was specified to achieve an adequate seal given the low bolt stress. Low stress spirals are less dense, are limited to Class 150 and 300, and do not seal as tightly as standard spirals. LS spirals cannot handle thermal cycling or the differential thermal contraction upon cooling of a 304SS flange mated to Aluminum.

Although on paper there was adequate gasket stress – 14,625 psi across the entire gasket surface compared to the 5,000 psi required – when targeting 25ksi bolt stress, the combination of bolt-stress-limited flanges with an inferior gasket caused this connection to leak regularly.

In July 2016, the facility began using the Dynamic Recovery Gasket type in the application with complete success. The flanges described above were Cold Box inlet and outlet connections, specifically ASME B16.47 NPS 40" Class 300 Series A, sealing methane at 450 psi at -150°F. Since then, the site continues to convert similar troublesome connections.

Ethylene Cold Box

The cold section of an ethylene manufacturing facility uses the same type of cryogenic cooling equipment, a cold box, to cool the ethylene mixture prior to separating secondary low carbon molecules like methane and ethane. Although bolt type, bolt stress, and gasket stress must be evaluated for each leaking connection, the DRG's inherent flexibility and spring-like elastic seal will behave the same in all cold boxes mating stainless to aluminum.

Other Contributing Factors

Bolting

L7s have a minimum working temperature of -150°F (5). If the bolts will see lower temperatures than this, B8s (304SS) or B8Ms (316SS) are recommended as they have a minimum temperature of -425 to -325°F (5). Stainless bolting comes in Class 1 and Class 2. Class 2 is strain hardened. Class 1 minimum yield strength at room temperature is 30,000 psi. Unless there is a flange limiting situation similar to the aluminum flanged cold box described above, Class 1 is undesirable because 30,000 psi bolt stress is often not enough to maintain a seal on most gaskets. Whenever possible, upgrade to Class 2 stainless. Yield strength of Class 2 bolts decreases with increasing diameter.

Installation

Never underestimate the benefit of a proper installation. Even the best gasket can fail if flange condition, alignment, fasteners, torque value and procedures are faulty and/or if inaccurate tools are used. ASME B31.3 and PCC-1 (6) recommend using a calibrated clicker torque wrench on process connections unless upgrading to hydraulic torquing or tensioning. Incorporating proper installation detailed in ASME PCC-1 will generally improve the performance of all gasket types.

Nut Factor

A common pitfall of all installations, including those considered proper, is inaccurate estimation of friction. Calculating a torque value requires an estimation of friction for the bearing surfaces of the bolts and nuts. K is the dimensionless factor used to represent that friction. Also referred to as the Nut Factor, K is often confused with the Coefficient of Friction, or Friction Factor. They are not the same but one can be determined from the other.

$$K = \text{Friction Factor} + 0.04 \quad \textbf{(6)}$$

K is used in the short equation for Torque:

$$T = K \cdot \frac{D}{12} \cdot F,$$

Where T = Torque, ft-lbs;
K = Nut Factor;
D = Nominal Bolt Diameter, inches;
F = Bolt Load, lbs_f.

K ranges from roughly 0.11 (low friction) to 0.23 (high friction) for bolted flanged connections. The magnitude of the value of K itself is small. Conversely and deceptively, a change in 'K' yields a significant impact on preload (bolt load). For example, applying 120 ft-lbs torque to a 5/8" stud bolt would generate 20,945 lbs preload with a K of 0.11 but produce only 10,017 lbs preload with a K of 0.23.

To complicate things further, the following conditions will cause a change in the Nut Factor: type of anti-seize, how much anti-seize is applied, type of stud (alloy vs. stainless), coatings and platings. Many lubricants do not publish K factors. Even if they do, they must be verified because test conditions may not mimic actual or simply because they are grossly inaccurate.

Pertinent to cold service, if bolt type is stainless, nickel-based lubricants are not recommended as they can promote galling. Also, when using stainless bolts, a general rule of thumb is that friction is 20% higher than with the equivalent B7 type bolt. Therefore, a torque used for a B7 condition should be increased by roughly 20% for a B8(M) assuming it is below yield stress.

The only way to verify K's accuracy is by measuring bolt preload in the field after torquing and then back-calculating K. This exercise is highly recommended.

Spring Washers

Spring washers are conical in shape and intended to help maintain bolt stretch. Over the last decade, use of spring washers on gasketed connections has declined for several reasons. First, they add an additional level of complexity to the joint where most prefer to simplify and reduce chance for error. Secondly, spring washers are directional. If installed upside-down, they are ineffective and could detract from an effective installation. Also, they are designed to activate at a certain level of compression. Not only does said compression require precise installation, but the load required to activate the washer must also be adequate to compress the gasket and contain the internal pressure. An increasing trend is to forego spring washers in favor of the Dynamic Recovery Gasket.

Conclusions

Cold service gasketed connections are challenged to maintain sealability during cooling, thermal cycling and throughout operation to the next maintenance turnaround. Options are available to achieve leak-free joints with low emissions. Best results are always achieved with better gaskets and proper installation. Proper installation includes the elimination of impacts, full application of anti-seize and a more accurate estimation of friction leading to better torque values. That in combination with quality spiral wound gaskets should survive slow to moderate rates of cooling and minor thermal cycling. Gaskets typically need to be upgraded when faced with rapid or flash cooling, frequent thermal cycling, or mating flanges of dissimilar metals. Kammprofile gaskets lack recovery and flexibility and are not recommended for these more difficult services. However, a Dynamic Recovery Gasket (DRG) has tremendous amounts of usable recovery and does not suffer plastic deformation upon compression like a spiral wound. The DRG is the preferred type to overcome obstacles associated with flexibility and recovery and is the only type to seal extreme cases like mating aluminum and steel flanges like that of BAHX inlet/outlet connections. A newer way of comparing gasket sealability is an EN 13555 Leakage test. Comparing leak rates at stresses above minimum recommended seating requirements and also comparing unloading curves can provide insight into gasket performance.

References:

1. EN 13555 - Flanges and their joints - Gasket parameters and test procedures relevant to the design rules for gasketed circular flange connections

2. EN 1591-1 - Flanges and their joints - Design rules for gasketed circular flange connections - Part 1: Calculation

3. ASME Boiler & Pressure Vessel Code (BPVC) - Section VIII – Division 1 – Mandatory Appendix 2

4. ASME B16.20 Metallic Gaskets for Pipe Flanges

5. ASME B31.3 Process Piping

6. ASME PCC-1 Guidelines for Pressure Boundary Bolted Flange Joint Assembly

AIChE Paper Number: 128g

SPECIAL VENTURI TECHNOLOGY FOR CONDENSATE REMOVAL

Marc Froeyen
Director
Steamloc® International

Prepared for Presentation at the 2018 Spring National Meeting
Orlando, Florida, April 22-26, 2018

AIChE Paper Number: 128g

SPECIAL VENTURI TECHNOLOGY FOR CONDENSATE REMOVAL

Marc Froeyen
Director
Steamloc® International

Abstract

Steam powered process and utility equipment need condensate removal, and the decision to install mechanical steam traps leads to ever-returning costs. This paper looks beyond the annual budget and gives insight into all the steam trap related costs during the lifetime of a facility. This information allows one to judge potential alternatives to mechanical traps by their true merits.

Special Venturi Technology ensures complete and continuous condensate removal without any moving mechanical parts. It is innovative and has a respectable 30+ year track record in the top 100 (petro)-chemical companies worldwide. In addition to inherent reliability, this technology brings energy savings for your plant and reductions on greenhouse gases.

SPECIAL VENTURI TECHNOLOGY FOR CONDENSATE REMOVAL

Table of contents

1. Introduction

The total cost of condensate removal is widely spread in organisations. It is difficult to group them and appoint their initial cause of what we all know is bound to happen:
"*The irrevocable nature of the eventual failure of all mechanical items.*"

Step by step, this paper nominates all steam trap related costs during the life time of a facility. Some costs are visible in yearly budgets, others have no real budget. The company in general will pay for these hidden costs – often unnoticed (such as the extra fuel to compensate for leaking steam traps).

Special venturi technology satisfies the concerns of both such budgets while guaranteeing complete condensate removal under all process conditions.

This technology offers advantages compared to older technologies on 4 proven spear points:

1. Extreme reliability to minimize maintenance costs
2. Stabilise/increase your process output
3. Avoid unnecessary energy losses and greenhouse gases
4. Very economical in a total cost comparison

All easily achieved by simply avoiding mechanical items in your steam traps.

2. Background

2.1 Purpose of steam traps

Steam is still a great way to transport energy to your production processes. This transport goes along with heat losses that convert a corresponding amount of steam into condensate. Also, during the heat transfer in process exchangers, where steam gives up its latent heat in favour of the process, steam continuously converts into condensate.

All condensate should be removed completely and as fast as possible to ensure dry steam to be transported to all steam users, and to ensure the total surface of a heater is available to exchange heat at all times. Remaining condensate can be a reason for water hammering with all related risks. A few technologies are used to do this job, but the most common way to do this is mechanical steam traps.

2.2 Process or Utility steam traps

Process steam traps serve the heart of your production (reboilers, autoclaves, heaters, etc....). Their operational outcome is directly linked to your production output. Average sized (petro)chemical plants have only some 15 to 30 of such process steam traps. Not much in numbers, but yet, they handle about 90% of total condensate.

Utility steam traps drain the steam distribution network in addition to smaller heating needs such as tracing, double pipe jackets and other heat loss compensating systems.

Average sized (petro)chemical plants have some 500 to 5,000 of such steam traps. Pretty high numbers, but they only handle about 10% of total condensate.

2.3 Basic principle of steam traps

Based on their operating principle, mechanical steam traps usually are classified in 4 groups:

- Floating ball

- Inverted bucket
- Thermostatic element
- Thermodynamic

Detailed operating principles of traps are not in the scope of this paper. However, as a short summary: all the above traps are based on mechanical elements that open or close internal valves at certain moments to drain condensate and non-condensable gasses. Most of these traps work intermittently as they are not always in sequence with the continuous condensate flow from heat transfer equipment. Venturi type traps have no such mechanical parts. Their continuous drain ensures immediate and total condensate removal at all times.

2.4 Failure positions of steam traps

The two primary ways steam traps fail is in the 'open' or 'closed' position:

- Closed failure will flood the condensate and heat transfer comes to a stop. A bypass might help, but often results into less efficiency, steam loss, erosion, safety issues, etc....
- Open failure will drain all condensate, so heat transfer continues, but not at full output because the fast steam is not fully condensing. As a consequence, you accept live steam losses and the potential for erosion of downstream piping.

Most steam traps mechanically fail in the open position. It makes sense to build traps this way since a closed failure requires immediate action to avoid process stops or safety issues like water hammering for which nobody wants to take credit. The steam leak from an open failure is also unwanted but holds less immediate risk, only energy cost. This allows extra time for corrective actions, although it might be tempting to stretch this time line in favour of other more urgent jobs.

> *"The 'Enbridge steam trap survey' tested the condition of 41,124 traps in 216 plants. This study showed a total of 24% defective traps (9,870 pieces) of which 68% 'blew through' (6,719 pieces) and 32% 'blocked' (3,151 pieces)."*
> ***B Griffin, Enbridge Steam Saver/Steam boiler plant efficiency, March 1, 2006.***

Debris in the piping is a general problem for all steam traps. This is also true for small diameter venturi traps. It is advised to properly flush the steam system. Where necessary, traps should be provided with internal strainers to avoid closed failures as best as possible.

2.5 Average life time and failure rates of Steam traps

The average trap lifetime is calculated as the total number by the broken traps per year. For example, 150 yearly failures on a population of 1000 traps result in an average life time of 6.7 years. The failure rate for that year is 15%. Further calculations will use this number.

> *"Average-quality traps may have just a 4-year life expectancy (which implies a 25% failure rate), while higher-quality steam traps may have an 8-yr life expectancy (12.5% average failure rate)."* ***Risko, J., Understanding Steam Traps, Chemical Eng. Progress, Feb 2011***

2.6 Average failure time of Steam traps

Most steam trap inspections are performed on a yearly basis. Traps can fail from the day after the previous or on the day before the present check. This brings the average failure time to 6 months. Raising the frequency will lower the failure time, but will increase the internal or external costs to perform these inspections.

2.7 Methods to check Steam traps

There are several ways to audit steam traps. Of course, a prerequisite is that each trap is in service. With the numerous amount of different trap principles and applied check technologies, it requires some skill to arrive at a correct assessment.

- Plants report very good results with ultrasonic tools, although it is labour intensive to first collect and upload all the necessary data prior to the field measurement.
- Thermal infrared cameras are also a good option. Close attention should be payed to the emissivity factor of the different materials. It is also important to keep in mind that the pipe size can influence the flow pattern and judgements on corresponding temperatures.
- More recently, there are real time and continuous monitoring systems for steam traps. Transmitters are built in for each (group of) trap(s) and connected to a central computer or mobile App. This can be helpful if you prefer to know the trap's condition instantly.

2.8 Steam loss of Steam traps

Open failing traps still drain condensate but simultaneously blow steam through their orifice. Napier provided the widely used equation to estimate the steam flow passing through the internal orifice of failing steam traps while draining condensate:

Imperial		Metric	
W (lb/h)	**= 24.24 x P_{abs} x D^2**	**W(kg/h)**	**= 0,247175 x P_{abs} x D^2**
Lb/h	= c x psia x inch2	kg/h	= c x bara x mm2
Here in:			
W	= steam loss in lb/h	1 kg/h	= 2.20 lb/h
P_{abs}	= delta steam pressure in psia	1 bara	= 14.50 psia
D	= diameter internal orifice in inch	1 mm	= 39.37x10-3 inch
24.24	= constant	0.24712	= constant

As an example, we worked out the daily cost for two leaks on average steam pressure: (Steam cost 15 $/ton - 8600 h/yr – this calculation is further used to determine total costs)

1. Utility steam trap 1/2" – internal orifice 1/8" (3.18 mm) - Blow-through failure
   Steam pressure 5 barg to back pressure 0 barg – condensate load 15 kg/h

$W = 24.24 \times 72{,}5 \times (1/8)^2$ $= 27.5$ lb/h ~ **4.5 $/day**	|	$W = 0.247175 * 5 * (3.18)^2$ $= 12.5$ kg/h ~ **4.5 $/day**

2. Process steam trap 3" - internal orifice 5/16" (7,9 mm) - Blow-through failure
   Steam pressure 16 barg to back pressure 2 barg – condensate load 16000 kg/h

$W = 24.24 \times 203 \times (5/16)^2$ $= 480.5$ lb/h ~ **77.50 $/day**	|	$W = 0.247175 * 14 * (7.9)^2$ $= 216$ kg/h ~ **77.50 $/day**

Venturi steam traps cannot fail open to waste energy. They are open the appropriate degree from the start.

2.9. Impact of leaking traps on your boiler load

Unless a plant benefits from exothermal processes to generate steam, it is the fuel-driven steam boiler that produces the energy that feeds all process installations. The energy that is lost from broken steam traps or cracked bypasses is not available for the process. The extra steam to compensate these losses is called the 'phantom load' of the boiler. This extra cost is silently absorbed by the overall plant budget.

3. Impact of failing traps during the next 30 years plant operation

3.1. Financial impact

The total cost from traps is built from following actions and events:
Initial Purchase / Audits / Energy loss / Spare parts / Labour to fix.

Other costs relate to: *Production impact / Shutdown preparation / ...* and minor costs such as repeated actions from *Purchasing / Warehouse / Accounting / etc...*

The impact on Operation can be a very high cost, but this is plant-specific and cannot be generalized.

The calculation below only highlights the costs to quantify the total impact of the main actions. We tried to work with average and conservative field values, however, we encourage each plant to apply their data to find their relevant outcome.

Starting points:	
Plant life time:	Expected to run for 30 years
Trap numbers:	Suppose 1000 utility and 25 process steam traps
Process trap (P)	1500 $/piece average
Utility trap (U)	150 $/piece average

Open/closed fail	68% / 32% (from "Enbridge survey")
Napier loss	Utility steam trap = 4.5 $/day
	Process steam trap = 77.50 $/day
Leak time:	A yearly audit results in average 6 months
Trap life time:	An average failure ratio 15% = 6.7 years life time
	= 1 original + 3 extra replacements for each trap
Audit trap	15 traps/hour = 50$ hour wage/15 = 3.4 $/location
Labour cost	50 $/trap change

Calculations:			
Initial purchase	(1000 x 150 $/piece) + (25 x 1500 $/piece)	=	187,500 $
Initial installation	(1025 x 50 $)	=	51,250 $
Audit traps:	1025 traps/yr x 30 yr x 3.4 $/location	=	104,550 $
Energy loss U:	1000 x 4 pc/30yr x 4.5 $/day x 365/2 days x 68%	=	2,233,800 $
Energy loss P:	25 x 4 pc/30yr x 77.50 $/day x 365/2 days x 68%	=	961,775 $
Extra traps U+P:	(3 x 1000 x 150 $) + (3 x 25 x 1500 $)	=	562,500 $
Extra labour:	(3 x 1000 + 3 x 25) hours x 50 $	=	153,750 $

TOTAL COST cumulatively generated from steam traps / next 30 years = **4,255,125 $**
(this is a multiplication of 17.8 to the initial purchase)

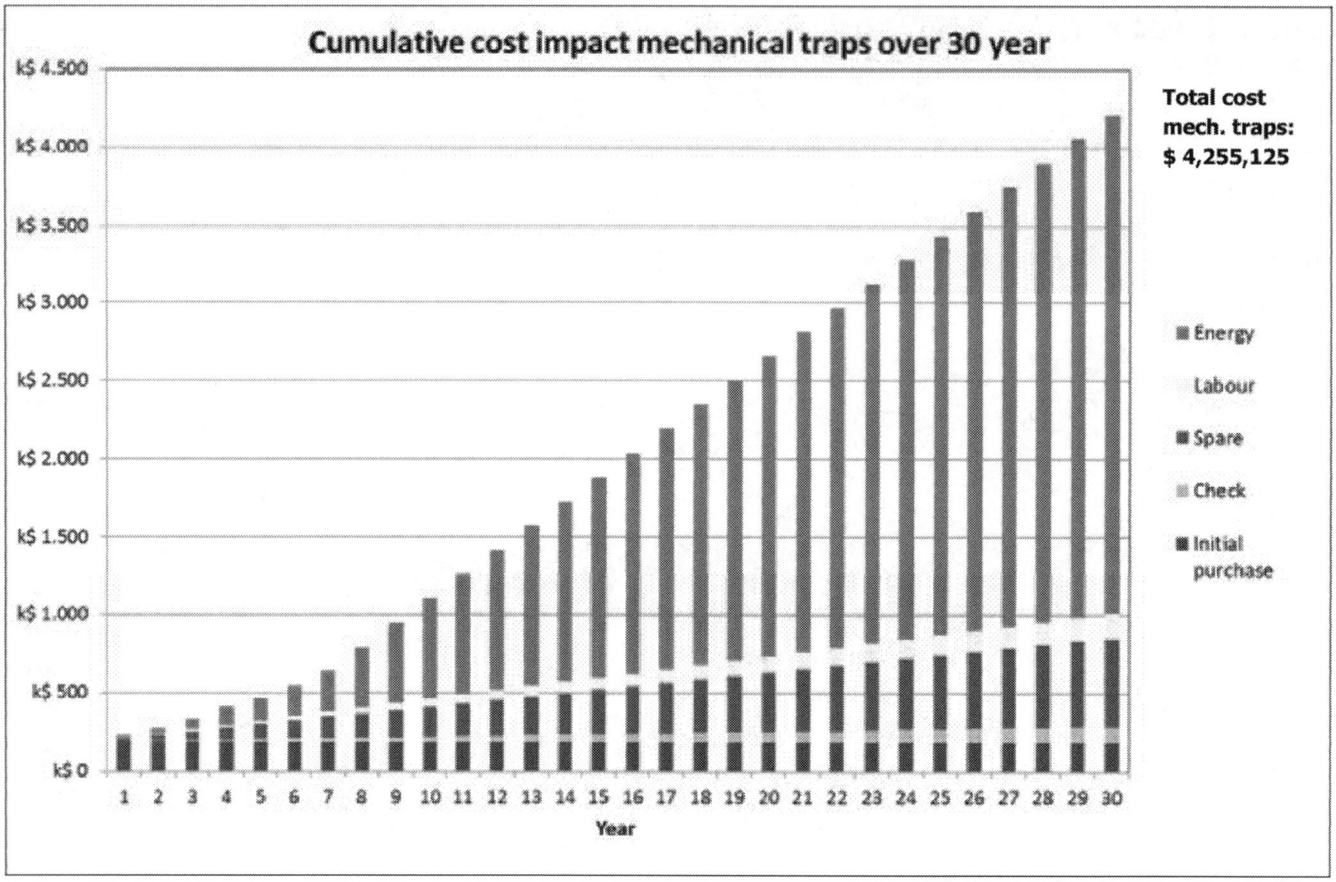

3.2. Environmental impact

Aside from the maintenance and energy impact, there is also an environmental impact. Burning coal, fuel or gas generates a certain amount of GHG (greenhouse gasses).

"Gasoline fired boiler MPSteam = 0.117 ton GHG/ton steam (Carbon Dioxide Equivalent) Greenhouse Gas Equivalencies Calculator – www.epa.gov "

"SCC - Social cost of greenhouse gases – Feb 2017 – 41$/ton CO2 - www.edf.org "

For the above calculation:	
Total steam loss in 30 years:	= 213.253 ton steam
Generated from gasoline, this represent an extra exhaust of:	= 25.033 ton GHG
Equivalent to	= 5360 cars driven for 1 yr
SCC (Social cost of carbon) estimated cost for CO2	= 1,026,353 $

This impact might require a correction for the fuel (natural gas, coal, etc ...) and the overall efficiency of the steam boiler installation.

4. What are we doing? And Why?

It is undisputed that steam installations require condensate drain principles. For over 100 years the market only offered mechanical solutions for this job and a complete industry was developed based on this technology.

Of course, this had benefits. People were educated on general steam knowledge and simultaneously trained about traps, how they work, what to do at failure, etc. Suppliers developed several kinds of inspection tools, and even help to organise trap audits, supply spare traps, etc.

All of this still results in a peculiar win-win where a facility can temporarily mitigate energy losses with a device which it is known in advance will fail again, while a supplier can preserve a steady turnover. For this reason, it is logical that trap suppliers are not incentivized to switch to non-mechanical steam traps.

It is time to step out of this vicious circle and look for a permanent solution that continually saves and at the same helps to limit unnecessary burning of fossil fuels. The 'Special Venturi Technology' will keep installations and facilities state-of-the-art.

All it takes is to study your costs and the will to alter deep-rooted habits in the organisation.

5. Venturi technology for condensate removal

5.1 What is it?

Venturi technology can be used to reach a highly effective liquid/gas separation. The 'active' element of this type of device is a well-designed and unique venturi construction founded on simple principles without any moving parts. It meets all requirements for a correct and effective condensate removal:

- drain all condensate
- drain it continuously
- drain without excessive steam consumptions, even under changing process conditions
- meet the typical 30 years design lifetime of the plant

5.2 How does it work?

At first sight, it seems a venturi will drain condensate and simultaneously pass lots of steam. This is a logical reflex if we only look at densities and velocities of the liquid and gas. Other features play a role for which the venturi design itself becomes paramount. A hole in a blind flange, or a pinched valve, or a simple orifice, etc ... they all belong to the general 'venturi-orifice' family. But only a proper venturi design will ensure the removal of all condensate and simultaneously inhibit the expected loss of live steam. The combination of differences in physical conditions such as density, velocity, altering of laminar to turbulent flows and compressing to expanding flows play a role in the combined action to limit steam passage in favour of the condensate flow.

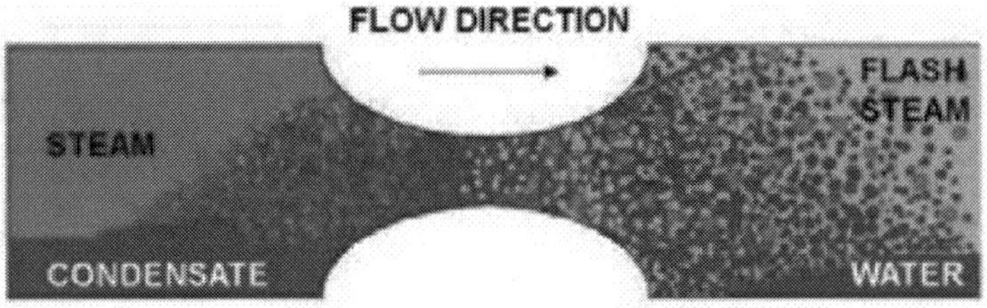

5.2.a Steamloc® principle design

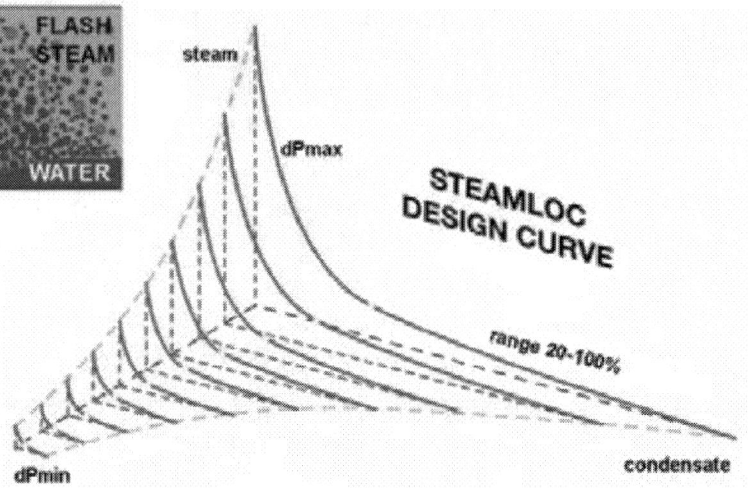

5.2.b Steamloc® design curve

Although one might expect the opposite, a well dimensioned venturi device has a very wide active range. It effortlessly handles big condensate load changes as they appear in batch processes. And since the capacity through the venturi is pressure dependent, the device is also fully self-regulating towards changing process conditions with respect to control valves on the steam inlet. Even air and other non-condensables will be removed without any additional mechanical element.

5.3 Common misconceptions about venturi steam traps

It may be thought that because a venturi is continuously open it will consume more steam than mechanical traps. Flow meters on heavy processes have shown that a properly designed venturi trap consumes steam in the same range or lower as correctly functioning mechanical traps. The real consumption happens when mechanical traps fail open which venturi traps cannot. Venturi are open from the beginning.

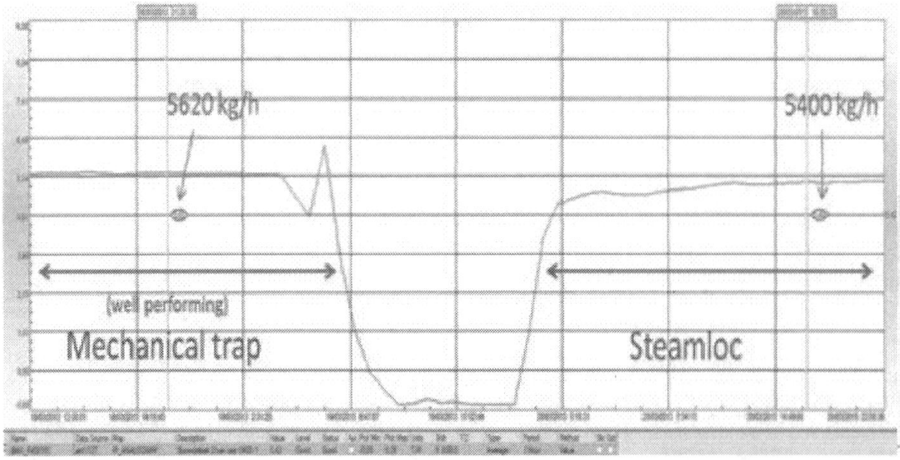

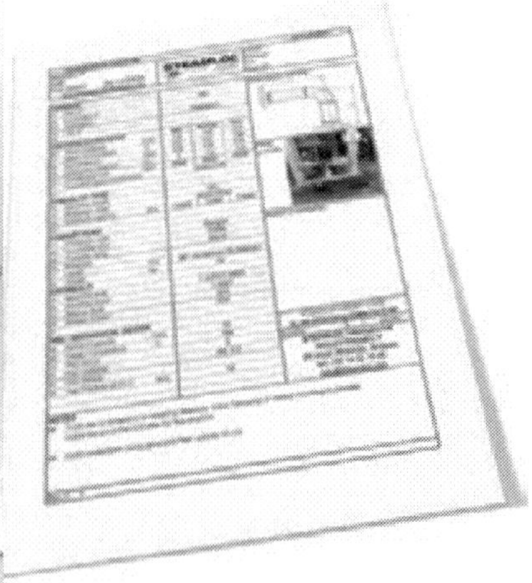

5.3a Compare consumption of mechanical trap and Steamloc®

Similarly, mechanical traps are closed when there is no condensate, but a venturi trap is always open, therefore it may be thought that a venturi would leak. In actuality, a normal operation has a continuous flow of condensate, whether big or small, there will be a flow. Steam consumption is calculated in trap design and can be compared to existing operations to evaluate the potential savings.

It is often said venturi traps can only be designed for unchanged load and pressures. A change in operational conditions would make them or too big and cause steam loss, or too small and flood the condensate.

Correctly designed venturi traps however have a wide operational range for cases where pressures remain unchanged and loads would vary. When control valves control the pressure, venturi capacities follow each variation immediately. Pressure controls always fall within the 80% working range, and also fouling cases are perfectly handled within this range.

5.4 Where to apply?

Venturi technology can drain condensate from any steam application. Process applications will give the highest return on investment because energy losses are bigger and often remain unnoticed longer, while the conversion cost is limited. Utilities are interesting spots because of the high numbers. Both a maintenance-based replacement program or an all-at-once conversion at the next shutdown could be considered.

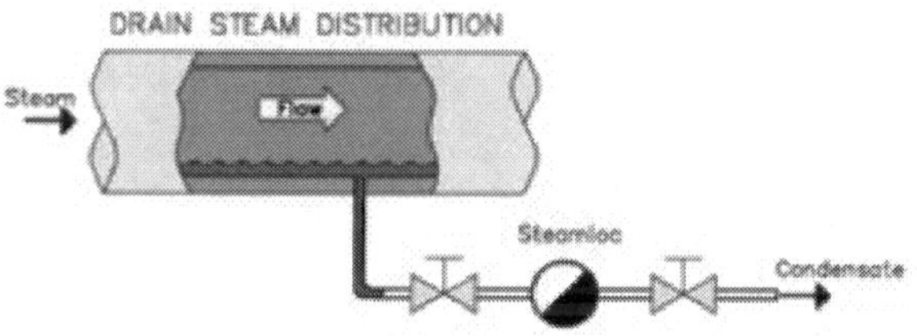

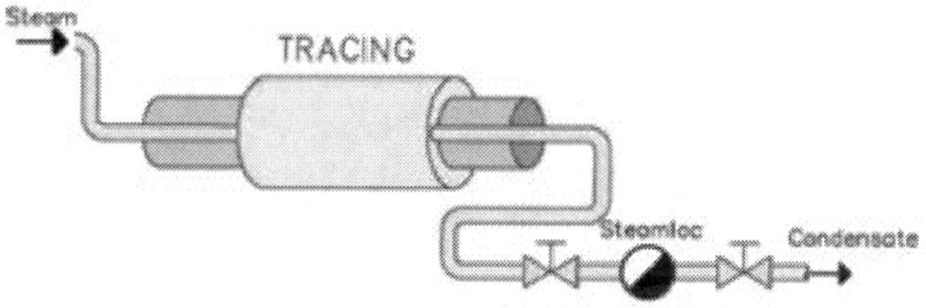

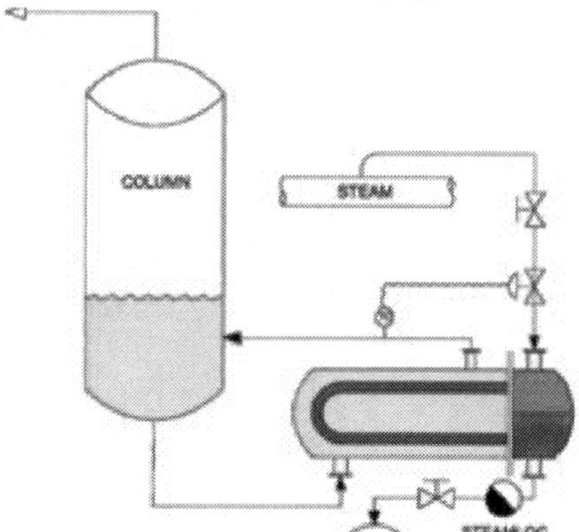

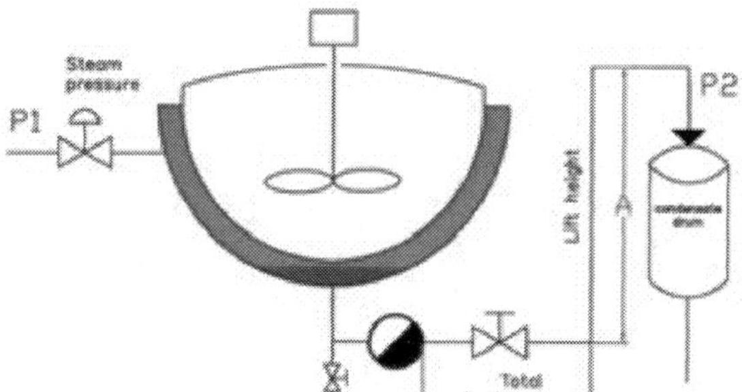

5.5 Major designs

Venturi technology for processes is typically built as a flanged spool piece with standard or customised lengths, all according to general equipment specifications. Customised shapes help to avoid complicated piping modifications. Materials are standard in the group of austenitic stainless.

For utility, a range is available to meet the widely varying shapes of mechanical traps. Connections are NPT, SW, Flanged, Quick Fit, ... and materials are also austenitic stainless. Custom made models are also possible.

5.5.a Process Venturi

5.5.b Utility Venturi

5.6 How to install?

The operation of venturi traps does not depend on gravity force. They are installed horizontal, vertical downwards, and upwards. They come in any shape, size, pressure rating and length to minimize piping modifications.

5.6.a Special Venturi Technology on reboilers

Venturi traps allow to simplify the piping layout to one single trap, one block valve and one decompression drain.

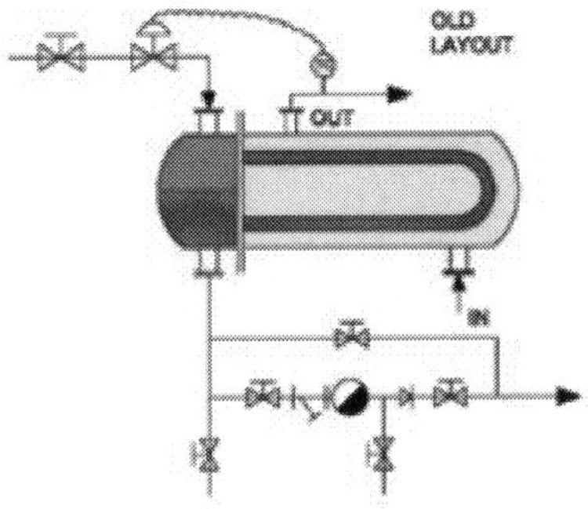

5.6.a Traditional layout

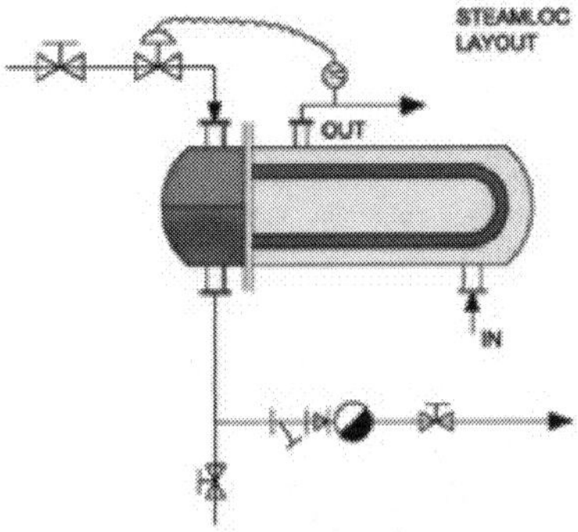

5.6.b Venturi layout

5.7 How to verify proper working

Inspection of a steam trap results in 4 options. Open failure generates the highest cost. Venturi type traps only have 2 options. Open failure is impossible. It is open from the start. The diagram implies that a simple temperature inspection for Steamloc is sufficient (hot is OK).

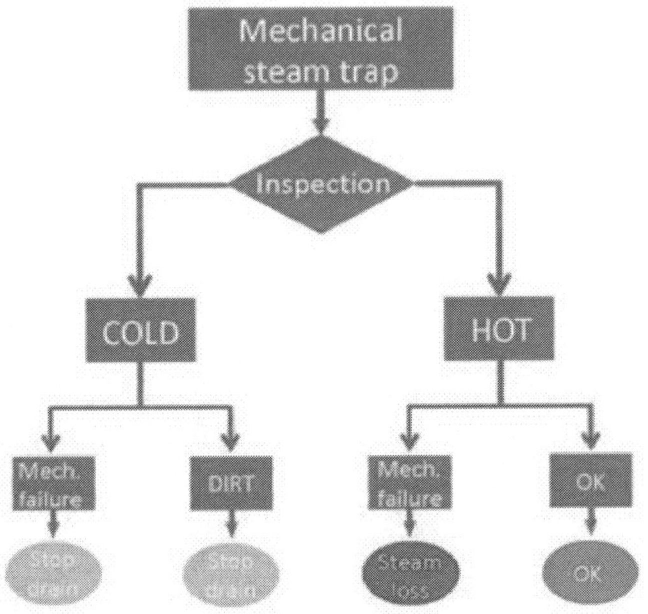

5.7.a Mechanical traps

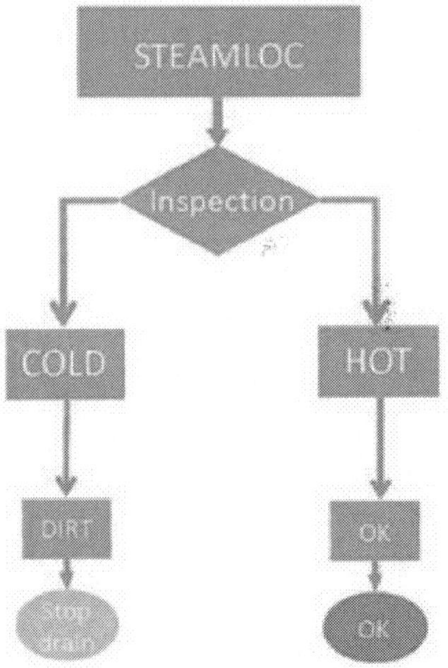

5.7.b Venturi traps

Infrared (IR) camera inspections may be effective for process and lined up utility.

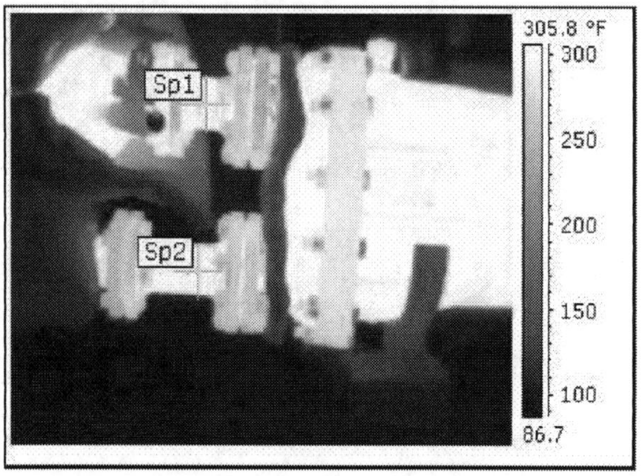

5.7.c Blowing process steam trap

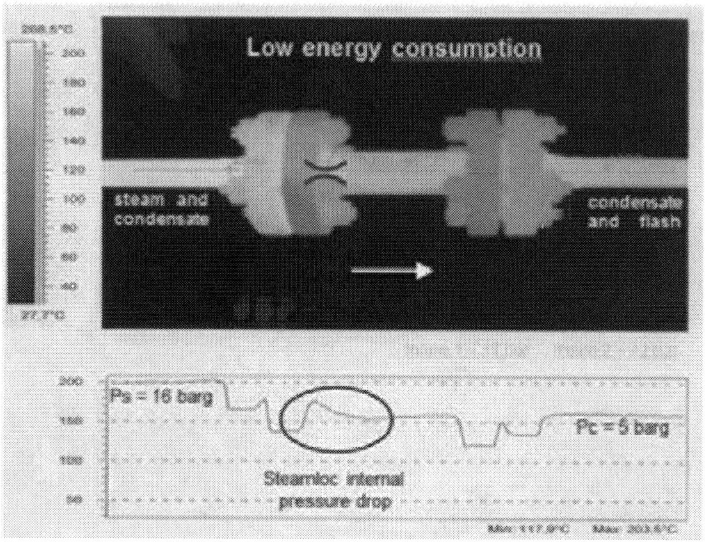

5.7.d

Special Venturi Technology

Utility drains to atmosphere can be inspected visually. Please pay attention to judge what you see correctly. Photo 5.7.e shows condensate, flash and leak. Photo 5.7.f shows condensate, flash and a small consumption.

5.7.e Blowing small utility steam trap

Special Venturi Technology

5.7.f

5.8 Compare Life-cycle Cost in a new project

Let's look at the previous costs under 3.1 (traps) and the impact of special venturi technology (to show price is not the restraining factor, we hypothetically assume venturi type are double priced as mechanical traps, which is conservative but unlikely)

Calculations:	**Steam traps**	**Special Venturi**
Initial purchase	= 187,500 $	= 375,000 $
Initial installation	= 51,250 $	= 51,250 $
Audit traps:	= 104,550 $	= 104,550 $
Energy loss U:	= 2,233,800 $	= 0 $
Energy loss P:	= 961,775 $	= 0 $
Extra traps U+P:	= 562.500 $	= 0 $
Extra labour:	= 153.750 $	= 0 $
TOTAL COST / next 30 years	= **4,255,125 $**	= **530,800 $**

5.9 How to implement this technology in an existing facility

A few process applications make plants familiar with the concept and build necessary confidence. Securing the rest of the process is a logical next step. Their limited numbers allow an easy follow-up for the performance. Utility is a logical continuation since the big numbers make their total energy losses relevant.

Let us look at the impact of a few possibilities (based on numbers from paragraph 3.1) (Initial purchase and installation is not taken in account – the plant exists already)

Choice 1: - Like-for-like mechanical steam trap replacements today

- Temporarily stop of energy loss at every replacement
- One accepts 4x energy losses during 30 years = 3,195,575 $
- + inspections + 3x new extra traps + installation = 1,008,300 $
- Total extra cost / 30 years = 4,203,875 $

Choice 2 – replace each failure by Venturi type (double priced)

- Stops energy loss at every replacement but requires one last energy loss to reach position on the replacement list
  At failure ratio 15% = 6,7/30 years energy loss = 713,678 $
- + inspections + 1x Venturi + installation = 449,600 $
- Total extra cost / 30 years = 1,163,278 $

Choice 3 – a full conversion into Venturi type today (double priced)

- Stops every future energy loss immediately = 0 $
- 1x new Venturi + installation + later inspections = 530,800 $
- Total extra cost / 30 years = 530,800 $

6. Case studies

Following are case studies of successful application of Special Venturi Technologies.

6.1 Stability on the steam inlet valve

The Amine Plant Operation team complained about the swing of a steam inlet control valve on a 40 barg heater for unknown reason. The intermittent operating principle of the inverted bucket steam trap was suspected.

A venturi trap was installed and immediately the swing disappeared. The explanation was found in the fact that a heat exchanger produces condensate continuously while the mechanical trap works intermittently. Venturi traps drain continuously in cadence with the heater output which improved the stability and reliability of the application.

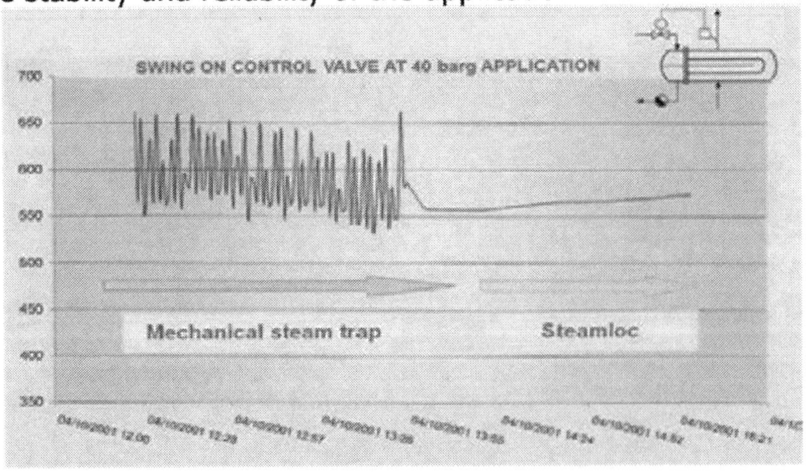

6.1.a behaviour of a steam inlet control valve with traps or venturi

6.2 Improved production output

This Distillation tower was equipped with two mechanical traps. One for the normal run, one for spare, just in case. Capacity problems forced the team to open both traps for which an instability appeared on the output product temperature. Once the two traps were replaced by one single venturi trap, a stable product temperature was achieved. Although it was not fully clear why, the Operations Team suspected one of the traps had failed open and created the instability.

The venturi trap also allowed to simplify the piping to straight in-out without the T-split.

6.2.a Two parallel steam traps

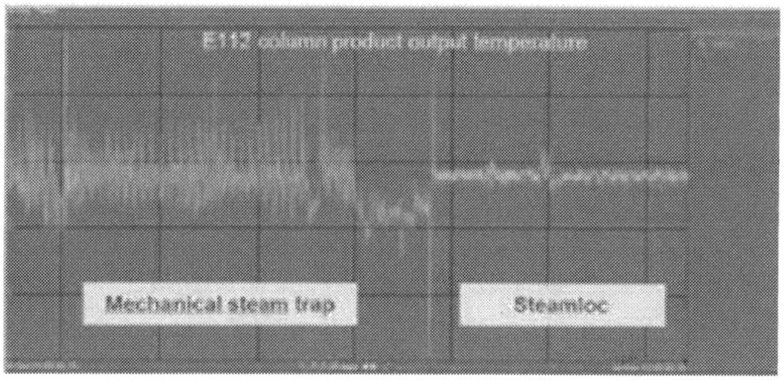

6.2.b Special Venturi Tech reaches stability

6.3 Reliability – traps and condensate piping

A Debutanizer Reboiler struggled with a recurring short service life of the mechanical trap resulting in erosion of downstream piping and water hammer in the condensate return line. A switch to another traditional trap resolved the short service life somewhat but the water hammer remained an issue. It was decided to replace the trap with a venturi trap. Water hammer disappeared, and no corrosion/erosion showed in time. Steam consumption, output temperature and control stability remained the same as before which is good, because the advantage is found in the fact venturi traps are not sensitive to erosion or leaking, and their operation prevents water hammer.

7. Summary

The steam trap's irrevocable nature to fail makes it a consumable that causes ever returning extra costs.

The venturi type steam traps can reduce these cost to the bare minimum of installation, thanks to the absence of any mechanical parts. Key-words are 'maintenance-free', 'reliability', and 'stability'. But the biggest win is the reduction of energy losses which also helps the environment.

Cost savings are within reach if new technologies are considered over traditional methods.

References

1. John/Robert/Frank Napier - "Napier formula" for steam loss through an orifice.
2. B Griffin (March 2006), Enbridge Steam Saver/Steam boiler plant efficiency
3. Risko, J. (Feb 2011), Understanding Steam Traps, Chemical Eng. Progress
4. Greenhouse Gas Equivalencies Calculator – www.epa.gov
5. Social cost of greenhouse gases (Feb 2017) – www.edf.org

AIChE Paper 126a

RDR - Revolution in Ethylene Yield

Harri Johannesdahl

CEO

Coolbrook Oy

Prepared for Presentation at the 2018 Spring National Meeting
Orlando, FL, April 24, 2018

RDR - Revolution in Ethylene Yield

Harri Johannesdahl

CEO

Coolbrook Oy

Abstract: Coolbrook is developing a revolutionary cracking technology, Roto Dynamic Reactor ("RDR"), a turbomachine for cracking hydrocarbons. The RDR technology provides the following benefits to the olefin production process compared to traditional technology,:

- Significantly higher ethylene yield (at least 20 % more ethylene+propylene)
- Lower energy consumption
- Sicnificantly less flaring
- Smaller plant investment costs

Coolbrook Oy, registered in Finland, which is a part of the European Union, owns three patent portfolios related to the invention. The key aspects of invention has been patented in countries that are noteworthy in respect of ethylene production. The RDR –reactor technology was originally invented by a hydrodynamic expert engineer, Mr. Bushuev, who performed tests with a model RDR prototype in 2007.

In 2012 current development project was started in Finland. Three existing patent families, all IP rights and related businesses are owned by Coolbrook Oy and thus governed by European Union Law.

The RDR heating method to crack feedstocks differs completely from traditional thermal furnaces. In a traditional furnace, naphtha or ethane or other feedstocks, break down under intense external heating of coil surfaces. In contrast, the RDR uses kinetic energy of high-speed spinning rotor to heat the feed gas close to 1000°C.

The RDR innovation has unique features for optimizing and balancing the chemical reaction and RDR can be designed to optimize selectivity and conversion. The controllability of the reactor and the process provides maximum yield of the most valuable and desired products and thus increases the operational flexibility and economical profitability of an ethylene plant.

The RDR technology can be installed in existing as well as greenfield ethylene plants. The cost of installing RDR technology in new plants is estimated to be slightly below or similar to the cost of installing existing cracking technology. Construction philosophy is to provide technology by using offsite modular construction minimizing turn around time and lowering cost of installation.

Feed composition, reactor gas temperature, steam ratio and residence time are the main parameters affecting the product distribution in cracking of hydrocarbons. RDR technology derives its biggest benefit from short residence time combined with high temperature both resulting unique energy addition method.

Current situation

Currently the key drivers for the development of the new chemical technologies are energy savings and reduced consumption of non-renewable feedstocks.

The existing technology of lower olefin production by cracking of hydrocarbons in the tubular furnace was created more than 70 years ago and doesn't satisfy modern requirements of efficient use of feedstock. Slow rate to increase temperature of feed mixture in tubular cracking furnaces results in non-optimum residence time. Thus, initially formed olefins remain excessively long period of time under conditions of intensive secondary reactions resulting in lower olefins yields. This disadvantage cannot be eliminated with the traditional technology because the rate of heat transfer through the tube walls to the reaction zone in radiant section of conventional cracking furnaces already has reached its physical limits.

RDR -New cracking technology

The major feature of the new process is non-traditional method of heat supply to the reacting mixture. In the rotodynamic reactor the heat required for cracking is not transferred through the tube walls like in the conventional tubular cracking furnaces. In RDR heat is generated directly and instantly in the reacting mixture by creating high velocity by rotor blades and converting high velocity to heat by static blades.

Due to such heat supply method, the temperature of the walls casing the reaction zone follows gas temperature and simultaneously the heat transfer limitations are avoided. The yield of desired cracked products is increased due to the higher temperature and shorter residence time in the reaction zone and also due to the uniform temperature profile in cross section of the process stream. Constant pressure in reaction zone also favours higher yields. Feed temperature to RDR is similar to the temperature to radiant section in conventional cracker.

Reactor design for the new cracking technology is shown in Figure 1. The reactor is a regenerative axial turbomachine. The reactor comprises of rotor equipped with blades (2) which form an axial-flow blade cascade, a fixed guide toroidal hoop (3) which adjoins to tips of the blades, and a casing (4) and encloses periphery of the rotor and the hoop forming a duct. This donut shaped duct (10) is between exit from the diffusing cascade and entry into the nozzle cascade. Port (6) for inlet stream is located after each partition and port (7) for outlet stream is located just before each partition. Upstream of the blade cascade the nozzle vanes (8) are installed forming a nozzle cascade, and downstream of the blade cascade the diffusing vanes (9) are installed forming a diffusing cascade.

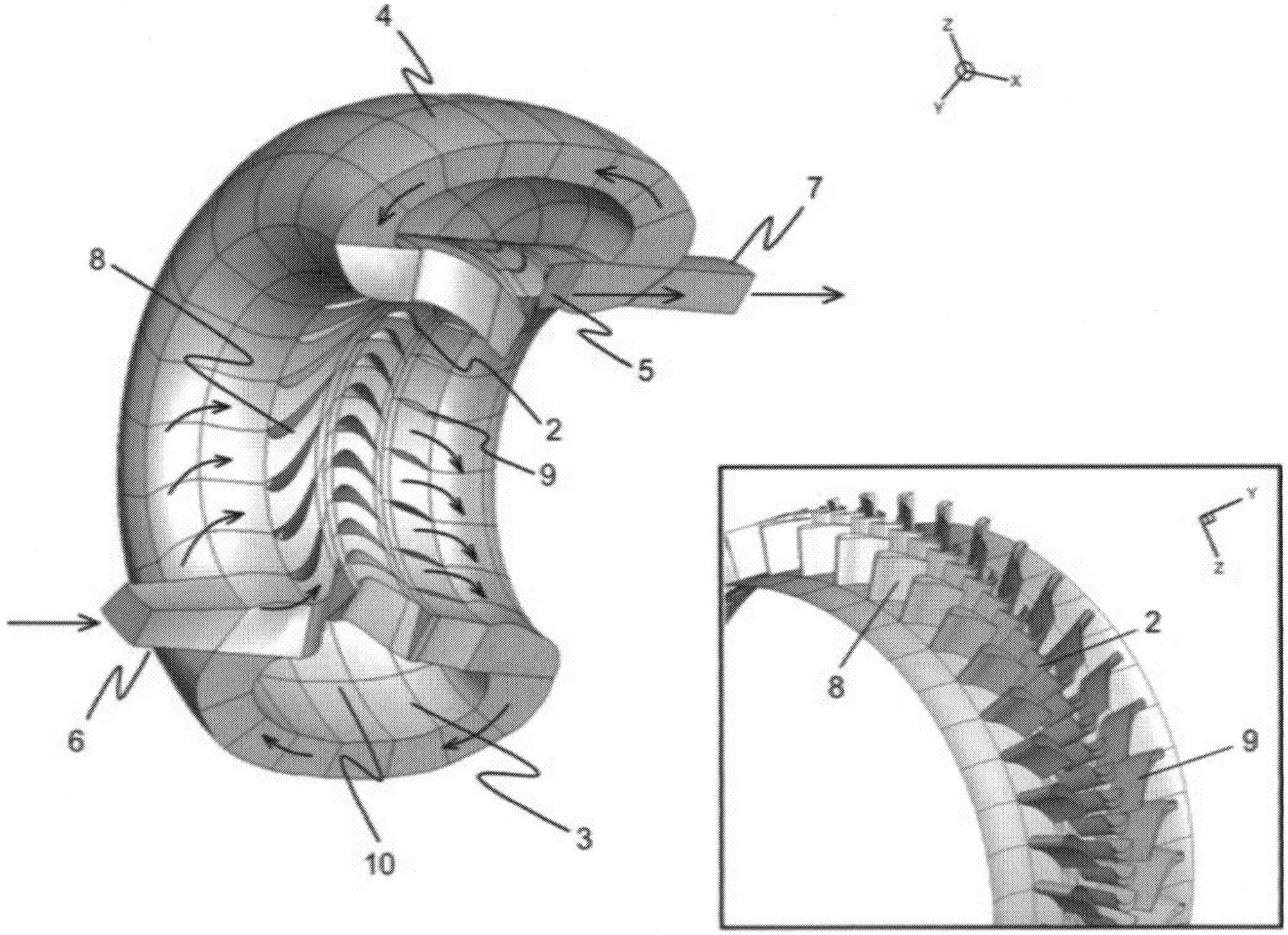

Figure 1. *Rotodynamic reactor for hydrocarbon cracking.*

When the rotor is rotated, a fluid in each working cavity is forced to move from the inlet port to the outlet port along a helical trajectory so that stream repeatedly cross the nozzle cascade, blade cascade, diffusing cascade and the vaneless space. The flow velocity in the vaneless space is subsonic. The flow velocities in the nozzle cascade, blade cascade and diffusing cascade are supersonic. When crossing the rotor blade cascade the flow receives the kinetic energy, which is converted into the heat when flow speed is reduced in the diffusing cascade. Heating of the stream is performed by impulses when this stream goes through the shock waves of high intensity .Thus the temperature-time curve has the sawtooth configuration. All stream particles have essentially the same temperature histories and so the reactor characteristics are close to the characteristics of the perfect flow-displacement reactor. The temperature history of stream particles is shown in Figure 2.

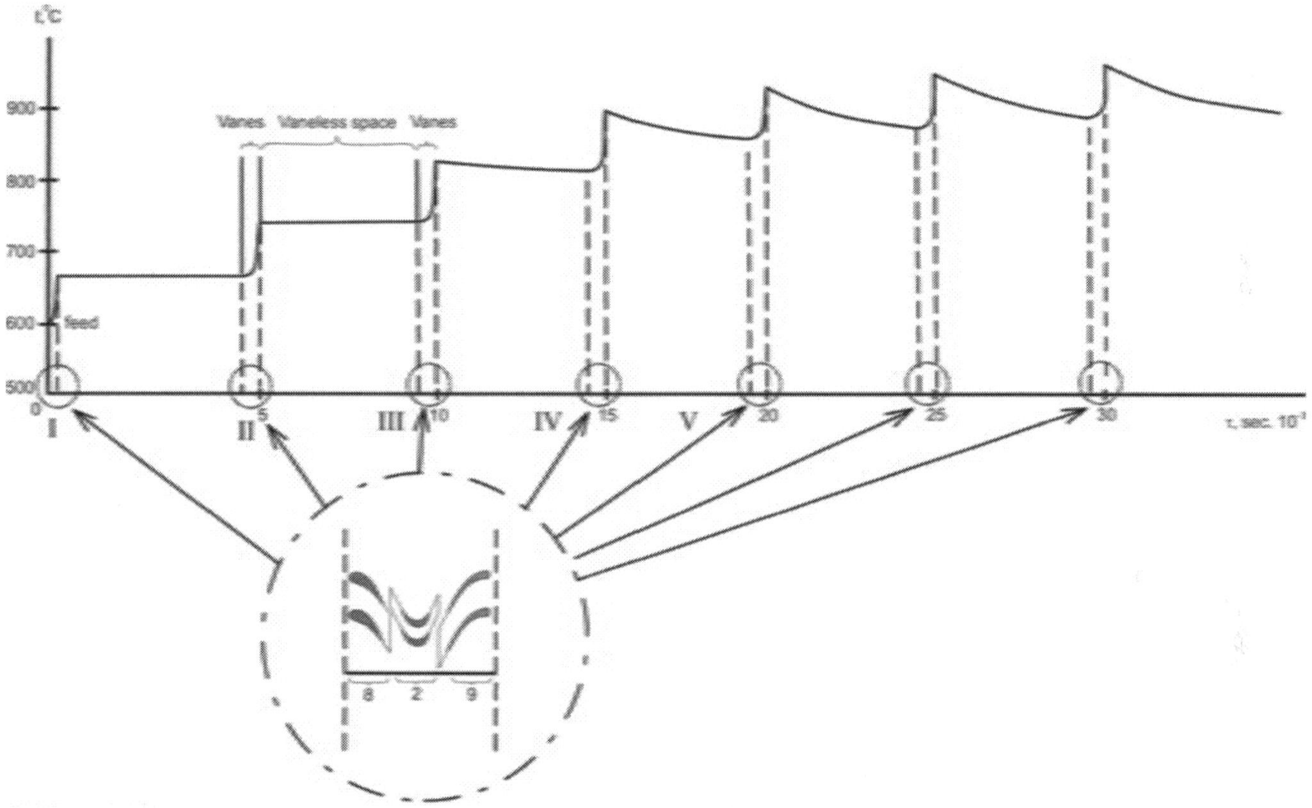

Figure 2. *Temperature jumps in RDR.*

Tests results of RDR-model rotodynamic reactor

A rotodynamic demonstration unit of small capacity was developed and manufactured to assess the technical feasibility of the new cracking process in 2007. The main test objective was to demonstrate olefins yield of the RDR reactor. The electric motor of 90 kW in power equipped with the step-up gearing was used to drive the reactor rotor.

The process conditions of the model reactor were similar to the process conditions of the commercial scale reactor: the temperature was 902°C at the end of transfere line, the residence time in reaction zone was 25-30 ms, and operation pressure was about 2.0 bar abs. Two naphtha feedstocks were tested. The feed flow rate was about 40 kg/h; the dilution of steam/naphtha ratio was 0.5.

The test results confirmed higher yields of main products compared to the best specimens of the conventional tubular cracking furnaces (Table 1).

Table 1. *Comparison to current technology*

Product	Designer, type of reactor, cracked gas outlet temperature (°C)						
	CB&I Lummus	Technip		KBR	[1)]COOLBROOK RDR		
	SRT-VI	GK-4	GK-5	Millisecond	Model rotary blade reactor		
	862 °C	848 °C	835 °C	900 °C	882 °C	891 °C	902 °C
Ethylene	30.30	29.03	30.50	34.40	**42.06**	**42.64**	**43.32**
Propylene	13.90	11.59	15.50	16.60	**13.53**	**13.10**	**12.34**
Sum	44.20	40.62	46.00	51.00	**55.59**	**55.74**	**55.66**

[1)]Temperature at the end of transfere line

These test results clearly confirm substantially better yield for for the RDR when compared with conventional furnaces. One important feature is that the highest temperature is at the exit of the reactor so reducing the risk of reactor coking.

RDR Pilot

Coolbrook has started a two-year RDR -pilot program. Targets of the program are to build a pilot reactor with auxiliary equipment to further develop and confirm technical aspects of this new technology, build a relevant partner network for commercialization, create RDR design software, and to demonstrate the invention to the petrochemical industry. The tasks and components for the Pilot phase:

- Test and confirm yields and conversion of different feeds
- Study coke formation and decoking methods

- Design basis for commercial reactor
- Validate reactor models
- Validate CFD modelling
- Test reliability of reactor

During the pilot one of the key objects is to study coking phenomena. The purpose is to understand effect of process conditions in RDR on coking and differences to conventional cracking. Due to shorter residence and higher temperature and without heating walls (coils) RDR may be less prone to coking than conventional furnaces.

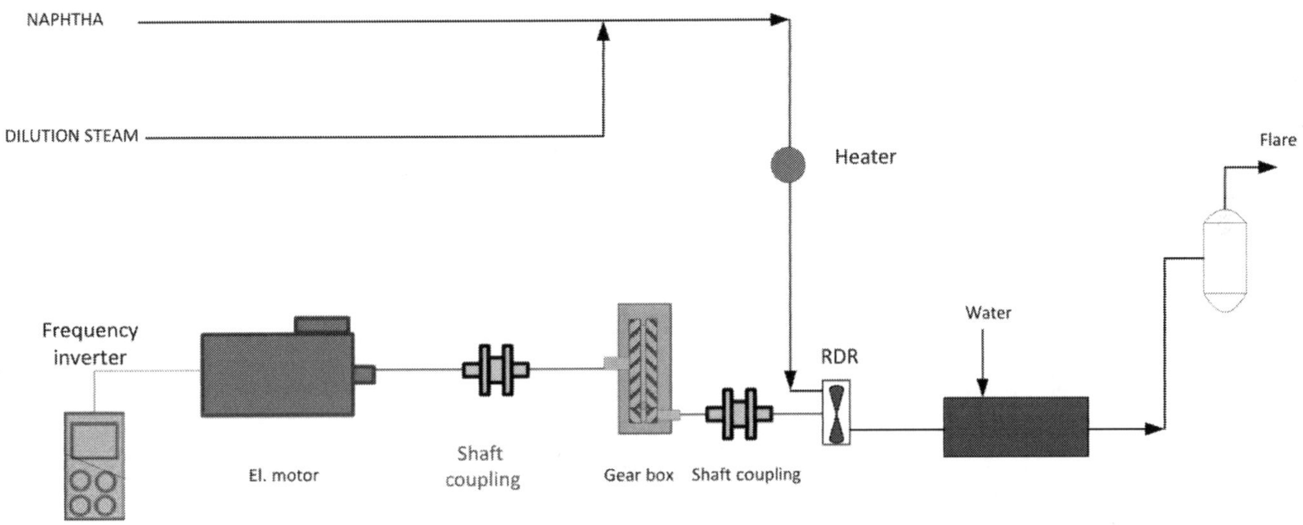

Figure 3. *RDR Pilot concept.*

Coolbrook has already several partners for the development work including Neste Engineering Solutions Oy for overall and chemical process design, Ferchau GmbH and Präwest GmbH from Germany for mechanical design and manufacturing, Arvos GmbH to design and build a quench cooler for the pilot. Experts from Universities of Cambridge and Oxford have been responsible for aerodynamic calculations and CFD modelling.

Kinetic model

Coolbrook has, together with Neste Engineering Solutions Oy, developed a–reactor model for the RDR –reactor. Sensitivity studies show that optimal temperature range is 960-990 °C. By rotodynamic reactor it is possible to get extemingly low residence time < 20 ms. Expected typical residence time for high Propylene and Ethylene yield is about 20-50 ms (naphtha feed). Optimum temperature depends on feedstocks, residence time in reactor and transfer line and target products. Below there are some examples of the simulation results with different feed stocks (Figure 4) and simulation of a Virgin naptha feed for RDR-pilot (Figure 5).

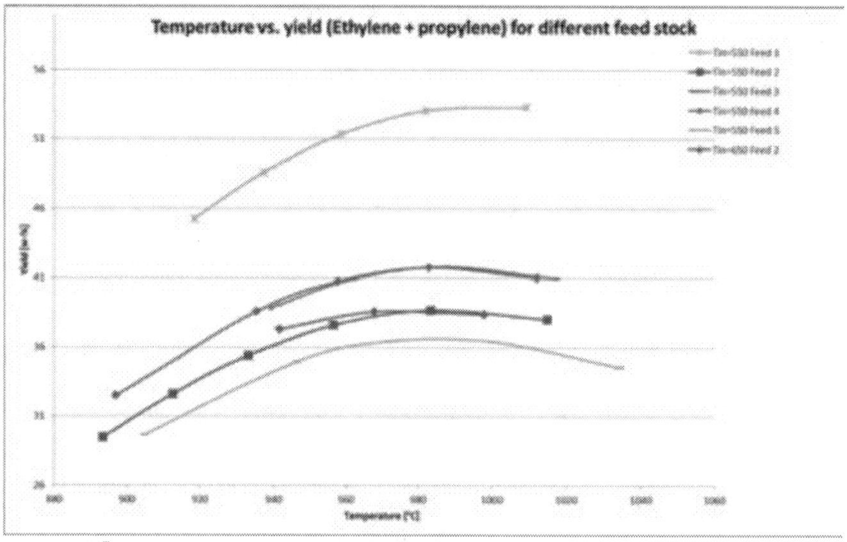

Figure 4. *Sensitivity study – different feed stock.*

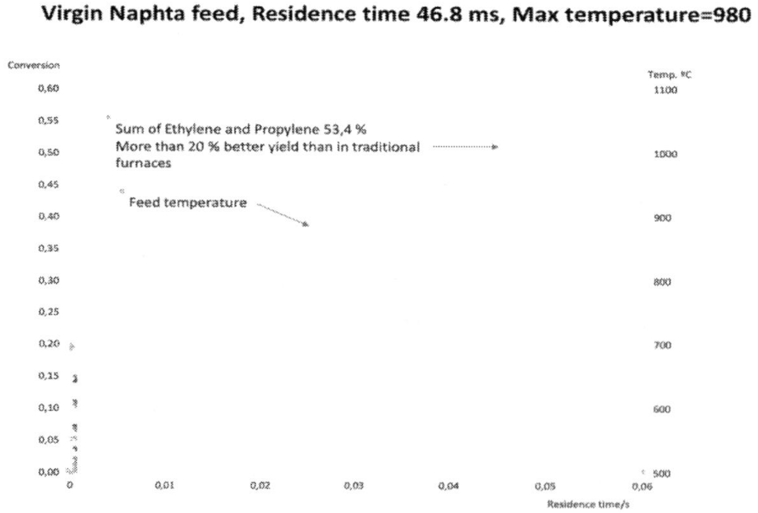

Figure 5. *Simulation of Virgin naphtha.*

Conclusion

RDR technology enables cracking at higher temperature with shorter residence time than in conventional thermal cracking (Figure 6). As a result of these conditions the combined Ethylene and Propylene yield improves by at least 20% resulting in significantly higher operating margins.

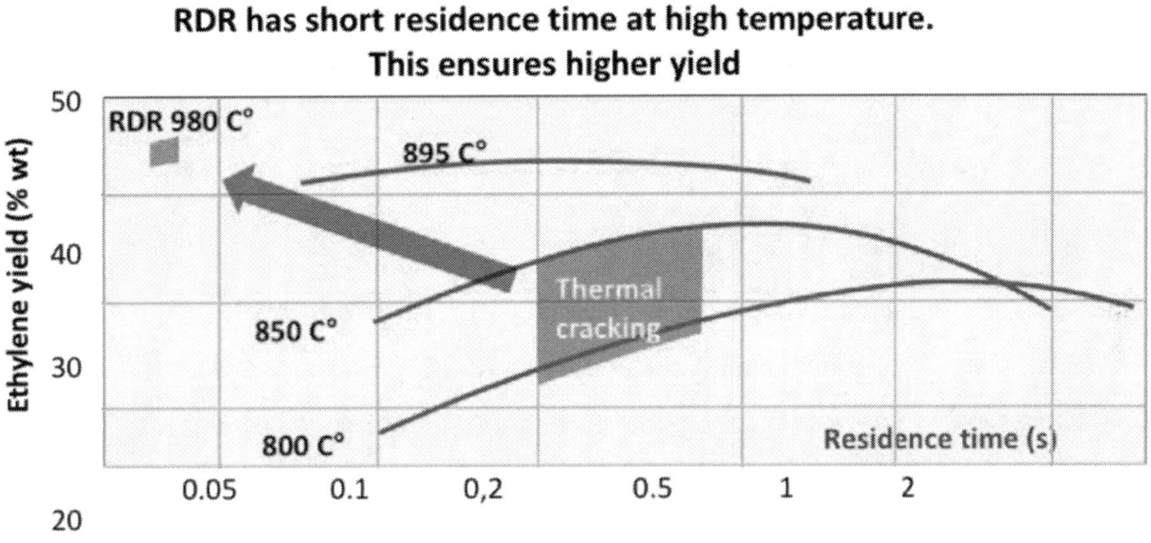

Naphta pyrolysis. Influence of temperature and residence time on ethylene yield.

Figure 6.

Other technical benefits include: steam ratio may be set from 50 % to 30%, smaller downstream investments because of product slate (less co-products), and decreased product losses by flaring due to-very short start up and shut down times. RDR is also feasible for cracking ethane, propane and butanes for high yields.

Most important economical benefits are increased yield, lower energy consumption, smaller investment costs and better controllability of the cracking process.

126B

FLAMELESS OXIDATION FOR HEATING OF ETHYLENE FURNACES

Dr. Sebastian Muschelknautz

CEO and Owner of TAKEOFF Engineering GmbH (Ltd.)

Prepared for Presentation at the 2018 Spring National Meeting

Orlando, Florida, April 22 – 26, 2018

ABSTRACT

Flameless Oxidation for Heating of Ethylene Furnaces

Dr. Sebastian Muschelknautz (Speaker), CEO of TakeOff Engineering GmbH,
Dr. Joachim Wünning, CEO of WS Wärmeprozesstechnik GmbH, President of WS Inc..

Flameless Oxidation (FLOX®) is a well established combustion method which was invented and developed 30 years ago by WS Wärmeprozesstechnik GmbH in Renningen, Germany. The application was targeted to limit the NOx-formation in furnaces where combustion air preheat is common. This holds true for steel heat treatment furnaces where in the meantime more than 50000 FLOX®-burners are in operation, which have been developed and fabricated by WS. TakeOff Engineering (TE) as an experienced engineering company in the Process Industry is cooperating with WS for the marketing and sales of FLOX®-burners in the chemical and petrochemical area. At the same time TE provides the furnace engineering for such projects.

More stringent NOx emissions will require more efforts in industries where little or no combustion air preheat is applied. Cracking furnaces in ethylene plants are such examples where conventional flame burners reach their limits, especially when fuel gases with fluctuating heating values are applied. FLOX®-burners can be installed in new furnaces or be retrofitted in existing furnaces without the need for fluegas NOx abatement (DeNOx/SCR). They offer a guaranteed maximum of 60 mg/Nm3 (29 ppm) NOx-emissions and an expected level of 30mg/Nm3 (15 ppm). Additional benefits of FLOX®-burners are very low turndown ratios and temperature uniformity or customized firing acc. to process heat requirements with increased overall heatflux density. Further on they have a broad fuelgas flexibility incl. high hydrogen concentrations as well as low noise emission levels. FLOX®-burners have also been successfully tested in coal dust firing mode.

This overview shows the fundamental working principles of FLOX®-technology and examples of existing FLOX®-burner installations in various industries and as well as the application potential in the process industry. A first commercial installation for FLOX®-burners in old cracking furnaces of an Ethylene plant is under preparation and the first furnace out of eight - equipped with FLOX®-burners - will be put back in operation in October 2018.

Further information on the principle of FLOX® and its applications can be found under: http://www.bine.info/publikationen/publikation/flammenlose-verbrennung/

1. Flameless Oxidation (FLOX®)

The FLOX®-technology developed by WS in Renningen/Germany refers to special industrial burner types which are characterized by a highly homogeneous reaction zone with even temperature distribution and with very low NOx-emissions. The unique feature is a nonvisible flame which occurs when the combustion air and internally recirculated combustion gases are intensively mixed into the fuelgas in a special way at high velocities. A so-called FLOX®-burner is started up in a conventional mode with a visible flame and is automatically switched to FLOX®-mode above appr. 850 deg C, which is considered to be a safe self-ignition temperature (fig. 1).

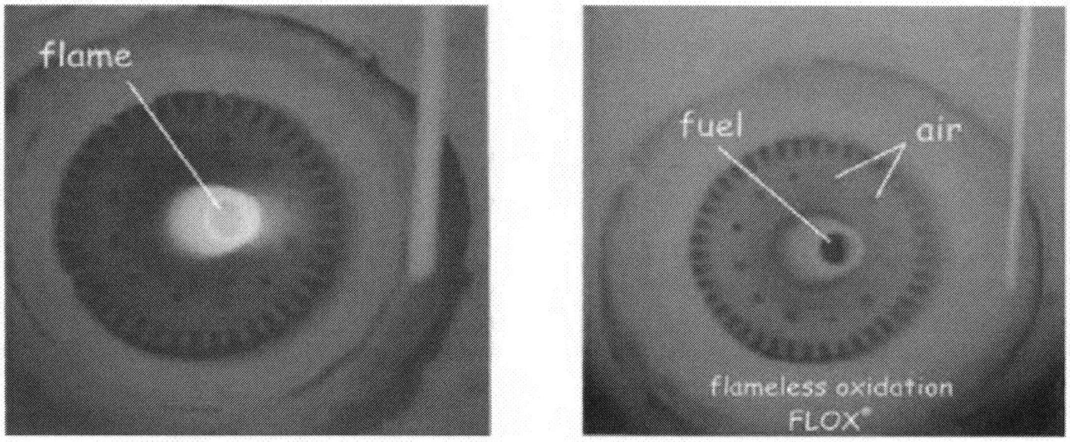

Fig. 1: FLOX®-burner in flame-mode for start-up (left) and in FLOX®-mode for normal operation (right).

Fig. 2 shows CFD-calculations of the flame-mode and the FLOX®-mode of a standard FLOX®-burner. The flame-mode is normally used for start-up of a furnace up to appr. 850 degC where the mode is switched to FLOX®. The peak-temperature in the FLOX®-mode is several hundred deg C lower than in the flame-mode which is due to the intense mixing of fuelgas and combustion air in the FLOX®-mode. A safe and stable combustion of the fuelgas is maintained without the need of a flameholding device. Due to the intense mixing FLOX®-burners are not sensitive to fluctuating fuelgas compositions and fuelgas is burned with very low noise emissions.

In order to achieve the high combustion air velocities a conventional low-pressure blower is sufficient. The combustion efficiency is similar as with conventional low NOx-burner as further elaborated in chapter 5.

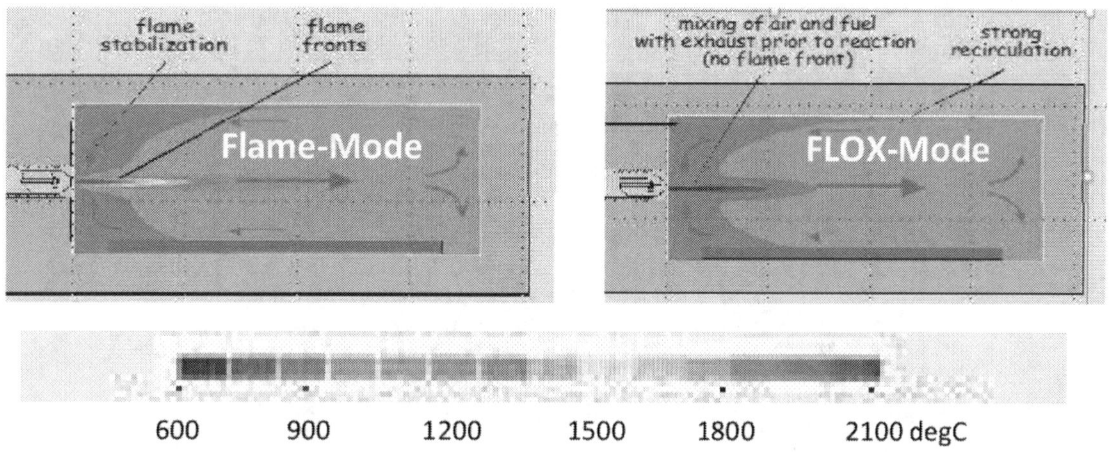

Fig. 2: CFD-calculation of the flame-mode and the FLOX®-mode of a standard FLOX®-burner

2. Industrial applications of FLOX®-burners to date

Some decades ago the steel heattreatment industry started to use waste heat to preheat combustion air for energy efficiency gains. Combustion air preheat in turn leads to high NOx emissions which typically cannot be fully compensated with conventional low-NOx-burners.

During tests at the WS-labs with air staging in the late 1980's, Flameless Oxidation (FLOX®) was accidentially discovered as a sudden drop in NOx-emissions upon largely increased combustion air velocities in the absence of a visible flame. The FLOX®-technology turned out to be a method of safe and stable combustion in a kind of 'overblown flame' upon conventional air-gas-ratios. This was up to that time widely considered as an uncontrollable burner operation upon industrial applications. Since then many patents around the FLOX®-technology and its burner applications have been granted to WS. Ten's of thousands of FLOX®-burners have been designed and manufactured since then and are in successful, reliable operation mainly in heattreatment furnaces (fig. 3) and to some extent in glass melting furnaces.

Fig 3: Steel strip annealing furnace with 116 tons of steel throughput per hour, equipped with 220 FLOX-burners with 160 kW each.

With regard to the process industry FLOX®-burners have been successfully installed in the following plants:

- Cantype-Reformer of Hydrochem (Linde) with 500 Nm3/h H2 production capacity.
- Large scale steamreformer in Germany (Linde) – 1 single 2 MW FLOX®-burner in parallel to conventional burners for reliability tests.
- Hydroprime productline of Linde as a standardized skid-mounted and truckable steamrefomer for 300 Nm3/h H2-production (see fig. 4).

In all these installations FLOX®-burners have shown a reliable operation. With respect to conventional top fired steamrefomers the FLOX®-technology has particular advantages with regard to NOx-emissions.

Fig. 4: Truckable Linde Hydroprime plant for 300 Nm3/h H2-production equipped with FLOX®-burners.

3. Types and performance of FLOX®-burners

The admission of combustion air in FLOX®-burners is achieved either through a simple rings lot around the central fuelgas injection tube or by a multiple of nozzles for intensified mixing with the fuelgas and the recirculating combustion gas (fig. 5).

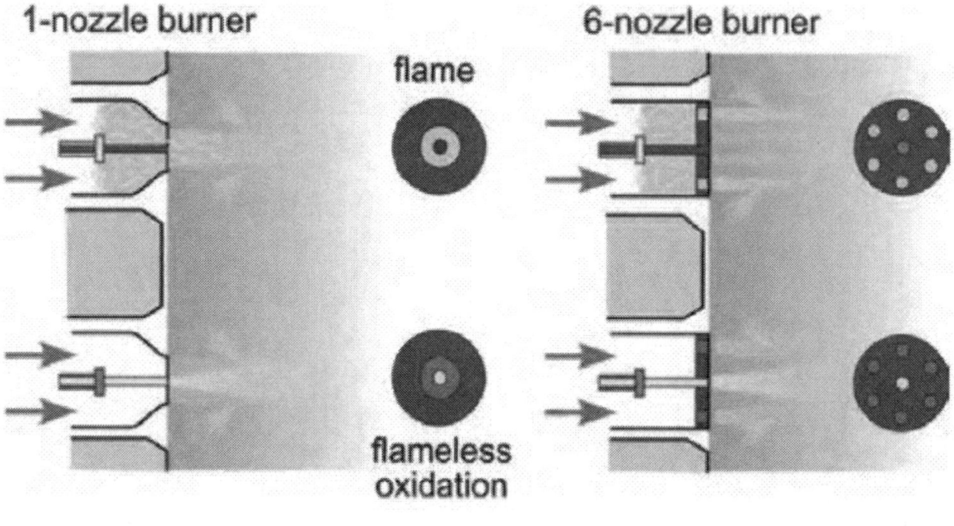

Fig. 5: Mixing of fuelgas (yellow) and combustion air (blue arrows) in FLOX-burners with single nozzle (=ring slot, left) and with a 6-nozzle design for intensified mixing (right).

Naturally NOx emissions increase with higher air preheat as well as with higher furnace temperature. Whereas with conventional low NOx-burners NOx emissions just slightly below 100 mg/Nm3 (49ppm) can be achieved, the FLOX®-technology can bring down the NOx-emissions to a guaranteed 60 mg/Nm3 (29ppm) with the expectation of getting as low as 30 mg/Nm3 (15ppm) under optimum operating conditions.

The FLOX®-burner standard design – called LUVOMAT - is shown in fig. 6. The model shown in the figure is for applications with external waste heat recovery as a dual-mode burner with flame-mode for startup and FLOX®-mode for normal operation. Each LUVOMAT has its own ignition and flame-supervision system as well as a local certified control unit for starting up the burner and automatically switching to FLOX®-mode and back. In heat treatment furnaces those burners are switched on and off in a kind of 'digital mode' in order to control the furnace temperature. This common practice in that industry turned out to be simple and very reliable despite the large number of mode-switching.

The drawing in fig. 5 on the left side shows a multi-bore design with 4 straight nozzles for FLOX-mode and 1 central nozzle for flame-mode. The combustion air is supplied at appr. 80 mbarg by an external blower and through an air-ducting system. The blower and the ducting are relatively inexpensive.

In its most simple form a FLOX®-burner like the LUVOMAT for the bulk of the heat duty in a furnace has no flame mode arrangement. For the start-up of the furnace there will be a small number of dual-mode FLOX® -burners. By this the overall costs of a FLOX® -burner arrangement can be kept low and are competitive to conventional low-NOx-burners.

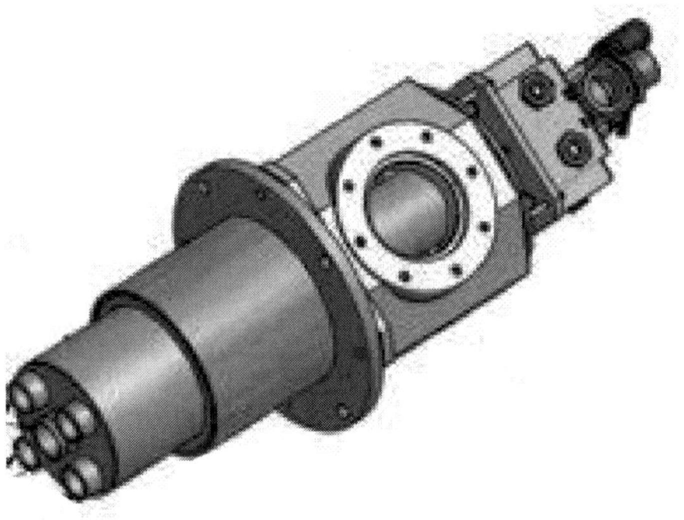

Fig. 6: Standard FLOX®-burner with the WS-tradename LUVOMAT

Fig. 7 shows a REKUMAT dual-mode FLOX®-burner with integrated air-preheat and a single straight nozzle design. In the photograph the inner tube of the fluegas-takeout duct is seen with a rippled surface for intensified heat exchange with the incoming air. The outer shell tube is not shown here.

Fig. 7: REKUMAT dual-mode FLOX®-burner with integrated air-preheat and a single straight nozzle design. The outer shell tube is not shown in the photograph.

Fig. 8 shows such REKUMAT-burners in the flame-mode during start-up of a heattreatment furnace. With this installation a straight flame is possible due to sufficient spacing to the opposite firebox wall.

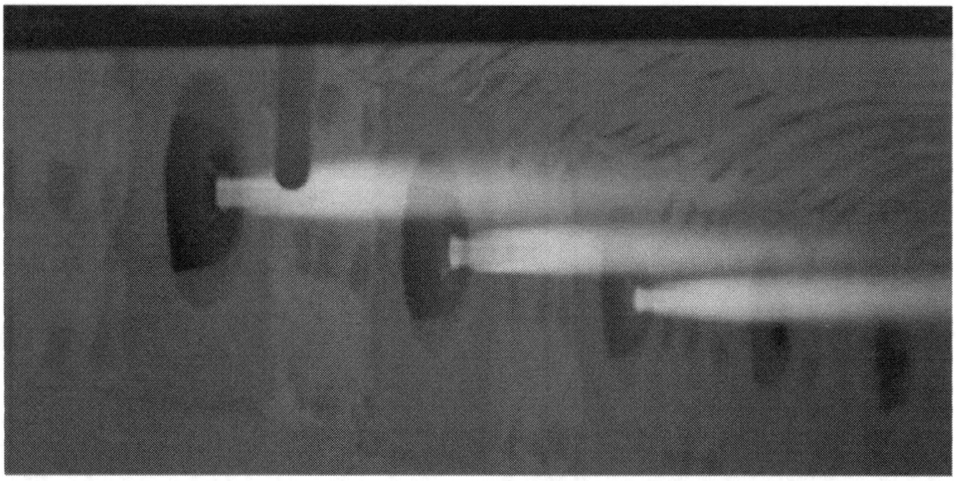

Fig. 8: REKUMAT-burners during start-up of a heattreatment furnace.

Upon applications where an external waste heat recovery is not available or does not make sense, the REKUMAT series is the FLOX®-burner of choice. Fig. 9 shows a REKUMAT FLOX®-burner with integrated air preheat by internal fluegas heat recovery. Such burners use a fraction of the combustion air in order to drive an eductor (injector) which takes out a certain percentage of the flue gas through a separate ringslot duct (compare photo in fig. 7). The outgoing fluegas heat is exchanged with the central incoming combustion air. With this system an integral waste heat recovery of up to 80% is achievable.

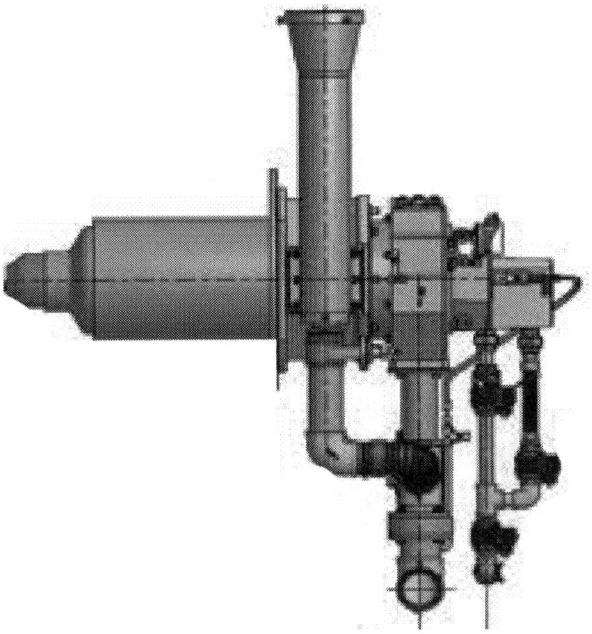

Fig.9: RECUMAT FLOX®-burner with integrated air preheat by fluegas heat recovery.

For FLOX®-burner applications in the process industry where the combustion heat shall be distributed mainly in parallel to the firebox wall a flat-wall FLOX®-burner is available. Here the burner cap has radial, somewhat angled openings which can be designed in a way that the hot gas is directed towards where the heat in the process tubes (coils) is needed. This represents an interesting solution for sidewall burners with customized firing.

Due to even temperature distribution and the low NOX emission of the FLOX®-technology it is attractive for many applications in the process industry, e.g. for cracking furnaces in ethylene plants.

4. FLOX®-burner for cracking furnaces in ethylene plants

Modern cracking furnaces in ethylene plants (fig. 10) are very large units with a very high thermal energy demand. They have 30 to 50 burners per furnace. Typically these burners are either floorburners with 2 to 3 MW each or a combination of floor- and sidewall-burners.

Fig. 10: Cracking furnaces in an ethylene plant.

The total fired duty of such a unit can be as high as 800 MW. Typically modern cracking furnaces have duties of 100 MW to 150 MW each and are supplied with floor burners or with a combination of floor- and sidewall-burners.

Usually the burner manufacturers guarantee NOx emissions acc. to the regulatory standards with conventional burners, which are in Europe currently 80 to 100 mg/Nm3 (39 to 49 ppm) for new furnaces and 60 mg/Nm3 (29 ppm) for furnaces with catalytic DeNox systems (SCR).

With low-NOx-burners one can achieve appr. 90 mg/Nm3 (44 ppm). For lower requirements a catalytic DeNOx system (SCR) is typically required which is installed as an integral part of the waste heat recovery. This adds significant costs and sometimes requires reinforcements of the furnace steelstructure upon furnace revamps due to the added weight of the SCR. Further on an SCR-catalyst has to be replaced typically every 5 years and the process consumes ammonia or urea which increases the costs further.

FLOX®-burners offer NOx values of 60 mg/Nm3 (29 ppm) and lower and are therefore a true low cost and less complex alternative to a SCR-system. Another advantage of the FLOX®-technology is the stability of the combustion due to high velocities of fuelgas and combustion air at the burner tip. Typical noise emission limit requirements for a cracking furnace are 86 dbA or sometimes even 80 dbA. FLOX®-burner expose very low noise emissions due to the smooth combustion process. Therefore silencers in the air duct are normally not required.

Fig.11 shows a typical heatflux distribution on the coils with conventional burners, in this case the combination of floor- and sidewall-burners. An optimum cracking process of the feedgas hydrocarbons inside the coils would require an almost even heatflux distribution. As shown the highest heatflux occurs at the lower end of the coils which is suboptimal.

With FLOX®-burners the overall heatflux to the coils will be more even distributed due to the higher flow velocities of the combustion gases which improves not only the mixing of fuelgas and combustion air but also the convective heat transfer. As a result one can expect an optimum cracking process which improves the lifetime of the coils and eventually also the ethylene yield. Besides the very low NOx-emissions FLOX®-burners have the advantage of a very high flexibility of fuelgas compositions. Even with hydrogen concentrations >20% the NOx-emissions will remain low. Tests which are currently undertaken will likely show that hydrogen concentrations upon stable burner operation can be much higher than 20%.

Additionally FLOX®-burners offer a high turndown ratio which can be almost zero if the combustion air ratio is kept at 50% load at the same time. This operation is of cause not very efficient. However it could enable the operator –instead of going to 'hot steam standby - to keep the furnace in operation upon short term plant shutdowns with very limited flaring from the warm section of the plant, e.g. upon a trip of the cracked gas compressor.

A typical FLOX® -burner arrangement for cracking furnaces consists of more than 70% of FLOX® -only-burners (just FLOX® -mode) while having the remaining burners as dual-mode FLOX® -burners. This is considered as sufficient for the start-up of the furnaces as well as for hot-steam-standby operation if the latter is desired to be below the self-ignition firebox-temperature (typically a conservative 850 deg C). It can also be considered to perform the decoking of the furnace coils with the dual-mode FLOX® -

burners only if the operator choses to do that temporary operation in the temperature range below 850 deg C.

By such arrangement the overall CAPEX of a FLOX® -burner system for a cracking furnace become very competitive and attractive for ethylene plant operators and investors while having additionally the OPEX-advantages of the FLOX® -burners compare to conventional burners.

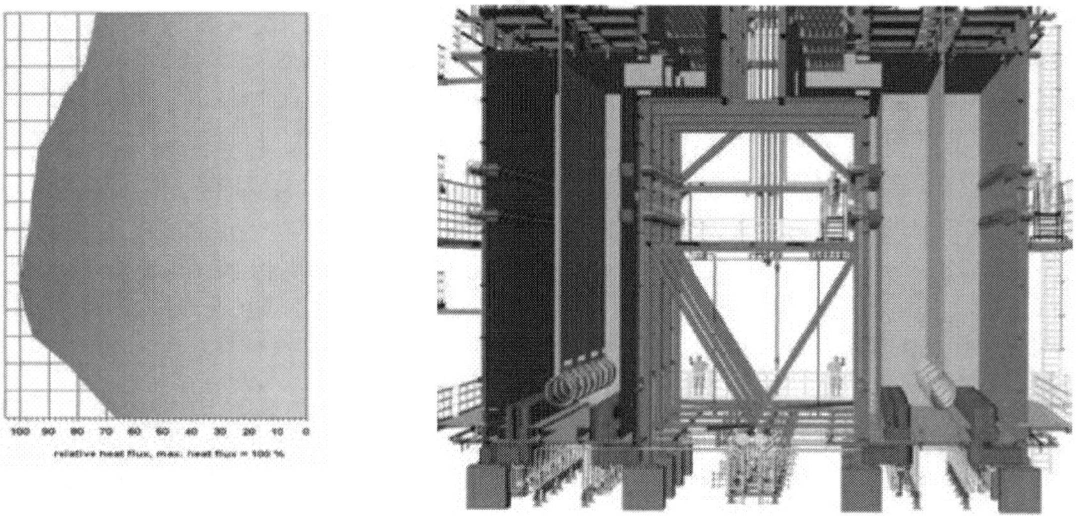

Fig. 11: Heatflux distribution to the coils with a mixed set of conventional floor- and sidewall burners.

5. Firing efficiency tests

WS has extensive lab- and test facilities within its R&D-center. Amongst those is a Mega-FLOX®-test furnace shown in fig. 12. Here FLOX®-burners with up to 1 MW can be tested in a floor burner arrangement with regard to NOx-emissions and heattransfer efficiency. The firebox temperature can be adjusted up to 1200 deg C and the heat generated by the burner will be removed with air-cooled tubes at the circular furnace wall in order to simulate various heat load profiles. The burners can be operated with or without combustion air preheat.

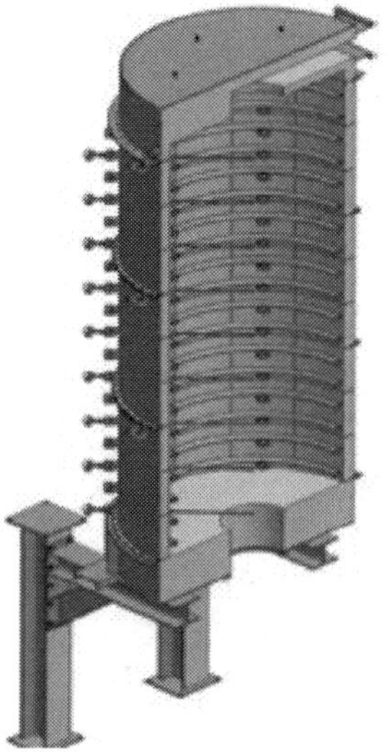

- For temperatures up to ~ 1200 degC
- Air cooled cooling tubes for simulation of load profiles from floor burners
- For burner capacities up to 1 MW
- With or without air preheat
- Optical and probe access

Fig. 12: Mega- FLOX® test furnace for floor burner tests up to 1 MW

For sidewall burner applications e.g. in old cracking furnaces flatwall FLOX®-burners as schematically shown in fig. 13 are available. The indicated dimensions are for 250 kW FLOX®-burners which are suitable for old and small cracking furnaces with sidewall burners. The number and the diameter of the radial bores in the burner caps are selected in a way that the flame jets do not interfere with each other thus safely avoiding any potential attachment to the firebox wall. At the same time direct impingement of the jets onto the coils is prevented.

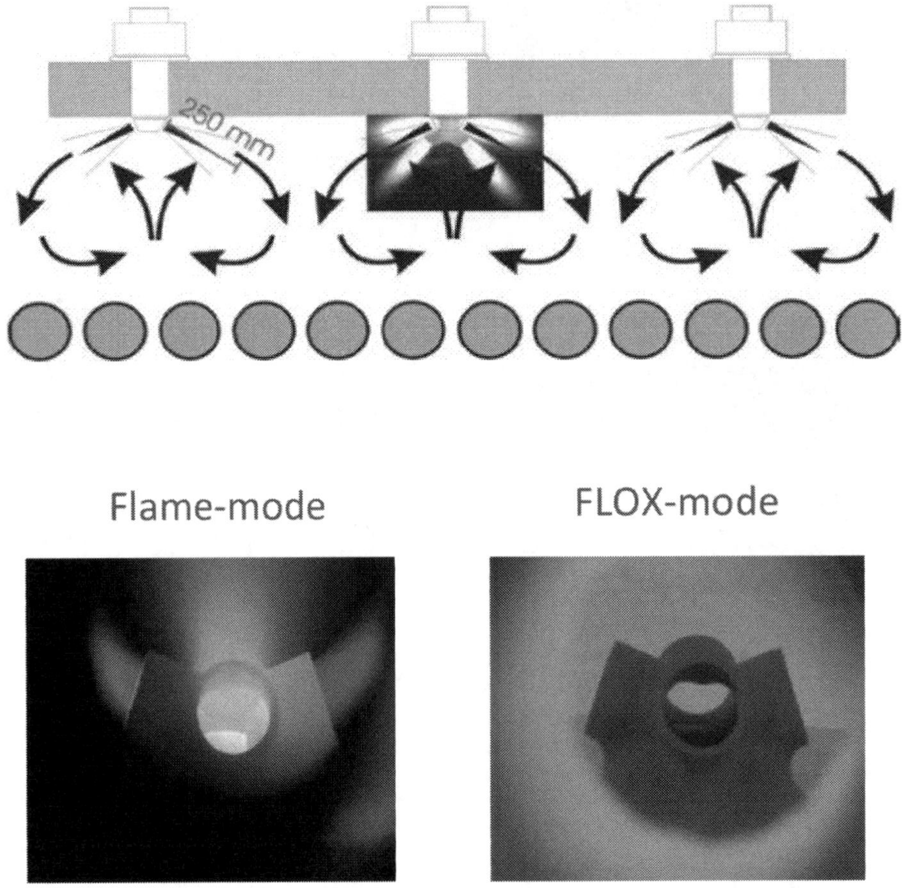

Fig. 13: Flatwall FLOX®-burners with specially designed endcaps (indicated spacing is for 250 kW burners)

In the lower part of fig. 13 the burner endcaps are shown in flame-mode and in FLOX®-mode operation. The geometry of these endcaps can be customized in many ways acc. to the needs of the heat distribution. They are manufactured either from high temperature metal cast or 3d-printed with ceramic material. The reliability of such designs is proven with long-term experience of such burner endcaps as part of FLOX®-burners in other industrial applications.

A recurring concern is whether FLOX®-burners can match the heattransfer efficieny of conventional burners since the peak temperature and consequently the heat radiation is assumed to be lower.

Upon a first application of FLOX®-sidewall burners in eight aged cracking furnaces of a large petrochemical operator this concern was addressed by conducting comparison tests in the WS testfacilities with a custom-made test furnace (fig. 14). It was designed in a way that the firebox represents exactly the environment of one full scale sidewall burner incl. the spacing between the wall and the coils, which are aircooled.

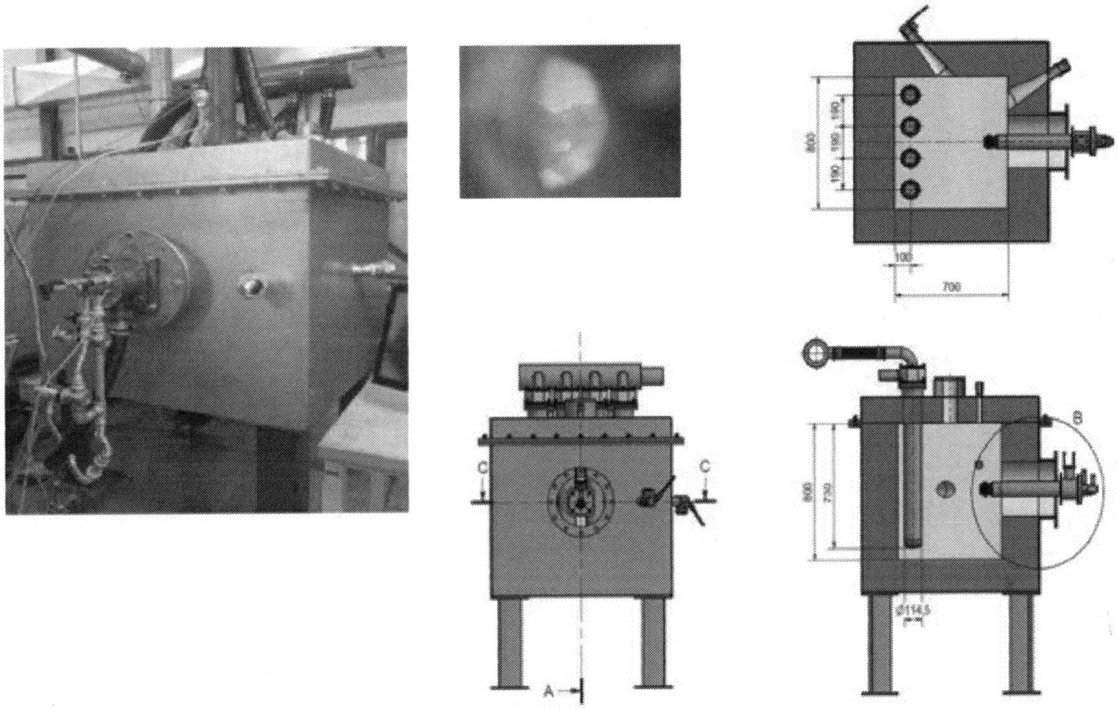

Fig. 14: Test furnace for comparison of full-scale sidewall FLOX®-burners with conventional low NOx-burners of similar capacity.

In these tests the FLOX®-burner operation was very smooth and stable even under extreme and adverse fuelgas compositions. There was no sign of impingement of unburnt fuel and the NOx emissions were much lower than the a.m. 60mg/Nm3 (29 ppm). The operation of the conventional low NOx-burner as a benchmark to the FLOX®-burner was by far not as stable as of the FLOX®-burner and the NOx-emissions were much higher.

It was most important to note that the measured heattransfer efficiency of the FLOX®-burner was similar to the one with the conventional burner. All in all these tests allowed to draw the conclusion that FLOX®-burners offer similar heattransfer efficiencies at much lower NOx-emissions than conventional low NOx-burners (see fig. 15).

This test result can be explained by the significantly increased convective heattransfer due to the high combustion gas velocities. This fact combined with a quite substantial heat radiation coming from the hot recirculating combustion gas compensates an eventually existing lack of radiation heattransfer. On the other hand the heatflux efficiency in terms of [kW/m2] could be higher due to the more even temperature profile. This could eventually increase the throughput in the coils, which might be of interest in revamp projects.

Ultimately these tests convinced the petrochemical operator to proceed with the FLOX®-burner-retrofit project and the first out of eight cracking furnaces equipped with FLOX®-burners is scheduled to come into operation in October 2018.

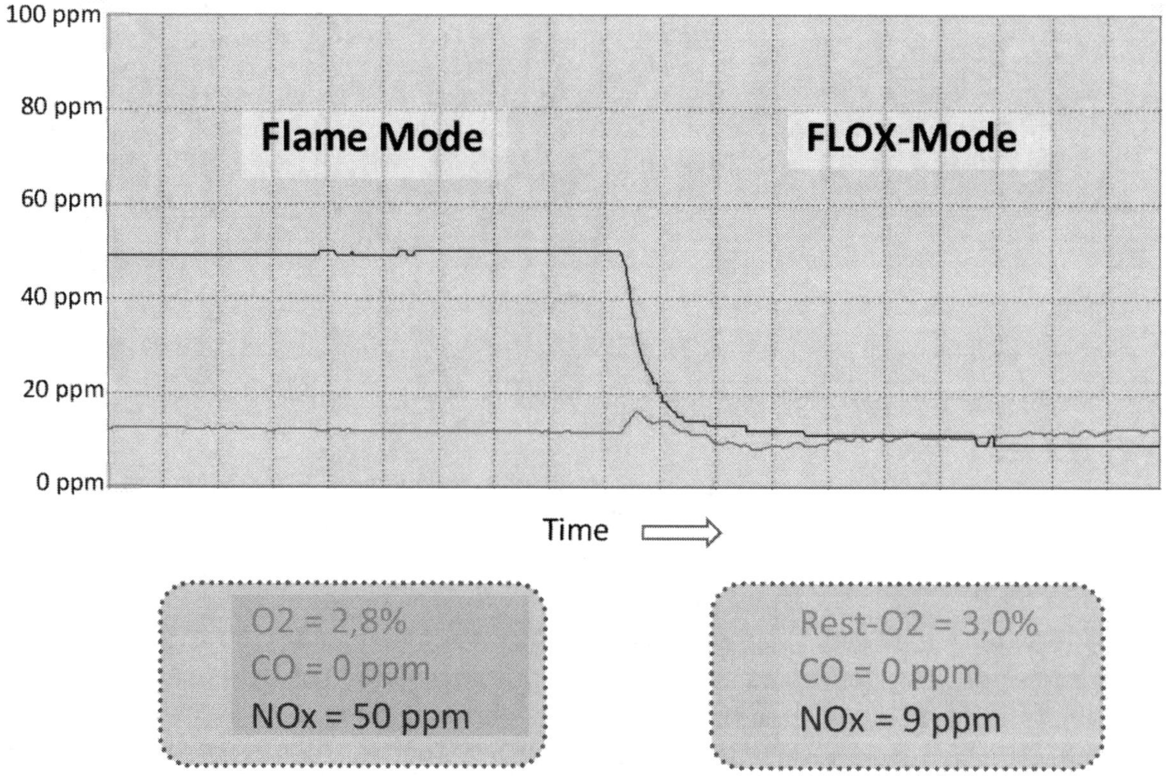

Fig. 15: Measured NOx-emission reduction upon shifting from flame-mode to FLOX®-mode (235 kW, lambda=1,15)

6. General conclusions for FLOX®-burner applications in the process industry

Applications of FLOX®-burners in the process industry have particular advantages with regard to the following points:

- Low NOx emission requirements which would normally ask for a DeNOx (SCR) unit in the fluegas stream.
- Fuel gases with normal compositions incl. low heating values, but also with fluctuating and extreme compositions including high hydrogen concentrations -> FLOX®-burners are very tolerant to such gases.
- Uniform heat transfer to the process coils and even temperature distribution in the firebox which extends the lifetime of the coils.
- Heat transfer efficiency is similar to conventional low-NOx-burners.
- Optimized heat transfer to the process with FLOX®-burners with increased heatflux density [kW/m2].
- Low noise emission requirements – with FLOX®-burners silencers in the airduct system can normally be avoided.
- Central air preheat is not installed but would be beneficial -> FLOX®-RECUMAT burner are a low cost solution for such applications.
- Very low turndown ratios.

126C

THERMAL DECOMPOSITION OF SULFUR COMPOUNDS AND THEIR ROLE IN COKE FORMATION DURING STEAM CRACKING OF HYDROCARBONS

Marko R. Djokic
Dr. ir.
Laboratory for Chemical Technology, Ghent University, Ghent, Belgium

Kevin M. Van Geem
Prof. Dr. ir.
Laboratory for Chemical Technology, Ghent University, Ghent, Belgium

Ruben Van de Vijver
Dr. ir.
Laboratory for Chemical Technology, Ghent University, Ghent, Belgium

Cato A.R. Pappijn
ir.
Laboratory for Chemical Technology, Ghent University, Ghent, Belgium

Natalia Olahova
Dr. ir.
Laboratory for Chemical Technology, Ghent University, Ghent, Belgium

Guy B. Marin
Prof. Dr. ir.
Laboratory for Chemical Technology, Ghent University, Ghent, Belgium

Prepared for Presentation at the 2018 Spring National Meeting
Orlando, Florida, April 22 – 26, 2018

THERMAL DECOMPOSITION OF SULFUR COMPOUNDS AND THEIR ROLE IN COKE FORMATION DURING STEAM CRACKING OF HYDROCARBONS

Marko R. Djokic
Dr. ir.
Laboratory for Chemical Technology, Ghent University, Ghent, Belgium

Kevin M. Van Geem
Prof. Dr. ir.
Laboratory for Chemical Technology, Ghent University, Ghent, Belgium

Ruben Van de Vijver
Dr. ir.
Laboratory for Chemical Technology, Ghent University, Ghent, Belgium

Cato A.R. Pappijn
ir.
Laboratory for Chemical Technology, Ghent University, Ghent, Belgium

Natalia Olahova
Dr. ir.
Laboratory for Chemical Technology, Ghent University, Ghent, Belgium

Guy B. Marin
Prof. Dr. ir.
Laboratory for Chemical Technology, Ghent University, Ghent, Belgium

Abstract: Sulfur-containing compounds play a key role in many industrial processes. Particularly for the steam cracking process, they have been linked with increased olefin selectivity, CO formation, and coke inhibition. This makes it imperative to quantify sulfur containing compounds to ppm and even ppb levels. On-line sulfur detection is a difficult task because of the adsorption of sulfur containing compounds on metal surfaces (tubings), as well as due to absorption and destruction in the GC system and capillary columns. In this work the influence of four different sulfur-containing additives, methanedithione (CS_2), (methyldisulfanyl)methane (DMDS), (methylsulfanyl)methane (DMS), and dimethyl sulfoxide (DMSO), on product selectivity, coke deposition, and CO production during steam cracking of hydrocarbons. The use of online comprehensive two-dimensional gas chromatography with sulfur chemiluminescence detection (GC × GC-SCD) is the key enabling technology to characterize the sulfur compounds. In addition to the GC × GC-SCD, the low molecular weight sulfur containing compounds such as hydrogen sulfide (H_2S) are analyzed using a thermal conductivity detector of a so-called refinery gas analyzer (RGA). Steam cracking in a pilot-plant unit revealed that all studied sulfur compounds are efficient in reducing the CO yield. Simultaneously, they strongly promote coke.

Introduction

Steam cracking is still the leading technology applied for the production of light olefins [1]. The lightest but the most important of them, namely ethylene and propylene, are the largest-volume platform chemicals of the petrochemical industry, produced at continuously increasing rates of more than 140 and 90 million metric tons per year, respectively [2-5]. Steam crackers predominantly consume both ethane and LPG from natural gas processing, or light and heavy naphthas, gas oils and atmospheric residues produced by atmospheric crude oil distillation. During cracking an undesired carbonaceous residue is formed on the inner wall of the reactor tubes and the transfer line heat-exchanger (TLE), which has a deleterious effects on both the energy efficiency and the economics of the process [6]. Coke deposition reduces the inner diameter of the coil, thus increasing the pressure drop and lowering the ethene selectivity. It also hampers heat transfer from the furnace side to the process gas in the reactor coil [7, 8]. Despite all the precautions taken to minimize the formation of coke, it is impossible to completely eliminate this issue at present. The last decades have witnessed considerable progress towards developing an effective technology to reduce coke formation. These technologies can be divided in three groups, namely the surface technologies [6], the use of three-dimensional reactor technologies [9, 10] and the use of additives [11-17]. Although the first two groups show promising results, using additives still remains the dominant and practical method for reducing coke formation during steam cracking. There are three generally accepted methods, by which additives can be applied: pretreatment of the inner surface of the cracking coils prior to the introduction of hydrocarbon feed to the reactor, continuous addition to the feed during cracking, and pretreatment of the inner surface followed by continuous addition to the feed during cracking.

In industrial olefin plants, sulfur-containing additives are commonly used to control CO production and also to what is assumed by most of the ethylene plant operators suppress coke deposition. It is not proven so far, however, whether there is a straightforward relation between CO production and the amount of coke unless it is assumed that CO is produced solely from the gasification of coke [11]. The role of sulfur-containing additives on diminishing the CO formation by passivating active nickel sites and thus forming nickel sulfides is proven [1], but their effect on coke formation is still complex and far from clear as evidenced by the contradictory reports with regard to this aspect [11, 13-19].

To improve understanding about the role of sulfur compounds on the steam cracking process four different sulfur additives have been experimentally studied during steam cracking of heptane: namely (methyldisulfanyl)methane (DMDS), (methylsulfanyl)methane (DMS), dimethyl sulfoxide (DMSO) and methanedithione (CS_2). The use of GC × GC-SCD has been applied to characterize the sulfur compounds.

Materials and Methods

Chemicals and Standards

Heptane, DMDS and CS_2 were obtained from Acros Organics. DMS, DMSO and 3-chlorothiophene, used as internal standard, were purchased from Sigma-Aldrich. Analytical gases were provided from Air Liquid. For producing steam, deionized water was used. All the chemicals and gasses were procured at the highest purities.

Pilot Plant for Steam Cracking

The experimental work was performed using the steam cracking pilot plant unit available at the Laboratory for Chemical Technology and has been extensively discussed elsewhere [20]. Here only specific details related to the experiments discussed in this study are mentioned. For a more detailed description of the pilot plant, the reader is referred to Supporting Information. As can be seen in Figure 1, the set-up consists of three parts: the feed, reaction and the analysis section. The feed section consists of different tanks from which both a liquid and a gaseous feed are fed to the reactor using computer controlled pumps and/or Cori-flow™ mass flow controllers. The furnace built of silica/alumina bricks is about 4m long, 0.7m wide and 2.6m high and divided into seven separate cells. The reactor coil is made of Incoloy 800HT and has a length of 12.8 m, with an internal diameter of 9 mm.

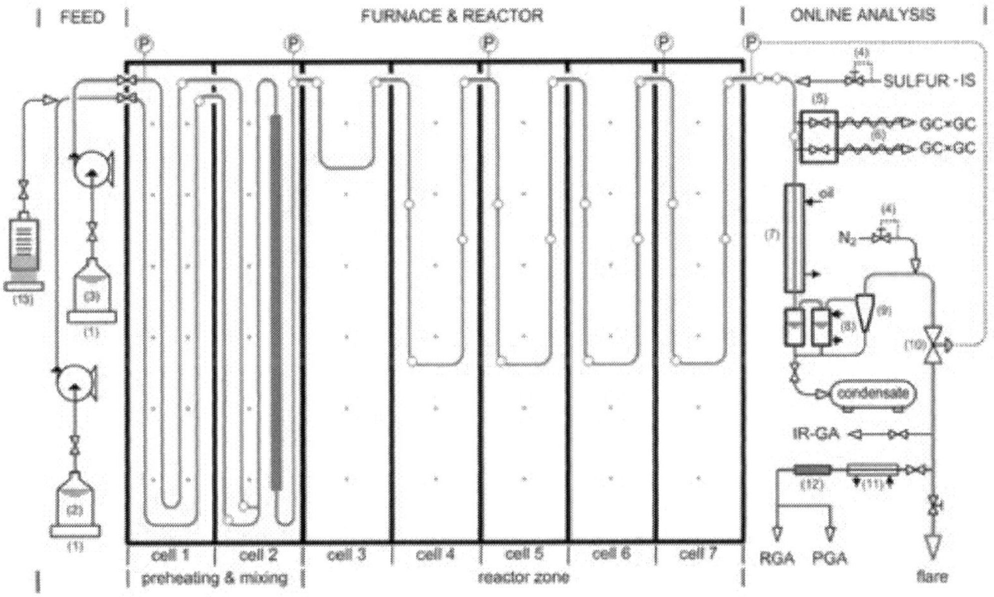

Figure 1: Pilot plant for steam cracking

The on-line identification and quantification of the entire product effluent was conducted using an up-graded comprehensive analysis section [21]. This approach allows analysis of the entire product stream, a wide boiling mixture containing all permanent gases, such as H_2, N_2, CO and CO_2, present in the effluent, and hydrocarbons ranging from methane to polycyclic aromatic hydrocarbons (PAHs), as well as sulfur containing compounds. The latter incorporates an infrared CO/CO_2 gas analyzer (IR-GA), a permanent gas analyzer (PGA), a refinery gas analyzer (RGA) equipped with a thermal conductivity detector (TCD) and flame ionization detector (FID), and 2 two-dimensional gas chromatographs (GC×GC), both equipped with an FID and one of them additionally with an SCD [22].

The use of an SCD on-line is of particular importance for the present study and allows obtaining a unique dataset of the decomposition products of the sulfur additives and also their reaction products with the heptane cracking products. The GC × GC-FID/-SCD is specifically designed in order to detect sulfur-containing compounds that are present in complex olefin-rich hydrocarbon matrices at very low concentrations. Both an FID and an SCD detector are connected in series which allows simultaneous FID and SCD analysis during a single run [22]. Since sulfur-containing compounds are highly reactive, they have a tendency to adsorb onto metal surfaces. For this purpose sulfur inert coated tubing were applied before and after the reactor coil. To determine absolute flow rates of all detected sulfur-containing compounds, a known amount of 3-chlorothiophene is continuously added to the effluent. 3-Chlorothiophene was chosen as an internal standard for sulfur compounds due to its easy quantification and absence in the product stream.

Quantification procedure for pure hydrocarbons as being developed Van Geem et al [21] and is given in Supporting Information. A similar approach was applied for the on-line SCD quantification of sulfur based species and the absolute flow rates of sulfur containing compounds present in the reactor effluent were determined using the following equation:

$$\dot{m}_i = \frac{V_i}{V_{st}} . \dot{m}_{st} . \frac{M_i}{\mathrm{N}_i . M_{st}} \quad (1)$$

With V_i and V_{st}: peak volumes of compound i and sulfur internal standard (3-chlorothiophene) respectively; N_i: the number of sulfur atoms in compound i and M_{st}: molar mass of the sulfur internal standard.

The flowrates were then transformed into weight percentages based on ppmw of elemental sulfur originates from sulfur-containing compound.

$$g_i = \frac{\dot{m}_i}{\dot{m}_{feed}} . \mathrm{N}_i \frac{M_{sulfur}}{M_i} \quad (2)$$

Where $\dot{m}_{feed}$ represents the know mass flowrate of the feed and M_{sulfur} is molar mass of sulfur.

Steam Cracking Experiments

All the cracking experiments with or without sulfur-containing compounds were carried out with heptane as a surrogate feedstock for light naphtha and steam as diluent. To test the influence of a broad range of sulfur concentrations on coking rates as well as on product yields, four sulfur-based additives were added to the feed in amounts varying from 4 to 851 ppmw of S relative to the heptane feed. Each concentration was kept fixed throughout the cracking run which typically lasts 6 hours. The cracking conditions used in this study are summarized in Table 1.

Table 1: Overview of the experimental conditions.

Flowrate of heptane (10^{-3} kg.s^{-1})	1.11
Flowrate of steam (10^{-3} kg.s^{-1})	0.56
Dilution (kg_{steam} / $kg_{heptane}$)	0.5
Duration (hours)	6
Coil inlet temperature (K)	823
Coil outlet temperature (K)	1123
Coil outlet pressure (10^3 Pa)	170

A typical coking experiment conducted in the pilot plant unit comprises continuous cracking and decoking of the reactor [20]. Before every cracking run the required temperature profile along the coil has to be established under a constant steam flow rate. Then the heptane is fed to the reactor and flows of both reactants were switched to flows specified in Table 1. The sulfur-containing additive is introduced at the inlet of the reactor by means of a ISCO 500D syringe pump. In a single experiment, and more specifically for the experiment with the continuous addition of 5 ppmw CS_2, presulfidation by DMDS of the reactor was conducted before cracking. The applied conditions, previously described by Dhuyvetter et al. [12], are specified in the Supporting Information.

Since the steam cracking process is endothermic in nature and involves high temperatures, the cracking and coke formation occur first and foremost in cell 3-7 where the temperature is higher than 873 K. Coke accumulated on the inner wall of the reactor is subsequently removed by burning off with a steam/air mixture. CO and CO_2 are continuously measured by means of an infrared detector. These values are subsequently used to calculate the amount of coke deposited in the reactor. For more detailed information on the conditions applied during decoking see Supporting Information.

Results and Discussion

Effect of sulfur compounds on product distribution

The effect of applying four sulfur-containing additives on product yields and coke formation was assessed in the pilot plant unit during steam cracking of heptane. Four different sulfur containing compounds, namely CS_2, DMDS, DMS and DMSO are added to the feed in different concentrations, varying from 4 to 851 ppmw of elemental S relative to the heptane feed. The main interest of these experiments is both the qualitative and quantitative analysis of the sulfur compounds in the effluent as well as coking rates and their effect on the product yields during heptane steam cracking. Figure 2 shows one of the two-dimensional chromatograms of the effluent acquired by means of GC × GC – FID. That the major products observed during steam cracking of heptane are: ethene, propene, methane, but-1-ene, buta-1,3-diene and to a lesser extent the aromatic compounds. The measured product yields for the experiments are summarized in Table 2. As steam cannot be considered inert because it takes part in reactions of steam reforming and coke gasification, normalization was performed based on the carbon balance instead of total mass balance.

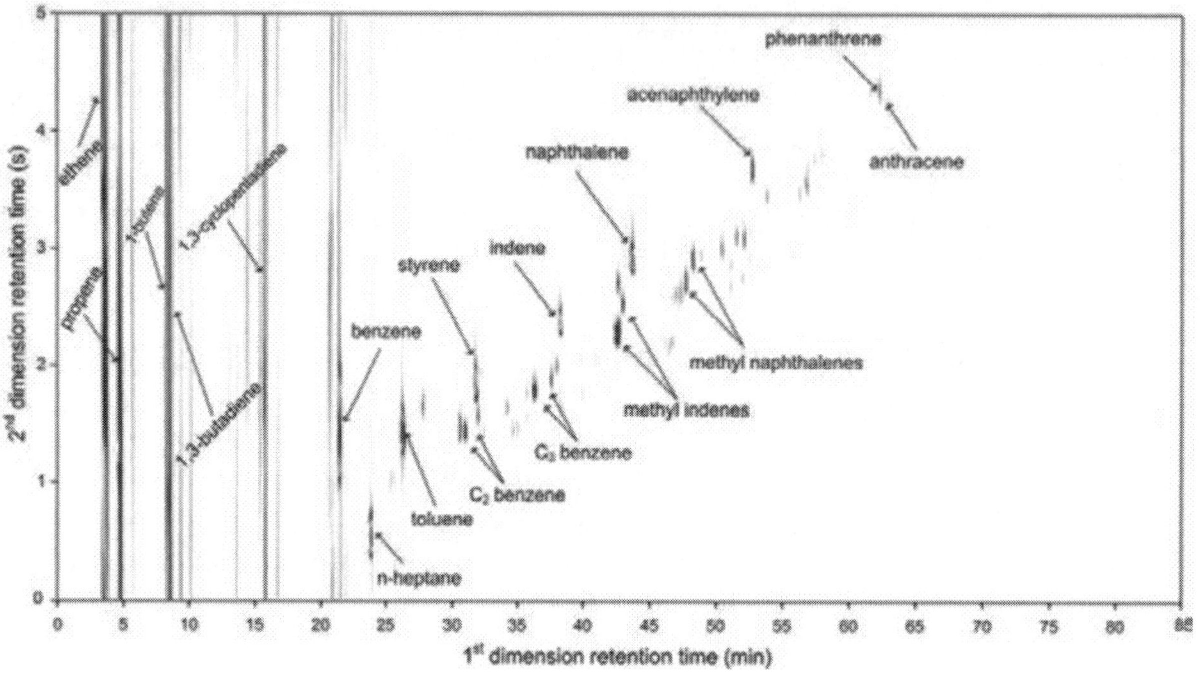

Figure 2: GC·GC-FID chromatogram of the effluent obtained during steam cracking of heptane. COT = 1123 K; COP = 170 ·103 Pa; dilution of 0.5 $kg_{steam}kg_{HC}^{-1}$.

Table 2: Summary of the measured product distribution (wt%) for steam cracking of heptane with/without the continuous addition of sulfur.

Experiment	Blank run	Addition of CS_2	Addition of DMDS	Addition of DMS	Addition of DMSO
HC-flow [10^{-3} kg.s^{-1}]			1.1		
H_2O flow [10^{-3} kg.s^{-1}]			0.6		
Dilution [kg_{steaam}/kg_{HC}]			0.5		
Additive concentrations [ppmw/HC]	0	5, 50, 200, 500, 1000	10, 200, 1235	10, 300, 1600	10, 200, 2000
Sulfur concentrations [ppmwS/HC]	0	4, 42, 168, 421, 842	7, 136, 841	5, 155, 826	4, 164, 821
		Yields (wt%)[b]			
ΣC4-	89.0	89.0	89.0	89.0	89.0
ΣC5+	11.0	11.0	11.0	11.0	11.0
P/E[a]	0.3	0.3	0.3	0.3	0.3
		C_{4-} Species			
H_2	0.98 ± 0.02	0.82 ± 0.01	0.80 ± 0.01	0.84 ± 0.0	0.83 ± 0.01
H_2S	0.00 ± 0.00	0.03 ± 0.01	0.07 ± 0.01	0.07 ± 0.00	0.04 ± 0.00
CO	1.38 ± 0.12	0.05 ± 0.00	0.06 ± 0.02	0.07 ± 0.01	0.10 ± 0.01
CH_4	13.44 ± 0.39	13.89 ± 0.12	13.85 ± 0.17	14.06 ± 0.18	14.09 ± 0.11

C_2H_6	5.25 ± 0.16	5.60 ± 0.05	5.73 ± 0.07	5.55 ± 0.17	5.75 ± 0.09
C_2H_4	43.97 ± 0.39	44.54 ± 0.31	44.64 ± 0.26	44.28 ± 0.16	44.24 ± 0.27
C_2H_2	0.51 ± 0.01	0.53 ± 0.00	0.52 ± 0.03	0.54 ± 0.03	0.53 ± 0.01
C_3H_8	0.51 ± 0.03	0.51 ± 0.01	0.52 ± 0.02	0.54 ± 0.01	0.5 ± 0.02
C_3H_6	14.89 ± 0.28	14.87 ± 0.17	14.84 ± 0.29	15.11 ± 0.31	15.03 ± 0.13
1,3-C_4H_6	5.12 ± 0.21	5.45 ± 0.09	5.32 ± 0.08	5.22 ± 0.10	5.17 ± 0.10
C_{5+} Species					
Benzene	3.81 ± 0.30	4.07 ± 0.26	4.12 ± 0.34	4.16 ± 0.27	4.03 ± 0.18
Toluene	0.91 ± 0.06	0.90 ± 0.11	0.90 ± 0.10	1.12 ± 0.06	0.91 ± 0.04
Indene	0.23 ± 0.01	0.26 ± 0.06	0.24 ± 0.04	0.23 ± 0.0	0.27 ± 0.02
Naphthalene	0.24 ± 0.07	0.29 ± 0.06	0.19 ± 0.05	0.18 ± 0.01	0.23 ± 0.01
Biphenyl	0.01 ± 0.00	0.01 ± 0.00	0.02 ± 0.00	0.01 ± 0.00	0.01 ± 0.01
Anthracene	0.01 ± 0.01	0.00 ± 0.00	0.01 ± 0.00	0.00 ± 0.00	0.00 ± 0.00

[a] P/E ratio of propene yield over ethene yield (kg/kg) [b] Product yields averaged over all cycles.

The conversion of heptane amounted to ~98 % and no significant differences in conversion were observed when sulfur compounds were added to the feed. The influence on the selectivity of the hydrocarbon species was minor. However, the decrease in the yields of hydrogen and carbon oxides is much more apparent when the continuous addition of sulfur compounds is evaluated.

Due to undesirable reactions on the surface catalyzed by nickel located in the cracking coil, small amounts of carbon oxide and carbon dioxide are produced. From a mechanistic point of view, two processes are important in describing steam/coke interaction, namely the direct gasification of coke with steam and metal-catalyzed removal of carbonaceous deposits by steam reforming [23]. As can be seen in Table 2 the blank experiment without continuous sulfur addition gives a higher yield of hydrogen and carbon monoxide (CO) compared to the other runs with continuous addition of sulfur. This is obviously associated with the catalytic influence of the wall in the absence of sulfur. It is well known that sulfur compounds act upon the catalytic gasification of the steam reforming process responsible for the production of CO. This can be explained by adsorption of sulfur containing moieties on the inner metal surface of the reactor where metal sulfides are formed. These do not catalyze the coke gasification, in contrast to nickel metal and nickel oxides [1].

When a sulfur-containing compound is added to the feed, a significant decrease (more than 90 %) of the CO amount is observed, as can be seen in Figure 3. The reason is not only the absorption of sulfur species on the inner wall of the coil but also coverage of the surface by coke, which blocks the catalytic active sites for steam reforming reactions. A similar trend was observed in work of Dhuyvetter et al. [12] and Reyniers et al. [11] where the addition of 50 ppmw of DMDS and CS_2 resulted in an important reduction of CO in the effluent. In addition, no further decrease of CO production with increasing quantities of DMDS was observed.

The main sulfur containing compound present in the reactor effluent is hydrogen sulfide (H_2S), which is the major product of the decomposition of DMDS [24, 25], as well as DMS, CS_2 and DMSO. Figure 4 reports yields of H_2S measured in the effluent during heptane steam cracking with continuous addition of CS_2, DMDS, DMS and DMSO by means of RGA-TCD.

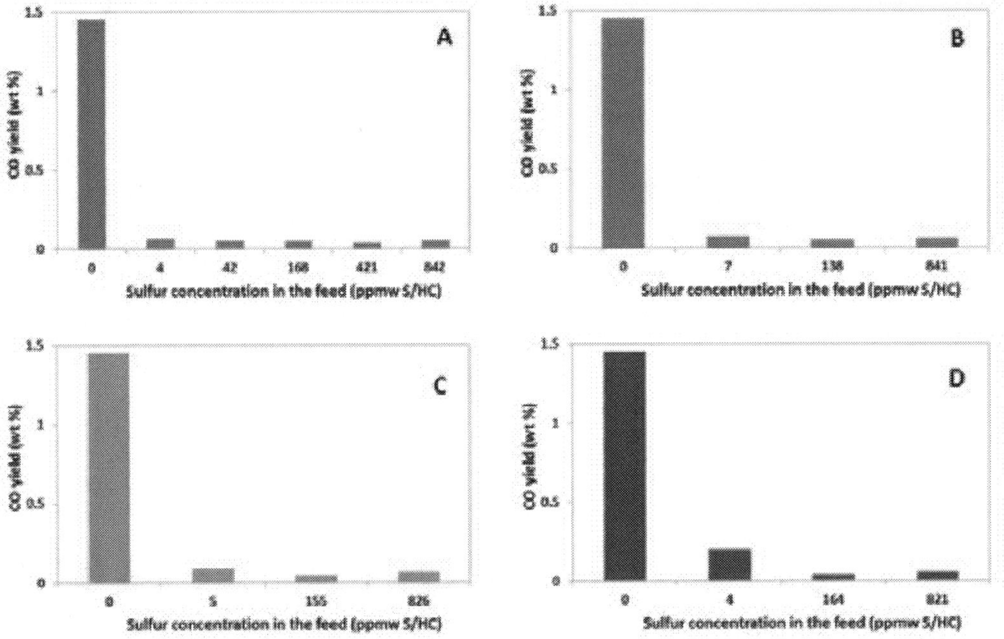

Figure 3: Asymptotic yield of CO as function of sulfur concentration.

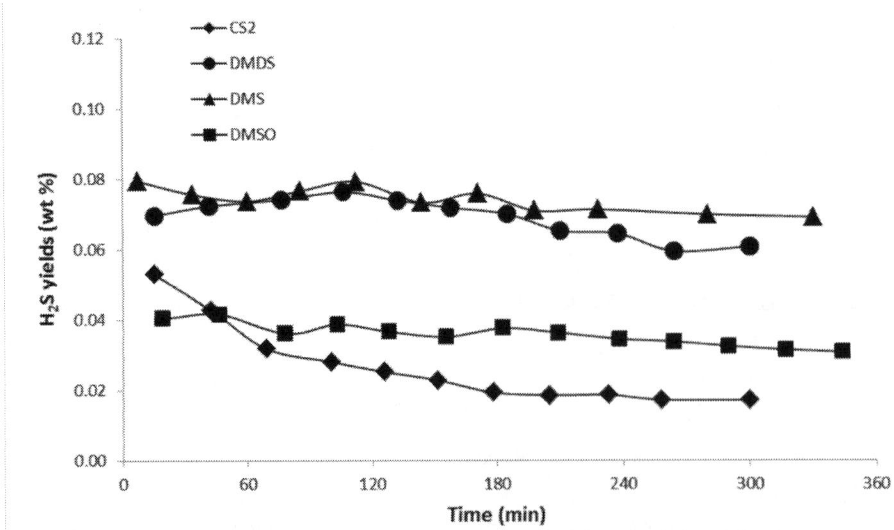

Figure 4: Yield of H2S as a function of time-on-stream at COT = 1123 K when 842 ppmw of S from CS_2, 841 ppmw of S from DMDS, 826ppmw of S from DMS, and 821ppmw of S from DMSO with respect to the feed were added.

As can be seen from Figure 4 and Table 3, most of the sulfur coming from DMDS and DMS ends up in the form of H_2S. In case of DMDS and DMS, respectively 94 and 86 % of the detected sulfur consists of H_2S. For DMSO, H_2S represents only 45 % of the measured sulfur. Even though CS_2 does not decompose completely under the studied conditions, H_2S remains the dominant sulfur containing product in the reactor effluent, with a selectivity of 38 %.

Besides H_2S, CS_2, DMS, DMDS, thiophene, methyl-substituted thiophenes (C_5H_6S), dimethyl(or ethyl)-substituted thiophenes (C_6H_7S) and benzothiophene are also detected in the effluent. This can also be seen from Figure 5 which presents the two-dimensional chromatogram of sulfur containing compounds in the effluent obtained using the GC×GC-SCD.

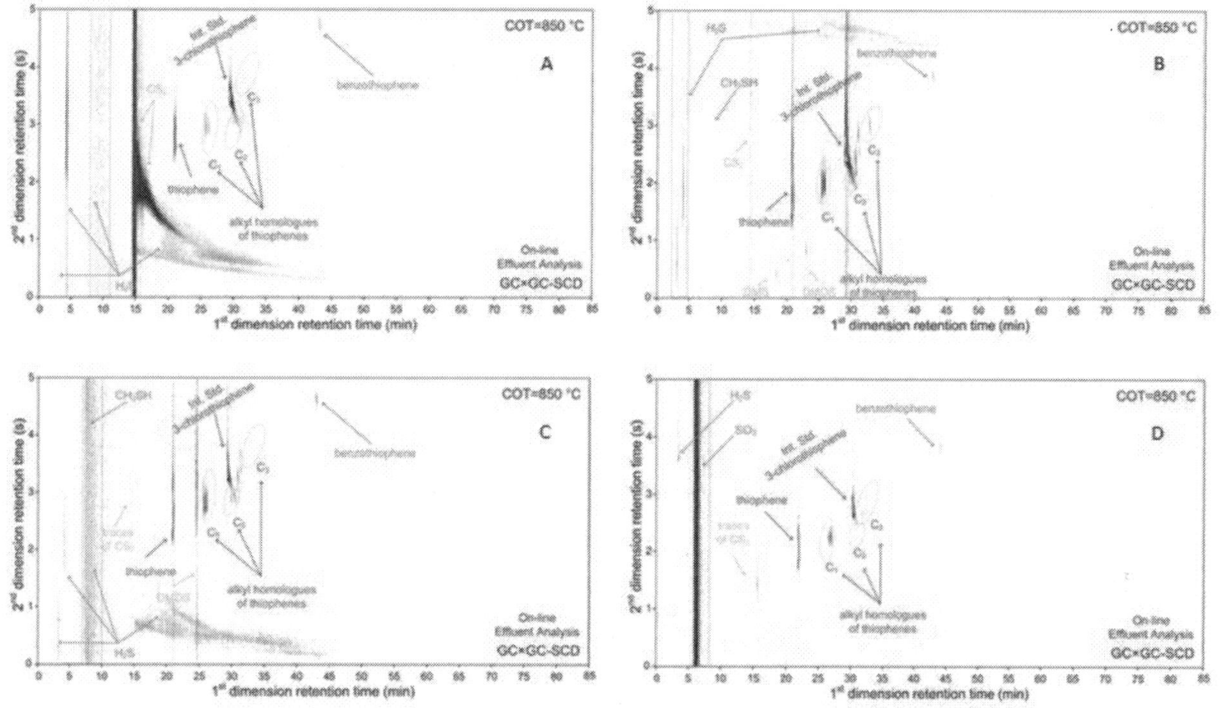

Figure 5: GC × GC-SCD chromatogram of sulfur-containing compounds contained in a cracking effluent at COT = 1123 K when (a) 842 ppmw of S from CS_2, (b) 841 ppmw of S from DMDS, (c) 826 ppmw of S from DMS, and (d) 821 ppmw of S from DMSO with respect to the feed were added.

Table 3 reports the yields of the sulfur containing products in the effluent for heptane steam cracking with continuous addition of all the sulfur-containing compounds added at the highest concentrations. Sulfur recovery refers to a ratio of the total amount of elemental sulfur detected in the effluent and the amount of elemental sulfur that is fed to the reactor. For all the additives the sulfur recovery reaches around 90 %, which is unprecedented.

Table 3: Yields of sulfur containing compounds in heptane cracking with the continuous addition of CS_2, DMDS, DMS and DMSO.

Experiment	**Addition of CS_2**	**Addition of DMDS**	**Addition of DMS**	**Addition of DMSO**
HC-flow [10^{-3} kg.s^{-1}]	1.1			
H_2O flow [10^{-3} kg.s^{-1}]	0.6			
Dilution [kg_{steaam}/kg_{HC}]	0.5	0.5	0.5	0.5
Additive concentrations [ppmw/HC]	1000	1235	1600	2000
Sulfur concentrations [ppmwS/HC]	842	841	826	821
Yields (ppmw S/HC)				
H_2S^a	317.0	671.9	619.0	344.0
SO_2	n.a.	n.a.	n.a.	370.0
CS_2	418.7	10.4	13.7	8.4
DMS	n.a.	1.2	25.7	n.a.
DMDS	n.a.	0.8	1.5	n.a.
Thiophene	23.4	24.7	46.1	28.6
C1 Thiophene	7.3	7.1	11.6	6.1
C2 Thiophene	2.2	0.4	0.3	0.1
Benzothiophene	4.7	0.4	1.2	1.4
Sum (ppmw S/HC)	773.3	716.9	719.1	758.6
Sulfur recovery (%)	91.8	85.2	87.1	92.4

[a] detected by means of RGA

CS_2 does not decompose to a large extent under these conditions, especially when added in these low concentrations. Besides CS_2, H_2S, thiophene together with its alkyl-substituted homologues and benzothiophene are detected in the effluent. Shum et al. [26] proposed that CS_2 decomposes by the radical mechanism giving primarily methanethione (CH_2S).

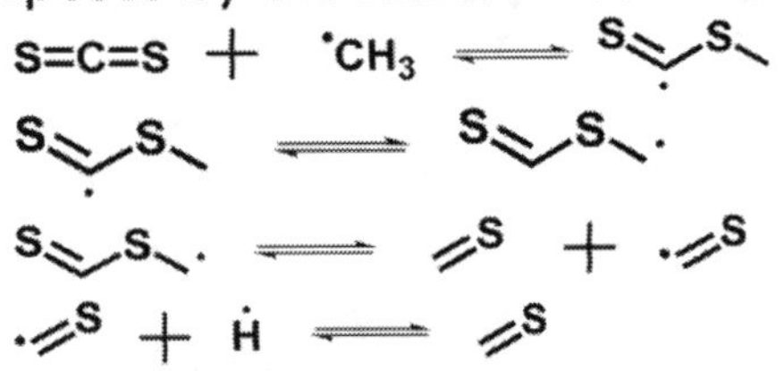

(1)

(2)

(3)

(4)

The thermal decomposition proceeds among others via the radical addition of methyl radicals (Eq. (3)). The formed radical can than isomerize via a hydrogen shift (Eq. (4)). The proposed reaction path leads mainly to the formation of CH_2S by β-scission (Eq. (5) and (6)), while no amount of this sulfur compound was detected in the effluent. However, CH_2S is a very reactive compound with a strong tendency to polymerize which makes it extremely difficult to detect, certainly by gas chromatography [27]. This could explain the unclosed mass balance for sulfur.

On the other hand, DMDS almost completely decomposes under the cracking conditions. Vandeputte [25] suggested that DMDS decomposes mainly to CH_3SH, H_2S, ethene and CS_2. This agrees with the observations in this work, but it is clear that because of the presence of substantial amounts of heavier products also heavier sulfur compounds can be formed by addition reactions or recombination reactions. As can be seen from Table 3, the main products from DMDS decomposition during this study are H_2S, CS_2, DMS, thiophene, methyl-, ethyl or dimethyl-substituted thiophenes and benzothiophene. At low temperature, DMDS already decomposes via substitutions (Eq. (7)), homolytic S-C (Eq. (8)) and S-S scission reactions (Eq. (9)) [25].

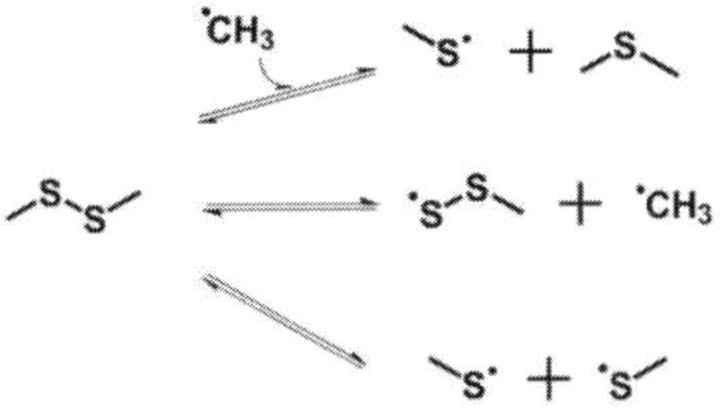

(5)

(6)

(7)

The resulting radicals can react via hydrogen abstraction, radical additions or recombination to form other sulfur-containing compounds, out of which H_2S, CH_2S and thiophene are the most important ones. The thyil radicals are also formed, mainly via substitutions (Eq. (10)), homolytic S-C scissions (Eq.(11)) and β-scissions (Eq. (12)) [25].

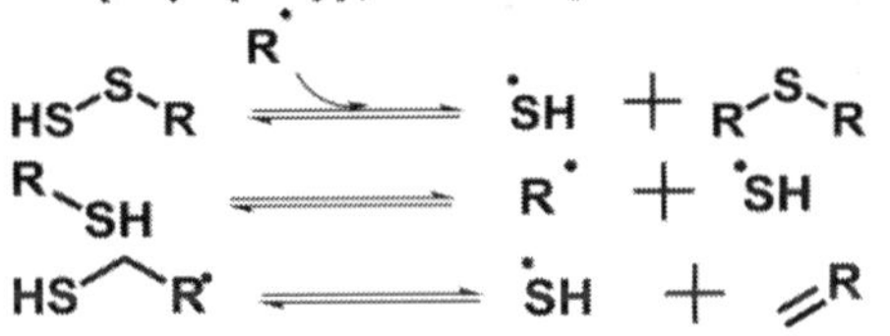

(8)

(9)

(10)

Furthermore, the unstable thyil or hydrogen radicals can react via substitutions to H_2S, Eq. (13) and Eq. (14) respectively.

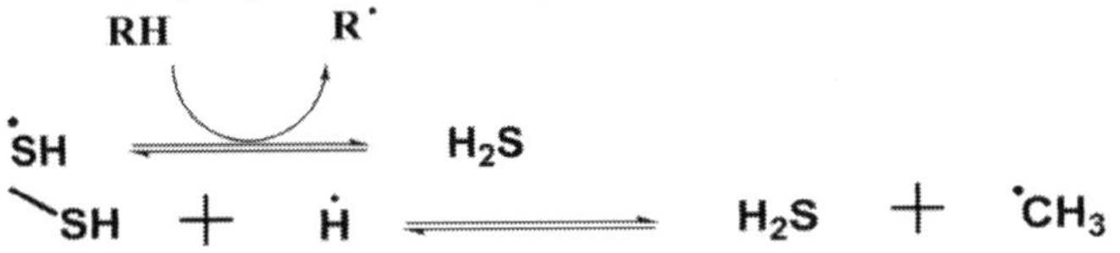

(11)

(12)

Bajus et al. [18] proposed a mechanism responsible for the formation of thiophene (Eq. (20) via a thiobutadienyl radical (Eq. (16). This can be formed as a reactive intermediate by reactions between 1,3-butadiene and a thyil radical (Eq. (15). Also under the present conditions this pathway seems the potential dominant route to thiophene:

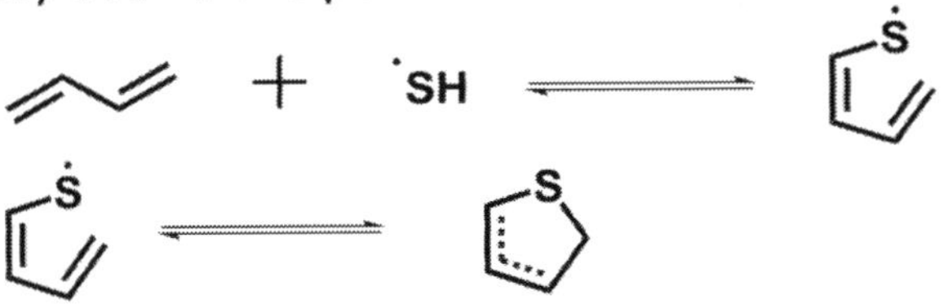

(13)

(14)

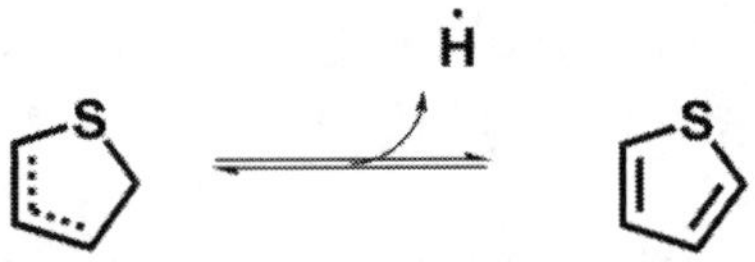

(15)

Formation of benzothiophene can be explained among others by Diels-Alder self-addition reactions of thiophene which also release H_2S as a product (Eq.18 and Eq. (19)).

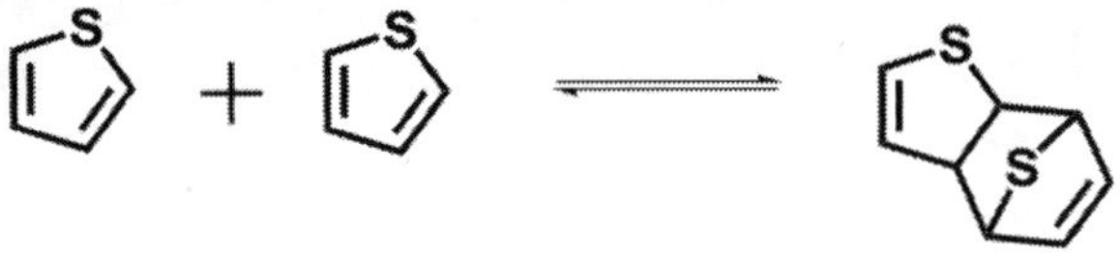

(16)

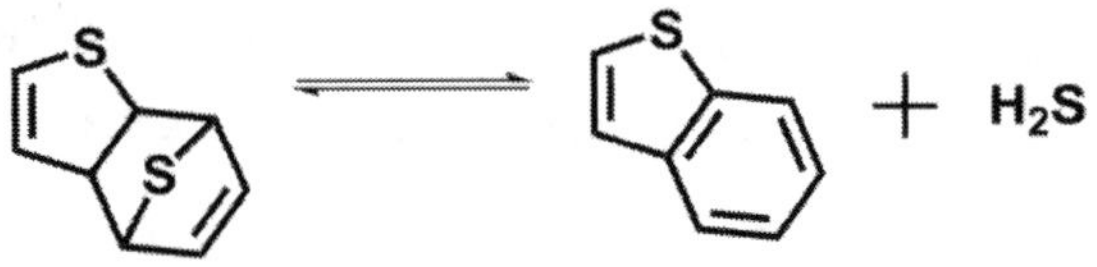

(17)

Benzothiophene can be also formed via reaction of a thiophene molecule with thiophyne (Eq. (20)) which is a very reactive intermediate formed by intramolecular dehydrogenation of thiophene [18].

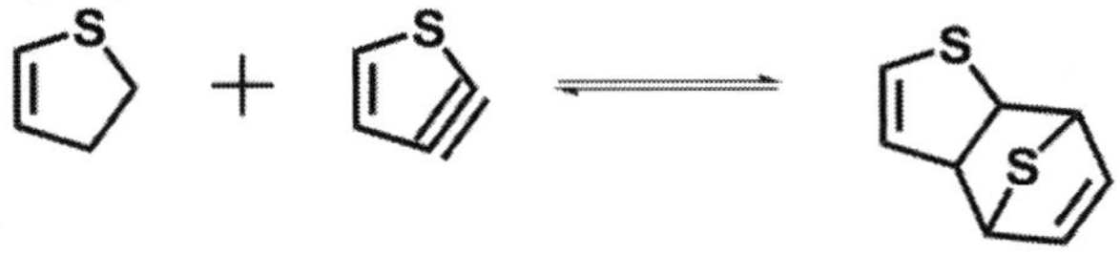

(18)

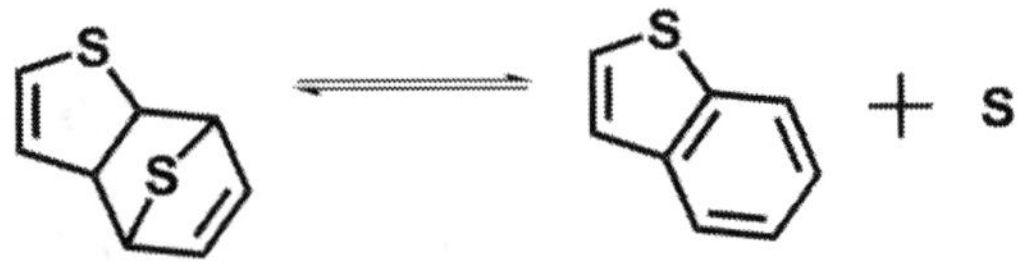

(19)

Nevertheless radical addition reactions of resonantly stabilized radicals will surely contribute to benzothiophene formation.

Sulfur eliminated from thiophene can enter into reactions in gaseous phase that lead to the formation of heptyl and thyil radical or can also react with the metal surface of the reactor (see Eq. (22)).

(20)

DMS was added in concentration of 850 ppmw of S relative to the heptane feed and 97% was converted into other sulfur-containing species, such as H_2S, CS_2, thiophene and its alkyl-substituted homologues together with benzothiophene. Shum and Benson [26] proposed in their work a small radical network built up of approximately 10 reactions responsible for the decomposition of DMS (see Eq. (23)-(33)).

(21)

(22)

(23)

(24)

(25)

(26)

(27)

(28)

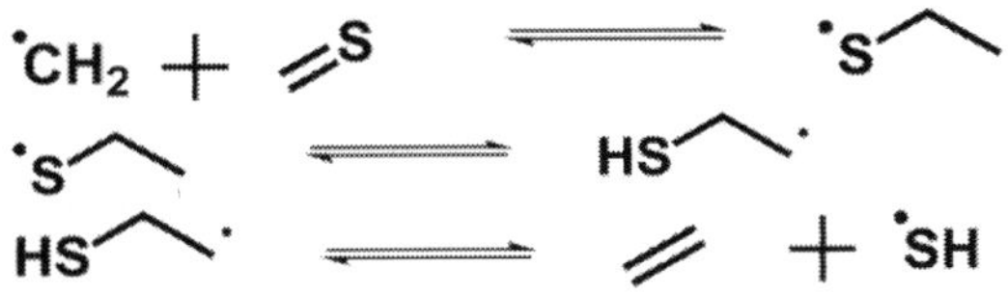

(29)

(30)

(31)

For the DMSO experiments, no DMSO was detected in the effluent which indicates that it is complete decomposed. From the 2-dimensional GC×GC SCD chromatogram of sulfur containing compounds (Figure 5 D) in the effluent the main product was sulfur dioxide (SO_2) when adding DMSO. Besides SO_2 H_2S, thiophene with its homologues and benzothiophene were detected. A trace amount of CS_2 was presented as well. Thyrion and Debecker [28] reported that the major products from the thermal decomposition of DMSO are methane, ethene and SO_2. However, the formation of some sulfur species such as DMDS, DMS and CS_2 indicates that thermal decomposition does not proceed via a simple mechanism.

Effect of sulfur compounds on coke formation

At the end of every cracking run the reactor was decoked with air/steam mixture at temperatures of 1073-1173 K. It is important to mention that amount of coke measured corresponds to the sum of the catalytically formed coke and pyrolytic coke deposited in radiant coil because it is determined by measuring the amount of CO and CO continuously when burning of the coke using an the infrared CO/CO_2 meter. Figure 6 shows the amounts of coke deposited in the reactor during 6 hours for all experiments as a function of the amount of sulfur in the feed. For the heptane coking run [COT = 1123 K; $\delta = 0.5\ kg_{steam}/kg_{HC}$; COP = 170 x 10^3 Pa] without any sulfur addition (blank experiment) 2.9 g of cokes is deposited in the reactor over 6 hours. As expected, the increase of sulfur concentration in the feed results in higher coke formation.

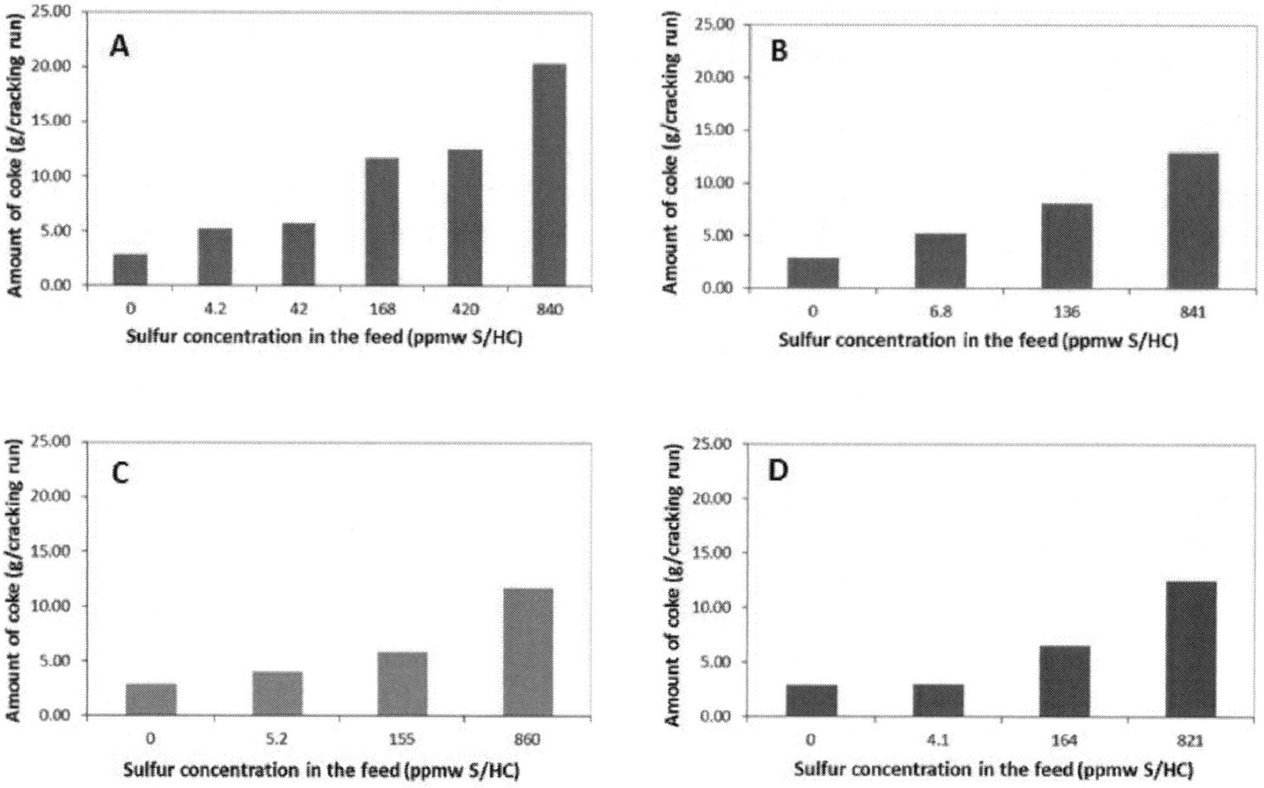

Figure 6: Amount of coke formed during steam cracking of heptane at the pilot-plant unit as a function of the sulfur concentration in the feed: A) CS_2, B) DMDS, C) DMS, and (d) DMSO.

Surprisingly, the most coke is formed during the continuous addition of CS_2 which does not decompose completely under cracking conditions. With continuous addition of 5-1000 ppmw CS_2, the amount of coke increased. Reyniers et al. [11] observed that the thermal stability of sulfur-based compounds follows the same order as their capacity to affect coke formation. This indicates that their detrimental effect may originate from the thermal decomposition leading to H_2S. Although the thermal decomposition of CS_2 yields mainly H_2S, sulfur centered radicals are also produced. These small reactive radicals from the gas phase can effect coking in two ways. Hydrogen and methyl radicals can enhance the number of radical sites in the coke matrix via hydrogen abstractions. Simultaneously they can also react with CS_2 which leads to the formation of radicals. The radicals thus formed from CS_2 by hydrogen abstraction can interact with the coke layer and create even more reactive centers in the coke matrix. The presence of another sulfur atom in this matrix can lead to additional reactions with other radicals such as hydrogen abstraction reactions and substitution reactions [26]. The observed higher coking rates thus can be caused by an increased reactivity of the coke matrix which is associated with the number of active sites. Starting from the addition of 6.8 ppmw of elemental sulfur from DMDS per heptane S the between total coke and amount of sulfur becomes linear, reaching the highest value of almost 13 grams of total coke when 841 ppmw of sulfur was added. For DMS, adding 5 ppmw of elemental sulfur relative to heptane led to an increase of the total coke of roughly 4 grams, and adding 155 ppmw increased the total amount of coke to approximately 6 grams. The highest value of total coke was almost 12 grams when 860 ppmw of elemental sulfur per hydrocarbon feed was added. Finally, the addition of 4.1 ppmw of sulfur originated from DMSO related to the feed resulted in an enlargement of the total amount of coke similar to that of the blank run, and increasing the amount of added DMSO to 164 ppmw of elemental sulfur per feed resulted in an increase of the total cokes to more than 6 grams. Further increase of the added DMSO increased the total amount of cokes to 12 grams at 821 ppmw of elemental sulfur relative to the hydrocarbon feed. In contrast with the contradictory results found in literature, several researcher observed the same effect. Reyniers and Froment [11] pointed out that with the continuous addition of increasing quantities of DMDS (25, 75 and 200 ppmw), CS_2 (1, 5, 20, 50, 150 and 200 ppmw), thiophene (100, 500 and 2500 ppmw) and benzothiophene (250 and 650 ppmw) to the feed, both the initial and the asymptotic coking rate increased. A promotional effect was also reported by Wang et al. [16]. With the continuous addition of DMDS, coke formation in the reactor and in the TLE increased as well.

Conclusions

The results demonstrate that the influence of ppm levels of sulfur compounds results mainly in an increase of coke formation and a decrease of CO formation. However, the influence on the product distribution, and more specifically on the ethylene selectivity, is not significant. Online GC·GC-SCD indicates that these compounds not only decompose but they also result in the formation of heavier sulfur compounds such as benzothiophenes. The heptane coking runs (COT= 1123 K; d = 0.5 $kg_{steam}kg_{HC}^{-1}$; COP= 170 ·103 Pa) with the continuous addition of DMDS, DMS,DMSO, and CS_2 always resulted in a reduction in CO yields of more than 90% compared to the experiment in the absence of sulfur-based additive, even when added in amounts lower than 5 ppm. However, compared to experiments when no sulfur was added, coke formation was substantially increased up to a factor of 7.

Acknowledgment

The research leading to these results has received funding from the Long Term Structural Methusalem Funding by the Flemish Government.

References

1. H. Zimmermann, R. Walzl, Ethylene, in: Ullmann's Encyclopedia of Industrial Chemistry, Wiley-VCH Verlag GmbH & Co. KGaA, 2000.

2. S.M. Sadrameli, Thermal/catalytic cracking of hydrocarbons for the production of olefins: A state-of-the-art review I: Thermal cracking review, Fuel, 140 (2015) 102-115.

3. J. Baker, GPCA: Ethylene continues expansion in 2012, in, ICIS Chemical Business, ICIS Chemical Business, 2012.

4. R. Koster, Global PP Outlook. ME Gas, U.S. Shale, China Coal But what about Europe?, in, IHS, 2014.

5. L. Koottungal, International survey of ethylene from steam crackers — 2015, in, Oil&Gas journal, 2015.

6. A.E. Muñoz, K.M. Van Geem, M.-F. Reyniers, G.B. Marin, Influence of the Reactor Material Composition on Coke Formation During Ethane Steam Cracking, Industrial & Engineering Chemistry Research, (2014).

7. G.C. Reyniers, G.F. Froment, F.-D. Kopinke, G. Zimmermann, Coke Formation in the Thermal Cracking of Hydrocarbons. 4. Modeling of Coke Formation in Naphtha Cracking, Industrial & Engineering Chemistry Research, 33 (1994) 2584-2590.

8. P.M. Plehiers, G.C. Reyniers, G.F. Froment, Simulation of the run length of an ethane cracking furnace, Industrial & Engineering Chemistry Research, 29 (1990) 636-641.

9. C.M. Schietekat, D.J. Van Cauwenberge, K.M. Van Geem, G.B. Marin, Computational fluid dynamics-based design of finned steam cracking reactors, AIChE Journal, 60 (2014) 794-808.

10. C.M. Schietekat, M.W.M. van Goethem, K.M. Van Geem, G.B. Marin, Swirl flow tube reactor technology: An experimental and computational fluid dynamics study, Chemical Engineering Journal, 238 (2014) 56-65.

11. M.-F.S.G. Reyniers, G.F. Froment, Influence of Metal Surface and Sulfur Addition on Coke Deposition in the Thermal Cracking of Hydrocarbons, Industrial & Engineering Chemistry Research, 34 (1995) 773-785.

12. I. Dhuyvetter, M.-F. Reyniers, G.F. Froment, G.B. Marin, D. Viennet, The Influence of Dimethyl Disulfide on Naphtha Steam Cracking, Ind. Eng. Chem. Res., 40 (2001) 4353-4362.

13. N. Rahimi, R. Karimzadeh, S.M. Jazayeri, K.D. Nia, An empirical investigation of the influence of sulfur additives on the catalytic rate of coke deposition and CO formation in the steam cracking of LPG over Incoloy 600 and stainless steel, Chemical Engineering Journal, 238 (2014) 210-218.

14. M. Bajus, V. Veselý, Pyrolysis of hydrocarbons in the presence of elemental sulfur, Collect. Czech. Chem. Commun., 45 (1980) 238-254.

15. M. Bajus, V. Veselý, J. Baxa, P.A. Leclercq, J.A. Rijks, Steam Cracking of Hydrocarbons. 5. Effect of Thiophene on Reaction Kinetics and Coking, Ind. Eng. Chem. Res. Dev., 20 (1981) 4.

16. J. Wang, M.-F. Reyniers, G.B. Marin, Influence of Dimethyl Disulfide on Coke Formation during Steam Cracking of Hydrocarbons, Industrial & Engineering Chemistry Research, 46 (2007) 4134-4148.

17. S.M. Jazayeri, R. Karimzadeh, Experimental Investigation of Initial Coke Formation over Stainless Steel, Chromium, and Iron in Thermal Cracking of Ethane with Hydrogen Sulfide as an Additive, Energy & Fuels, 25 (2011) 4235-4247.

18. M. Bajus, Sulfur compounds in hydrocarbon pyrolysis, Sulfur Reports, 9 (1989) 46.

19. M. Bajus, J. Baxa, Coke formation during the pyrolysis of hydrocarbons in the presence of sulfur compounds, Collect. Czech. Chem. Commun., 50 (1985) 16.

20. S.P. Pyl, T. Dijkmans, J.M. Antonykutty, M.-F. Reyniers, A. Harlin, K.M. Van Geem, G.B. Marin, Wood-derived olefins by steam cracking of hydrodeoxygenated tall oils, Bioresour. Technol., 126 (2012) 48-55.

21. K.M. Van Geem, S.P. Pyl, M.-F. Reyniers, J. Vercammen, J. Beens, G.B. Marin, On-line analysis of complex hydrocarbon mixtures using comprehensive two-dimensional gas chromatography, J. Chromatogr. A, 1217 (2010) 6623-6633.

22. T. Dijkmans, M.R. Djokic, K.M. Van Geem, G.B. Marin, Comprehensive compositional analysis of sulfur and nitrogen containing compounds in shale oil using GC × GC – FID/SCD/NCD/TOF-MS, Fuel, 140 (2015) 398-406.

23. D.L. Trimm, Fundamental Aspects of the Formation and Gasification of Coke, in: L.F. Allbright, B.L. Crynes, W.H. Corcoran (Eds.) Pyrolysis - Theory and Industrial Practice, Academic Press, Inc., New York, 1983, pp. 203-232.

24. D. Salari, A. Niaei, R. Shoja Mohammad, R. Nabavi, Coke Formation Reduction in the Steam Cracking of Naphtha on Industrial Alloy Steels Using Sulfur-Based Inhibitors, ijcre, 8 (2010) 20.

25. A.G. Vandeputte, M.F. Reyniers, G.B. Marin, Theoretical study of the thermal decomposition of dimethyl disulfide, J Phys Chem A, 114 (2010) 10531-10549.

26. L.G.S. Shum, S.W. Benson, The pyrolysis of dimethyl sulfide, kinetics and mechanism, International Journal of Chemical Kinetics, 17 (1985) 749-761.

27. M.E. Peach, E. Weissflog, N. Pelz, Mass spectra of polymeric thioformaldehydes (CH2S)3, (CH2S)4, (CH2S)5 and (CH2S)x, JOURNAL OF MASS SPECTROMETRY, 10 (1975) 6.

28. F.C. Thyrion, G. Debecker, Thermolysis of dimethyl sulfoxide, International Journal of Chemical Kinetics, 5 (1973) 583-592.

AIChE Paper 129d

Separation of Olefin-Paraffin Mixtures with Optiperm™ Customized Amorphous Fluoropolymer Membranes

Dr. Hannah Murnen

VP Business Development

Compact Membrane Systems

Dr. Sudip Majumdar

Chief Technology Officer

Compact Membrane Systems

William Charlton

VP Separations

Compact Membrane Systems

Prepared for Presentation at the 2018 Spring National Meeting

Orlando, FL, April 22, 2018

Separation of Olefin-Paraffin Mixtures with Optiperm™ Customized Amorphous Fluoropolymer Membranes

Hannah Murnen

VP Business Development

Compact Membrane Systems

Sudip Majumdar

Chief Technology Officer

Compact Membrane Systems

William Charlton

VP Separations

Compact Membrane Systems

Abstract: The separations of olefins from paraffins are some of the costliest, most energy intensive, and most technically difficult separations in the chemical industry due to the very similar size and nature of the molecules being separated. The technologies currently available for olefin-paraffin separation include distillation, adsorption, and absorption. Olefin-paraffin mixtures are conventionally separated mainly by distillation which is highly energy and capital intensive. A membrane based olefin-paraffin separation technology has the potential to provide substantial economic benefit to petrochemical industry and drastically reduce the energy required. While separations with membranes have been demonstrated in the laboratory, problems with membrane stability have prevented development of commercial systems.

Compact Membrane Systems (CMS) has developed Optiperm™ membrane, a customized amorphous fluoropolymer (CAF) containing silver ions that selectively transport olefin molecules from a mixture of olefin and paraffin. Optiperm™ membrane has shown stable performance over 12 months with high propylene flux and propylene/propane selectivity in the laboratory. Similar results were also obtained with ethylene and ethane gas mixtures. A broad range of tests have been conducted to characterize the membrane resistance to common poisons (such as hydrogen, acetylene, carbon dioxide) in refinery gas streams.

These silver based Optiperm™ membranes have been tested in the laboratory with actual refinery gas mixture from a distillation column. A pilot Optiperm™ system has been installed recently at the Delaware City Refining Company (DCRC) site in Delaware City, DE. The testing of the pilot unit has been initiated to provide real-world evidence of membrane stability in realistic conditions validating our laboratory testing. Pilot unit test results as well as some of our economic modeling results showing very low payback times and positive IRRs for a variety of scenarios governing petrochemical processes will be discussed in this paper.

Background

The separation of ethylene and propylene, two of the major building blocks for the chemical industry, is one of the most energy-intensive. It is estimated that separating alkanes from alkenes through distillation uses 0.3% of energy globally. In addition, distillation columns to separate these molecules are very capital-intensive. A membrane based olefin-paraffin separation technology has the potential to provide substantial economic benefit to the petrochemical industry and to drastically reduce the energy required. In addition, membrane systems are scalable, allowing their implementation for smaller, stranded streams where distillation columns have not previously been installed due to limited economic value. Membrane processes utilizing facilitated transport membranes have been extensively studied for separating olefin/paraffin mixtures. While separations have been demonstrated in the laboratory, problems with membrane stability have prevented development of commercial systems.

Membrane Technology

Compact Membrane Systems (CMS) has developed Optiperm™ membrane, a customized amorphous fluoropolymer (CAF) containing silver ions that selectively transport olefin molecules from a mixture of olefin and paraffin. CMS Optiperm™ membranes are based on a silver-containing fluoropolymer material. The technology exploits a mass-transfer mechanism known as facilitated transport. Silver incorporated into the membrane polymer backbone as a binding site for the double bond in the olefins transporting them across the membrane as depicted in the cartoon below. Saturated paraffins do not have the same interaction. This difference in affinity allows for olefins to pass through the membrane at a significantly higher rate than do paraffins.

Adapted from Cussler E.L.: Facilitated Transport. In: Membrane Separation Systems, vol. 2, US DOE Report, DOE/ER/30133-H1 (1990)

Figure 1. *Illustration of Facilitated Transport mechanism – the vines represent the silver cations in the polymer backbone and Tarzan represents the olefin molecule*

The two key membrane properties that characterize the effects of facilitated transport are **permeance** and **selectivity**.

- **Permeance** is defined as the rate at which one component travels through a given surface area of membrane for a given driving force. This performance metric is typically expressed in GPU (gas permeation unit). Optiperm™ membranes have olefin permeances greater than 150. This translates into a propylene flux of 0.25-0.85 ton/ft^2/year under normal operating conditions (e.g., 30-200 psi, 30°C, and 20-90% olefin)
- The **selectivity** of one component to another is simply the ratio of the permeance of the two components. This tells you how much olefin is passing through the membrane as compared to paraffin. Optiperm™ has selectivity ranging from 20-80 depending on the operating conditions and feed composition.

The Optiperm™ membrane has both high permeance and high selectivity, meaning high purities can be reached with a reasonable membrane area. These characteristics yield very attractive process economics.

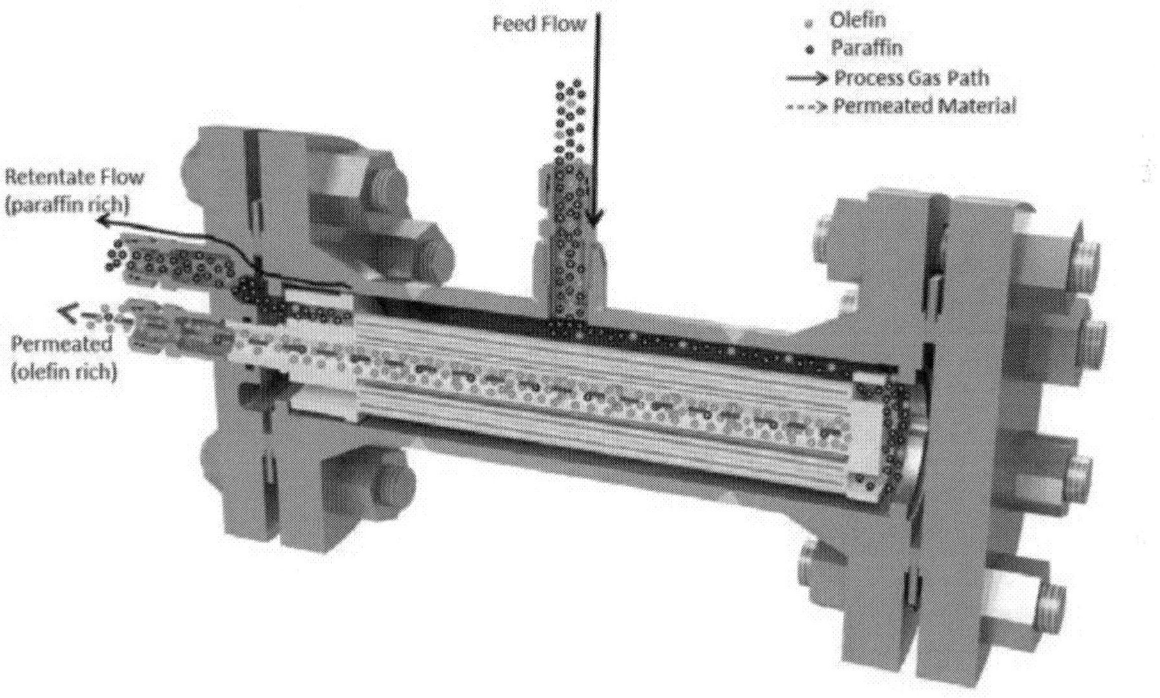

Figure 2. *Illustration of Optiperm™ membrane module*

The membrane can perform separations of olefins from paraffins at a wide range of temperatures, pressures, and compositions. CMS has demonstrated olefin removal at temperatures up to 70°C (158°F) and pressures up to 250 psig. For an efficient separation to occur, the olefin-paraffin mixture must be in the gaseous state. Separation performance requires humidification as water vapor improves the mobility of the olefin through the membrane. The membrane can remove olefin in both concentrated and dilute streams for compositions ranging 5-95% by volume of olefin. The membrane can process mixtures of C_2, C_3, and C_4 olefin and paraffin. Typical lab evaluations involve exposing the membrane to a stream of fixed pressure, temperature, and composition. The chart displayed on the next page shows a performance profile for a membrane separating a 20% propylene/80% propane mixture at 60 psig and room temperature. As you can see, the membrane has shown stable performance over >13 months, a vast improvement over current state of the art materials.

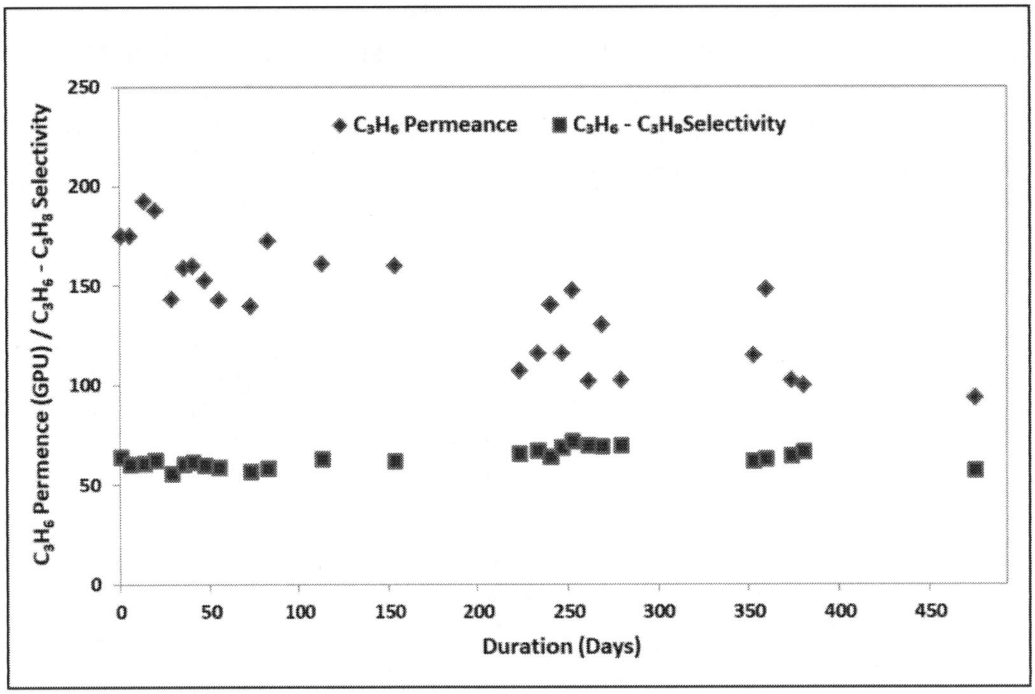

Chart 1. *Optiperm™ membrane aging study*

A broad range of tests have been conducted to characterize the membrane resistance to common poisons in process gas streams. These include testing under conditions such as 5% hydrogen or 500 ppm acetylene. We have seen good resistance to both hydrogen and acetylene from the Optiperm™ membranes. Based on our testing, we recommend process streams contain:

1. <100 ppm acetylene
2. <2% hydrogen
3. <100 ppb total sulfur

We generally include a sulfur pretreatment unit operation as part of the full membrane upgrading system to eliminate the concern that sulfur will detrimentally affect membrane lifetime.

Economics

CMS has performed multiple evaluations to assess the economic viability of the Optiperm™ membrane. These exercises rigorously examine the operational, capital, administrative, and associated auxiliary equipment (e.g. compression and drying) to return a pressurized, dry product or recycle stream. The assessments looks at a representative process and uses the membrane to add value by recovering and purifying an olefin-rich reactor feed stream and for enriching a paraffin rich product stream.

Example Application – Polypropylene Reactor Purge

Domestic polypropylene production capped near 4.4 million US tons in 2014 and resin demand continues to grow at 2 – 2.5% annually. Production methods vary in terms of catalysts employed, vapor or liquid phase reactors, and operating temperature and pressure. In general, polymer grade propylene (99.5+% C_3H_6, 0.5% C_3H_8, balance impurities) is fed to a reactor at moderate temperature (66-87°C) and high pressure (440-590 psi). After reaction, the polypropylene product is sent to a separator and transferred to a degassing bin. The unreacted propylene is recompressed and recycled back to the reactor. This recycling of the propylene monomer results in the buildup of inert propane in the reactor. Left untouched the propane takes up space in the reactor and reduces catalyst efficiency. As a result, a certain amount of material must be continually purged from the reactor circuit, In the process of removing undesired propane valuable propylene is returned to the splitter column for reprocessing, sent to plant fuel system for energy recovery or is lost to the plant flare.

CMS's membrane technology gives polypropylene producers a direct method for recovering propylene conventionally lost as purged material while still preventing accumulation of inert propane. A simple process flow diagram illustrates the integration of an Optiperm™ membrane system into a polypropylene manufacturing process.

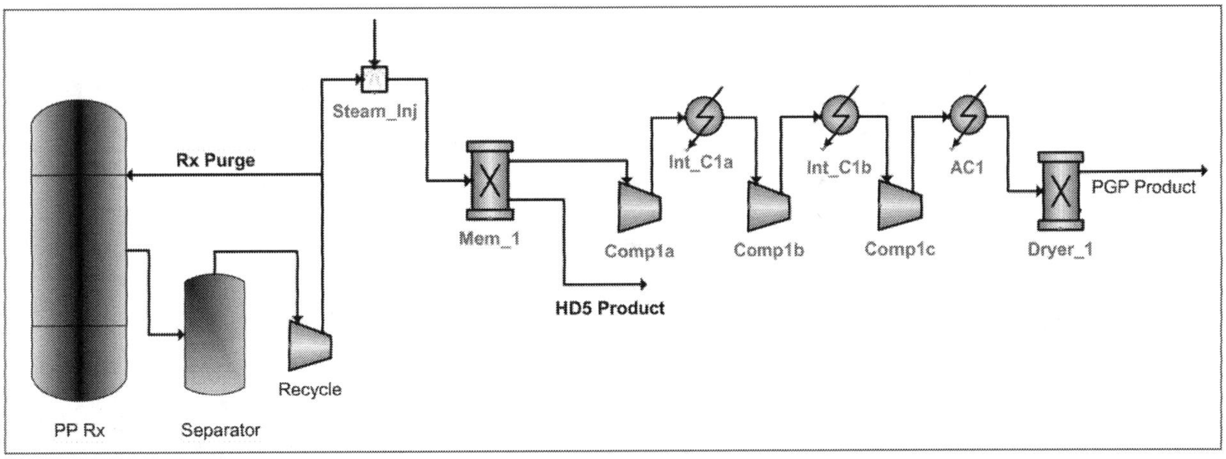

Figure 3. *Process Flow Diagram for polypropylene reactor purge recovery*

In this case study where the Optiperm™ membrane is used to recover propylene from a polypropylene reactor purge stream. The study examined a 31.5 KTA polypropylene reactor with a purge content of 10% propane. Using standard capital cost estimations for equipment, a calculated annual operating expense, the current market values for polymer grade propylene and HD5 propane, to calculate an investment return on a membrane system requiring a $0.7MM capital investment and $0.2MM in annual operating expenses. The reduced feed stream cost quickly offsets the capital expenditure and operating costs giving an IRR of 155% and payback period of around 7 months.

These economics are robust over a wide range of conditions. The sensitivity analysis for a number of variables is shown below. The biggest sensitivity to the economics comes from the permeance of the membrane and the value of the product streams. Higher permeance through the membrane minimizes the required area for the separation and therefore improves the capital cost for the project. CMS Optiperm™ membranes have very high permeances, contributing to positive economic outcomes across a wide variety of applications.

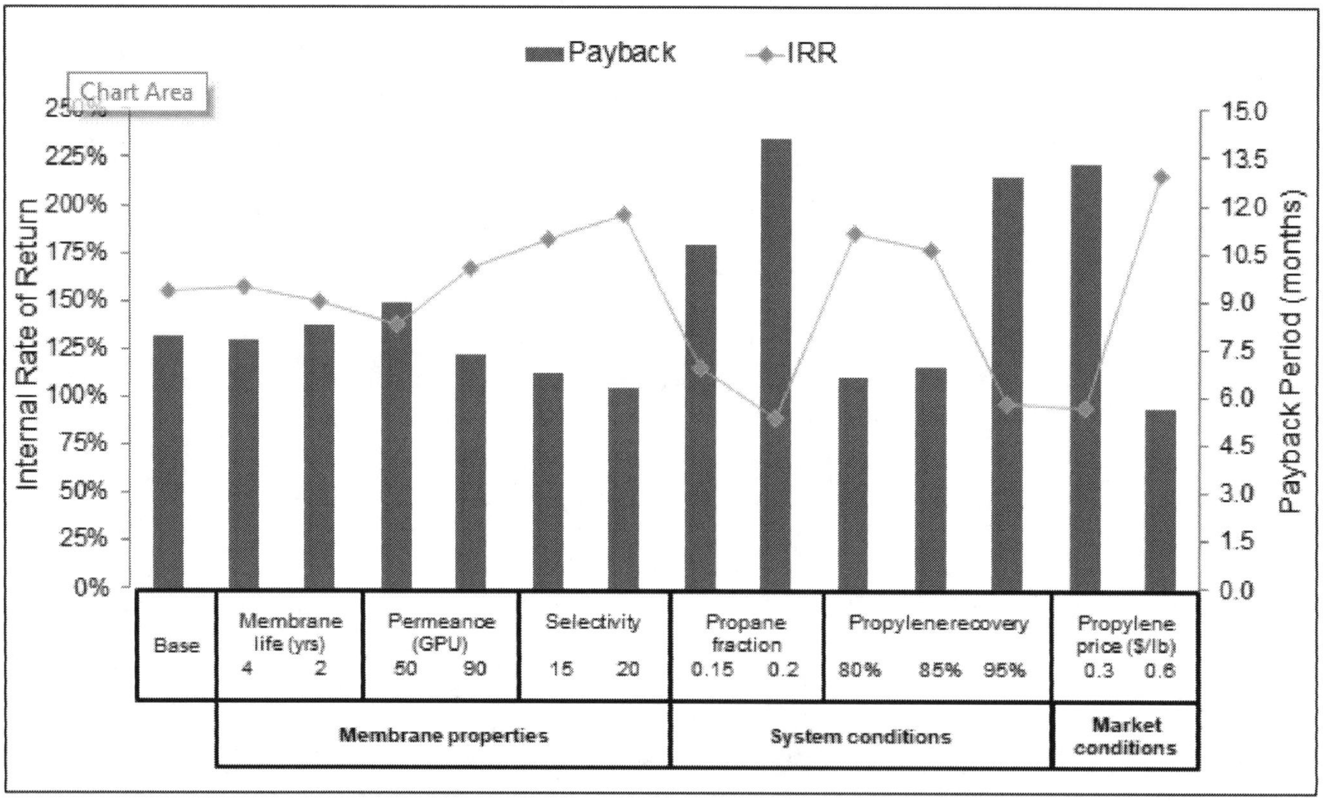

Chart 2. *Sensitivity analysis for Polypropylene Rx purge study*

As expected, more frequent membrane replacement, lower olefin permeance, and inexpensive market prices for propane and propylene increase the payback period of the membrane system. However, in all cases evaluated across a range of likely scenarios, the investment is still strongly positive with the IRR remaining greater than 90% for all scenarios evaluated.

A listing of our base case assumptions used in the economic sensitivity analysis is shown here in Table 1.

Table 1. *Modeling parameters for polypropylene reactor economic evaluation*

Modeling Parameter	**Value**
Membrane Permeance (Propylene)	70 GPU
Membrane Selectivity	10
Feed Pressure	240 psia
Permeate Pressure	47 psia
Membrane Lifetime	3 years
Polymer Grade Propylene Value	$0.45/lb
Reactor Propane Content	10%
Propylene Recovery	90%

Membrane Production and Scale up

CMS is currently scaling up manufacturing to meet expected demand for the Optiperm™ membrane. These scale up activities include membrane module scale-up, supply-chain diversification, QC development, procurement of large scale coating equipment and expansion of manufacturing space. An example of a prototype Optiperm™ spiral wound membrane is shown below.

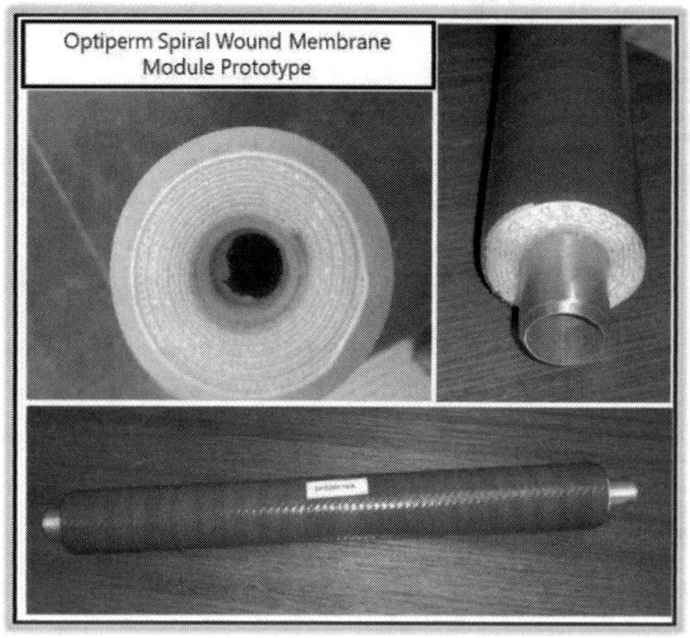

Figure 4. *Optiperm™ spiral wound membrane modules*

We continue to develop and perfect our large scale membrane coating capabilities utilizing techniques learned and honed over 20 years of fabricating membranes in our Newport De facilities.

A performance profile map of a 42" wide membrane sheet separating a 20% propylene/80% propane mixture at 60 psig and room temperature shows the uniformity of our coating process with average propylene permeance of 222 ± 14 GPU and a propylene selectivity of 46 ± 12

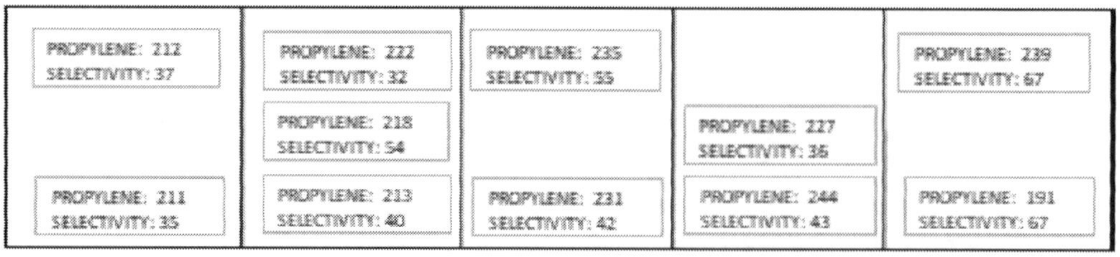

Figure 5. *Performance profile map for 42" wide membrane sheet*

This work will allow CMS to be producing its Optiperm™ membrane at significant quantities by 4Q 2018.

Pilot tests

Our pilot unit at DCRC was started up in late August, 2017 utilizing a paraffin-rich mixed C_3 feed stream to gauge the technical performance and economic viability of a full scale system. The pilot trial will provide real-world evidence of membrane stability in realistic conditions validating our laboratory testing.

Our field testing has delivered valuable insight on the key operating parameters that influence membrane performance and has also challenged the operating team to work through several challenging issues including startup and steady state operation of a very small unit operating in a refinery environment.

The key learnings:

- Field membrane performance is stable. Pre-field membrane qualification testing and post-field performance tests confirm steady membrane performance with permeance and selectivity matching the pre field testing values.
- When optimal operating conditions are reached (including high humidity), membrane performance matches laboratory data at similar conditions.
- To date six membrane configurations have been field-tested to identify the optimal combinations of polymers, membrane thickness and materials of construction to enhance membrane performance
- Pilot rig temperature control and relative humidity continues to challenge field operations and limits the repeatability of membrane field performance data
- The low designed feed rates and extreme atmospheric conditions are the main contributors to poor temperature and relative humidity control of the process feed to the membrane. While challenging at the pilot scale, temperature and humidity control at commercial scale are not expected to be issues.
- An updated humidification column design is currently underway, we are optimistic the redesigned column and improved heat tracing of the process piping will dampen the atmospheric effects allowing for steady temperature and humidity control resulting in consistent membrane performance.
- Feed pretreatment catalyst are performing as designed to remove sulfur and provide the desired feed quality to the membrane

Figure 6. *Pilot rig on-site at DCRC Refinery*

Full-Scale System

As developments with interested parties progress CMS has looked forward to design full-size membrane systems for use at a refinery or polymer production operation. Some key features of a full scale system include:

- Use of steam for temperature maintenance and feed humidification
- Recompression of recycle stream to system pressure
- Bundling of multiple membranes into one housing to reduce capital expense
- Modular construction of membrane systems to facilitate efficient field erection
- Flexible housing orientation to reduce systems footprint

The benefit of the membrane form factor is that a large amount of material can be processed without taking up an excessive amount of refinery space. An example system was developed for a high capacity C_3 splitter. The splitter generates 233,000 ton/year of paraffin rich bottoms with an average composition of 90% propane at 200 psig. A membrane system is used to recover olefin from this bottoms stream. The system recycles 58,000 ton/year of permeated material to the column, enriched in propylene to 54%. The system simultaneously generates 175,000 ton/year of HD5 propane. A proposed membrane bank contains 8 housings each holding 42 membrane cartridges. This design takes up a 28′ by 28′ area of refinery, or 784 ft^2. In the projected design the membranes are oriented horizontally and the housings are not stacked. Vertical orientation and stacking can both be implemented in order to reduce the footprint of the full scale system. A cross-sectional view of such a system is shown below with operators shown for scale:

Figure 7. *Illustration of an Optiperm™ membrane system to debottleneck a C_3 Splitter*

On-Purpose Propylene Production from Light Hydrocarbons with MFI-Zeolite/Metal-Oxide Composites

*Dr. Shinya Hodoshima, Azusa Motomiya, Shuhei Wakamatsu, Takenori Kanda, Ryuichi Kanai and Dr. Fuyuki Yagi**
Research & Development Center, Chiyoda Corporation, Yokohama, Japan

Introduction

Recently, propylene and ethylene demand is steadily increasing, especially, propylene, being a raw material in the petrochemical industry, has been increasingly significant because the global demand for propylene has increased at growth rate of 4-5 wt% annually [1]. Since the majority of propylene (ca. 60%) is currently produced as a byproduct by thermal steam-cracking of hydrocarbon feedstock such as naphtha, the thermal crackers can no longer meet the propylene demand because of the low weight ratio of propylene to ethylene (ca. 0.5) in its product. In addition, this technology requires a huge amount of thermal energy (temperature: 850-900°C). The recent shift to ethane-based feedstock for producing ethylene also could not affect the propylene supply, because this technology provides little yield of propylene. Any new method for producing propylene efficiently from light-naphtha, being widely available feedstock, should be thus developed to meet future demand for propylene and reduce energy-consumption in olefin production. Moreover, the conversion of various light hydrocarbons (e.g., shale oil, condensates recovered from shale gas and GTL-naphtha) into valuable light olefins is also significant task in terms of the efficient use of resources.

Catalytic cracking of naphtha-based hydrocarbons over solid-acid catalysts, e.g., ZSM-5, have been actively investigated [2,3] as an alternative method for on-purpose propylene production to conventional thermal steam-cracking. In particular, catalytic cracking using fixed-bed reactor has the following advantages compared with steam-cracking: (1) high weight-ratio of propylene to ethylene; (2) being energy-saving. Furthermore, the cracking process equipped with fixed-bed reactor is superior to fluidized catalytic cracking (FCC) processes in terms of operational simplicity in reaction unit. However, this type of catalytic cracking process has not been commercialized, because practical cracking catalyst with long lifetime, being applicable to fixed-bed reactor, has not been still developed.

From the above-mentioned viewpoints, we have focused on the development of highly active and stable catalyst for light-naphtha cracking in order to produce propylene efficiently under mild conditions at around 550°C. The proprietary zeolite-based catalysts for cracking, consisting of ZSM-5 zeolites containing Fe and Ga (Fe-Ga-Al-MFI) and metal oxides (e.g., aluminum oxide and silicon oxide) [4-7], were developed in the present study. This paper presents excellent performance of these proprietary catalysts for cracking.

Experimental procedure

Synthesis of MFI-type zeolites and characterization

MFI-type zeolites containing some heteroatom such as aluminum, gallium and iron (Al-MFI and Fe-Ga-Al-MFI) were hydrothermally synthesized [4-7]. Desired aqueous solution containing Si and Al source, Si, Al and Ga source and Si, Al, Ga and Fe source were prepared as mother gel with the following molar compositions: (1) Si/Al = 25; (2) Si/(Fe+Ga+Al) = 25, Fe/(Fe+Ga+Al) = 0.4, Ga/(Fe+Ga+Al) = 0.2, Al/(Fe+Ga+Al) = 0.4; (3) Si/(Fe+Ga+Al) = 200, Fe/(Fe+Ga+Al) = 0.4, Ga/(Fe+Ga+Al) = 0.2, Al/(Fe+Ga+Al) = 0.4. Each mother gel was charged into autoclave and then heated at 150°C for 24 h under adequate stirring conditions. Both the Al-MFI (Si/Al = 25) and the Fe-Ga-Al-MFI (Si/(Fe+Ga+Al) = 25) zeolites were characterized by XRD, XRF ^{27}Al-NMR, ^{71}Ga-NMR and UV-VIS/DRS measurements. The Fe-Ga-Al-MFI zeolite (Si/(Fe+Ga+Al) = 200) was characterized by both X-ray diffraction and X-ray fluorescence measurements.

Preparation of Fe-Ga-Al-MFI zeolite/metal-oxide composite catalysts

Composite catalysts, consisting of the Fe-Ga-Al-MFI zeolite (Si/(Fe+Ga+Al) = 200, Fe/(Fe+Ga+Al) = 0.4, Ga/(Fe+Ga+Al) = 0.2, Al/(Fe+Ga+Al) = 0.4) and metal oxides (aluminum oxide or silicon oxide), were prepared as follows. Firstly, the sodium-type Fe-Ga-Al-MFI zeolite (Si/(Fe+Ga+Al) = 200, Fe/(Fe+Ga+Al) = 0.4, Ga/(Fe+Ga+Al) = 0.2, Al/(Fe+Ga+Al) = 0.4) was hydrothermally synthesized by the same method described in the above section. Secondly, after both the zeolite and metal-oxide (γ-alumina (Cataloid AP-1, JGC Catalysts and Chemicals Ltd.) or fumed silica (AEROSIL 200, Evonik Industries AG)) were mixed at different weight ratios (Zeolite/Al_2O_3: 65%/35%, Zeolite/SiO_2: 65%/35% and 85%/15%), powdery mixture was molded into cylindrical shape with diameter of 1.0 mm by extrusion [4-7].

Catalytic cracking of *n*-hexane over zeolites and zeolite/metal-oxide composites

Catalytic cracking of *n*-hexane (*n*-C_6H_{14}) diluted with nitrogen over the synthesized zeolites (Al-MFI and Fe-Ga-Al-MFI, Si/T ratio in mother gel: 25, T = Al, Fe+Ga+Al) were performed for 5 h under atmospheric pressure by use of a flow-type reactor in fixed-bed mode. The normal hexane was employed for reaction test as a model reactant representing naphtha feedstock. The *n*-hexane diluted with nitrogen (molar ratio of N_2 to *n*-C_6H_{14}: 15/1) was supplied to the catalyst-bed at 550°C with the Liquid Hourly Space Velocity (LHSV) of *n*-hexane at 1.0 h^{-1} [4,5]. Catalytic cracking of *n*-hexane without dilution over Fe-Ga-Al-MFI/Al_2O_3 (Si/(Fe+Ga+Al) ratio in mother gel: 200, Mixed weight ratio in preparation step: 65%-Zeolite/35%-Al_2O_3) and Fe-Ga-Al-MFI/SiO_2 composites (Si/(Fe+Ga+Al) ratio in mother gel: 200, Mixed weight ratio in preparation step: 65%-Zeolite/35%-SiO_2) were performed for 100-350 h in the fixed-bed reactor under the following conditions: Temperature; 565°C, Pressure; 0.1 MPa, Charged amount of catalyst: 2.0 mL, Feedstock; *n*-hexane without dilution, LHSV of *n*-hexane; 5.0 h^{-1}. Both *n*-hexane conversion and product yields were evaluated in the same manner as cracking over the MFI-zeolite only. In addition, a long-term reaction with the Fe-Ga-Al-MFI/SiO_2 catalyst (Si/(Fe+Ga+Al) ratio in mother gel: 200, Mixed weight ratio in preparation step: 85%-Zeolite/15%-SiO_2) was carried out to evaluate its catalytic stability under the following conditions: 565-595°C; 0.1 MPa; LHSV of *n*-hexane at 4.5-7.0 h^{-1} [6].

Results and discussion

Fe-Ga-Al-MFI zeolite as main component for catalytic cracking

All of zeolites synthesized in the present study were confirmed by the XRD measurement to show the diffraction patterns corresponding to MFI-type framework. Most of aluminum and gallium species in the Al-MFI (Si/Al = 25) and the Fe-Ga-Al-MFI (Si/(Fe+Ga+Al) = 25) zeolites were confirmed to be incorporated in the zeolite framework by ^{27}Al-NMR and ^{71}Ga-NMR measurements (**Fig. 1**). In addition, most of iron species in the Fe-Ga-Al-MFI zeolite (Si/(Fe+Ga+Al) = 25) were also found to be incorporated in the zeolite framework by UV-VIS/DRS measurement (**Fig. 2**).

Comparison of catalytic performance between protonated Al-MFI zeolite (H-Al-MFI) and protonated Fe-Ga-Al-MFI zeolite (H-Fe-Ga-Al-MFI) with the almost same acid density (Si/T ratio: 25, T=Al, Fe+Ga+Al) was carried out under the same conditions.

Both **Fig. 3** and **Table 1** show product selectivity and catalytic activities in cracking of *n*-hexane diluted with nitrogen at 550°C over these zeolites synthesized in the present study, respectively. As shown in **Fig. 3**, the product distribution was obviously changed by introducing heteroatom into the H-Al-MFI zeolite as a reference sample. Though the conventional H-Al-MFI exhibited comparatively high selectivity to light olefins, it also showed high selectivity to aromatics. Iron atoms and gallium atoms in the zeolite have been recognized to reduce acid strength of zeolites [8] and accelerate dehydrogenation of alkanes [9], respectively, the combination of Fe and Ga atoms was considered to cause a drastic change of acidity. As a result, the H-Fe-Ga-Al-MFI-zeolite with optimized composition (Al/(Fe+Ga+Al)=0.3, Ga/(Fe+Ga+Al)=0.3, Fe/(Fe+Ga+Al)=0.4) showed the excellent catalytic performance for light-hydrocarbon cracking compared to the conventional Al-MFI(ZSM-5).

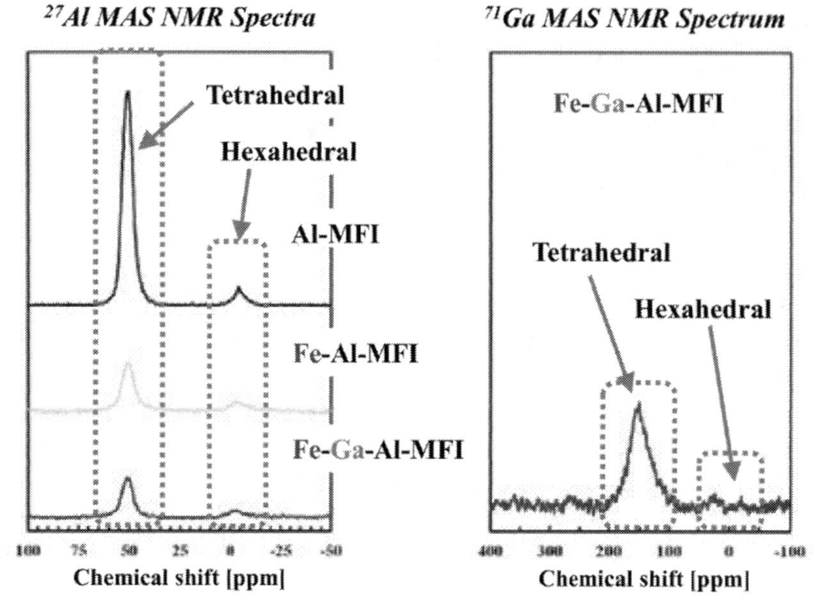

Fig. 1 Results of ^{27}Al-NMR and ^{71}Ga-NMR measurements over various MFI zeolites

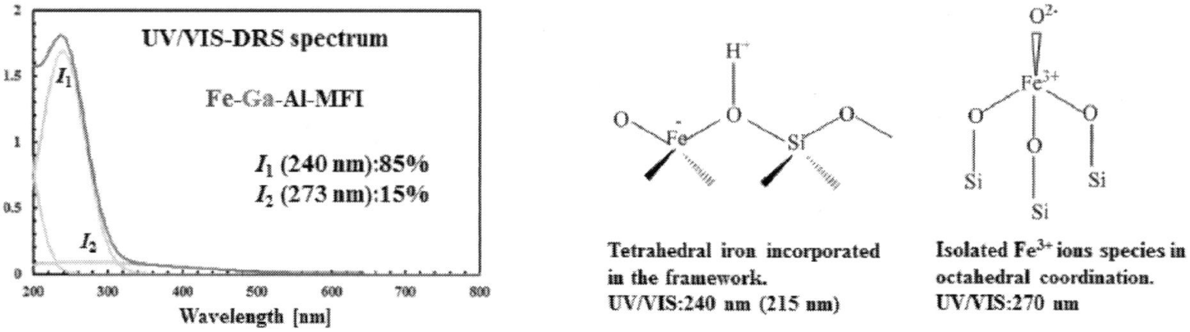

Fig. 2 Result of UV-VIS/DRS measurement over H-Fe-Ga-Al-MFI zeolite

Fe-Ga-Al-MFI/metal-oxide composites for catalytic light-hydrocarbon cracking

Zeolite catalysts are generally formed into shapes suitable for industrial application by using metal-oxide binder (e.g., aluminum oxide) to improve mechanical

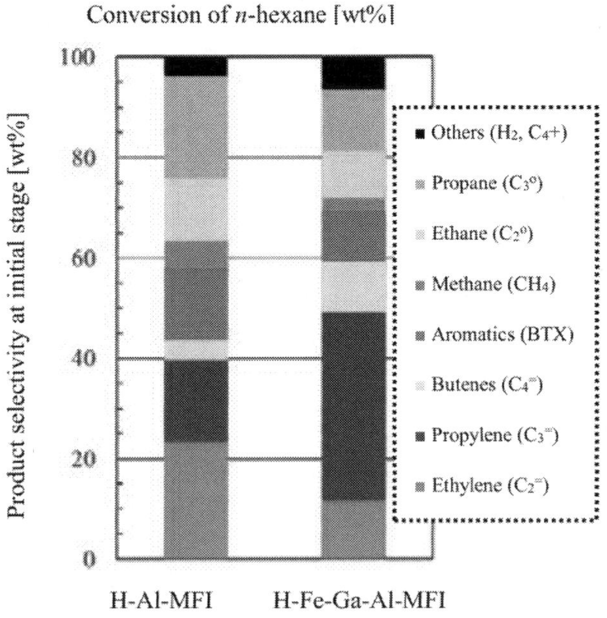

Catalyst(zeolite species);
H-Al-MFI(Si/Al=20.0), H-Fe-Ga-Al-MFI (Si/(Fe+Ga+Al)=19.4, Al/(Fe+Ga+Al)=0.3, Ga/(Fe+Ga+Al)=0.3, Fe/(Fe+Ga+Al)=0.4)
Reaction conditions; Temp.: 550°C, Press.: 0.1 MPa, Cat. vol.: 1.0 mL, Feedstock: n-C_6 diluted with N_2, Molar ratio of n-C_6/N_2: 1/15, LHSV of n-C_6: 1.0 h^{-1}

Fig. 3 Initial selectivity in *n*-hexane cracking over H-Al-MFI and H-Fe-Ga-Al-MFI zeolites

Table 1. Initial activities for n-C_6 cracking over H-Al-MFI and H-Fe-Ga-Al-MFI zeolites

Zeolite species	Acid density* [mol/mol]	Conversion [wt%]	Yield [wt%]	
			$C_3^=$	Aromatics
Al-MFI	20.0	99.9	16.3	14.2
FeGaAl-MFI	19.4	71.8	27.0	7.1

*Si/T ratio of zeolite species (T=Al, Fe+Ga+Al)
Temp.: 550°C, Press.: 0.1 MPa, Feedstock: n-C_6 diluted with N_2, Mixed molar ratio of n-C_6/N_2:15/1, LHSV of n-C_6: 1.0 h^{-1}

strength and reduce pressure drop across the catalyst-bed. In the present study, cylindrical composites with diameter of 1.0 mm, consisting of the Fe-Ga-Al-MFI zeolite (Si/(Fe+Ga+Al) = 25, Fe/(Fe+Ga+Al) = 0.4, Ga/(Fe+Ga+Al) = 0.3, Al/(Fe+Ga+Al) = 0.3) and aluminum oxide, were prepared by extrusion. **Table 2** compares catalytic activities for *n*-hexane cracking with respect to the Fe-Ga-Al-MFI and the Fe-Ga-Al-MFI/Al_2O_3 composites (Zeolite/Al_2O_3 mixed ratio: 65 wt%/35 wt%). Both composite catalysts exhibited higher *n*-hexane conversions and apparent yields of light olefins than those in the case of the zeolite alone. Weight of light olefins generated per one hour with zeolite of one gram, i.e. space time yield of light olefins, was also enhanced by use of the composite catalyst, suggesting that zeolite species combined with aluminum-oxide binder were well dispersed [5].

Excellent properties of Fe-Ga-Al-MFI/metal-oxide composites under practical conditions

For the purpose of applying the present catalyst to light-hydrocarbon cracking under practical conditions, cracking of *n*-hexane without dilution was examined at 565°C. Compared to the cases using *n*-hexane feedstock diluted with nitrogen, catalytic cracking of *n*-hexane without dilution is advantageous from the following viewpoints: (1) Saving thermal energy required for heating diluent; (2) No irreversible catalyst deactivation derived from the dealumination. However, hydrocarbon cracking in the absence of diluent requires catalyst with high resistance to coking. **Table 2** shows catalytic activities for *n*-hexane cracking in the presence and absence of diluent over Fe-Ga-Al-MFI/Al_2O_3 composites (Si/(Fe+Ga+Al) molar ratio: 25 and 200, Zeolite/Al_2O_3 mixed ratio: 65 wt%/35 wt%) and Fe-Ga-Al-MFI/SiO_2 composite

Table.2 Activities for catalytic *n*-hexane cracking over Fe-Ga-Al-MFI zeolite and Fe-Ga-Al-MFI-based composites

Catalyst (Si/T ratio)	Feedstock (Mixed molar ratio)	LHSV / Temp. $[h^{-1}]$ / [°C]	Conv. [wt%]	Yield [wt%]			Amount of carbon deposit [wt%]	Olefin productivity $[g\text{-}(C_2^{=}+C_3^{=})/g\text{-zeolite h}]$
				$C_2^{=}$	$C_3^{=}$	BTX		
Fe-Ga-Al-MFI(25) only	n-C_6 / N_2 (1 / 15)	1.0 / 550	71.8	8.3	27.0	7.1	-	0.53
Fe-Ga-Al-MFI(25)/Al_2O_3	n-C_6 / N_2 (1 / 15)	1.0 / 550	80.9	10.0	28.3	10.8	-	0.62
Fe-Ga-Al-MFI(25)/Al_2O_3	n-C_6 only	5.0 / 565	81.4	5.4	10.6	19.4	-	0.26
Fe-Ga-Al-MFI(200)/Al_2O_3	n-C_6 only	5.0 / 565	67.3	8.2	18.1	7.3	24.6 (100 h)	1.4
Fe-Ga-Al-MFI(200)/SiO_2	n-C_6 only	5.0 / 565	64.2	7.4	15.3	4.3	4.8 (340 h)	1.5

Catalyst; Fe-Ga-Al-MFI/Al_2O_3 composites, Zeolite/Al_2O_3 ratio: 65 wt%/35 wt%, Fe-Ga-Al-MFI/SiO_2 composite, Zeolite/SiO_2 ratio: 65 wt%/35 wt%,
Reaction conditions; Pressure: 0.1 MPa, Charged volume of catalyst: 1.0 mL (n-C_6/N_2), 2.0 mL (n-C_6 only),
Reaction time: 5 h (n-C_6/N_2, n-C_6 only, Fe-Ga-Al-MFI(25)/Al_2O_3), 100 h (n-C_6 only, Fe-Ga-Al-MFI(200)/Al_2O_3), 350 h (n-C_6 only, Fe-Ga-Al-MFI(200)/SiO_2)

(Si/(Fe+Ga+Al) molar ratio: 200, Zeolite/SiO_2 mixed ratio: 65 wt%/35 wt%). In contrast to the cracking of *n*-hexane diluted with nitrogen, the composite catalyst with the Si/(Fe+Ga+Al) ratio of 25 resulted in rapid deactivation due to aromatics or coke formation in cracking of *n*-hexane without dilution at 550°C and the LHSV of *n*-hexane of 1.0 h^{-1}. By applying the zeolite catalyst with higher Si/(Fe+Ga+Al) ratio of 200 (i.e., smaller amount of acid sites) at higher temperature (565°C) and higher LHSV (5.0 h^{-1}), i.e. shorter contact time, selectivity to light olefins was obviously enhanced, giving high propylene yield of 18.1 wt%. Here the space time yield of light olefins was ca. 1.4 $[g\text{-}(C_2^{=}+C_3^{=})/g\text{-zeolite h}]$, being ca. twice of the value (0.62) in nitrogen dilution. In cracking of *n*-hexane without dilution, aromatization proceeds much easier than cracking of *n*-hexane diluted with nitrogen or steam due to bimolecular reaction of *n*-hexane and/or re-adsorption of light olefins on acid sites, indicating that aromatics would be generated quite selectively by using zeolites with high acid densities. In other words, zeolites with low acid densities would give high selectivity to light olefins by suppressing aromatics formation.

Catalytic stability of the Fe-Ga-Al-MFI/Al_2O_3 composite with the Si/(Fe+Ga+Al) ratio of 200 were evaluated by performing *n*-hexane cracking without dilution at 565°C for long time. **Fig. 4(A)** shows time courses of yields of ethylene and propylene. Both ethylene yield of ca. 8.0 wt% and propylene yield of ca. 18 wt% obtained at initial stage were maintained for ca. 80 h in spite of severe conditions to cause coke formation easily. In order to improve catalyst lifetime further, the Fe-Ga-Al-MFI-zeolites combined with silicon oxide binder (SiO_2), being highly resistant to coke formation, were employed for catalytic cracking of *n*-hexane under the same conditions as the reaction with the Fe-Ga-Al-MFI/Al_2O_3 composite. As shown in **Table 2** and **Fig. 4(B)**, by using the Fe-Ga-Al-MFI/SiO_2 catalyst with the mixed ratio of 65 wt%/35 wt%, aromatics yield was diminished (4.3 wt%) and its lifetime was much longer (ca. 340 h) than that (ca. 100 h) of the Fe-Ga-Al-MFI/Al_2O_3 catalyst. Moreover, the Fe-Ga-Al-MFI/SiO_2 catalyst with the mixed ratio of 85 wt%/15 wt% performed longer its lifetime than 1,000 h in *n*-hexane cracking without dilution at 565-595°C with the LHSV of *n*-hexane at 4.5-7.0 h^{-1}, as shown in **Fig. 4(C)**. The long lifetime was applicable to catalytic hydrocarbon cracking in fixed-bed. These results indicate that it is necessary to suppress the formation of aromatics to prolong lifetime.

Since the difference of binder species resulted in remarkable improvement of catalyst durability, it is considered that a kind of interaction between zeolite and silicon oxide and/or coverage of strong acid sites on external surfaces by silicon oxide might give excellent resistance to coke formation. We also made sure that proprietary catalyst could be easily regenerated by the coke combustion with diluted air.

Simplified process flow diagram of proprietary process is shown in Scheme 1. First of all,

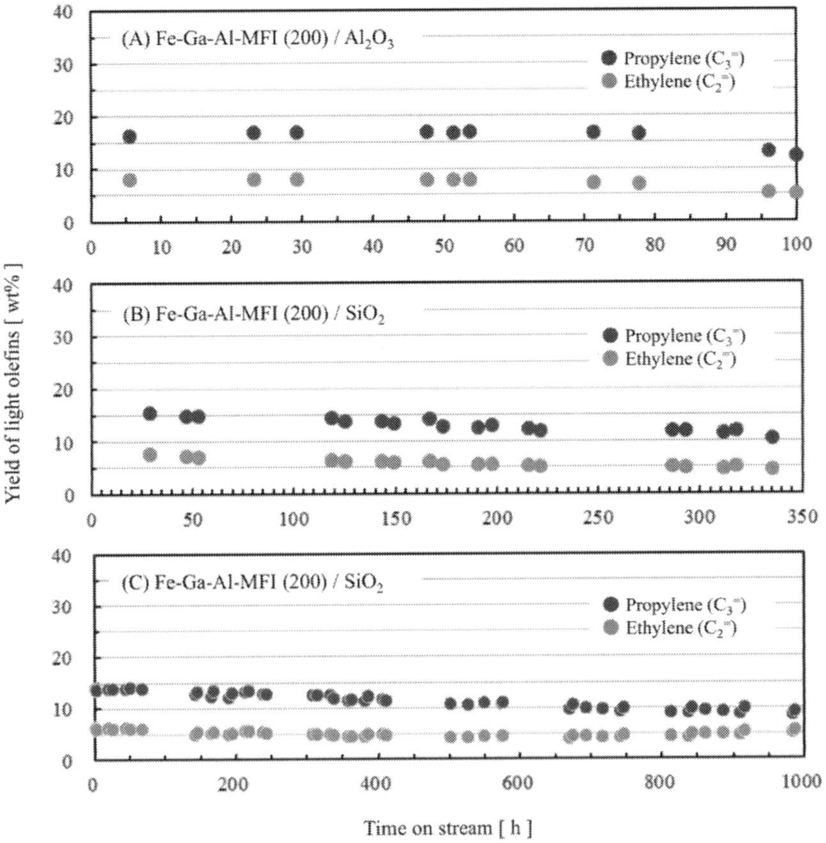

Fig. 4 Time courses of *n*-hexane cracking without dilution over H-Fe-Ga-Al-MFI/Al_2O_3 catalyst

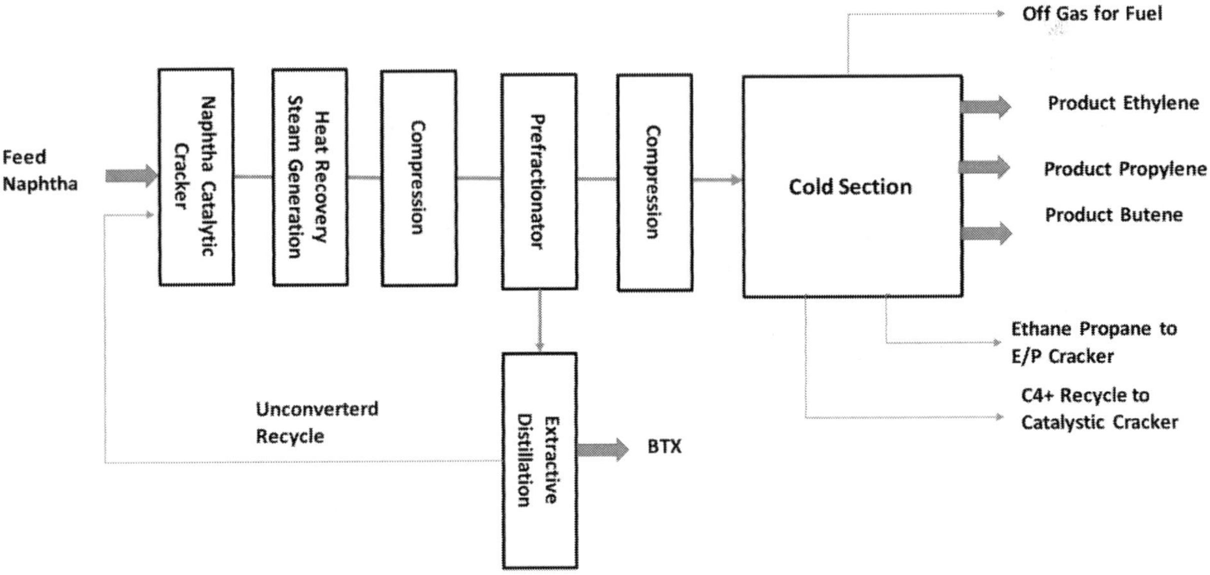

Scheme.1 Simplified process flow diagram of proprietary process

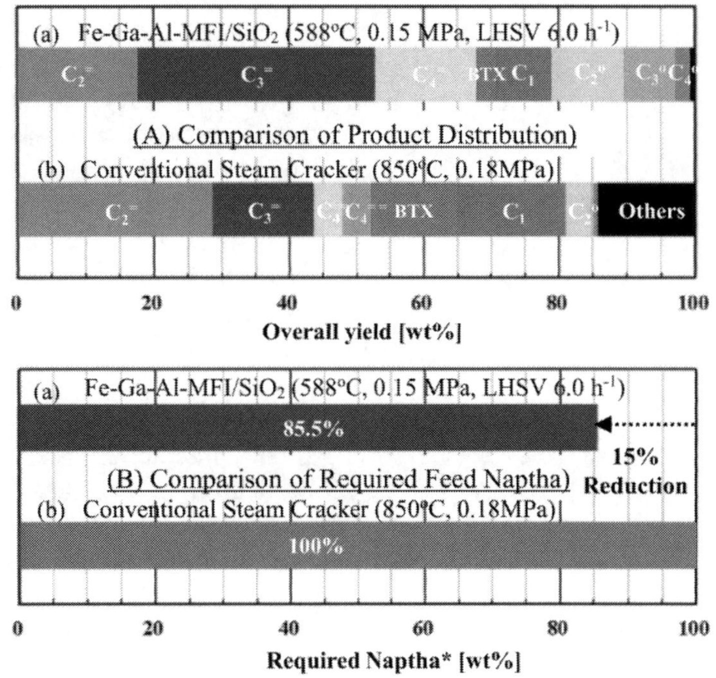

Fig. 5 Comparison of product distribution (A) and required feed naptha (B) between conventional naptha cracking and our catalytic cracking

raw naptha is fed to the reactor of naptha catalytic cracker, and subsequently, outlet products from the reactor are also fed though heat recovery steam generation. At this section high pressure steam is generated and utilized in this process. At the following pre-fractionator, BTX is separated as final product, and unconverted light hydrocarbons such as C5-C8 paraffins are recycled to catalytic cracker. Then, separated products are compressed, and fed to cold section. At the cold section, product olefins are divided into final products (ethylene, propylene and butene) by using 3 distribution columns. Off gas such as metane is utilized as fuel, and separated ethane and propane are fed to E/P (ethane/propane) cracker to obtain ethylene and propylene. Separated butane from the bottom of debutanizer is also recycled to catalytic cracker.

Fig. 5 (A) compares product distribution between conventional naptha cracker and proprietary catalytic cracker. As shown in Fig. 5 (A), propylene, ethylene, butane and BTX yields of catalytic cracker (a) are 35.1, 17.7, 15.1 and 3.0 wt%, respectively. As for Fig. 5 (A)-(a), olefin yields were estimated by the iteration calculation taking into consideration of the unconverted feed recycle. On the other hand, propylene, ethylene, butane, butadiene and BTX on conventional steam cracker (b) are 15.0, 28.6, 4.2, 4.4 and 13.0 wt%, respectively. Remarkable increase of propylene, and corresponding decrease of ethylene and BTX was observed on proprietary catalytic cracker. The amount of valuable olefins (ethylene + propylene + butane + butadiene + BTX) is 70.9%, and such olefins on steam cracker is 65.2%.

Fig. 5 (B) also compares required feed naptha between our catalytic cracking (a) and conventional naptha cracker (b). In these case, required naptha is calculated to obtain unit weight of valuable olefins (ethylene + propylene + butane + butadiene + BTX). [Required naptha (wt%) = Feed naptha(wt) / Beneficial olefins (ethylene + propylene + butane + butadiene + BTX) (wt) x 100 .]

As shown in Fig.5, required feed naptha can be decreased with 15%, which will directly contribute to the reduction of operation cost. Under current development status, this technology is bench scale, so we are trying to scale up to demonstration scale within 2-3 years, and we would like to commercialize it without delay of market demand.

Conclusions

The proprietary composite catalysts, consisting of H-Fe-Ga-Al-MFI-zeolite and metal-oxide such as silicon oxide, developed in the present study made it possible to produce propylene quite efficiently from light hydrocarbons (e.g., light-naphtha) at around 565°C and suppress energy-consumption in reaction unit. It was also confirmed that these catalysts performed long lifetimes that is applicable to cracking process in simplified fixed-bed reactor. From these experimental results of those catalysts it was suggested an efficient process for increasing propylene yield definitely by combination with existing cracking technologies could be realized.

Conceptual process flow diagram and reactor design were proposed. Simple fix-bed reactor is available, and catalyst regeneration is also applicable. In our process, production of valuable olefins such as ethylene propylene, butane and BTX can be increased considerably, which results in the decrease of feed naptha with around 15%. So this technology will be contribute to saving energy and operating cost of chemical and petroleum fields all over the world, especially in Asia, Midle East, Europe and US. This technology is now a bench test phase, so we are trying to move on to a demonstration test phase within 2-3 years, and we would like to commercialize it in the closely near future.

Acknowledgements

A part of the present study was financially supported by the Japan Oil, Gas and Metals National Corporation (JOGMEC, 2012-2013) and the New Energy and Industrial Technology Development Organization (NEDO, 2015-2018). We would like to express our deep gratitude here. The authors acknowledge Prof. Sachio Asaoka (The University of Kitakyushu) for guidance regarding zeolite synthesis.

References

[1] Forecast of global demand and supply for petrochemical products for the period 2008 to 2021 complied (http://www.meti.go.jp/english/press/2017/0628_003.html)

[2] H. Konno, T. Okumura, Y. Nakasaka, T. Tago, T. Masuda, J. Jpn. Petrol. Inst., **55**(4), 267 (2012).

[3] A. Corma, J. Mengual, P.J. Miguel, Appl. Catal. A; General, **421-422**, 121 (2012).

[4] S. Hodoshima, A. Motomiya, S. Wakamatsu, R. Kanai, F. Yagi, Res. Chem. Intermed., 41(12) (2015) 9615-9626.

[5] S. Hodoshima, A. Motomiya, S. Wakamatsu, R. Kanai, F. Yagi, Micropor. Mesopor. Mater., 233 (2016) 125-132.

[6] S. Hodoshima, A. Motomiya, S. Wakamatsu, R. Kanai, F. Yagi, Topics in Catal., submitted.

[7] S. Hodoshima, A. Motomiya, S. Wakamatsu, R. Kanai, F. Yagi, Chem. Eng. J., submitted.

[8] A. Chatterjee, T. Iwasaki, T. Ebina, A. Miyamoto, Micropor. Mesopor. Mater., **21**, 421 (1998).

[9] T. Inui, F. Okazumi, J. Catal., **90**, 366 (1984).

126f

IMPROOF: INTEGRATED MODEL GUIDED PROCESS OPTIMIZATION OF STEAM CRACKING FURNACES

Steffen H. Symoens
ir.
Laboratory for Chemical Technology, Ghent University, Ghent, Belgium

Marko R. Djokic
Dr. ir.
Laboratory for Chemical Technology, Ghent University, Ghent, Belgium

Jia Zhang
Ms.
Laboratory for Chemical Technology, Ghent University, Ghent, Belgium

Georgios Bellos
Dr.
DOW Benelux B.V., Terneuzen, The Netherlands

Dietlinde Jakobi
Dr.
Schmidt + Clemens GmbH + Co. KG, Lindlar, Germany

Jörg Weigandt
Dr.
Schmidt + Clemens GmbH + Co. KG, Lindlar, Germany

Sebastian Klein
Mr.
Schmidt + Clemens GmbH + Co. KG, Lindlar, Germany

Frédérique Battin-Leclerc
Prof. Dr.
Centre National de la Recherche Scientifique, Nancy, France

Geraldine Heynderickx
Prof. Dr.
Laboratory for Chemical Technology, Ghent University, Ghent, Belgium

Joost Van Thielen
Mr.

CRESS B.V., Breskens, The Netherlands

Benedicte Cuenot
Dr.
European Centre for Research and Advanced Training in Scientific Computation, Toulouse, France

Tiziano Faravelli
Prof. Dr.
Politecnico di Milano, Milan, Italy

Gilles Theis
Mr.
John Zink International Luxembourg SARL, Luxembourg

Philippe Lenain
Mr.
Ayming France, Lyon, France

Andrés E. Muñoz G.
Dr.
AVGI, Ghent, Belgium

John Olver
Dr.
Emisshield Inc., Blacksburg, Virginia, USA

Kevin M. Van Geem
Prof. Dr. ir.
Laboratory for Chemical Technology, Ghent University, Ghent, Belgium

Prepared for Presentation at the 2018 Spring National Meeting
Orlando, Florida, April 24, 2018

IMPROOF: INTEGRATED MODEL GUIDED PROCESS OPTIMIZATION OF STEAM CRACKING FURNACES

Steffen H. Symoens
ir.
Laboratory for Chemical Technology, Ghent University, Ghent, Belgium

Marko R. Djokic
Dr. ir.
Laboratory for Chemical Technology, Ghent University, Ghent, Belgium

Jia Zhang
Ms.
Laboratory for Chemical Technology, Ghent University, Ghent, Belgium

Georgios Bellos
Dr.
DOW Benelux B.V., Terneuzen, The Netherlands

Dietlinde Jakobi
Dr.
Schmidt + Clemens GmbH +Co. KG, Lindlar, Germany

Jörg Weigandt
Dr.
Schmidt + Clemens GmbH +Co. KG, Lindlar, Germany

Sebastian Klein
Mr.
Schmidt + Clemens GmbH +Co. KG, Lindlar, Germany

Frédérique Battin-Leclerc
Prof. Dr.
Centre National de la Recherche Scientifique, Nancy, France

Geraldine Heynderickx
Prof. Dr.
Laboratory for Chemical Technology, Ghent University, Ghent, Belgium

Joost Van Thielen
Mr.
CRESS B.V., Breskens, The Netherlands

Benedicte Cuenot
Dr.
European Centre for Research and Advanced Training in Scientific Computation, Toulouse, France

Tiziano Faravelli
Prof. Dr.
Politecnico di Milano, Milan, Italy

Gilles Theis
Mr.
John Zink International Luxembourg SARL, Luxembourg

Philippe Lenain
Mr.
Ayming France, Lyon, France

Andrés E. Muñoz G.
Dr.
AVGI, Ghent, Belgium

John Olver
Dr.
Emisshield Inc., Blacksburg, Virginia, USA

Kevin M. Van Geem
Prof. Dr. ir.
Laboratory for Chemical Technology, Ghent University, Ghent, Belgium

Abstract: The IMPROOF project will demonstrate several of the latest technological innovations in the field of fouling minimization and energy efficiency improvement at pilot and industrial scale. These include the use of new advanced high temperature alloys, in combination with novel 3D reactor technologies and coated with high emissivity coatings. In this work, the coking resistance of two different industrial reactor materials, a chromia forming Centralloy® ET 45 Micro and an alumina forming Centralloy® HT E, was compared during steam cracking of propane in a pilot plant furnace. In addition, the Centralloy® HT E reactor material was combined with the 3D SCOPE® reactor technology in order to bench mark it to bare tubes. Four out of five coking cycles were performed at the SOR industrial conditions, while during the 4th coking cycle the reactors were exposed to high-temperature EOR industrial conditions in which tube metal temperatures (TMTs) of 1050 °C were reached. While the ET 45 Micro reactor was shown to deteriorate after the 4th coking cycles, alumina forming HT E reactors proved stable with an even improved coking resistance. Applying 3D SCOPE® technology reduced the temperature gradients from the reactor wall towards the center, resulting in decreased coke formation due to lower TMTs compared to bare reactors. After exposing all the reactors to EOR conditions, Centralloy®

HTE SCOPE® had an outstanding coking performance, producing nearly 3 times lower amounts of radiant coke compared to the bare ET 45 Micro and HT E reactors. To further improve the energy efficiency of olefin furnaces a high emissivity coating will be applied to the furnace walls in future experiments. In addition to the application of a high emissivity coating applied on the external surface of the radiant coils.

Introduction

Steam cracking is considered a mature technology and the most dominant process for olefin production [1]. Nevertheless, a lot of research is still conducted towards steam cracking optimization. The growing demand of light olefins ensures maximization of the output of existing plants and novel technological developments towards reducing the energy input per ton of ethene and propene. One of the most important contributions to the total energy consumption per ton is the unavoidable decoking step. Decoking of steam cracking reactors is necessary to burn off the deposited coke on the inside reactor wall of the long tubular reactors, suspended in the furnaces, as a result of unwanted side reactions [2]. That is primarily why steam is added to the feed, also improving the selectivity towards the desired light olefins, i.e. ethylene and propylene. Typically, the outlet temperature of the tubular reactor is very high, around 820 – 890 °C. In modern cracking furnaces, the residence time is reduced to a few hundred milliseconds in order to improve the yield of desired light olefins. After the cracking temperature has been reached, the gas is quickly quenched to stop the cracking reaction in a transfer line heat exchanger to recover as much energy as possible. Approximately 8 % of the chemical industry's total primary energy is attributed to steam cracking and is therefore the most energy-consuming process of the sector [3]. In addition, the process is responsible for massive amounts of CO_2 emissions [4]. Massive amounts of economical revenue can be achieved by improving the energy efficiency of the steam cracking plant as the needed energy counts for an important part of the light olefin production in typical ethane and naphtha based plants [4]. Ethylene furnaces have to decoked regularly, typically after run lengths in the order of 30-60 days, to remove the coke that was build up in the coil. During decoking, mixture of steam-air is sent through the reactor to combust and gasify the coke, while the production of the desired products is halted for roughly 48 hours [5]. The heat-transfer efficiency of the firebox decreases by 1-2 % from start-of-running to end-of-run cracking conditions, due to the extra thermal insulation of the deposited coke layer. This efficiency decrease results in an increase of the fuel consumption of the furnace by 5 % [6].

Several technological innovations exist to reduce coke formation including the application of high-performance alloys [2, 7-16], coatings [17-19], pretreatments [20, 21], feed additives [22, 23], and 3D reactor technologies [5, 24-26]. 3D reactor designs manufactured out cutting-edge reactor alloy, combined with optimized process control and a more uniform heat transfer could increase run lengths [27], with the result of an increased lifetime of the furnaces and a reduction in CO_2 emissions. Munoz et al. [2] indicated that better performing radiant coils can reduce catalytic coking, while it was mentioned by Schietekat et al. [25] that advanced 3D coil design such as the swirl flow reactor design, and presumably other 3D reactors such as the SCOPE® design [28] (by Schmidt + Clemens GmbH + CO. KG), can mitigate coke deposition. Optimal geometries reduce the radial gradients inside the reactor by increasing mixing, and thus lowering the wall temperatures where coke formation happens. Lower wall tube temperatures will extend the run length, by lowering the coking rate exponentially, resulting in lower energy consumption for decoking on an annual basis. An incredible useful tool to optimize these geometries is advanced 3D CFD reactor simulations. And most of the time it is the only instrument available to investigate the performance of these 3D profiles.

In this work a 3D reactor design, SCOPE® [28], is, for the first time, been experimentally evaluated under industrial relevant operating conditions. In total three different reactors were studied in a pilot plant steam cracking unit: a reference material Centralloy® ET 45 Micro, a reactor manufactured from Centralloy® HT E as an advanced coil material and a reactor that combines the advanced coil material with the SCOPE® 3D reactor design. Five cracking cycles were conducted of varying durations, to investigate both the catalytic and asymptotic coking behavior. All cycles except for the fourth resemble industrial propane cracking conditions, while the conditions for cycle 4 are chosen to resemble high temperature carburization conditions that are present in industry during end-of-run conditions. At end-of-run conditions, the material undergoes aging, after which it is observed in industry that coking behavior of the material behaves different.

In the future the energy efficiency of the pilot plant will be evaluated before and after the application of a high emissivity coating on the furnace refractories. In a subsequent step a high emissivity coating will be applied on the external surface of the radiant coils to further improve the overall energy efficiency.

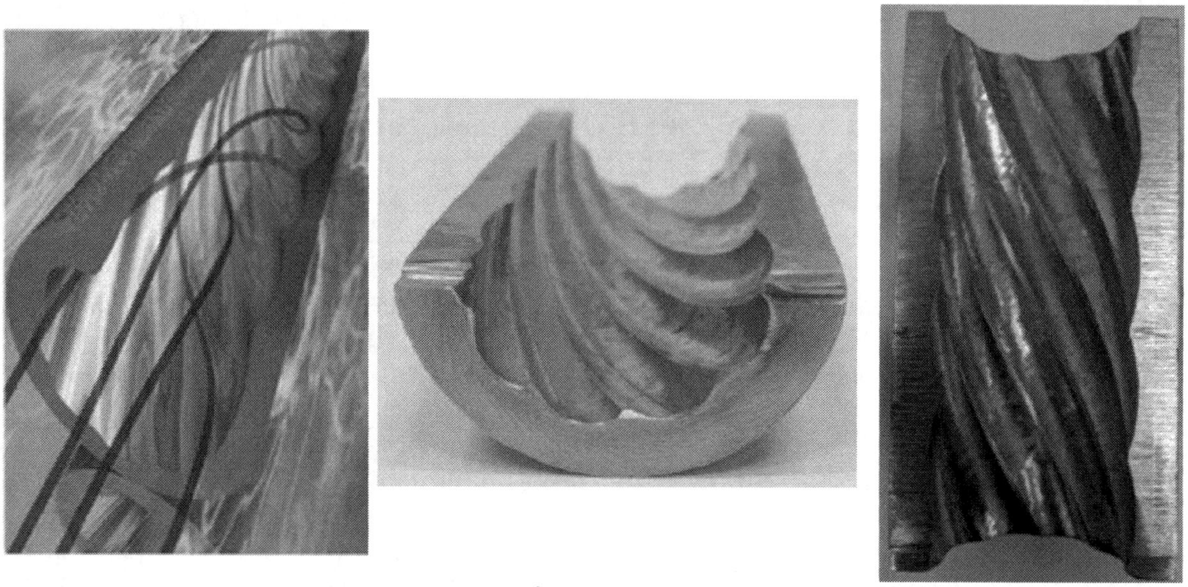

Figure 1: SCOPE® 3D reactor coil technology

Pilot plant

The experiments were performed on the pilot plant for steam cracking setup available at the Laboratory for Chemical Technology at the University of Ghent. A schematic representation of the setup is given in Figure 2. The setup is been described in numerous previous articles [17, 18, 22, 29-34], and therefore, only a short description of the setup is given.

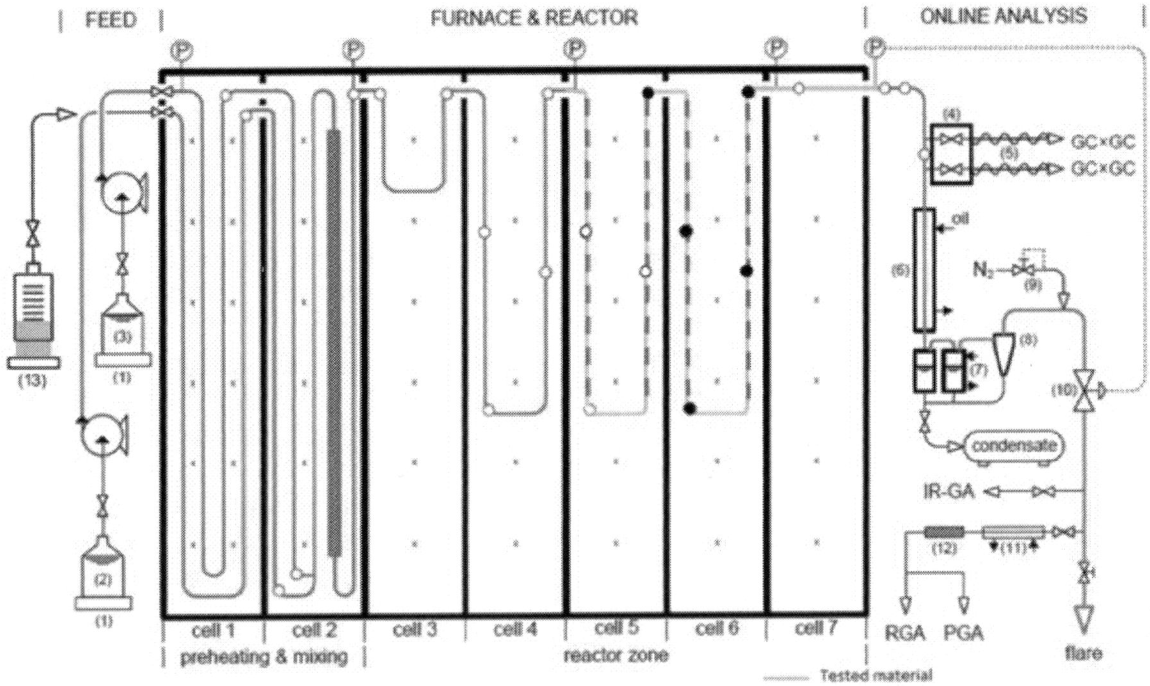

Figure 2: Schematic overview of the pilot plant setup during the experiments with the three different reactors: ET 45 Micro, HT E and HT E SCOPE® (green) (blue dashed line: SCOPE® profile), with the same inlet section in Incoloy 800HT (red). ((○): process gas temperature, (•): process gas temperature and, for the reactors with HT E material, reactor outer wall temperature, 1: electronic balance, 2: demineralized water reservoir, 3: liquid hydrocarbons reservoir, 4: heating sampling oven, 5: heated transfer lines, 6: oil cooled heated exchanger, 7: water cooled condenser, 8: cyclone, 9: thermal mass flow controller, 10: outlet pressure regulation valve, 11: water cooled heat exchanger, 12: dehydrator, 13: ISCO 500D syringe pump, (P): process gas measurements, GC×GC: two-dimensional gaschromatogry, RGA: refinery gas analyzer, PGA: permanent gas analyzer, IR-GA: infrared gas analyzer)

The setup consists out of three main parts: a feeding section, a reactor section and an analysis section.

The feeding section consists out of different vessels and bottles from which a wide variety of both gaseous and liquid hydrocarbon feedstocks can be fed towards the reactor.

The reactor section comprises the reactor and the furnace. The furnace has seven separate cells for which the temperature can be controlled individually, in order to impose a certain temperature profile. In the first two cells, the feedstock is preheated, evaporated (if necessary) and mixed with the diluent (steam), allowing homogeneous conditions at the beginning of the reactor. Steam cracking and coke deposition only occur in the last four cells, where the temperature is above 600 °C. Cell three, in this study, acts as a connection between the convection and the radiant section of the setup. The gas phase temperature is measured by means of twenty thermocouples and four pressure transducers that are mounted along the coil.

An additional five thermocouples were welded on the outer surface of the reactor, in order to measure the skin tube temperatures during operation. The three studied reactors were made out of two heat-resistant tube alloys which were welded together. Incoloy 800HT in the first two radiant cells and ET 45 Micro (reactor 1) or HT E (reactor 2 and 3) in cell 5 to 7. The first part of the reactor is almost 4.8 m long with an internal diameter of 9 mm and is placed in cells 3 and 4. During the experiments the temperature in these cells is kept below 644 °C to keep cracking and thus coking in these cells to a bare minimum. The second part is located in cell 5 and 6, bypasses cell 7 and is manufactured out of ET 45 micro or HT E (green reactor line in Figure 2). The 3D technology 'SCOPE®' is only applied in the vertical straight tubes and is marked with a blue dashed line. The internal diameter of the ET 45 Micro and HT E bare coils is 37.4 mm, while the reactor coil with the SCOPE® profile was specifically designed to have an equal cross sectional area as the bare reactors. This design resulted in an internal diameter ranging from 35 to 39.1 mm from top-to-top and valley-to-valley, respectively, and nine fins. Table 1 shows the alloy composition as it is reported by the manufacturer.

The analysis section allows on-line identification and quantification of the entire product stream, a wide boiling point mixture, containing all permanent gases, such as H_2, N_2, CO and CO_2 and hydrocarbons ranging from methane to polycyclic aromatic hydrocarbons.

Table 1: Typical composition of the Reactor Material INCOLOY® 800HT® and the Commercial Centralloy® ET 45 Micro and Centralloy® HT E [35-37]

Alloy	Composition [wt %]								
	C	Si	Mn	Cr	Fe	Ni	Al	Nb	Additions
INCOLOY® 800HT®	0.06-0.10	< 1.0	< 1.5	19.0-23.0	39.5 min	30.0-35.0	0.25-0.60	-	Ti, Al+Ti 0.85-1.20
ET 45 Micro	0.45	1.6	1.0	35	bal.	45	-	1.0	MAE, RE
HT E	0.45	-	-	30	bal.	45	4.0	0.5	MAE, RE

RE: Addition of reactive elements; MAE: Addition of micro-alloying elements

Experimental procedure

The experiments performed on the pilot plant setup consisted of several steps: pre-sulfiding (pre-S), cracking and decoking.

A total of 5 coking/decoking cycles were performed for each reactor. During the initial heating up phase, steam was fed to the reactor while the temperature of the reactor was gradually increased from 300 to 800 °C with a heating ramp of 50 °C/h. This slow heating rate is specifically chosen in order to reflect the typical heating-up process of the industrial process where thus a thin but very uniform protective oxide layer is formed. Prior to every coking step a pre-sulfiding step with DMDS was performed. When the temperature profile remains stable, sulfur is introduced to the reactor to passivate it. Before introducing the hydrocarbon feed to the reactor, the reactor is heated under steam atmosphere, until the desired temperature profile is reached. Then, the flow rate of propane and steam are set to the desired value.

Simultaneously with the introduction of the hydrocarbon feed, hexane diluted DMDS is added to the gas stream.

Three different coking durations were employed for the standard temperature profile, namely 2, 6 and 12 hours, while the end-of-run (EOR), high tube metal temperature (TMT), experiment was run for 1 hour and 40 minutes. After every coking cycle, the deposited coke was burned off with a steam-air mixture, while the CO and CO_2 formed during the combustion and gasification of the coke was continuously measured with an infrared analyzer. With the measured amounts of CO and CO_2 the total amount of burnt off coke can be calculated. To start a decoking cycle, the reactor is heated under an inert flow of helium, to preserve the coke layer. After reaching the desired temperatures (cell 3 to 7 at 800 °C), the infrared measurement is started and steam is added to the reactor together with a flow of helium. Once helium flow is stopped air is admitted to the reactor. When the measured CO_2 concentration drops below 1.0 vol%, the temperature in cells 5 to 7 are increased to 900 °C. When practically all the coke is gasified and/or combusted, the flow of steam is switched off and the decoking procedure continues with a pure air atmosphere.

All experimental runs were carried out at a constant coil outlet pressure, while maintaining a coil residence time of ~0.9 s and Reynolds numbers in the order of $4.2 - 5.4 \cdot 10^3$. During the experiment 'TMT max', the COT is set at 110 °C (bare tubes) / 160 °C (SCOPE®) higher than baseline COT. Although this is not industrially realistic, the experiment is included to mimic the high TMT values at the end-of-run conditions. Metallurgical aging of the material is hereby simulated as it is shown that at higher metal temperatures the Fe-Ni-Cr alloys carburize faster than their alumina forming alloys [16]. From the second industrial coking cycles onwards the material has been previously exposed to these harsh end-of-run conditions, i.e. the alloy experienced metallurgical aging. The order of the coking cycles is presented in Table 2. These different durations are employed to have a better distinction between catalytic and asymptotic coking rates, where the standard coking cycle of 2 hours gives an indication for the catalytic coking rate. Based on previous experiments the stage of catalytic coke formation was estimated to last approximately 3 hours [29].

Table 2: Order of the experiments with duration and COT/TMT set point

Coking cycle	Duration [h]	COT [°C]
1	6	Baseline
2	2	Baseline
3	6	Baseline
4*	1.67	Baseline + 110/ + 160**
5	12	Baseline

*Experiment TMT max, EOR condition

** HT E + SCOPE®

Tube metal temperatures

During the experiments the tube metal temperatures (TMT) were continuously measured via weld on thermocouples for the reactors involving HT E. For the ET 45 Micro reactor, no weld-on thermocouples were utilized. The TMTs were in this case measured via an infrared dual wavelength pyrometer from Sensortherm® Metis M322 and a manually inserted thermocouple type K that was touching the outer tube surface of the coil at various positions along the axial direction. These two measuring means are depicted in Figure 3.

The position that resulted in the highest TMT is reported and used for the control of the 'TMT max' experiment.

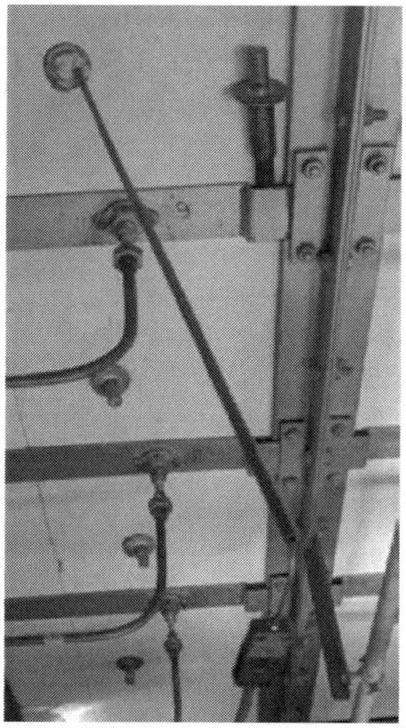

Figure 3: Picture of the setup for the temperature measurements of the outer surface of the ET45 micro reactor. (Left) Infrared temperature measurement; (right) manual thermocouple type K touching the reactor surface through an ignition hole

Results and discussion

Product yields

The goal of this study is to investigate the performance of the three different reactors on their coke mitigation abilities while maintaining the same cracking severity.

Figure 4 shows the yields of ethene and propene averaged over different analyses taken during a given experiment.

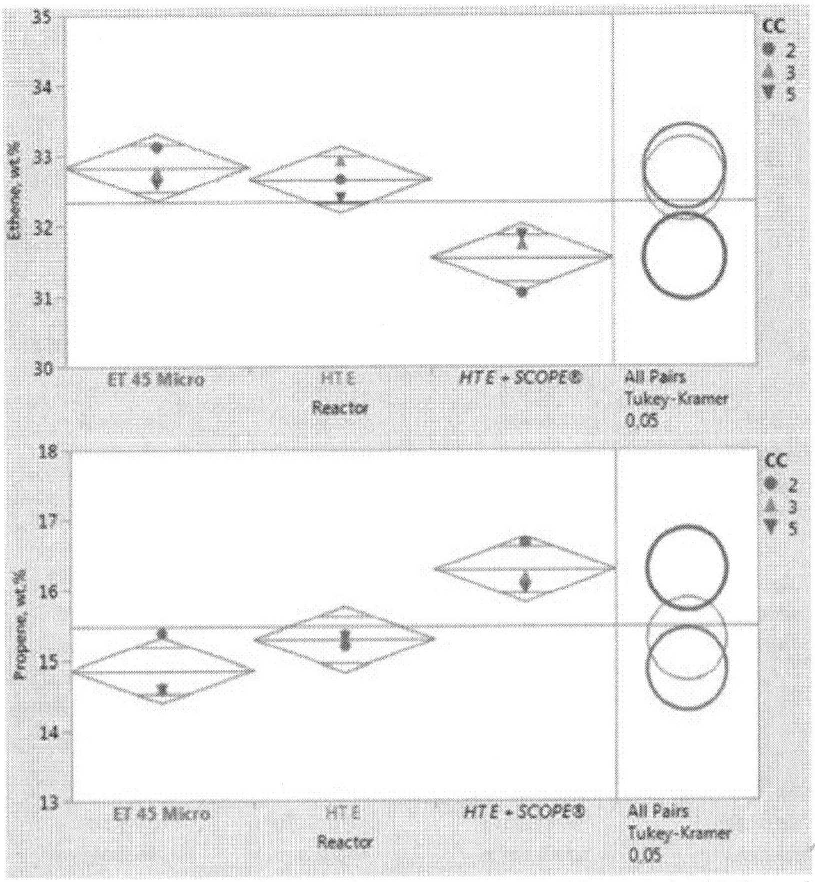

Figure 4: Ethene (top) and propene (bottom) variance per coking cycle (CC) and per reactor

The statistical analysis demonstrates that there is no significant change in ethene formation between ET 45 Micro and HT E, but HT E + SCOPE® obtains lower ethene yield. While the propene yield increased for HT E + SCOPE® in comparison with both ET 45 Micro and HT E. The former is caused by the reduced over-cracking in the near wall region due to lower tube metal temperatures.

Coke formation

A comparison of the coking rates is depicted in Figure 5.

As the coking rate of the second cycle is giving a representation of the catalytic coking rate, its values are obviously larger. This explanation can also be applied for the fourth coking cycle, although the influence of the applied extreme cracking severities will share a bigger contribution to the overall coking rate. This is due to the fact that a higher severity or cracking temperature results in higher coking rates [38, 39].

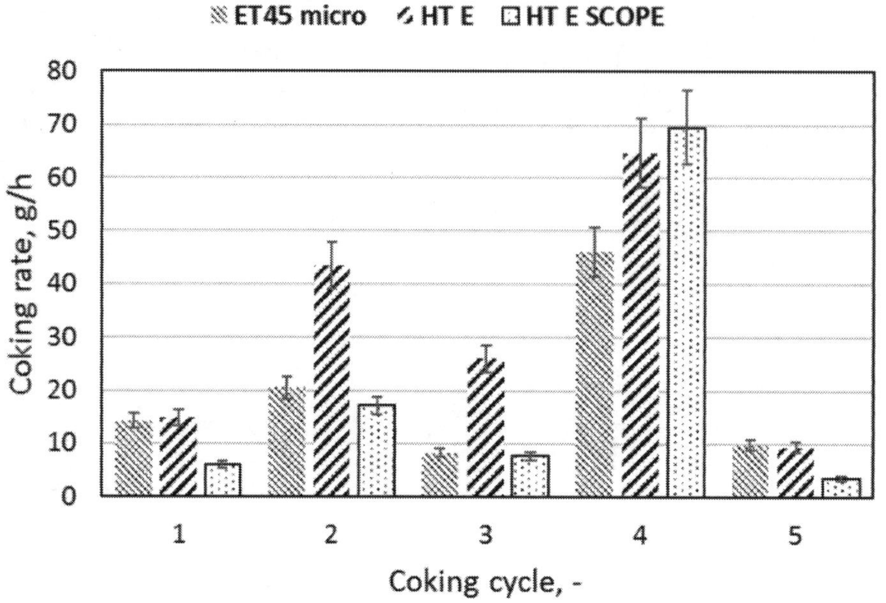

Figure 5: Coking rates as a function of the coking cycles for the three different reactors: ET 45 Micro, HT E and HT E + SCOPE®

Considering the fact that in an industrial steam cracker the coils experience maximal TMT values during the EOR of the first coking cycle, aging the alloy at the start of its lifetime. And adding the fact that the asymptotic coking rate has most influence on the overall run length [40], makes the coking rate from cycle five the most interesting value of the study. In fact, the difference between the third and fifth coking rate holds most information on how the actual reactor will perform in an actual steam cracker.

For ET 45 Micro the coking rate increased when comparing cycle 3 with cycle 5. An increase of 1.2 was observed. The increase is even larger when the observed coking rates are deconvoluted in a purely catalytic and an asymptotic contribution. This can be achieved by subtracting the amount of coke obtained during the 2 hour coking cycle (CC2) from the amount of coke measured during the third and fifth cycle and attributing the remaining coke to an asymptotic rate. The factor then becomes 3.4. The coking rate increase is attributed to the change in the oxide scale of the chromia forming alloy. ET 45 Micro is protected by a complex system of oxide layers with chromium-manganese spinel on top of a chromia layer and with a thin layer of SiO_2 at the oxide-metal interface [16], visualised in Figure 6. During cycle four, the metal was aged metallurgical to mimic extreme end-of-run conditions, for which the TMT reaches values above 1050 °C. During those harsh conditions the protective oxides tend to transform into chromium carbides. Furthermore, the growing coke layer on top of a chromia- or Cr-Mn-spinel scale incorporates significant amounts of Cr and Mn as nanoscale sized (worm-like shaped) oxides. The structure of these oxides suggests a kind of oxide intercalation into the coke. This phenomenon is corrosive since it extracts Cr and to a lesser extent Mn from the base material and by that, contributes to the depletion of oxide forming elements in the alloy.

Although the oxides are restored during decoking, the decoking is only partially reestablishing the original oxides. In other words, the oxide to carbide transformation initiates a non-reversible damage to the protecting scale resulting in a rough, ash-like oxide with an increased surface area that accelerates subsequent coking rates significantly [16]. This finding is in stark contrast with the coking rate decrease observed for the reactors manufactured out of HT E. After the high temperature EOR coking cycle, a factor 2.8 and 2.1 reduced coking rates were obtained for HT E and HT E + SCOPE®, respectively. When assigning a catalytic and asymptotic coking rate the coke reduction factors become even higher, i.e. 6.7 and 3.4 for HT E and HT E SCOPE® respectively. The protective scale formed after the heating up phase on the HT E reactors is expected to be composed out of Al_2O_3 and an oxide with a combination of Cr and Al, as it is shown in Figure 7 [16].

Due to a much better thermodynamic stability alumina scales are favorable as protection at extreme conditions typical for a steam cracking furnace provided that the kinetics ensures an unbroken and tight scale growth and that internal oxidation is suppressed. The latter is achieved by a suitable alloy composition [16].

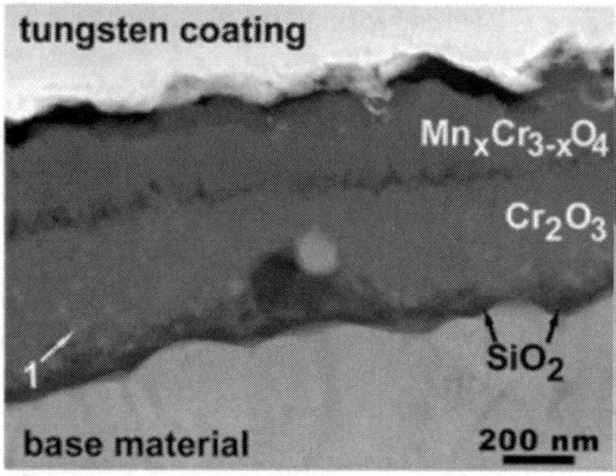

Figure 6: STEM-HAADF micrograph of the oxide scale on the ET 45 Micro material after the heating-up phase (1 : particles with Fe and Ni) [16]

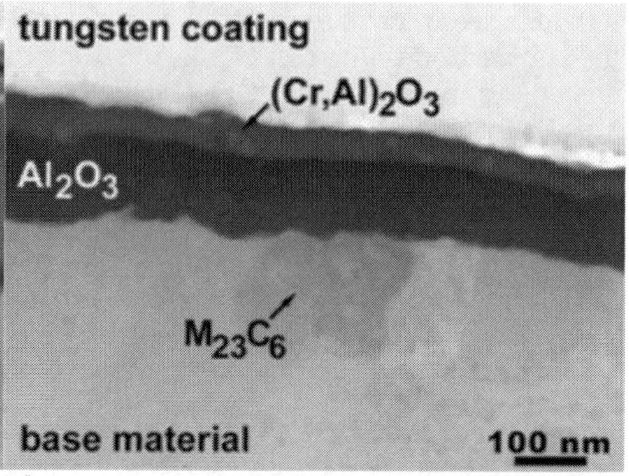

Figure 7: STEM-HAADF micrograph of the oxide that can be found on the HT E reactor surface after the heating-up phase [16]

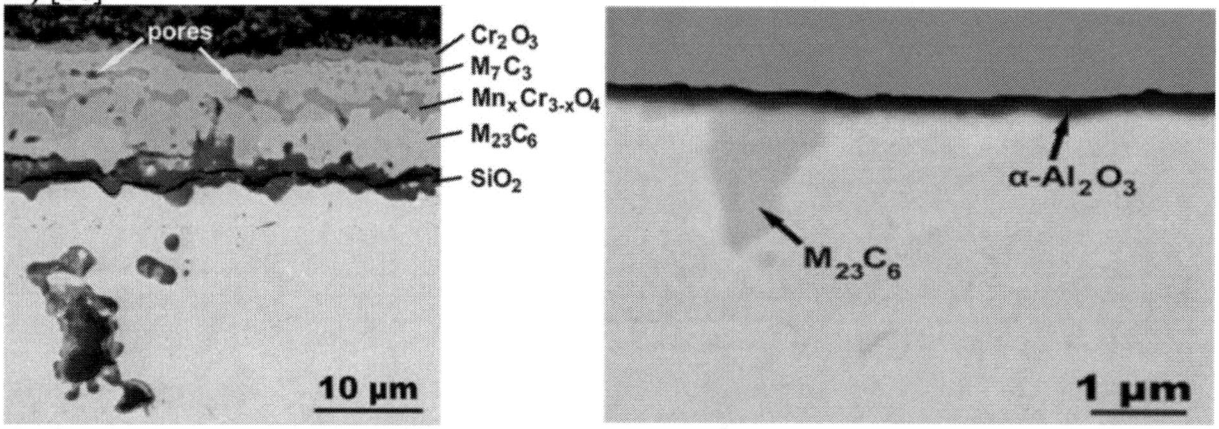

Figure 8: SEM-micro-graphs (BSE) of the reaction layers in cross section: Test conditions: 1 coking-decoking cycle with 1 h decoking; max temperature 1100 °C. (left) ET 45 Micro; (right) HT E [16]

In general, the HT E + SCOPE® reactor had an outstanding performance with regard to its coking tendency, except for cycle 4, the 'TMT max' experiment. It has to be mentioned that the COT was increased by another 50 °C in comparison with the bare reactors, to compensate for the lower tube wall temperatures.

The excellent coking performance during the standard tests can be immediately attributed to the (expected) lower radial gradients of the SCOPE® 3D profile. The profile enhances mixing, reduces the average cup temperature and thereby reducing the inner surface temperature.

Figure 9 shows the deposited coke as a function of time for the different reactors. Although HT E has a higher coking rate at the beginning, its rate drops after even one single higher temperature cycle of short duration faster and eventually drops below the coking rate of ET 45 Micro. The combination of HT E + SCOPE® shows a better coking performance than the other two tested reactors over the presented time regime.

Figure 9 further shows a better performance for the reactors manufactured out of HT E when the estimated coke deposition (based on the tests prior to the high TMT experiment) is compared with the one measured after the high temperature experiment. The coke deposition reduction is less pronounced for HT E + SCOPE®, because it is more difficult to push to higher temperatures. During the tests prior to the high temperature exposure, the α-alumina scale is not fully established, Cr,Al-oxides are still present (see also Figure 7). This mixed oxide scale changes when the material is exposed to higher temperatures, during which the α-Al_2O_3 is fully developed (Figure 8 right). The complete opposite trend can be observed for the reactor manufactured out of ET 45 Micro. The reactor performs worse in comparison with the estimated coke deposition. This can be attributed to the chromia scale that suffers deterioration during the high TMT experiment (Figure 6 versus Figure 8 left). The difference between the HT E reactor and HT E + SCOPE® can be solely explained by the reduction in tube wall temperature.

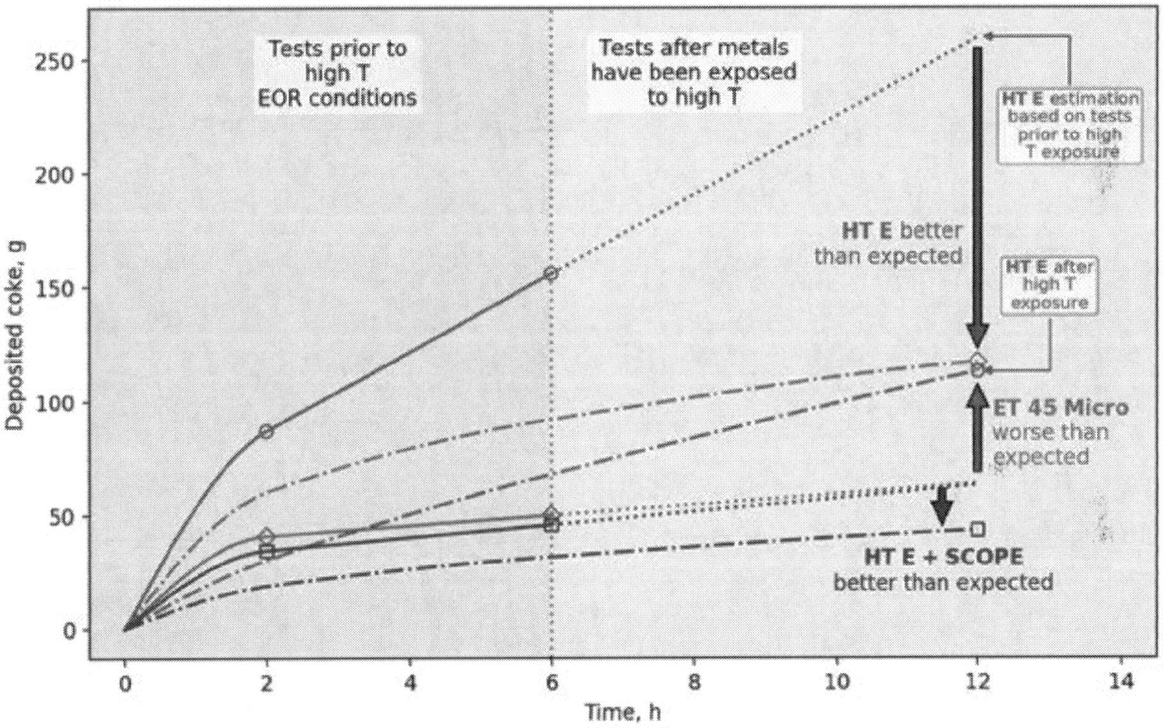

Figure 9: Deposited coke as a function of time for the three different reactors, ○ (red) HT E, ◇ (green) ET 45 Micro, and □ (blue) HT E + SCOPE®. Dotted lines are estimated values based on the tests prior to the exposure of the tubes to higher temperatures, while the dash-dotted line shows the estimated coke (based on cycle 5, the 12 h coking run) after the exposure at EOR high temperatures. Tests prior to high TMT conditions (blue haze), before 6 hours, tests after high TMT experiment (green haze), after 6 h.

Tube metal temperatures

When comparing the measured outer surface temperature during the standard experiments (Figure 10), it can be observed that the temperatures of the HT E + SCOPE® reactor are significantly lower than for the bare HT E reactor, a difference of at least 20 °C. This can be again explained by the lower radial temperature gradients of the reactor [5]. Which means that for the same inner gas phase temperature a lower metal tube temperature is required, which has, as a further consequence, that longer run lengths can be achieved because the metallurgical threshold temperature will be reached at a slower pace.

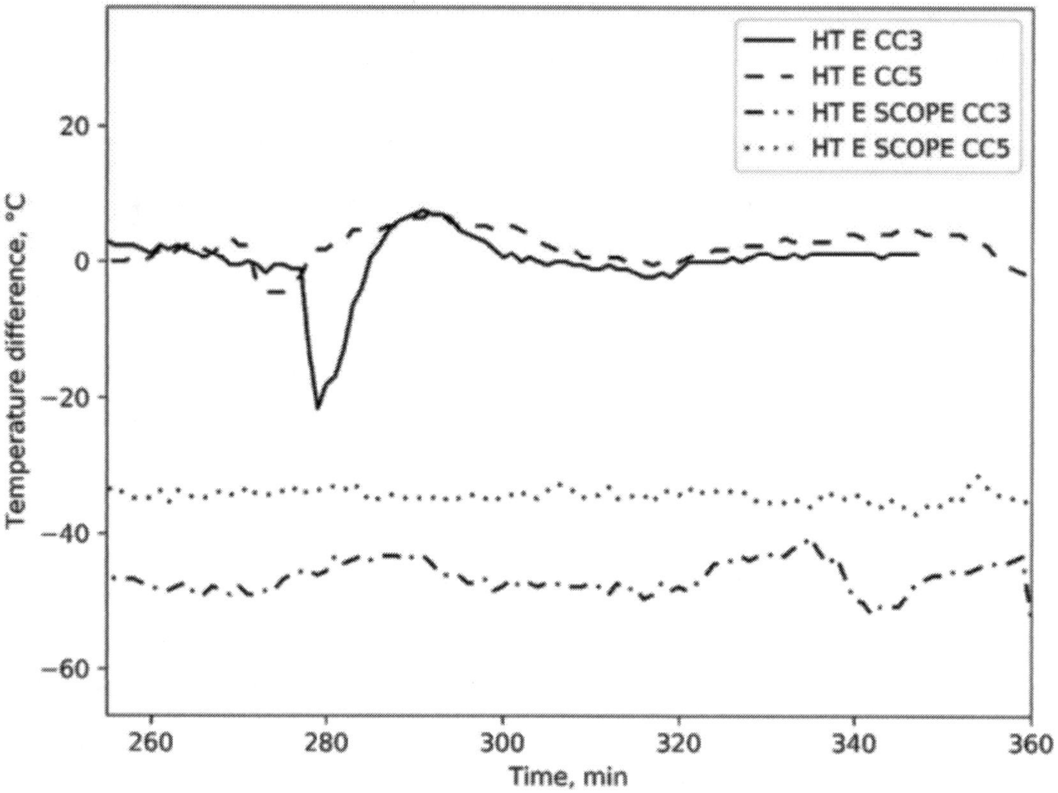

Figure 10: Relative difference of the measured tube metal temperatures during coking cycle 3 and cycle 5 for the reactors manufactured out of HT E, i.e. bare and SCOPE® (reference: average of HT E CC5)

Furnace fuel consumption

During every step of the experiments, the natural gas consumption was monitored. The natural gas serves as fuel for the furnace. It is mixed with air by the combustion controller and then distributed over the 90 wall burners. Its consumption is, therefore, a good measure for the fuel intake of the furnace and thus the energy efficiency. Because the operating conditions were kept identical for different reactors, the energy efficiency of the tested reactor can be derived. In general, roughly 10 % lower fuel consumptions were obtained for the HT E + SCOPE® reactor.

This, again, shows that due to a more uniform temperature distribution of the SCOPE® design, a better heat transport from the furnace towards the reacting gas phase is achieved. Less over- and under-cracking of the mixture can lead to the same yield output with a lower heat input.

Conclusions

Three different reactors were experimentally validated in a pilot plant setup under industrially relevant conditions for propane cracking. The first reactor is constructed from Centralloy® ET 45 Micro, a chromia forming alloy. The second and third reactor were manufactured from Centralloy® HT E, an alumina former. A 3D reactor technology, called SCOPE®, was installed in one of the Centralloy® HT E reactors. The cross sectional area of all three reactors were kept identical. Five coking cycles of different durations were imposed for all three reactors, to acquire knowledge in the catalytic and pyrolytic coking mitigation capabilities of the reactors.

ET 45 Micro shows a better catalytic coking behavior at the initial lifetime stages of the coil (before aging) than HT E. However, it appears that the asymptotic coking rate of HT E is lower than for ET 45 Micro. The transformation of the protective oxide scale to an agglomeration of carbides explains why ET 45 Micro worsens when the coil is subjected to high temperature end-of-run conditions, while the coking behavior of Centralloy® HT E even improves after aging. As the main contributor to the industrial run length is attributed to the asymptotic coking rate, it becomes clear that HT E will perform better in a typical industrial furnace.

The combination of the better performing material, Centralloy® HT E, with the 3D geometry, SCOPE®, outperforms the other two reactors regarding its coking behavior. A factor 2.7 lower coking rate after the first industrial metallurgical aging cycle can be achieved by combining alloy HT E with the 3D SCOPE® technology. The reason for lower coking rates can be attributed to the lower radial gradients inside the coil. These lower radial gradients result further in lower tube metal temperatures, meaning that the industrial plant can be operated longer because the temperature threshold of the material would be reached slower. The fuel consumption to fire the furnace was almost 10 % lower when comparing the bare HT E reactor with HT E + SCOPE®.

Lower coking rates, and lower tube metal temperatures at the start-of-run will both increase the run lengths. Add the fact that lower gas consumptions were measured for the combination, Centralloy® HT E + SCOPE®, makes a clear indication that the energy efficiency of an industrial steam cracker can be drastically increased by selecting the optimal reactor technologies.

Acknowledgment

The work leading to this intervention has received funding from the European Union H2020 (H2020-SPIRE-04-2016) under grant agreement n°723706 and from the COST Action CM1404 "Chemistry of smart energy carriers and technologies.

References

1. ExxonMobil, *The outlook for Energy: A view to 2040.* 2012.
2. Munoz Gandarillas, A.E., et al., *Coking Resistance of Specialized Coil Materials during Steam Cracking of Sulfur-Free Naphtha.* Industrial & Engineering Chemistry Research, 2014. **53**(35): p. 13644-13655.
3. Ren, T., M.K. Patel, and K. Blok, *Steam cracking and methane to olefins: Energy use, CO2 emissions and production costs.* Energy, 2008. **33**(5): p. 817-833.
4. Ren, T., M. Patel, and K. Blok, *Olefins from conventional and heavy feedstocks: Energy use in steam cracking and alternative processes.* Energy, 2006. **31**(4): p. 425-451.
5. Schietekat, C.M., et al., *Computational Fluid Dynamics-Based Design of Finned Steam Cracking Reactors.* Aiche Journal, 2014. **60**(2): p. 794-808.
6. Zimmermann, H. and R. Walzl, *Ethylene*, in *Ullmann.* 2012, Wiley: Weinheim.
7. Petrone, S.S.A., R.C. Mandyam, and A.G. Wysiekierski, *Surface alloyed high temperature alloys.* 2000, Google Patents.
8. Verdier, G. and F. Carpentier *Consider new materials for ethylene furnace applications. An innovative metallurgy solves maintance issues.* 2011. **90**, 61-62.
9. Industries, M. *Manaurite XM: Heat resistant alloys for hydrocarbon processing.* 2012 [cited 2015; Available from: http://www.manoir-industries.com/site/ref/scripts/FR_Brochures-Petrochem_122.html.
10. Industries, M. *Manaurite XTM: Heat resistant alloys for hydrocarbon processing.* 2012 [cited 2015; Available from: http://www.manoir-industries.com/site/en/ref/Manoir-Petrochem-Brochures_75-72.html.
11. Group, S.C. *Why choose S+C? Because you know what matters in the petrochemical industry.* 2016 [cited 2018 21/02/2018]; Available from: https://www.schmidt-clemens.com/fileadmin/sc-downloads/SC_Broschuere-Petrochemie_ENG.pdf.
12. Munoz Gandarillas, A.E., et al., *Influence of the Reactor Material Composition on Coke Formation during Ethane Steam Cracking.* Industrial & Engineering Chemistry Research, 2014. **53**(15): p. 6358-6371.
13. Crynes, L.L. and B.L. Crynes, *COKE FORMATION ON POLISHED AND UNPOLISHED INCOLOY-800 COUPONS DURING PYROLYSIS OF LIGHT-HYDROCARBONS.* Industrial & Engineering Chemistry Research, 1987. **26**(10): p. 2139-2144.
14. Jakobi, D., C.K. van Moesdijk, P., and A. von Richthofen, *Tailor - Made Materials for high temperature applications: New strategies for radiant coil material development*, in *NACE Corrosion 2009.* 2009.
15. Jakobi, D., P. Karduck, and V.R.A. Freiherr, *Nickel chromium alloy.* 2011, Google Patents.
16. Jakobi, D., P. Karduck, and V.R.A. Freiherr, *The High-Temperature Corrosion Resistance of Spun-Cast Materials for Steam-Cracker Furnaces - A Comparative Study of Alumina- and Chromia-Forming Alloys*, in *NACE Corrosion 2013.* 2013.
17. Schietekat, C.M., et al., *Catalytic Coating for Reduced Coke Formation in Steam Cracking Reactors.* Industrial & Engineering Chemistry Research, 2015. **54**(39): p. 9525-9535.

18. Olahova, N., et al., *CoatAlloy ™ Barrier Coating for Reduced Coke Formation in Steam Cracking Reactors: Experimental Validation and Simulations.* Industrial & Engineering Chemistry Research, 2017.
19. Petrone, S., et al., *CATALYZED-ASSISTED MANUFACTURE OF OLEFINS (CAMOL): YEAR-(4) UPDATE ON COMMERCIAL FURNACE INSTALLATIONS.* AIChE: Texas, 2010. **Vol. 88a**.
20. Sarris, S.A., et al., *Optimization of the in Situ Pretreatment of High Temperature Ni–Cr Alloys for Ethane Steam Cracking.* Industrial & Engineering Chemistry Research, 2017. **56**(6): p. 1424-1438.
21. Zhang, Z.B. and L.F. Albright, *Pretreatments of Coils to Minimize Coke Formation in Ethylene Furnaces.* Industrial & Engineering Chemistry Research, 2010. **49**(4): p. 1991-1994.
22. Wang, J., M.F. Reyniers, and G.B. Marin, *Influence of dimethyl disulfide on coke formation during steam cracking of hydrocarbons.* Industrial & Engineering Chemistry Research, 2007. **46**(12): p. 4134-4148.
23. Olahova, N., et al., *Coking tendency of 25Cr-35Ni alloys: influence of temperature, sulfur addition and cyclic aging.* Industrial & Engineering Chemistry Research, 2018.
24. DeSaegher, J.J., T. Detemmerman, and G.F. Froment, *Three dimensional simulation of high severity internally finned cracking coils for olefins production.* Revue De L Institut Francais Du Petrole, 1996. **51**(2): p. 245-260.
25. Schietekat, C.M., et al., *Swirl flow tube reactor technology: An experimental and computational fluid dynamics study.* Chemical Engineering Journal, 2014. **238**: p. 56-65.
26. Gyorffy, M., et al. *Mert performance and technology update.* in *AIChE Spring Meeting: Ethylene Producers Conference, Tampa Bay, USA.* 2009.
27. Zhang, Y., et al., *Impact of flue gas radiative properties and burner geometry in furnace simulations.* AIChE Journal, 2015. **61**(3): p. 936-954.
28. Wolpert, P., et al., *Finned tube for the thermal cracking of hydrocarbons, and process for producing a finned tube.* 2008, Google Patents.
29. Dhuyvetter, I., et al., *The influence of dimethyl disulfide on naphtha steam cracking.* Industrial & Engineering Chemistry Research, 2001. **40**(20): p. 4353-4362.
30. Pyl, S.P., et al., *Biomass to olefins: Cracking of renewable naphtha.* Chemical Engineering Journal, 2011. **176–177**: p. 178-187.
31. Van Geem, K.M., et al., *On-line analysis of complex hydrocarbon mixtures using comprehensive two-dimensional gas chromatography.* Journal of Chromatography A, 2010. **1217**(43): p. 6623-6633.
32. Dijkmans, T., et al., *Production of bio-ethene and propene: alternatives for bulk chemicals and polymers.* Green Chemistry, 2013. **15**(11): p. 3064-3076.
33. Olahova, N., et al., *Thermal Decomposition of Sulfur Compounds and their Role in Coke Formation during Steam Cracking of Heptane.* Chemical Engineering & Technology, 2016. **39**(11): p. 2096-2106.
34. Pyl, S.P., et al., *Wood-derived olefins by steam cracking of hydrodeoxygenated tall oils.* Bioresource Technology, 2012. **126**: p. 48-55.
35. Group, S.C., *Centralloy(r) ET 45 Micro Material Data Sheet.* 2009.
36. Group, S.C., *Centralloy(r) HT E Material Data Sheet.* 2014.

37. Xu, L.Y., et al., *Failure analysis of Incoloy 800HT pipe at high temperature.* Engineering Failure Analysis, 2013. **31**: p. 375-386.
38. Van Geem, K.M., et al., *Coke Formation in the Transfer Line Exchanger during Steam Cracking of Hydrocarbons.* Industrial & Engineering Chemistry Research, 2009. **48**(23): p. 10343-10358.
39. Froment, G.F., *COKE FORMATION IN THE THERMAL-CRACKING OF HYDROCARBONS.* Reviews in Chemical Engineering, 1990. **6**(4): p. 293-328.
40. Reyniers, G.C., et al., *COKE FORMATION IN THE THERMAL-CRACKING OF HYDROCARBONS .4. MODELING OF COKE FORMATION IN NAPHTHA CRACKING.* Industrial & Engineering Chemistry Research, 1994. **33**(11): p. 2584-2590.

ENTERPRISE MANUFACTURING INTELLIGENCE TO BRING DATA AND KNOWLEDGE TOGETHER FOR REAL TIME DECISION MAKING

Chaitanya Khare
Associate Analytical Manager
The Dow Chemical Company

Mary Beth Seasholtz
Associate Technology Director
The Dow Chemical Company

Richard Rolke
Global Improvement Leader
The Dow Chemical Company

Jim Petrusich
Vice President of Sales
Northwest Analytics

Prepared for Presentation at the 2018 Spring National Meeting

Orlando, FL, April 22-26, 2018

ENTERPRISE MANUFACTURING INTELLIGENCE TO BRING DATA AND KNOWLEDGE TOGETHER FOR REAL TIME DECISION MAKING

Chaitanya Khare
Associate Analytical Manager
The Dow Chemical Company
Mary Beth Seasholtz
Associate Technology Director
The Dow Chemical Company
Richard Rolke
Global Improvement Leader
The Dow Chemical Company
Jim Petrusich
Vice President of Sales
Northwest Analytics

Abstract:

Enterprise Manufacturing Intelligence (EMI) is a method which gives personnel quick access to critical parameters in real time and provides suggestions on pro-active actions to take to keep the plant running well, therefore avoiding plant crisis. Knowledge about a plant or unit operation is captured through EMI in two ways: (1) trending representative tags and calculations in real time and (2) providing suggestions for pro-active actions to take based on the deviation from normal conditions.

The future of the manufacturing industry is towards building intelligence into an organization through use of data -driven decisions. A significant part of this process is a onetime investment to capture current technical knowledge of the operation. This establishes a solid platform for ongoing continuous improvements. The value that can be gained by putting the data into context is through improved reliability of the process.

The future in the manufacturing industry will include a knowledge database which is connected to the EMI tool. When real time data are connected to Dow's immense collection of knowledge through different sources of literature, this should result in a system which pro-actively engages subject matter experts to prevent any unforeseen circumstances like a plant shutdown while at the same time evolves itself with the new learnings from each new situation that are encountered.

INTRODUCTION

The chemical industry has come a long way since its manufacturing inception in 1736 with Sulfuric acid production [1]. Since then manufacturing of chemicals has increased in capacity and efficiency in the conversion of feedstocks. Improvements are a constant in the chemical industry, with safety of the people, community and the environment remaining the number one priority.

While technology exists for process control and optimization, efficient troubleshooting of a problem/issue is still a challenge. Traditional troubleshooting projects start off with first understanding the plant status by pulling and assessing the plant data. This is followed by calculating pertinent metrics, and following them back in time to understand the series of events that caused the current problem.

Enterprise Manufacturing Intelligence (EMI) is a method which gives personnel quick access to critical parameters in real time and provides suggestions on pro-active actions to take to keep the plant running well, therefore avoiding plant crisis. Knowledge about a plant or unit operation is captured through EMI in two ways: (1) trending representative tags and calculations in real time and (2) providing suggestions for pro-active actions to take based on the deviation from normal conditions.

This report is a guiding document for how to bring data and knowledge together for real time decision making. This report will be particularly useful for the people who are charged with keeping plants running well; whether that be in a manufacturing plant, R&D, or the technology center.

WHAT IS EMI?

Enterprise manufacturing intelligence (EMI) is a term which applies to software used to bring a corporation's manufacturing-related data together for the purposes of reporting, analysis, visual summaries, and passing data between enterprise-level and plant-floor systems. As data are combined from multiple sources, it can be given a new structure or context that will help users find what they need regardless of where it came from. The primary goal is to turn large amounts of manufacturing data into real knowledge and drive business results based on that knowledge [2].

At Dow, EMI contains several pieces, (see Figure 1). It includes: (1) real-time status board about data and pertinent calculations, (2) easy drilldown to graphs, (3) suggested actions for keeping the plant in the "sweet spot", and (4) place to enter comments, ideas, collaboration which can be mined in the future. That is, it becomes an "Operational Guidance System" useful for collaboration between Plant, R&D and Technology Center personnel. The information readiness for decision making is key for helping any manufacturing plant by the technology center and R&D. The information readiness makes EMI a meeting place of all the real time decisions in a startup, normal, or shut down phase of an operating unit.

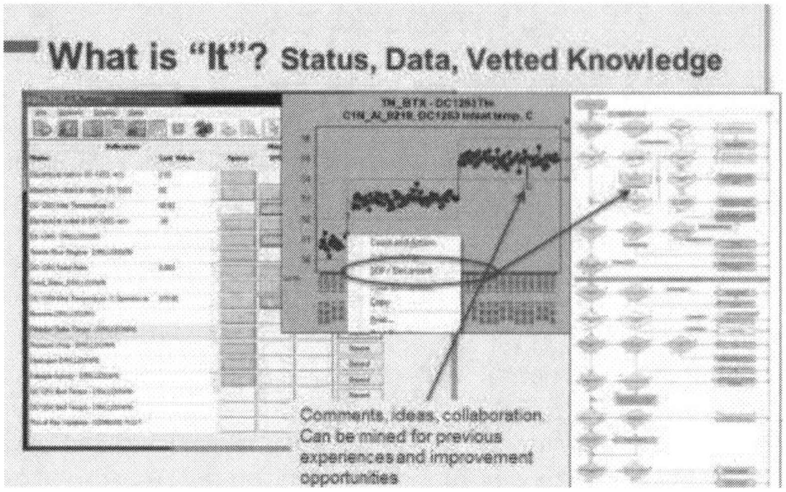

Figure 1: Components of an EMI Deployment

HOW TO IMPLEMENT EMI

To understand how to implement EMI it is good to first understand how plant support has been historically conducted.

Old paradigm:

Within the old paradigm, the plant would call out for help to either the tech center or R&D organization after they were already having problems. (see Figure 2). Then, a team would go through the rigor of first understanding the issue by speaking with the plant operations team. They would then start collecting the data for the plant. They would analyze this information while at the same time testing their different hypothesis. Eventually something would be learned and a report would be written about the knowledge gained. But, the learnings would likely not be sustainably incorporated into plant operations, so problems recurred.

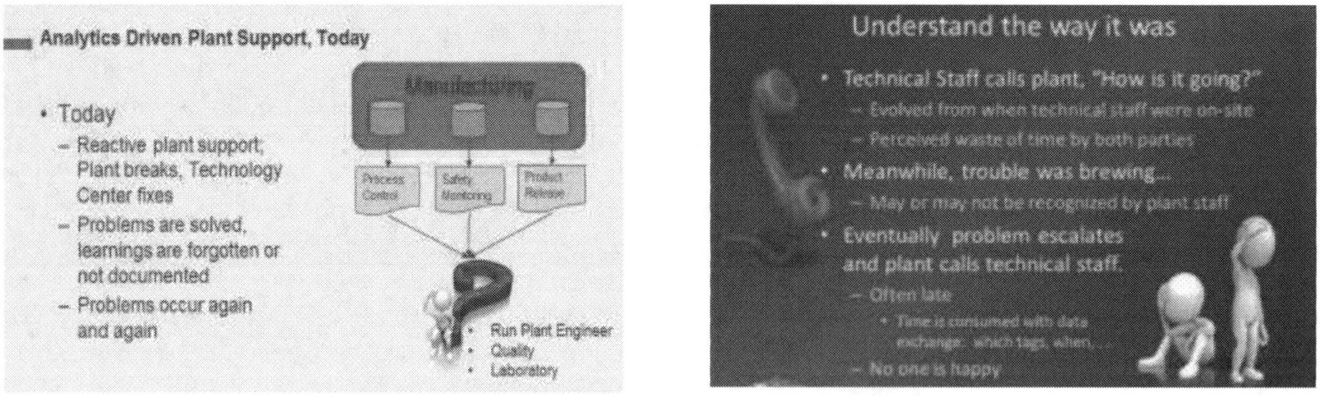

Figure 2: Old paradigm of plant support

New paradigm:

The new paradigm is to implement EMI to provide an early warning system which will prevent issues from occurring and sustainably incorporate learnings into plant operations. (see Figure 3). The target selection process for EMI implementation is based on a business's prioritized AU gaps. Choosing projects in this way leads to good support from the leadership and the technical experts.

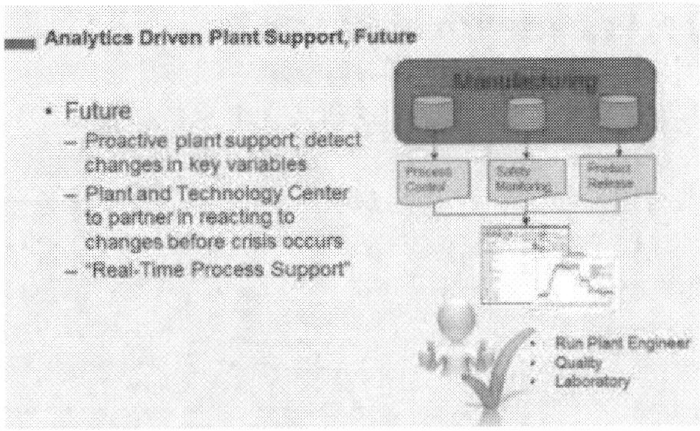

Figure 3: The new paradigm using EMI for plant support

Once a priority AU gap is identified, a team of subject matter expert's (SME's) from the technology center and/or business R&D and the operating plant is formed. This team is charged with building a "logic decision map". Rigorous brainstorming by this team is facilitated to have a healthy and positive conversation. During the initial step the goal is to draw a simplified process flow diagram and collect the most important parameters in the plant. As discussions proceed about the different parameters, historical and real time data are brought into the conversation. This helps to test different hypothesis around the parameter of choice. These data-driven discussions help expose and if needed, debunk, tribal myths.

Once the process is understood by the team, and there is an appreciation for the varied metrics highlighted, the next step is to discuss the AU-gap. For example, "What, when not monitored, would shut the plant down?" This question can trigger a lot of different ideas from the SME's. Below is an example of the interactions described:

- Assume that the overall production goal in a plant is the conversion across a series of reactors. The target is to keep it above 90%.
- Next, ask the question, "What, when not monitored, would shut the plant down?" Discussion on the above question reveals that factors other than failing conversion can bring the plant down. That is, the overall conversion is not the most critical parameter as it can be tuned to go above 90% if some activity is lost in the reactors/catalysts.
- So what would be more relevant than the overall conversion? It could be quality of the feedstock, it could be impurities, or something like water or oxygen that may be a poison for the system.
- Where do we get the answers to the question "What shuts the plant down"? We get these answers quite often from a previous root cause analysis from a previous situation that led to a plant shutdown. These answers could be known from the catalyst provider or from the technology MET documents or R&D work.

Now the team should be able to build out the logic decision map as they now understand the process and what causes the AU Gaps. That is, the team needs to identify the most important parameters to monitor, what ranges those metrics need to be kept within, as well as what actions should be taken to keep the key metrics in range. When thinking through appropriate actions, the team must resolve how to balance competing phenomena. All this information is put into a tool which shows real-time data as well as suggested actions to take if metric(s) are out of range. The data should be arranged so that at first glance only the pertinent metrics are shown, but with a few clicks, other data which are helpful for understanding a situation more completely can be found.

What makes the new paradigm better than the old?:

The new paradigm of creating an EMI system helps the team focus on analysis of data and determination of pertinent metrics without a plant crisis looming over head. In the old process, the troubleshooting would start after the trouble started, which was too late. With regards to obtaining and interacting with the data in order to proactively prevent a crisis from developing, the EMI approach allows for more time to be spent analyzing the data than in the past (see Figure 4).

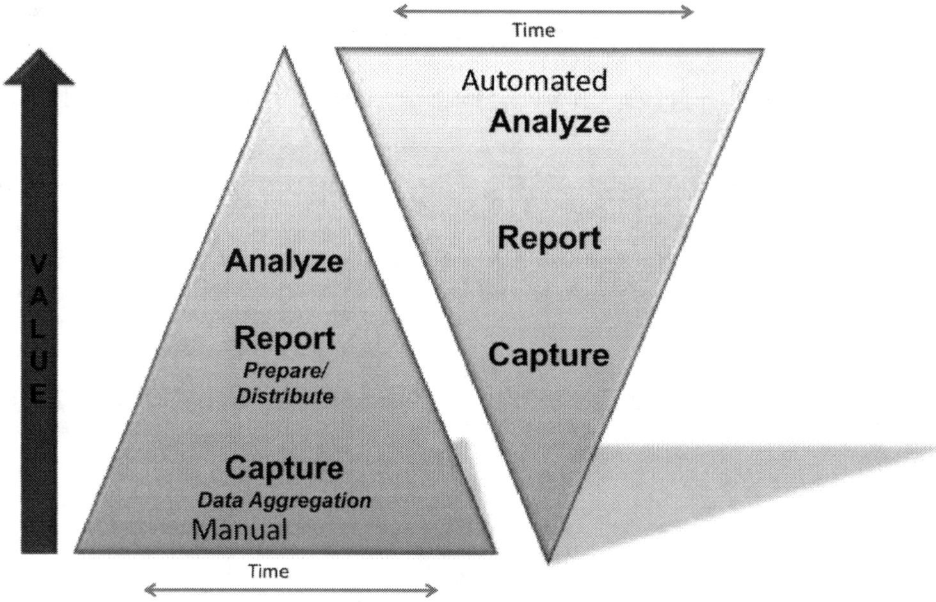

Figure 4: Difference between the old and the new paradigm in terms of time spent vs creating value

One can apply EMI for making decisions at different levels within an organization (See Figure 5). In terms of implementing EMI at the transactional level, one should be aware that at Dow a number of systems are already in place, and this is likely not an opportunity-rich environment.

The most valuable EMI implementations to date have been at the tactical level of operation. The tactical level is where plant support takes place and the different business functions work together to help the plant troubleshoot or avoid a crisis. At the tactical level decisions are made taking into consideration the details within the technology and the current operating condition of the facility. Having a logic decision map at this level of operation delivers the most promising results and facilitates collaboration between the support and operations functions. That is, EMI becomes the meeting place for R&D, technology center and the plant to collaborate to proactively prevent a problem.

The operational level EMI implementation shows the summary of the entire plant operation encompassing multiple technologies. Such an implementation is very useful during morning production meetings as it provides an efficient summary of the different plant areas. Time is spent talking about how to proactively prevent problems, not fussing over what data to look at. This helps to improve the efficiency of the plant staff meetings with data readily available for decision making. Another kind of operational level implementation is to show the status of same technology (ex. reactors) across a fleet of plants. Then decisions can be made about which locations operate more or less efficiently and learnings can be leveraged across the fleet.

The strategic level of EMI implementation is immensely valuable for making large scale capital decision across a business or an organization. The information presented can provide guidance in making decisions at the business level. Making such information available in real time is useful to the business leadership driving the decisions.

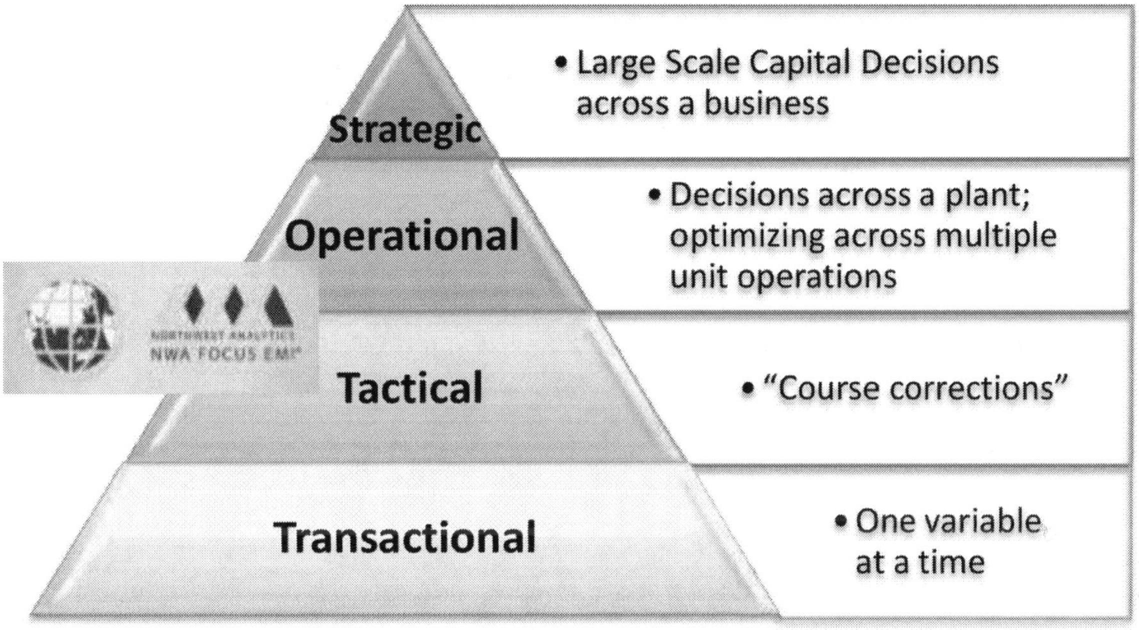

Figure 5: Decision matrix with different tools at different stages

Managing the change in culture:

An important question generally asked is: <u>"How does a typical implementation proceed?"</u> (See Figure 6). Building the logic decision map is a major step in implementing EMI, however it is only half the journey. The second half of the journey is more complex than debating the technical details. This part of the journey is the cultural transformation from reactive troubleshooting to proactive data-driven decision making.

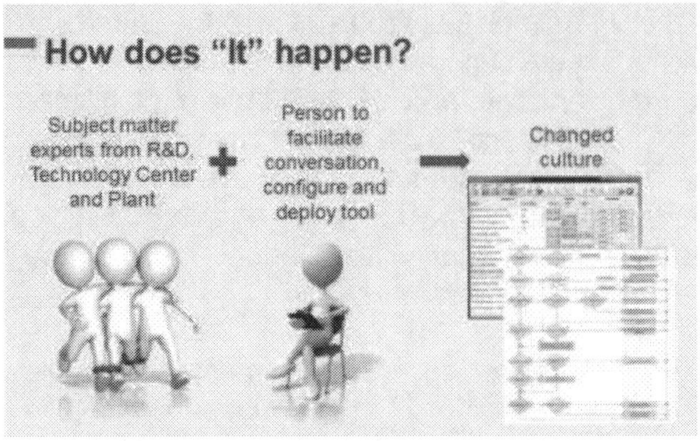

Figure 6: Typical method of EMI implementation

Users of EMI come from the Technology Center, Plant, Lab, Quality, and R&D. Implementing EMI helps each user from the various functions connected with the production facility. The production or improvement engineer in the plant using EMI gets the information about his/her plant, perhaps in the first 5 minutes of their day. This lets them head into meetings with knowledge about the plant status; therefore they can answer questions that may arise. In addition, knowing the current status of the plant helps ensure the plant prioritizes the day according to the situations that have been observed in EMI.

Users from the Technology Center and R&D are looking to understand what is ongoing at all the plants within a particular technology. With EMI in place across a business for that technology, a technology SME can have accurate information about all plants, even if he/she is located remotely from the manufacturing facilities. This information helps prioritize efforts in the technology center and R&D.

The Lab and Quality EMI implementations put in perspective all the online/offline measurements in the plant which maintains an accurate focus on helping understand the results from the process to maintain the expected product quality.

The first draft of the EMI deployment is often all red (See Figure 7 – Start). That is, many of the parameters are initially operating outside of the recommended ranges. As the users start using EMI every day, they monitor the newly established key parameters. They also learn what appropriate changes/improvements are needed to keep these key parameters in acceptable ranges and the system becomes green (See Figure 7– Now). This transformation from the start of the implementation to end of the implementation is the culture change taking place from reactive to proactive. But, it is not an easy step. This change takes time, energy, and a lot of commitment to doing the right thing from the plant, technology center and R&D. However, at this point in the implementation the data are starting to work for us, instead of us working to acquire and assemble the data. The critical parameter control charts will start informing users of any deviations from the normal and the decision map will guide users in making the appropriate decision in a proactive manner.

When EMI is implemented in plant operations, the above mentioned users benefit immensely. For example: In a morning production meeting, usually there is the main day staff which is required. The day staff in a large plant is around 30 people. Typically, to go through all the details in a big facility the morning production meetings could span around one hour. Discussions take place across several issues ongoing in the plant. However, we might be spending time from folks who are not really necessary to be present in those discussions. With the use of EMI, the morning production meetings can be reduced to a mere 15 minutes because the data are organized and readily available. EMI helps in maintaining focus on safety and the current issues that are alarming for the plant. After the larger team sees the issues, it is possible to take the issues offline to discuss in a small SME group. This approach can save a lot of time for several of the day staff in an operating facility. In simple terms, time saved would be 45 minutes for 30 people every day.

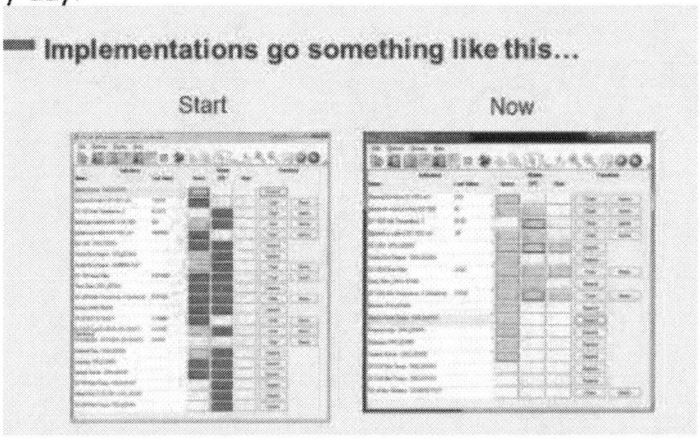

Figure 7: Natural progression of an EMI implementation from start to the current state

FUTURE OF EMI

The future of the manufacturing industry is towards building intelligence into an organization through use of data -driven decisions. A significant part of this process is a onetime investment to capture current technical knowledge of the operation. This establishes a solid platform for ongoing continuous improvements. The value that can be gained by putting the data into context is through improved reliability of the process (See Figure 8).

The future in the manufacturing industry will include a knowledge database which is connected to the EMI tool. When real time data are connected to Dow's immense collection of knowledge through different sources of literature, this should result in a system which pro-actively engages subject matter experts to prevent any unforeseen circumstances like a plant shutdown while at the same time evolves itself with the new learnings from each new situation that are encountered (see Figure 9).

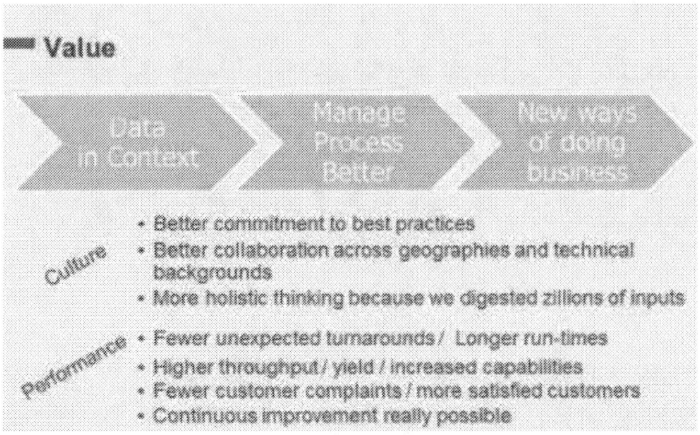

Figure 8: Return on investment from using our data through the intended culture and performance

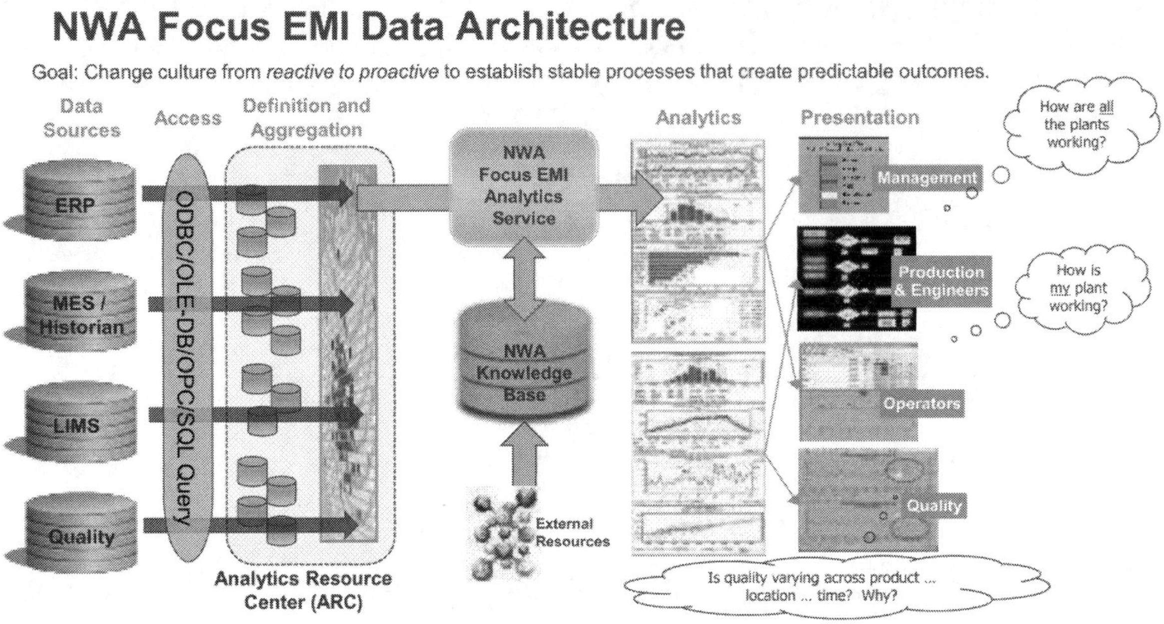

Figure 9: Architecture flow chart for generating the knowledgebase through the use of data in NWA Focus EMI

ACKNOWLEDGMENTS

We would like to acknowledge numerous discussions during the germination of EMI with Lloyd Colegrove, and Howard Clark. We would also like to acknowledge the Hydrocarbons business in helping us test this new concept as a pilot and staying on with us during the initial phase.

REFERENCES

1. First industrial revolution, https://en.wikipedia.org/wiki/Industrial_Revolution
2. EMI, https://en.wikipedia.org/wiki/Enterprise_manufacturing_intelligence
3. L. Colegrove, M. Seasholtz, C. Khare (2016), "Getting started on the Journey", Chemical Engineering Progress, 112(3), 41-45.
4. C. Khare, S. Chin (2017), "Potential for data analytics opportunities in SMART chemical industry", Chemistry Today, March/April, vol. 35(2).

AIChE Paper Number: 169B

Big Data Analytics for Improved Overall Asset Effectiveness of Ethylene plants

Dr. Pratap Nair
President & CEO
Ingenero

Prepared for Presentation at the 2018 Spring National Meeting
Orlando, Florida, April 22-26, 2018

Big Data Analytics for improved Overall Asset Effectiveness of Ethylene Plants

Starting up and sustaining production levels consistently from an ethylene plant does not always match expectations. Maintaining and managing furnace operations, the Cracked Gas compressors, Transfer Line Exchangers, chiller health and the distillation operations in the downstream cold section, to get the best Overall Asset Effectiveness (OAE), is a constant challenge. OAE is a combination of Throughput, Availability and Yield. Any one of these three parameters could impact performance of the plant. Big Data Analytics can provide insights and prescriptions that help balance these three parameters to sustainably maximize the OAE.

Data availability is not a constraint, but utilization of the available data is. Data capable of providing insights is available from diverse sources like, the DCS, operator logs, laboratory, maintenance and engineering documents. Traditionally about 90% of the collected data is archived and not even looked at. More complete utilization of this data, utilizing Big Data Analytics techniques modified to suit the process manufacturing application, yields valuable insights to help improve the overall asset effectiveness as well as individual equipment effectiveness.

While several Ethylene plants have used subsets of these big data analytical techniques quite successfully to enhance OAE, the full application of the big data techniques tailored to process manufacturing allows improved OAE levels beyond the current paradigms.

Successful implementations have shown that Big Data Analytics in Process Manufacturing is not just an off the shelf tool that can be picked up and applied. Success of the analytics initiative for Chemical Process Manufacturing entails several aspects, to ensure effective implementation and sustenance. Effective data handling, a synergistic use of Machine Learning and fundamental models for Descriptive, Predictive, Prescriptive and Diagnostic applications, an effective business intelligence interface to provide readily actionable visual information based on a solid data and analysis foundation and a work process are all important aspects to be considered.

Data quality issues and skills availability are two of the top challenges organizations typically face when trying to implement analytics systems. The paper shows how these challenges were successfully addressed using combined models based analysis and the Internet of Things (IoT).

This paper showcases ethylene facilities where a Big Data Analytics initiative has been successfully implemented at the process manufacturing level to improve OAE. The paper also covers the key ingredients and success factors based on lessons learnt from these applications.

The Need in Ethylene Plants

Starting up and sustaining production levels consistently from an ethylene plant does not always match expectations. Typical challenges faced when operating an ethylene plant:

- Furnace run lengths below expectation
- Unexpected tube failures and lower than expected tube life
- Rapid coking
- Inefficient furnace decokes
- Equipment reliability and lower than expected uptime
- Below capacity operation
- Recurring fouling in TLEs, chillers, CGC circuit
- Utility limitations
- Refrigerant and caustic losses
- Lower than expected product recovery
- Higher than expected specific energy consumption
- Off spec product
- Excessive flaring

The raison d' être for the ethylene plant asset is to get best possible production of ethylene, without excessive costs and without incurring safety and compliance incidents. Each of the challenges listed above, impact the production from the asset (the ethylene plant).

The ethylene plant is complicated to run, with all its parameter interactions. Most of the ethylene plants today, have large amounts of data and more, that is collected and available, indicating what is going on within the pipes. Making sense of this data can give insights into why the plant is behaving the way it is and hence enable actions to fix any anomalous behavior, that threatens a manufacturer's profitability and survival, be it efficiency, safety or regulatory compliance related.

So how does one use the data to address the need? While the underlying approach to address this need may sound familiar, the level to which the inaccuracies in the data are handled and gross approximations made and/or intuition relied upon to address these needs using data, would vary depending on the decade we are operating in. Digital technologies that are now maturing can be leveraged to help minimize inefficiencies and risks that may have been considered acceptable operations norms before.

Using data - The overall approach

The overall approach to define metrics using the data from the plant operation, to better understand and characterize the operation and serve as a basis for actions to saftely improve performance, may not very different from prevalent practice. The digitalized approach improves the timeliness and reliability of the information, allowing a much higher value to be derived than before. The overall approach entails:

- Defining metrics that help measure the effectivess of the asset (the plant) in meeting its production goal
- Identifying and defining key parameters that impact the effectiveness metric
- Calculating the metrics and key impact parameters
- Defining performance benchmarks
- Computing the gaps from possible performance benchmarks
- Monitoring the metrics and parameters
- Identifying gaps from benchmarks
- Zeroing in on the root causes for these gaps
- Developing actions to close the gaps
- Implementing the actions
- Tracking impact of actions on the gaps
- Fine tuning and keep up with changes caused by variables we have no control on, with the help of these metrics

The Overall Asset Effectiveness (OAE) is a useful metric that provides a high level view of the asset performance. We can drill down from this metric to details at individual equipment of device levels, to understand the root causes of below expected OAEs. The effectiveness of individual equipment or sections can also have an analogous metric: Overall Equipment Effectiveness (OEE). This metric has been widely used in discrete manufacturing and in Batch process manufacturing to some extent, but has not been used as much in contnuous process manufacturing, especially in bulk chemicals manufacturing.

Overall Asset Effectiveness (OAE) – A useful metric

The overall asset effectiveness (OAE) is a factor comparing the actual production of the product to a benchmark production. In case of the ethylene plant it is tons of ethylene produced over a specified period of time (could be an hour, a day, a week, a month, year,....), as compared to a benchmark production of the product over the same period. Product could also be defined as Ethylene + Propylene, depending on the key objective of the plant. Possible benchmarks:

- Design production
- Historical Best production
- Ideal possible production

It is more convenient to take design as the basis benchmark for calculating the OAE, since it is relatively more constant than the other two benchmarks. The other benchmarks can then be normailzied against the design basis.

There are three key parameters that impact the production:

- Throughput (total feed to the plant, e.g. kgs/hr of Ethane)
- Yield (kgs of product/kgs of feed)
- Availability (onstream time of the plant)

Each of these parameters when compared to the corresponding design benchmark, gives rise to three effectiveness factors: Throughput Effectiveness (TE), Yield Effectiveness (YE) and Availability Effectiveness (AE). The formulae and relationships are detailed in Annexure 1.

Two case examples of the OAE metric and its key impacting parameters (TE, YE and AE) are shown in Figure 1 below.

Figure 1

Overall Asset Effectiveness Case Examples

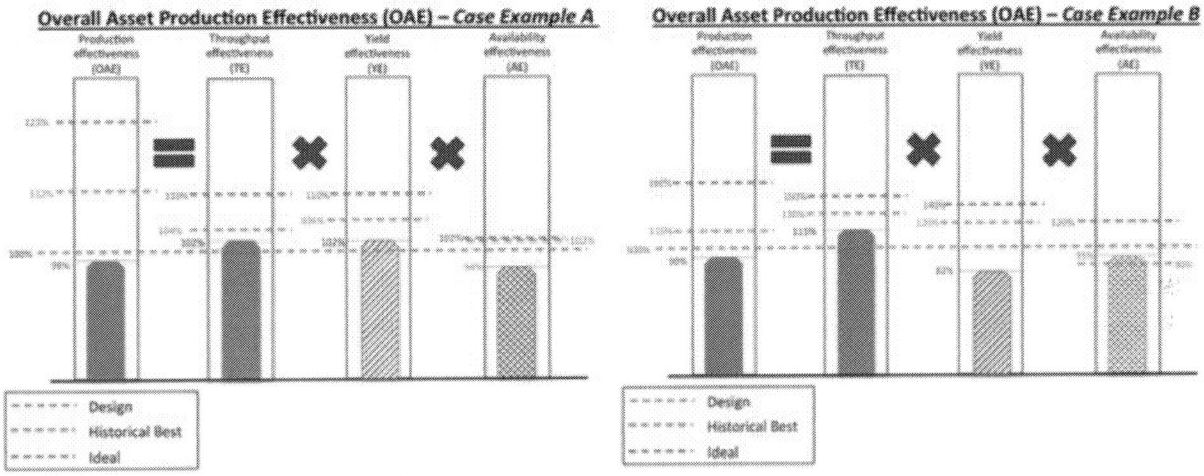

The gaps could also be shown as waterfall charts (Figure 2 below), indicating the extent to which each parameter (TE, YE and AE), impacts the OAE.

Figure 2

Overall Asset Effectiveness Gap Analysis

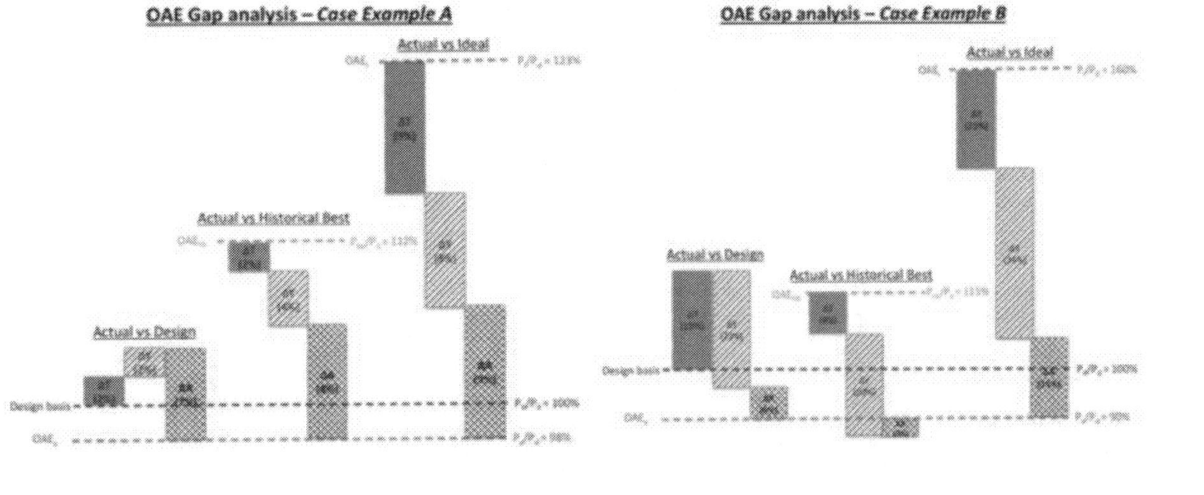

In the Case Example A shown in Figures 1 and 2, the OAE is 98%, meaning that the actual production for the period considered is at 98% of the production the plant has been designed for. The figure also shows that the plant has been operated before at 112% of design production levels (The Historical Best), for the very same conditions of the feed profile and furnace feed configuration. Also, at these feed and furnace configurations, calculations show that the plant could possibly be run at 123% of the design production levels, at the existing conditions of the plant equipment.

The resolution of OAE into its components (Throughput, Yield and Availability) shows that the main cause for production being below design levels, is on account of the plant availability during this time period considered, which is at 94% of design availability. The throughput and Yield are operating above design levels.

While the details have not been shown here in the diagram or as separate figures, further drill down in this case showed that this availability short fall came from lower than expected run-lengths of the furnaces. This was focused upon and addressed.

In Case example B, it is seen that the production is at 90% of the design level, on account of yield being at 82% of its design level and partly because of the availability being at 95% of design. Further drill down in this case showed that the yield gap was on account of an anomalous operation of a few furnaces, caused by instrument errors driving the APC incorrectly. Some of the yield gap in this case was also found to arise from losses in the recovery section of the plant. The Availability gap was again on account of furnace run-lengths.

The OAE helps us focus on the area to work on that impacts the bigger picture goal. As we observed above, in the Case Example A, the immediate area to focus on was the availability, since the plant seems to be doing well with respect to throughput and yield, atleast versus design, even though it is possible for them to be doing better on these parameters (when comparing with ideal possible) and they have done better before on these parameters. In Case Example B, Yield was the immediate focus area, followed by availability as a first step, while the plant could be doing better on all three prameters and hence production, versus what is ideally possible.

The OAE is a high level metric, which drills down to more detailed analogous defined metrics (OEE) at the section or equipment level, to identify the root case of the gaps and allowing focus on the area where the fix is required.

Key Challenges to be addressed

One of the key challenges that make it difficult to use the measured and stored data from the manufacturing process, as is, are Errors in data. Recent surveys have consistently shown this to be the top challenge[1], organizations face when trying to use the collected and archived data, as is.

---

[1] *Business Intelligence, Analytics and data management survey in late 2015 by TDWI showed data errors to be the number one challenge (@40%), followed by analytics skills (@38%) (further details covered in Reference 1).*

The data errors results in non-closure of the material balance around the plant.

The OAE metric, and the visual summary provided in Figures 1 and 2 cannot be meaningfully calculated with raw data. The data errors could give an erroneous picture of the relationship and gaps beteween the parameters. The ideal benchmark has to be calculated based on boundary conditions around the plant, like feed and product specs and offtake as well as equipment well being and status. The numbers for design and historical best have to be picked depending on the feed slate, the furnace configuration, etc., so we are comparing "apples to apples".

Technologies now available can be leveraged to address the data error and dynamic benchmark challenges.

Paradigm shift to leverage Industry 4.0 technologies to improve OAE

There has been a lot of media coverage on Industry 4.0. There is school of thought that believes it to be the fourth Industrial Revolution in the making (Figure 3a). Industry 4.0 is the cyber-physical transformation of manufacturing that promotes connected manufacturing and a digital convergence between industry, businesses and other processes.

The vision of Industry 4.0 will result in a smart factory and make full use of digital technologies for manufacturing. It is currently in its infacncy and beginning to take place in isolated ways. It is expected to be transformative, with end-to-end digitalization, resulting in a fully interoperable ecosystem of machines and partners across the supply chain, with data both informing and correcting the course of action.

Figure 3a

The Industrial Revolutions and where Industry 4.0 fits in historically

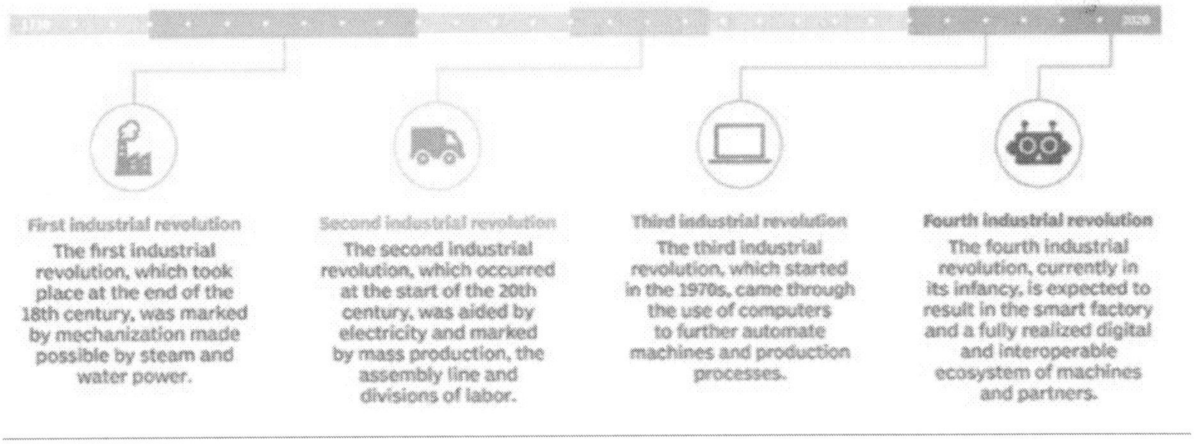

Source: Definition Industry 4.0, Tech Target, March 26, 2018

Industry 4.0 has its basis in Digital assets (technologies and concepts). Some of the technologies/concepts that are behind the Industry 4.0 discussion are:

- **Big Data Analytics**
- **Digital Twin**
- **Machine Learning**
- Cloud
- **Predictive Analytics**
- **IoT**
- **AI**
- OPC UA
- **Adaptive Analytics**
- **Everything-as-a-service**
- **Smart Sensors**
- Data Lakes

While all of the technologies may not have direct relevance to the process manufacturing application, right now, combining the applicable ones in the right way can add tremendous value to improve operational efficiency. Figure 3b below shows that within the digital asset life cycle management category, Operations performance effectiveness improvement has a significant value for the industry. The key is to leverage these technologies, without getting too far ahead on implementing bleeding edge techniques versus leading edge and feasible ones.

Figure 3b

The Significance of Operations Performance Improvement in Digitization

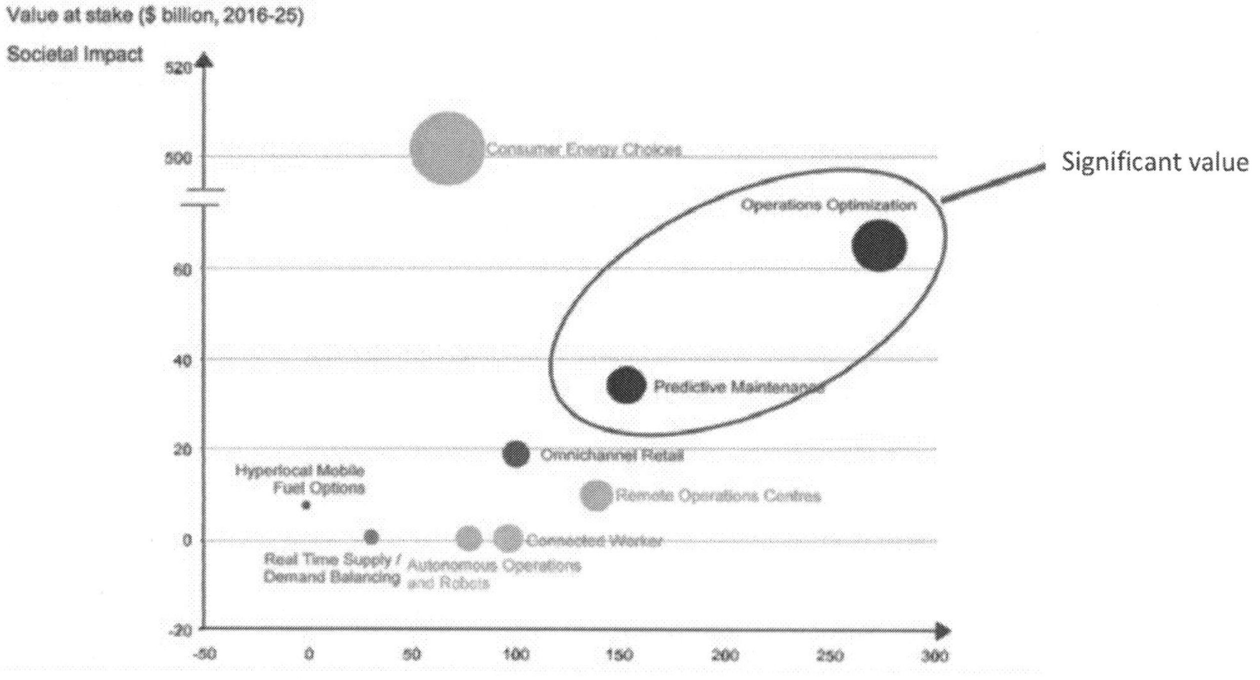

Source: World Economic Forum/Accenture analysis Industry impact $ billion

Table 1 below shows paradigms versus possible improvements leveraging Industry 4.0.

Table 1

Evolution of Managing Process Operations – Paradigm changes

Paradigm (since 1990's)	**Industry 4.0** -> **Current possible** (Digitalized application)	**Future** (Automated applications)

• Machine automation
• Advanced Process Control (DMC)
• Onsite operations support by process engineers
• Access to data (but overload)
• Access to modeling software (but typically unsused)
• Operator/engineer personal experience and intuition based operation
• Guided by licensor SOPs
	• Big Data analytics
	• Digital Twin
	  o Predictive
	  o Adaptive
	• Collaboration between Digital asset experts + process engineer team
	• IoT
	  o Collaborative remote operations engineering
	  o Operations decision support as a service
	• Interactive
		• Automate using AI
		• Auto adaptive analytics
		• Reduce human intervention in the analytics process

Outcome		
• Early detection of anomalies often missed due to inability to uncover truths and insights that aren't readily obvious or don't align with conventional process wisdom or intuition	• Insights not otherwise captured • Early attention to anomalies and actions to be taken that may otherwise slip attention • Improved efficiency • Improved Reliability	• Faster realization time and action

Companies that leverage digital technologies to simplify, automate and optimize operations, better than paradigms allow, will be at the forefront of transforming the industry to ensure they thrive in the volatile market conditons. Big Data Analytics is one of the technologies that is often discussed under Industry 4.0. Like many of the other technologies, stand alone Big Data Analytics may not deliver the expected value. It has to be combined with other techniques to be able to extract value for operations. As a result, the boundaries between some of these technologies often seem blurred. An effective utilization of Big Data Analytics entails:

- The development and adaptive maintenance of a Digital Twin
- The collaborative working between process engineers, data scientists and/or digital asset experts
- Effective use of IoT to leverage remote experts by taking the data to the experts, instead of the other way around to address the skills challenge and also lend focus
- The thought of using the analytics support-as-a-service
- Dashboard for effective visualization of information generated by the analysis

An overview of the Digitalized system for process manufacturing is shown in Figure 4 below.

Figure 4

Overview of Digitalized system for Process Manufacturing

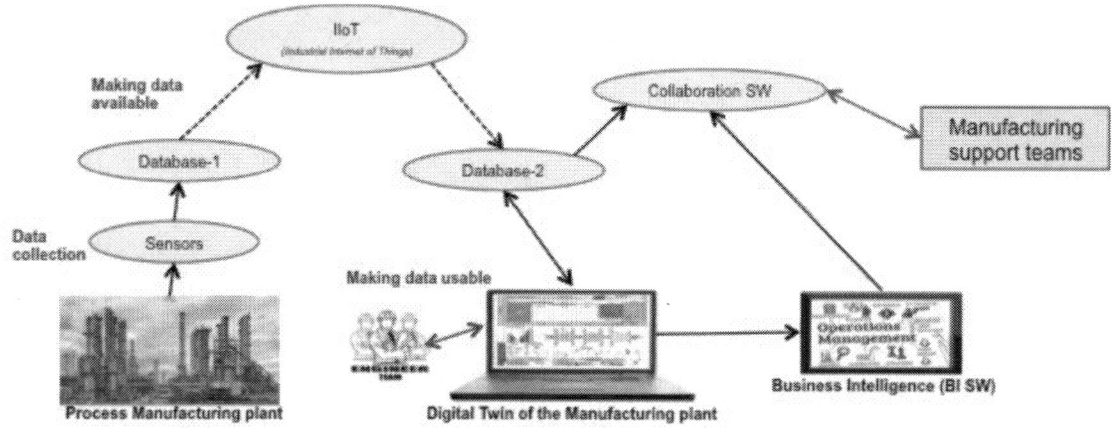

Big Data Analytics and the Digital Twin in Process Manufacturing

Data availability is not a contstraint in most Ethylene plants and process manufacturing plants in general. Plant data is collected at regular intervals in the order of seconds from the control sysytem and made available on plant historians. In additon Laboratory Information systems, operator logs provide additional, albeit less frequent data. Additional information about the process configuration is available in engineering drawings and data sheets. Business data that drives plant targets is available from the ERP system. There is no dearth of data, in fact there is an overload. Utilization of data is the constraint. As shown in Figure 5 below, more than 90% of the data collected is archived and not even looked at. The data available from a plant qualifies as Big Data.

Figure 5

Big Data in Process Manufacturing – limited use paradigm

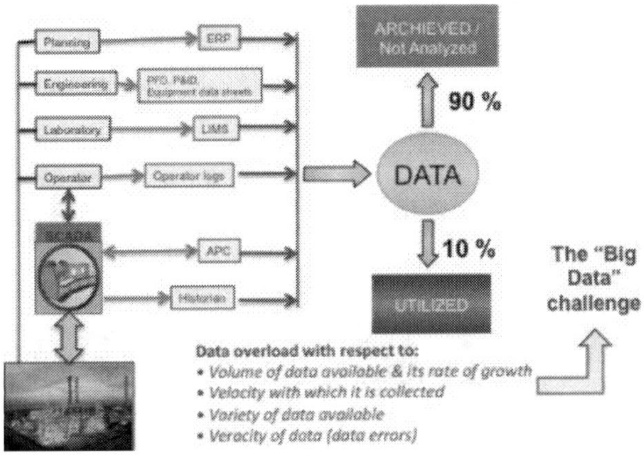

Big Data Analytics has been widely used of late in Social Media and Customer/Market analysis. However, Big Data Analytics in Process Manufacturing has to be handled differently from its current popular applications in other fields.

- In Process Analytics the "Why?" is as important as the "What?" as a result from the Analytics, unlike in Media and Customer analytics, where the "What?" is largely sufficient
  - ✓ Find and interpret patterns in data
  - ✓ Use for predictive purposes
  - ✓ Extract meaningful relationships to use to improve the process
- In Process Big Data Analytics, need combine Machine Learning models with Fundamental models
  - ✓ Machine Learning models
    - ➢ Predictive
  - ✓ Fundamental models
    - ➢ Surrogate models to explain
    - ➢ Predictive in a larger operating range

The Analysis engine for analyzing the big data from process manufacturing has to be a combination of Machine Learning models and fundamental models. A more detailed coverage of the application of Big Data Analytics to Process Manufacturing may be found in Reference 1. A case example of the type of models that go behind the Ethylene process manufacturing analysis and the benefits from its use, is shown in Figure 6. The Machine Learning models are driven purely by the data collected from the plant. The fundamental models help explain the physics and chemistry and the why, behind the observations from the data driven models.

Figure 6

Process Manufacturing Big Data Analytics requires a combination of models

Models and Analysis *(combination of models used)*

- Interactive dashboard
- Special Studies
- DMC Variable Tracker
- Coking Model: Furnace Run-length Predictor
- Machine learning and statistical model
- Performance Benchmarking
- Overall equipment Effectiveness Metric
- Equipment Health Monitoring (EHM)
- Safety Deviation Tracker
- Yield Model (COILSIM1D)
- Maintained Steady-state model (Hysys)
- Convection section model (HTRI)
- Exchanger HTRI models

Benefits

- Maximizing 13 furnace operation
- Improving furnace run lengths
- Optimizing furnace performance (i.e. conversion and yield with better health)
- Improved reliability
- Extending time between turnarounds
- Better control and less variability
- Cold section capacity increase

This basket of models integrated with one another, provides an accurate representation of the asset (the ethylene plant). Maintenance of these models is necessary to keep it mirroring the plant conditions as it continuously changes over time, on account of changing feed conditions, environmental conditions, equipment condition and product requirement, so that it can mimic the behaviour of the plant.

This system of models can predict how the plant will behave to changing conditions, it can explain the root cause of certain observations, it can help identify errors in measurements and it is a useful tool to run scenarios and what-ifs, to see how the plant will behave, without actually testing it on the plant.

In the Industry 4.0 parlance, this is referred to as the Digital twin, a term introduced by GE originally, albeit as a support for design. Gartner has included the Digital Twin in the top five digitization trends in 2017 and expects several things to be represented by digital twins over the next five years. It has been defined as "*a dynamic software model of a physical thing or system, that can be used to analyze and simulate real world conditions, response to changes, improve operations and add value, based on data provided by sensors in the physical world and the physics of how the components of a thing operate and respond to the environment*".

The utility of the digital twin in product design has been well appreciated. Its utility in process manufacturing is also very significant as was elaborated in detail in the Reference 1 paper. A significant utility of having a maintained digital twin (a compendium of tuned models) is required to address the significant data errors challenge as explained in greater detail in the paper in Reference 1. However, it is important to understand, what it takes to develop and maintain it and the processes that have to be structured around it to make it useful and deliver value. Developing a model or a set of models to mirror an ongoing operation and maintaining it is a far more complex exerecise than developing and using a design quality model. However, the benefits and value of doing this far exceeds the effort and cost.

Functionalizing the Digital Twin & Enabling Chemical Process Big Data Analytics

While the Industry 4.0 concepts and techniques can be very useful to improve performance benchmarks of process manufacturing, it requires changes to the way the operations is run and managed. Data Analytics is the base for value delivering decision making in this digitized paradigm.

Certain thoughts that are considered sacrosanct based on operating paradigms need to be overcome and changes made to based on certain realities, in order to make use of the Digital Twin and Big Data Analytics:

- The thought that by monitoring the data, insights will be gained and suggestions made to operations will generate lasting change is incorrect.
- Data availability and dashboards alone do not deliver results – sustained analysis and communication is key to knowledge and value capture
- It has to be appreciated that implementing changes to operating conditions in the plant is not automatic or easy
    - Change have to happen at the site of control of equipment and the changes require monitoring, correction, ongoing communication and convincing
    - Significant analysis is required to generate insights and actions that will consistently add value and provide the confidence to drive the change at the operations level
- Models built and maintained to reflect actual plant conditions and that can predict its behavior are essential to gain actionable insights based on analysis instead of intuitional feel and directional guidance
- Implementation requires communication with operations that is based on analytics to instill confidence and make it happen (i.e. "your thoughts are not better than my thoughts"...but if digital twin showed results from change, I can make it happen)

The typical operations analytics paradigm:

- A typical plant has an occasional converged process simulation of the plant run to check for general agreement with operation on an infrequent basis (i.e. monthly)
    - The ground work necessary to perform and sustain analysis is significant and requires a specialized engineer
        - ✓ Monthly runs planned but they rarely happen
    - Substantial effort is required with data and model preparation, maintenance and use, to get distilled decision quality information

- ✓ It requires application with expertise and persistence on a frequent basis to keep it pertinent for regular use
- ✓ An unmaintained model may provide "directional guidance", at best, because it is unable to get to decision quality specifics and answer questions and often not offer the confidence in the results
- ✓ The process simulation by itself does not characterize the end to end ethylene plant operation, but only portions of it

  - Engineers typically spend ~80% of their time finding and validating information versus running and analyzing result
    - ✓ As a result, little model based analytics ever get done

- The work and frequent nature of using the tuned models that mirror the plant is what results in the value, which is not the typical paradigm in the production and technical support environment

Even if data capture, storage and dashboard presentation is good, utilization requires knowing:

- How to clean and filter that data for use, identifying and weeding out bad data
  - Veracity of the collected data is a major challenge that restricts its use
- The analytics tools required for the desired result
  - The analytics and tools to apply requires knowledge of techniques as well as the manufacturing process (the parts of the digital twin engine)
- How to manage the analytics tools behind the digital twin
  - To build models to reflect reality and tweaking to ensure convergence
  - To manage the typical convergence issues of analytics tools and models
  - To go beyond just data driven analysis and use fundamental models required for process insights

A unique combination of expertise (digital asset experts) is required to address these issues that typically hinder utilization. This is often most efficiently done by having this done as a service, by an independent remote team, which allows the operating company teams to focus on prioritizing actions and getting them implemented instead of getting bogged down getting to a base of actionable information from which results can be delivered. The IoT concept makes this possible by having data taken to the relevant remote experts instead of having these experts come to the data, which is often not feasible and sustainable. This application of IoT addresses the skill availability challenge identified in the TDWI survey.

Figure 7 shows this division of effort and collaboration and the reality of keeping the digital twin functional and of ensuring conversion of data to usable and actionable information. The Manufacturer operations support team's productivity improves by being able to focus on the necessary and value added tasks, of understanding, prioritizing, ensuring implementation and thinking beyond instead of spending time on validating information and maintaining the digital twin.

Over time, several of these tasks will be automated using AI techniques (a combination of Machine Learning and validated rule based programs). While the AI algorithms and techniques to do this are available today, configuring these is more time consuming and requires the process shown in Figure 7 to have functioned for some time to capture the "learnings" and patterns from the work process and replace repeatable and predictable learning parts of it using the AI algorithms. These need to be rigorously tested and compared to a working system, before it can be implemented to go live.

The digital twin and this collaborative interaction between the Manufacturing operations support teams and the digital asset expert team (the digital twin is the key digital asset), as shown in Figure 8, offer a paradigm shift from typical current operation. Figure 9 describes the synergy between the various manufacturing support teams and the productivity benefits from the distributed collaborative operations engineering using the digital twin. Figure 10 shows the typical delivery from the decision support analytics-as-a-service center, that supports the digital twin.

Figure 7

Reality of keeping the Digital Twin alive – Distributed roles

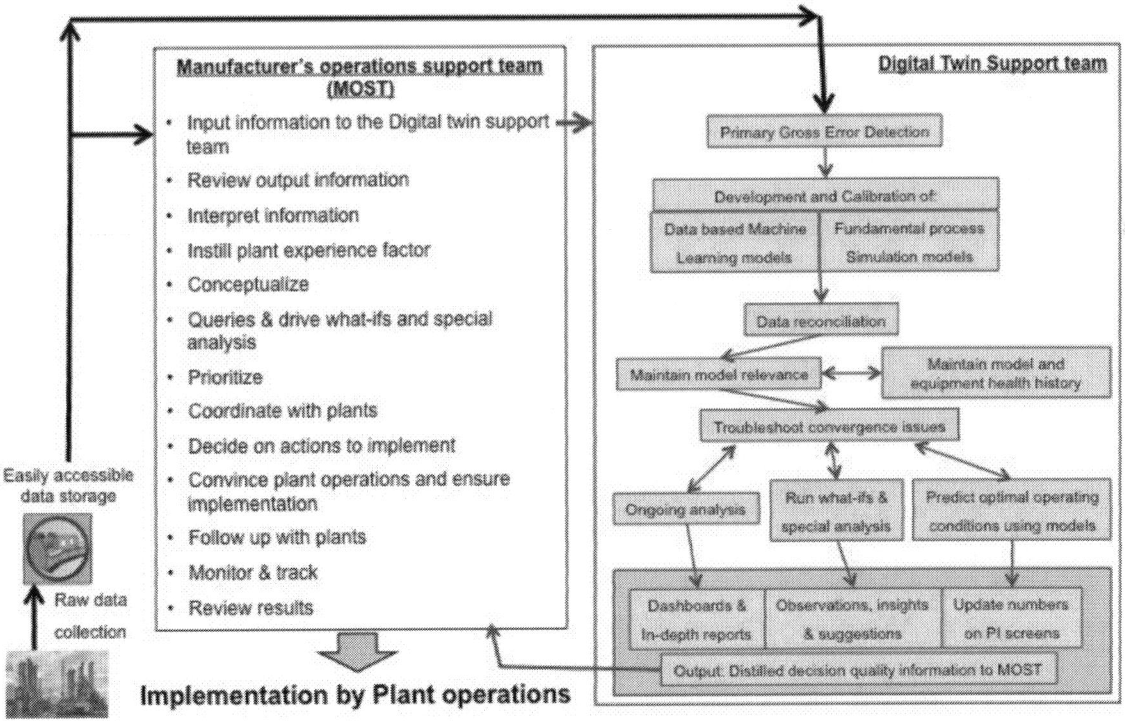

Figure 8

Collaborative operations engineering

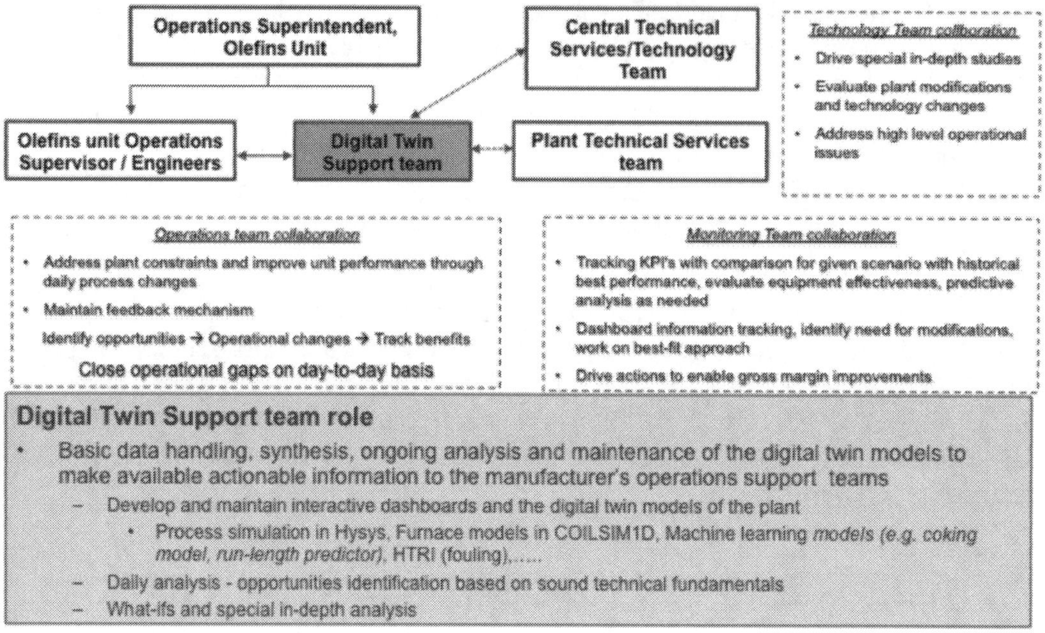

Figure 9

Facilitation of value added work by Operations support teams by the Digital Twin support team

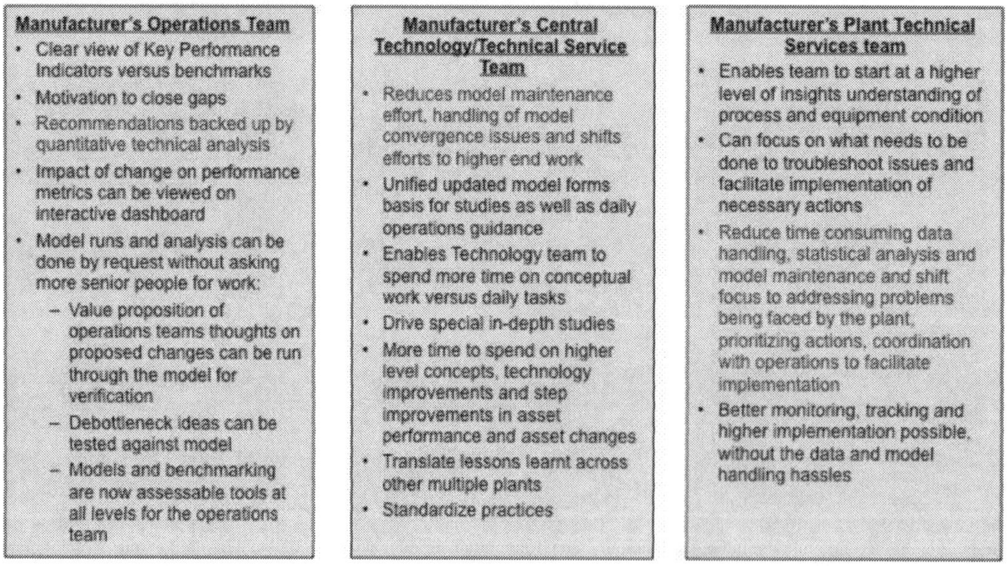

Figure 10

Typical delivery between the Collaborative Operations Engineering centers

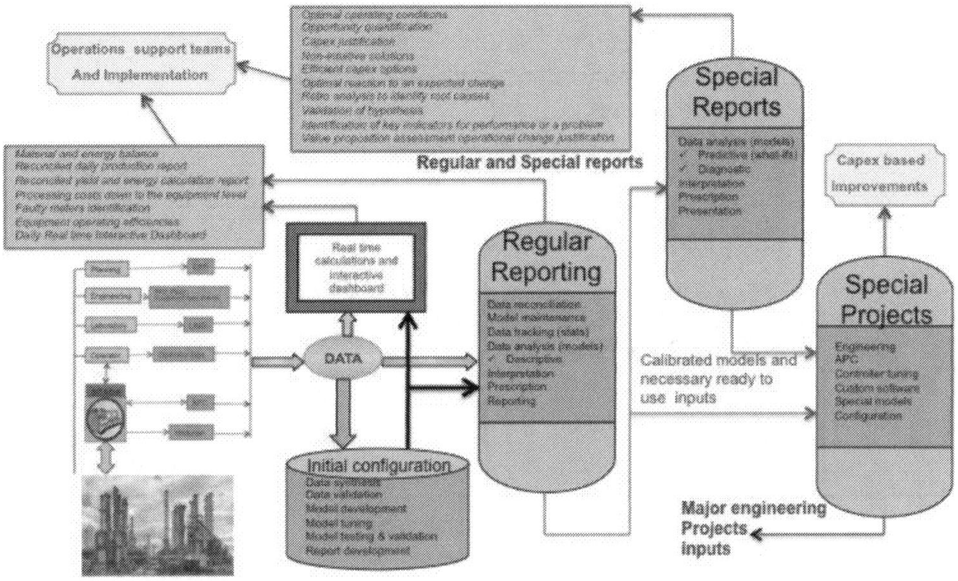

The future automated operation with AI, will utilize rules based on experience with current activity suporting the digital twin, simplified models where possible and machine learning to partially automate the digital twin support activities. The scenario over the next couple of years will likely be as shown in Figure 11, versus the current manned work process as shown in Figure 7.

Figure 11

The future scenario on operation of the Digital Twin

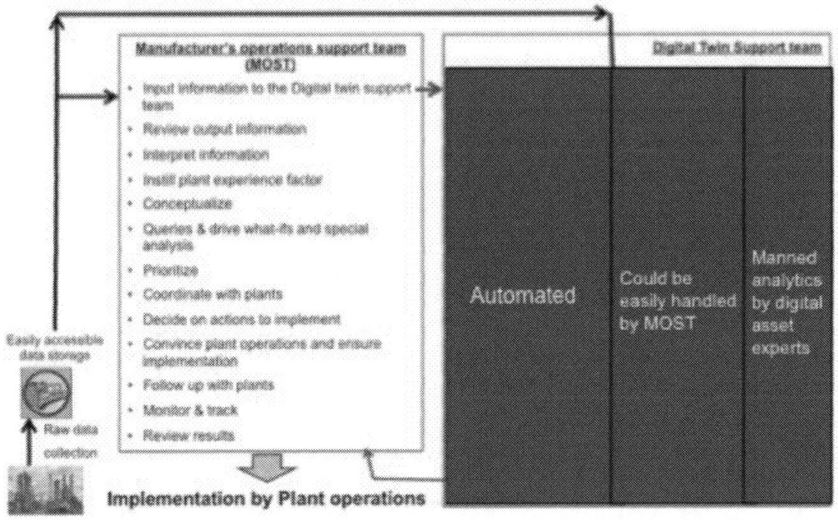

Case examples of results from application of Big Data Analysis using a Digital Twin at multiple ethylene plants

This Integrated analytics program was implemented at three ethylene sites of a major ethylene producer and has shown consistent value addition year after year. Table 2 shows a summary of some of the value derived.

Table 2

Results for three plants for a Major Ethylene Producer

Parameter	Facility A	Facility B	Facility C
Furnace Runlength	**2 x Design**	**+100%**	**+50%**
Yield	Increased	Increased	+2.3%
CGC Performance	+60 tpd	RCA solved problem	Improved
Capacity Increase	Debottleneck study	Best ever (@107%)	Max capability
Other	C2 Splitter optimized / Exchanger improve	Same + RCA	Decoke and C3 splitter improve
Significant What-if Results	25 proactive	8 proactive + 31 queries response	Numerous

The increased run-lengths helped improve the Availability Effectiveness of these assets and hence the OAE, in each of the above cases, besides lowering the costs due to fewer decokes.

Results of performing such analytics at another ethylene facility in North America are shown in Table 3.

Table 3

Improvement over previous year with Big Data Analysis using a Digital Twin

Parameter	Improvement	Remarks
Production		
Ethylene	6.8%	In spite of lighter feed slate operation
Propylene	10.1%	Record production
Furnaces overall		
Runlength	4.1%	Overall improvement
Availability	6.4%	
Variability	-1.8%	
C2" Fractionator		
Feed	5.2%	Capacity increased in spite of increased C2" in feed
C2/C2" ratio	7.1%	
Column ΔP	5.3%	
C2 spec	2.4%	

Key Success Factors – Lessons learned

Some of the key lessons learnt from the application of big data analytics to ethylene manufacturing, using the digital twin. This application has been evolved over time and with experience from its implementation and success at several ethylene sites.

- It is not just about the technology or tools. The process around the tools and application is equally, if not more important for success.
- The process and tools can be modified over time, to incorporate new technologies once they are mature enough to apply, to add efficiency to the existing work flow process, if it is able to deliver better results.
- Change management: Openness to new ways of doing things, made possible by technology. Sticking to the old process may not allow the new technology application to deliver results. For example, concepts like collaborative operations engineering, remote operations decision support-as-a-service are paradigm shifts for most operations.
- Top executive level buy-in and support is required for this application to be implemented successfully.

References:

1) P. Nair, (March 7, 2017) "Tailoring Big Data Analytics for Optimizing Ethylene Plant Asset Performance" 2017 Spring National AICHE Meeting, Paper 2B (Ethylene Producers Conference).
2) Cliff Saran, (Oct 26, 2017) "Digital Twins: Revolutionizing product business" Computer Weekly.
3) Diann Daniel, (Mar 26, 2018) "What is Indutry 4.0" Tech Target posting by Margaret Rouse.
4) L.F.Colegrove, M.B. Seasholtz, C. Khare., (March 2016) "Big Data – Getting Started on the Journey", CEP, pp. 41-45
5) Guide, (Feb 21, 2018) "Jargon Buster Guide to Artificial Intelligence" Computer Weekly.
6) P. Nair, (March 2007) "Large amounts of data requires an outside management system", Houston Business Journal.

Annexure 1

Overall Asset Effectiveness

$$P_a = T_a \times Y_a \times A_a \times \psi_{os} \times t$$
$$P_d = T_d \times Y_d \times A_d \times \psi_{os} \times t$$

$$P_a/P_d = (T_a/T_d) \times (Y_a/Y_d) \times (A_a/A_d)$$
$$OAE_a = TE_a \times YE_a \times AE_a$$

Where

P_a = Actual production (product e.g. C2", C2"+C3")
P_{hb} = Historical best production (product e.g. C2", C2"+C3")
P_I = Ideal production (product e.g. C2", C2"+C3")
P_d = Design production (product e.g. C2", C2"+C3")

T_a = Actual Feed Throughput (total feed to plant, e.g. C2, C2+C3)
T_{hb} = Historical best Feed Throughput (total feed to plant, e.g. C2, C2+C3)
T_I = Ideal Feed Throughput (total feed to plant, e.g. C2, C2+C3)
T_d = Design Feed Throughput (total feed to plant, e.g. C2, C2+C3)

Y_a = Overall actual yield of product (e.g. kgs C2"/kgs C2 fed)
Y_{hb} = Overall historical best yield of product (e.g. kgs C2"/kgs C2 fed)
Y_I = Overall ideal yield of product (e.g. kgs C2"/kgs C2 fed)
Y_d = Overall design yield of product (e.g. kgs C2"/kgs C2 fed)

A_a = Actual Availability of the Asset (the plant)
A_{hb} = Historical best Availability of the Asset (the plant)
A_I = Ideal Availability of the Asset (the plant)
A_d = Design Availability of the Asset (the plant)

ψ_{os} = On stream factor (function of time period being considered)
t = elapsed time

OAE_a = Actual Overall Asset Effectiveness = P_a/P_d
TE_a = Actual Throughput Effectiveness = T_a/T_d
YE_a = Actual Yield Effectiveness = Y_a/Y_d
AE_a = Actual Availability Effectiveness = A_a/A_d

Benchmarks:

$$OAE_{hb} = TE_{hb} \times YE_{hb} \times AE_{hb}$$
$$OAE_I = TE_I \times YE_I \times AE_I$$

OAE_{hb} = Historical best OAE = P_{hb} / P_d
OAE_I = Ideal OAE = P_I / P_d

TE_{hb}= Historical best Throughput effectiveness = T_{hb} / T_d
TE_I = Ideal Throughput effectiveness = T_I / T_d

YE_{hb}= Historical best Yield effectiveness = Y_{hb} / Y_d
YE_I = Ideal Yield effectiveness = Y_I / Y_d

AE_{hb}= Historical best Availability effectiveness = A_{hb} / A_d
AE_I = Ideal Availability effectiveness = A_I / A_d

Note:

Decoupling Availability & Throughput/Yield

- Availability at any point in time for any equipment is 0 or 1 (efficiency of working of the equipment is not reflected in the availability – this is reflected in the throughput)
- Throughput linked by downstream constraints of equipment and its efficiency (this enables delinking the Throughput and Availability factors)
- Throughput only considered over the time when the plant or equipment is available
- Availabilty of the plant = Min (availability of equipment in series and avg of equipment in parallel in the process flow)

Figure 12

Simple illustration: Concept of decoupling Throughput from Availability in the OAE calculations

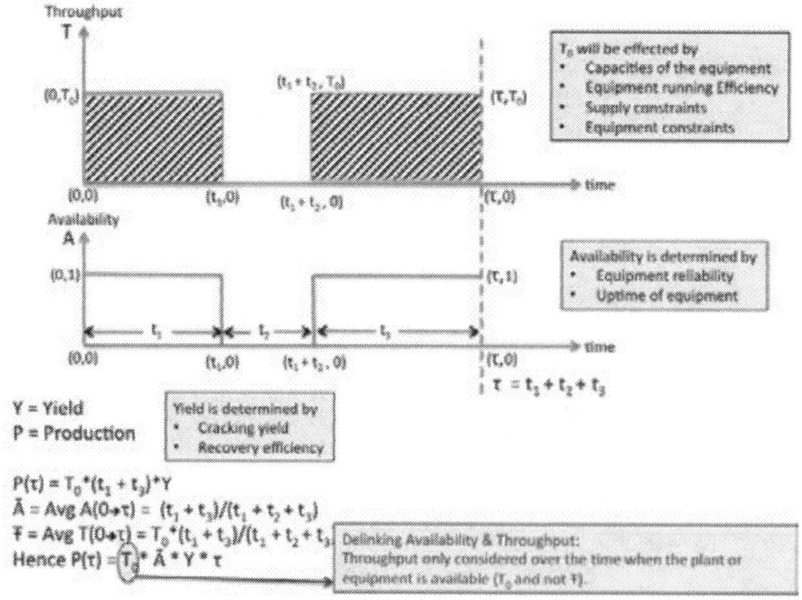

OAE Gap analysis

Gap between Design and Actual

Production gap = $1 - (P_a / P_d)$

Contribution of Throughput gap to the Production gap:
$(1 - (P_a / P_d))*(1 - (T_a/T_d))/(3 - ((T_a/T_d) + (Y_a/Y_d) + (A_a/A_d)))$

Contribution of Yield gap to the Production gap:
$(1 - (P_a / P_d))*(1 - (Y_a/Y_d))/(3 - ((T_a/T_d) + (Y_a/Y_d) + (A_a/A_d)))$

Contribution of Availability gap to the Production gap:
$(1 - (P_a / P_d))*(1 - (A_a/A_d))/(3 - ((T_a/T_d) + (Y_a/Y_d) + (A_a/A_d)))$

Gap between Historical Best and Actual

Production gap = $(P_{hb} / P_d) - (P_a / P_d)$

Contribution of Throughput gap to the Production gap:
$((P_{hb} / P_d) - (P_a / P_d))*((T_{hb}/T_d) - (T_a/T_d))/(((T_{hb}/T_d) + (Y_{hb}/Y_d) + (A_{hb}/A_d)) - ((T_a/T_d) + (Y_a/Y_d) + (A_a/A_d)))$

Contribution of Yield gap to the Production gap:
$((P_{hb} / P_d) - (P_a / P_d))*((Y_{hb}/Y_d) - (Y_a/Y_d))/(((T_{hb}/T_d) + (Y_{hb}/Y_d) + (A_{hb}/A_d)) - ((T_a/T_d) + (Y_a/Y_d) + (A_a/A_d)))$

Contribution of Availability gap to the Production gap:
$((P_{hb} / P_d) - (P_a / P_d))*((A_{hb}/A_d) - (A_a/A_d))/(((T_{hb}/T_d) + (Y_{hb}/Y_d) + (A_{hb}/A_d)) - ((T_a/T_d) + (Y_a/Y_d) + (A_a/A_d)))$

Gap between Ideal and Actual

Production gap = $(P_I / P_d) - (P_a / P_d)$

Contribution of Throughput gap to the Production gap:
$((P_I / P_d) - (P_a / P_d))*((T_I/T_d) - (T_a/T_d))/(((T_I/T_d) + (Y_I/Y_d) + (A_I/A_d)) - ((T_a/T_d) + (Y_a/Y_d) + (A_a/A_d)))$

Contribution of Yield gap to the Production gap:
$((P_I / P_d) - (P_a / P_d))*((Y_I/Y_d) - (Y_a/Y_d))/(((T_I/T_d) + (Y_I/Y_d) + (A_I/A_d)) - ((T_a/T_d) + (Y_a/Y_d) + (A_a/A_d)))$

Contribution of Availability gap to the Production gap:
$((P_I / P_d) - (P_a / P_d))*((A_I/A_d) - (A_a/A_d))/(((T_I/T_d) + (Y_I/Y_d) + (A_I/A_d)) - ((T_a/T_d) + (Y_a/Y_d) + (A_a/A_d)))$

Table 4: Ready reckoner table for OAE gap analysis calculations

Compare between	Multiplier	Numerator			Denominator
	P	T	Y	A	
Actual vs Design	$1 - (P_a/P_d)$	$1 - (T_a/T_d)$	$1 - (Y_a/Y_d)$	$1 - (A_a/A_d)$	$3 - ((T_a/T_d) + (Y_a/Y_d) + (A_a/A_d))$
Actual vs historical best	$(P_{hb}/P_d) - (P_a/P_d)$	$(T_{hb}/T_d) - (T_a/T_d)$	$(Y_{hb}/Y_d) - (Y_a/Y_d)$	$(A_{hb}/A_d) - (A_a/A_d)$	$((T_{hb}/T_d) + (Y_{hb}/Y_d) + (A_{hb}/A_d)) - ((T_a/T_d) + (Y_a/Y_d) + (A_a/A_d))$
Actual vs Ideal	$(P_I/P_d) - (P_a/P_d)$	$(T_I/T_d) - (T_a/T_d)$	$(Y_I/Y_d) - (Y_a/Y_d)$	$(A_I/A_d) - (A_a/A_d)$	$((T_I/T_d) + (Y_I/Y_d) + (A_I/A_d)) - ((T_a/T_d) + (Y_a/Y_d) + (A_a/A_d))$

AIChE Paper Number 169f

Start Listening to Your Heat Exchanger
- Smart Data Application Enables Insight into Process Equipment

Dr. Jörg Weidenfeller
Director RnD / Technology
Axel Kindgen
Director IT / Digitalisation
Karsten Stückrath
President & Managing Director
SCHMIDTSCHE SCHACK / ARVOS GmbH:

Prepared for Presentation at the 2018 Spring National Meeting
Orlando, Florida, April 22-26 2018

AIChE Paper Number 169f

Start Listening to Your Heat Exchanger
- Smart Data Application Enables Insight into Process Equipment

Dr. Jörg Weidenfeller
Director RnD / Technology
Axel Kindgen
Director IT / Digitalisation
Karsten Stückrath
President & Managing Director
SCHMIDTSCHE SCHACK | ARVOS GmbH:

Abstract:

Transfer line exchangers, their operation and reliability are critical components in the production of Ethylene and the economics of the plant.

Since the focus of the operator is onto the process itself, standard sensor installations are located to monitor and control the process flows. The status and condition of the heat exchanger might be derived by measured data up- and downstream the equipment and is based on individual experience of the operator. In addition regular inspections are conducted.

Transfer line exchangers equipped with additional sensors provide smart data for the operation and maintenance team. This Industrial Internet of Things (IIoT) solution is based on proven cloud technology and is adaptable/ flexible to various customer requirements. It also makes use of machine learning algorithms in order to allow predictions for different parameters.

All those smart data will be used to generate regular operation and health reports for the process heat exchanger, which provide the customer a deeper insight into his equipment and finally enhance the equipment availability.

Introduction

SCHMIDTSCHE SCHACK | ARVOS GmbH, formerly known as ALSTOM Power Energy Recovery is designer and manufacturer (OEM) for transfer line exchangers (TLE) in Ethylene industry and waste heat recovery systems in further chemical, petrochemical and metallurgical processes. The company was founded more than 100 years ago and was the first who introduced and established steam super-heaters for steam locomotives. Up to this time the boiler of steam locomotives produced only saturated steam with an appropriate poor efficiency. The founder of SCHMIDTSCHE SCHACK was driven by improving this product, increasing the efficiency of the steam process and implementing a new technology into the existing envelope of the steam locomotive.

This company philosophy and motivation have not changed. More than 50 years ago the transfer line exchanger was introduced and established in the Ethylene industry by replacing the quench. Today further improvements of the TLE are consequently introduced by keeping the existing envelope of the hardware.

The status quo in Ethylene plants is that sensors and instrumentation are installed to monitor and control the production process and the process flows. From those measurement locations upstream and downstream the transfer line exchanger the status and condition of the hardware might be derived. The interpretation of those data strongly depends on the knowledge and experience of the individual operator. The real status and condition of the transfer line exchanger is obtained during regular shut-down of the production line and inspection of the TLE. This indirect approach has uncertainties due to

- Data are not directly collected at the TLE and do not describe TLE condition
- inspection during shut-down only provides a snap-shot with limited confidence for future condition

The transfer line exchangers, their operation and reliability are critical components in an ethylene plant with an impact on the plant economics. In the transfer line exchangers the sensible heat in the cracked gas of an ethylene furnace is converted into high pressures steam (HPS). HPS is used to drive turbines, compressors and pumps in the separation train of an ethylene plant and is most important in the crackers energy system. Any loss of steam production by failure or decreased performance due to fouling of the TLE has to be compensated by HPS import from boilers requiring additional fuel gas or fuel oil with a negative effect on plant economics. A high reliability and performance of the TLEs are most important, as utilization of the TLEs have a significant impact on the plant economics and require a high quality of the TLE, for maximum utilization and performance.

In general availability and performance are independent of each other, as a heavily fouled TLE will operate at a low performance level. Maximizing the availability and performance of the TLEs by online monitoring and predictive maintenance increases the asset utilization and will improve the plant economics significantly.

Enabling The Smart TLE

McKinsey's digital compass describes that asset utilization drives value by making the best use of the hardware. Predictive maintenance thus can drive value by decreasing the planned and unplanned asset downtime. Based on sensor readings early indications of hardware problems can be detected and timely corrections can be performed. Additionally maintenance resources can be prioritized and availability can be increased. Typically, predictive maintenance decreases the total downtime by 30 to 50%.

To support the needs of Ethylene plant operators SCHMIDTSCHE SCHACK started to further improve TLE's functionality by applying IIoT technology and keeping TLE's envelope. This requires a stepwise approach as shown in Figure 1. At the base level the TLE is equipped with additional sensors. The measurement locations are at the points of interest for the TLE but not for the process. Those sensor readings are input for further evaluation and assessment of the status and condition of the TLE. The data are acquired and stored in a independent system. This approach ensures that the Data Control System (DCS) of the plant will not be impacted or disturbed. Optional operational data can be integrated by connecting to the DCS if the operator requests this. From the sensor readings meaningful values and equipment Key Performance Indicators (KPI's) are derived and monitored. Additionally the data are used for models which are based on OEM's domain knowledge with respect to thermal and mechanical design. In addition artificial intelligence (AI) is used to discover abnormalities in operation, to forecast the reliability and performance of the TLE and to provide diagnostics and future recommendations.

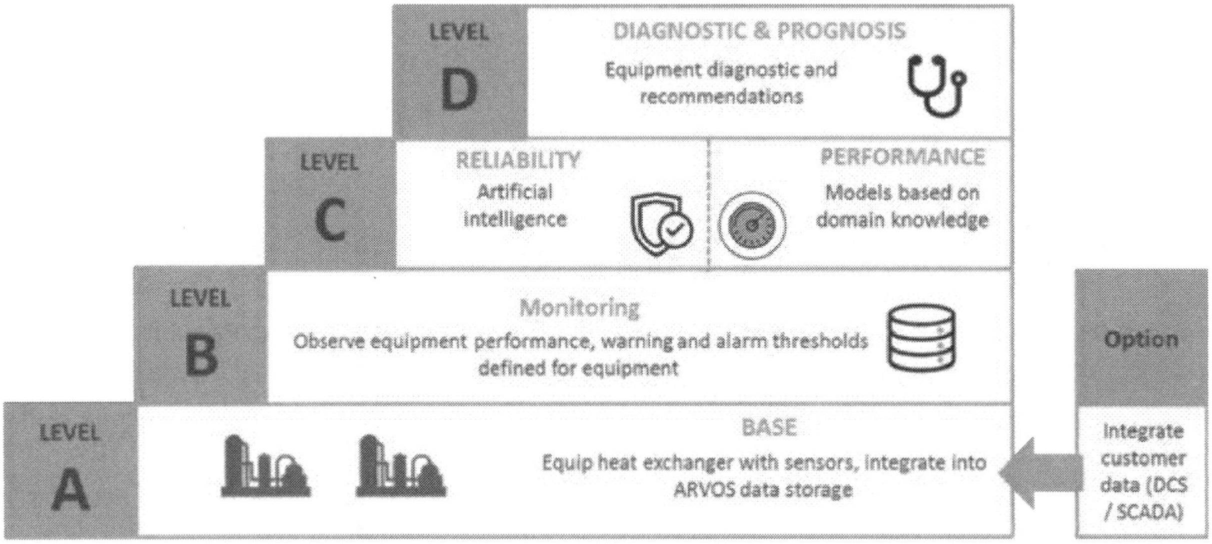

Figure 1: Stepwise approach to increase asset utilization

Technical Set-Up

The measurement locations for the additional sensors at the TLE will be installed during regular manufacturing process in the SCHMIDTSCHE SCHACK workshop. At site the sensors will be installed after erection of the TLE. The independent data acquisition system (DAQ) will also be installed nearby the TLE. The DAQ consists of Input/Output controllers connected to a edge-device which sends the live data via high-speed wireless network to the SCHMIDTSCHE

SCHACK data storage. The complete DAQ system is enclosed in an explosive protected (ATEX) control box.

The SCHMIDTSCHE SCHACK data storage manages the storage and data handling. Live data are evaluated and visualized, historical data are stored. Both types of data are used for further data manipulation by domain knowledge algorithms and machine learning. This enables diagnostic of the current TLE status, future condition prognosis and recommendation service. In summary this describes the new SCHMIDTSCHE SCHACK Zero.One platform for collection and analysis of data from transfer line exchangers, see also Figure 2. Zero.One is a cloud-based platform as a service and enables asset performance management and analytics by connecting equipment, data and people.

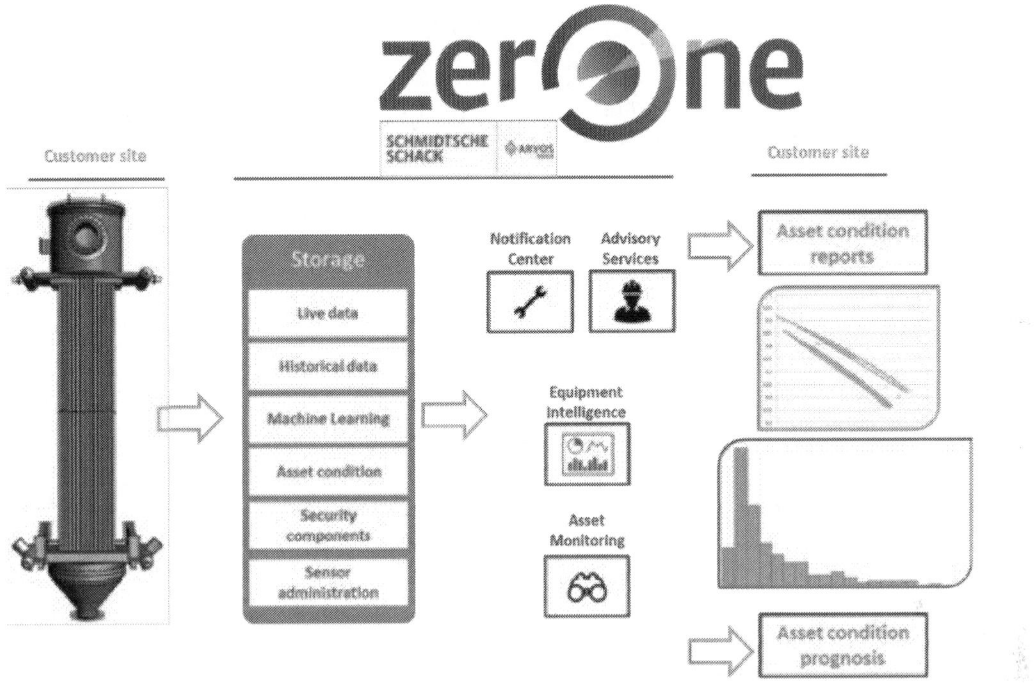

Figure 2: End to end set-up for TLE monitoring and prognosis

Operators can receive frequently condition and status reports of their TLE. Plant managers will get an overview about all TLEs in their plant. This information is available in asset, plant and fleet dashboards and reports (see Figure 3) which have different information levels.

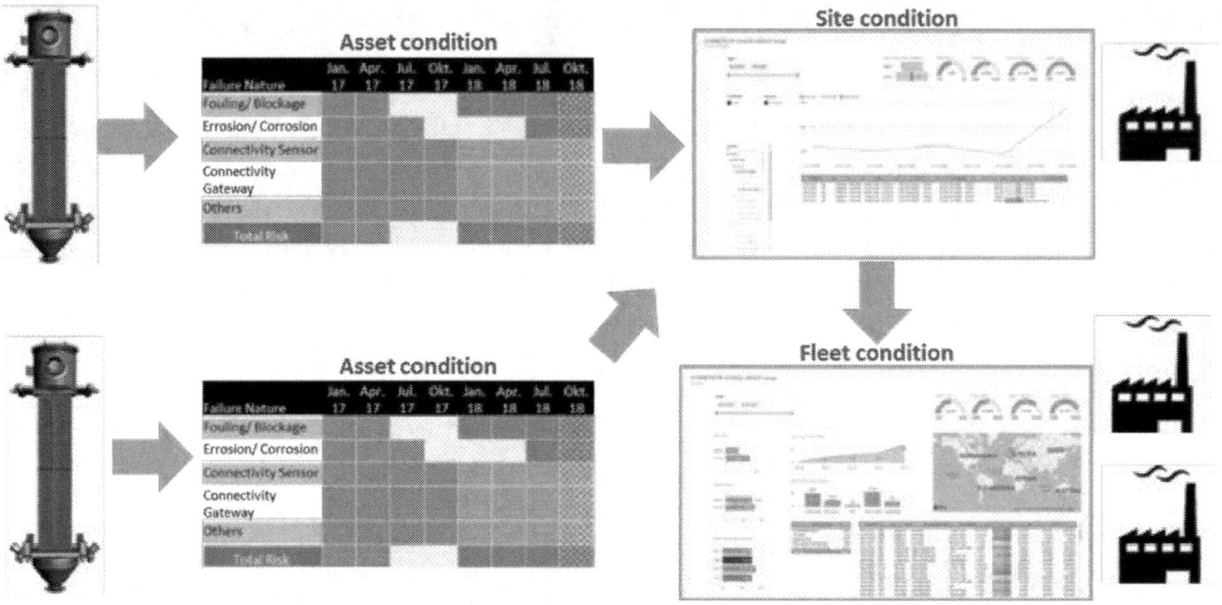

Figure 3: Dashboards for TLE condition

From Raw To Smart Data

Collecting and monitoring the operational data of a TLE, like gas inlet and outlet temperatures and pressure, water pressure etc. are an important requirement to come to a right conclusion about the equipment condition. However, only assessing the operational data is not sufficient since those do not directly reflect

- TLE design topics (thermal and mechanical),
- aging of the TLE and
- abnormalities during operation.

To address these subjects several methods for raw data post-processing are available. SCHMIDTSCHE SCHACK decided to apply an approach which combines OEM design knowledge and machine learning. In the following only examples of the current machine learning applications are discussed.

Machine Learning

To predict the future TLE performance by artificial intelligence the right approach and algorithm has to be identified. For that purpose operational data were simulated and random noise was overlaid to simulate fluctuations in the sensor readings. The principle of a machine learning algorithm is shown in Figure 4. Input values are modified by the model, which creates the predicted value. The model is trained in that way, that the error between output from the model and the target value is minimal. This will be achieved by adapting the algorithm in the model. The training of the model is finished, when the prediction achieves a pre-defined accuracy.

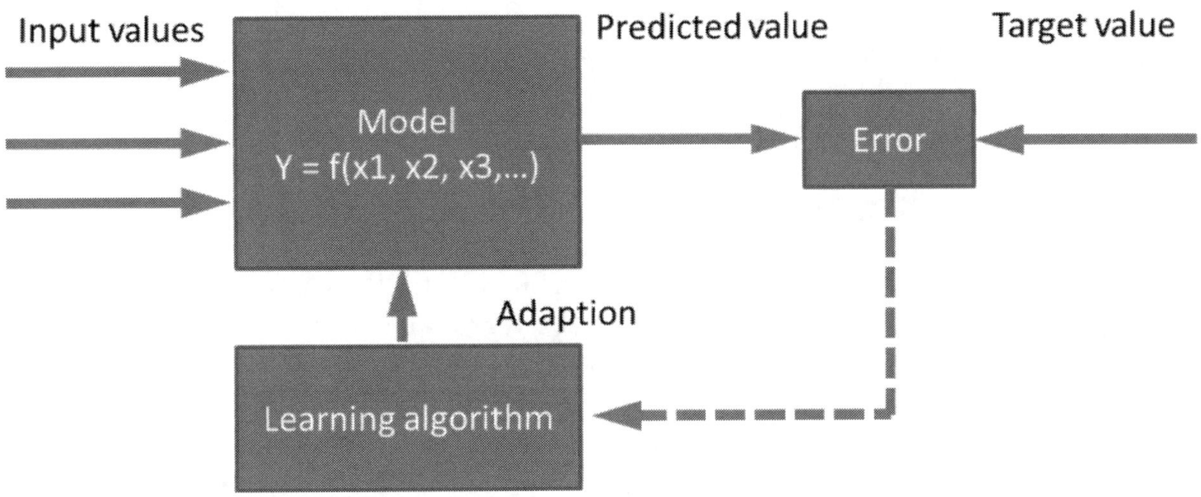

Figure 4: Principle of machine learning

When training a model, further verification is of the trained model is required to verify

- how the trained model reacts to new and unknown data
- if the training did not specialize the model in that way that accurate predictions are only possible with the training data

For that purpose a second and unknown data set has to be applied to the trained model to verify the training and the quality of prediction.

The use case for machine learning is to predict the time period to the next de-coking cycle. The signal of gas outlet temperature was used as training input. The TLE is operated first with heavy feed and after that the feed is changed to light feed until de-coking. After that the operational sequence starts again with heavy feed and again switching to light feed (see Figure 5). The training data describe 100% load and 100% heating surface.

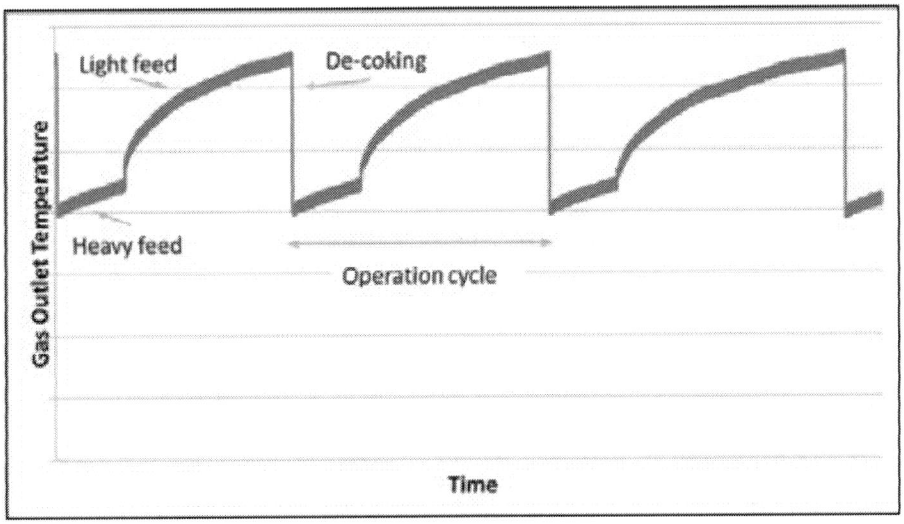

Figure 5: Training data

The training was performed only for three operation cycles. After the training the algorithm was applied to unknown test data to predict the remaining time until de-coking. In addition, several unknown boundary conditions were implemented (see Figure 6):

1. heating surface reduced by 15%, 100% load, reduced operation time
   (blockage of tubes reduces heating surface, for 100% load the gas outlet temperature increases faster, in consequence reduced operation time)
2. heating surface reduced by 15%, 90% load, normal operation time
   (operator reduces load to 90% to achieve the normal operation time)
3. no reduction of heating surface , 90% load, extended operation time
   (operator reduces load to extend operation time)
4. no reduction of heating surface , 100% load, normal operation time

The algorithm shows very good prediction with an error of ±2 days for normal light feed operation. The preceding heavy feed operation has a bigger error in prediction but the feed switch is well recognized. The algorithm also predicts the time until de-coking with a sufficient accuracy for the unknown boundary conditions of reduced heating surface, reduced load, extended or reduced operation time. In case that the training would consist also these events, a further accuracy improvement for the prediction is expected.

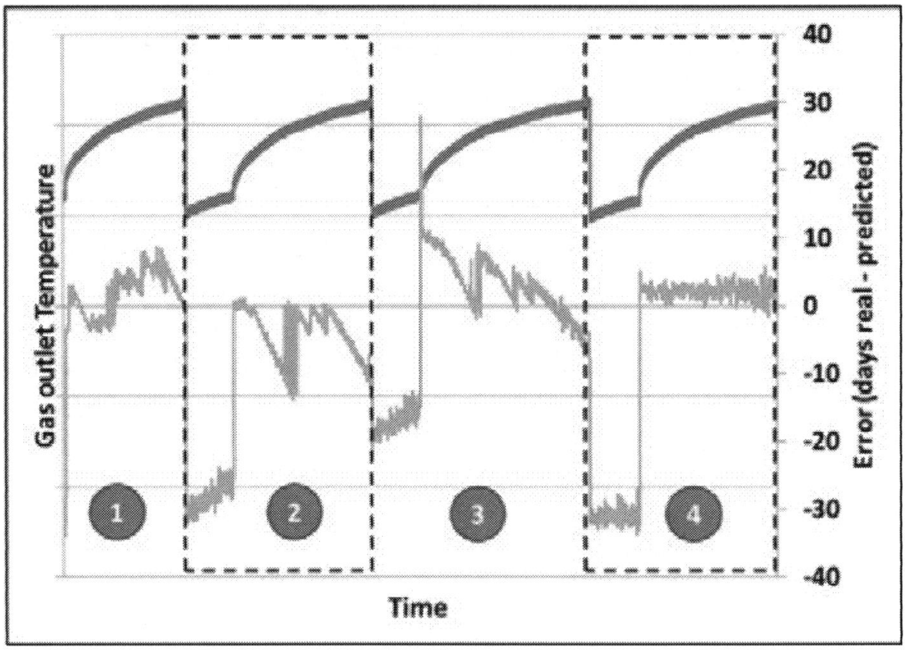

Figure 6: Test data, error between time until de-coking and prediction

Detection of Abnormality

During TLE operation several abnormalities can occur. For example single tubes can suddenly be blocked by refractory or coke particles. Blocked tubes reduce the overall heating surface of the TLE and therefore theoretically this should be visible by an increase in gas outlet temperature and pressure drop across the tubes. However, it will be difficult to detect this event by only monitoring these two signals. But with the help of artificial intelligence this will be possible.

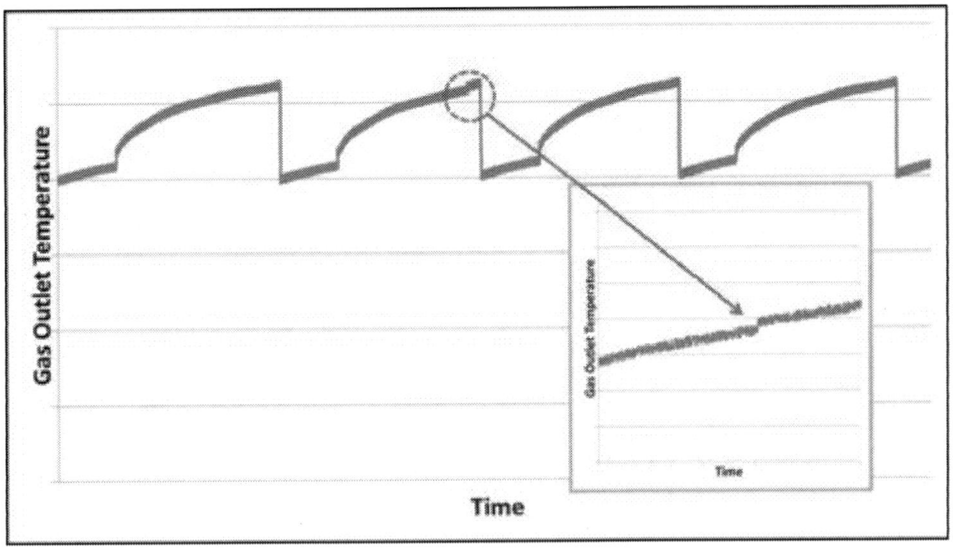

Figure 7: Temperature signal and abnormality

Figure 7 shows the gas outlet temperature which was used to train the appropriate algorithm. The small change in gas outlet temperature due to tube blockage is also shown. The algorithm was trained to detect this kind of blockage event. After the adjustment phase a second and unknown set of data was used to verify the algorithm. The result of the blockage prediction is shown in Figure 8. Here the algorithm clearly identifies the event.

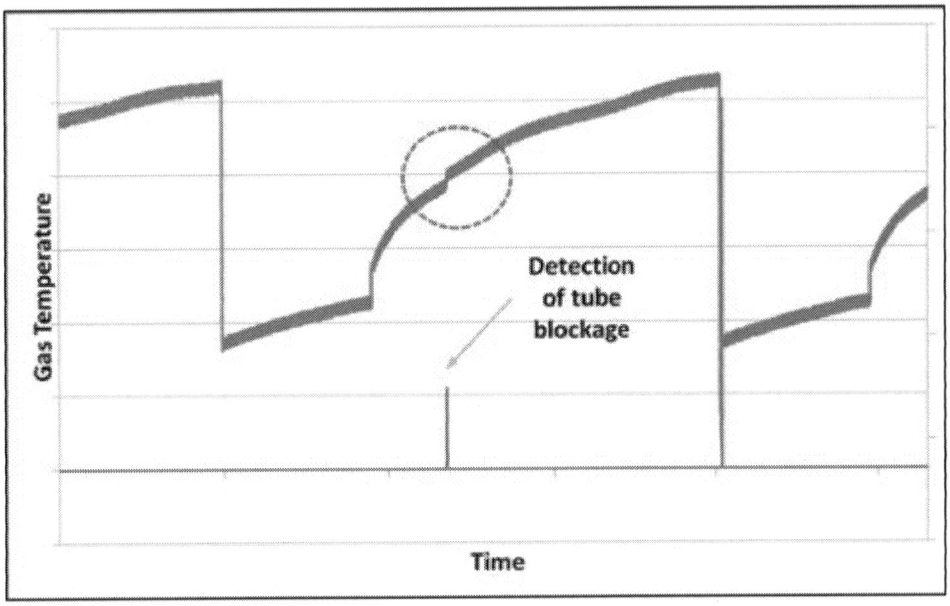

Figure 8: Tube blockage detection

Conclusion

This paper illustrates the next consequent step in further improving the functionality of SCHMIDTSCHE® transfer line exchanger by applying new technologies. Zero.One comprises additional sensors installed at the equipment and applying a cloud based platform to analyze and assess the asset. This service combines OEM design knowledge and artificial intelligence to detect abnormalities during operation and to predict future events. This will create transparency into the TLE condition, helps to optimize availability and productivity and improves planning reliability for the operator and the maintenance team.

513845 Incorporating Model-Based Technology with Big Data Concepts for Monitoring Ethylene Production

Abstract:

Data is becoming abundant and supplied in many different forms through mobile devices, wireless communications and big storage servers. With rich available data production facilities are applying "Big Data" concepts to analyze, identify and correlate parameters that indicate asset performance and health. Many model based solutions have been applied to Ethylene production providing significant benefits in monitoring process performance and optimizing production. Model based solutions feature intrinsic relationships characterized by first principle concepts or determined empirically that have been qualified for accuracy. Incorporating existing model based solutions and big data analysis offers potential for discovering process indicators that improve understanding of process and equipment performance. The resulting awareness advances process knowledge beyond "what is happening to the process" to "how did the process gets to this point" and "what other paths for the process to take" . Evaluating information generated by model based solutions and stored in large data sources can be cumbersome. Collaboration with analyzing large amounts of information and displaying discoveries in digestible form to operations and managers such as dashboards and high performance graphics often requires support from information technologist. Reliance on IT to provide a forum for data analysis and presentation of findings may restrain advances in process knowledge discoveries. Solutions that incorporate model based applications, big data analysis and provides engineers with a forum for easily analyzing information and displaying findings provides an opportunity for advancing asset performance and health.

513439 Improving Ethylene Furnace Performance with Big Data Analytics

Abstract:

Parameters around furnace performance include throughput, yield, runlength, coil life, and energy efficiency. Increasing the factors that allow maximization of any of one of these furnace performance criteria will reduce the level of performance of the other criteria. Big Data analytical techniques have been repeatedly proven to effectively find the optimal settings to meet the furnace operational objectives.

A key to successful implementation of Big Data to Ethylene furnaces is, however, to combine it with first principles reaction kinetic models. Pure empirical models work well with predicting equipment failure and thereby addressing availability issues. Availability is a variable for Ethylene furnaces to be managed. While one can argue that with sufficient data and time, empirical models can be made to replicate furnace performance over a range of feedstocks and conditions; however first principles models can also replicate performance closely with an enhanced ability to view relationships. The logical approach is to utilize the modeling approach (i.e. first principles versus empirical) that can best illustrate actual performance within the constraints of difficulty and time in building and maintaining the model. This approach can be very effectively utilized to increase production and runlength.

Adjustment variables for the furnaces include firing, draft, flow, feed balancing, and dilution steam and understanding the ultimate effect of these on coil inlet pressure, coil outlet pressure, tube metal temperature, pressure drop in TLE, firing duty, furnace efficiency, residence time, and coil outlet temperature is key to better operation of the furnace. By combining first principles modeling with predictive modeling using Big Data techniques results the ability to completely predict the performance of the furnace over the life of its run. The impact of adjustments is reflected in the performance and production of the furnace over the likewise variable runlength of the furnace. With this information furnaces can be set to optimize the production goals and targets that themselves change with feedstock variability and planned decoke scheduling.

In this presentation, models utilized will be described and their output provided. Use of this output to identify furnace anomalies at various stages in its life and the ability to take corrective actions will be shown. While dependent on feedstock availability and downstream limitations, achieving anywhere from 5% to 20% increase in production has occurred in numerous facilities.

514113 Ethylene Risk and Technology Review: What's the Outlook?

Abstract:

Per the court-mandated timeline, the EPA is currently working through the statutorily-required risk and technology review (RTR) for the ethylene NESHAP (40 CFR, Part 63, Subpart XX and Subpart YY). Industry trade groups have been engaged in meetings with EPA as this process has moved forward. It is anticipated the proposed rule, due in late 2018, will impact current ethylene operations. It is anticipated the rule will mirror certain aspects of the refinery rule, such as the flare and PRD language and likely will address furnace stack emissions and decoke emissions (both decoke to the fire box and decoke to an atmospheric pot) and potentially waste water emissions.

This paper will discuss the possible proposed rules and the potential impact on the ethylene facilities.

AIChE PAPER NUMBER

Delivering a Successful Ethylene Section 114 Information Collection Request Test Program

David L. Elam, Jr
Project Director and Vice President
TRC

Jeff Thomason
Air Source Testing Manager
John Zink Hamworthy Combustion

Prepared for Presentation at the Spring National Meeting
Orlando, FL April 22-26, 2018

Delivering a Successful Ethylene Section 114 Information Collection Request Test Program

David L. Elam, Jr
Project Director and Vice President
TRC
Jeff Thomason
Air Source Testing Manager
John Zink Hamworthy Combustion

Abstract:

Although every emission testing project is important, some are more important than others. Routine test programs help sources demonstrate compliance with permit limits. Engineering test programs help sources make fiscally responsible decisions regarding operations or the installation of air pollution control equipment. But Information Collection Request (ICR) test programs are different – the data from these programs will be used to evaluate emission levels and may result in the establishment of emission limits or work practices that can affect the economic viability of an individual facility or an entire industry. Accordingly, ICR test programs demand the highest level of test program planning and execution to deliver defensible data that fully conforms to the requirements of the ICR.

In April 2016, EPA issued an ICR for ethylene production facilities under the statutory authority of Section 114 of the Clean Air Act (CAA). This ICR was one of the most complex that EPA has issued and presented significant technical and logistical challenges. The decoking process itself presents numerous technical challenges including elevated moisture, stack temperature variability, operational availability, and source access.

This paper and associated presentation reflect the mantra of "prepare hard, test easy." The significant technical hurdles in sampling will be addressed, as well as the logistical and personnel side of the project. When the logistical aspects of the sampling program from glassware washing procedures to the need to have 4,000 pounds of ice delivered are handled before test team deployment, the full intellectual capabilities of the test team can be focused on the range of challenges associated with field operations.

Introduction

An Information Collection Request (ICR) is a set of documents that describes reporting, recordkeeping, survey, and other information collection requirements that a federal agency imposes on the public. The Paperwork Reduction Act (PRA) requires that every federal agency must obtain approval from the Office of Management and Budget (OMB) before collecting identical or similar information from 10 or more members of the public. ICR requirements can vary in complexity, ranging from a request for existing information that is readily available to one that requires the production of substantial historical data and the generation of new data. Agencies use ICR data to comply with statutes. For example, the U.S. Environmental Protection Agency (EPA), under the statutory authority of Section 114 of the Clean Air Act uses ICRs to fulfil its requirements associated with the residual risk and technology reviews of New Source Performance Standards (NSPS) and Maximum Achievable Control Technology (MACT) standards.

A cursory search of www.regulations.gov indicates that EPA renewed or initiated close to 200 ICRs in the 12 month period beginning on July 1, 2015. Many of these were renewals, simply requesting the continued collection of data associated with previously approved ICRs; however, some ICRs imposed substantial requirements on industry. Specifically, the ICR targeting Ethylene Production Facilities issued in April 2016 certainly cost targeted facilities several hundred thousand dollars to collect and report the required information. And perhaps more important than the near-term costs associated with completing the ICR are the long-term and ongoing costs associated with regulations that will be promulgated based on ICR results.

This paper describes an approach for end-to-end management of an ICR program, focusing on specific requirements of EPA's Section 114 ICR for Ethylene Production Facilities.

Managing an ICR as a Project

Responding to an ICR is an important undertaking for an organization that should be approached using project management best practices. Although responding to an ICR can involve several hundred individual tasks and span several months, six areas require particular attention to ensure that requested information satisfies its intended purpose, is responsive, qualified, technically sound, reported accurately, and properly incorporated in rulemaking.

1. Participate in Stakeholder Meetings that Shape the ICR. EPA typically seeks involvement of the affected community when developing ICRs. Often, key trade organizations and industry groups are invited to participate in the development of the ICR. It is important to use this opportunity to understand EPA's goals in conducting the ICR, the schedule for the ICR program, and how ICR data will be used. Trade organizations and industry groups have the relevant process and operational knowledge critical to guiding the development of a practical and realistic ICR that yields the required data.

2. Assemble the ICR Response Team. An ICR response often requires efforts from environmental, operational, and information technology staff. The legal department may need to be involved to ensure protection of confidential information. Contractors and consultants may be required to support the ICR with site preparation, data collection, data analysis, reduction, and interpretation. Engage these team members early and rely on their expertise to support the ICR response through the entire process.

3. Provide Feedback on Draft ICRs. It is important to review and understand draft ICRs and comment on them in the required timeframe. Although it is easy to focus on the cost of responding to the ICR when seeking changes, the most persuasive arguments for modifying an ICR are found in the areas of safety, technical feasibility, and data defensibility. When challenging a draft ICR in any of these areas, it is important to be clear about the specific issue, the limitations of the proposed approach, and provide an alternate approach to satisfying the data need. For example, if it is unsafe to conduct a particular sampling operation, provide a defensible engineering calculation or alternative sampling approach that satisfies the data need. Similarly, if a measurement yields a detection limit that will not guide a meaningful risk assessment, explain the sampling, analytical, or process operation issues that affect the detection limit and how that detection limit will drive interpretation and use of ICR data.

4. Perform the ICR Work. Once the ICR is issued, manage the effort as a project relying on a project manager for coordinating and tracking the various tasks. Although an ICR may be issued with incremental submittal requirements, it is likely that the ICR will require simultaneous work on multiple tasks instead of independent work on sequential tasks to meet the submittal schedule. Responding to a complex ICR will require the efforts of an overall project manager and a team of task managers, who must balance ICR response with ongoing operational responsibilities.

5. Report ICR Data on Time and in the Required Format. Reported ICR data are going to be used by EPA to evaluate the effectiveness of existing regulations and determine the need for new or revised ones. It is therefore important that ICR data be collected in time to allow adequate review of the information before it is submitted. Although conformance with the ICR requirements is important, it is also important to evaluate the collected data from a broader perspective, considering data quality, completeness, and representativeness. ICRs may require electronic reporting, which can be challenging to learn. Accordingly, it is important to allow enough time for both the reporting of electronic information and the review of electronically reported information.

6. Remain Engaged in the Regulatory Process as a Stakeholder. As EPA reviews, reduces, and analyzes ICR data, it is important for affected stakeholders to understand the dataset in terms of technical accuracy and quality and remain engaged with EPA to guide the interpretation of data and the application of practical pollutant control strategies and achievable emission limits.

Overview of the Ethylene Production Facility ICR

In April 2016, EPA issued an Information Collection Request (ICR) for ethylene production facilities under the statutory authority of Section 114 of the Clean Air Act (CAA). As required by Section 112 of the statute, EPA must examine Maximum Achievable Control Technology (MACT) standards with respect to residual risk and technological advancements every eight years and promulgate revised MACT standards if the Risk and Technology Review (RTR) indicates that changes are necessary. EPA uses the authority of Section 114 of the CAA to gather information that supports the RTR evaluation process.

MACT requirements for ethylene production plants were promulgated in 2002 (40 CFR Part 63, Subpart XY; Generic Maximum Achievable Control Technology Standards and 40 CFR Part 63, Subpart XX, Ethylene Manufacturing Process Units: Heat Exchange Systems and Waste Operations). The Ethylene MACT regulates emissions of the following organic hazardous air pollutants (HAPs):

- Benzene
- 1,3-Butadiene
- Cumene
- Ethylbenzene
- Hexane
- Naphthalene
- Styrene
- Toluene
- *o*-, *m*-, and *p*-Xylene

EPA's recent Ethylene Production ICR was issued to 20 separate facilities representing a universe of 191 ethylene cracking furnaces and 37 heat exchangers. Unless a facility conducted testing that met the requirements of the ICR in the preceding five years, each facility had to select **one** ethylene cracking furnace for emission testing during cracking and decoking operations and **all** identified heat exchange systems.

The ethylene cracking furnace test programs were particularly complex, requiring multiple sophisticated sampling trains to measure the aforementioned compounds and:

- Aldehydes (Acrolein, Acetaldehyde, Formaldehyde, Propionaldehyde)
- Ammonia
- Carbon Dioxide
- Carbon Monoxide
- Ethane
- Ethylene
- Filterable Particulate Matter
- Flow Rate
- Hydrogen Cyanide
- Metal HAPs (Sb, As, Be, Cd, Cr, Co, Pb, Ni, Mn, Hg, Se, And P)
- Methane
- Oxygen
- Polycyclic Aromatic Hydrocarbons
- Propylene
- Total Hydrocarbons

The heat exchanger test programs involved collection of water samples for laboratory analysis and the use of the El Paso method to determine the following strippable compounds:

- Benzene
- 1,3-Butadiene
- Chlorine

- Chloroform
- Cumene
- Dimethylformamide
- Ethylbenzene
- Ethylene
- Hexane
- Naphthalene
- Propylene
- Styrene
- Toluene
- Total Strippable Volatile Organic Compounds (VOC)
- *o*-, *m*-, and *p*-Xylene (and total)

Although EPA identified a range of compounds that had to be measured in connection with these test programs, it did not establish detection limits for the measurements. Instead, EPA identified sample methods and run times.

EPA required the use of FTIR for several of the target pollutants. In addition, continuous emission analyzers (CEMs), two isokinetic sample trains, and Reference Method 18 sampling procedures were required to quantify the numerous target compounds. Seven (7) test runs were required for the ethylene cracking furnaces. During decoking operations, which can span several weeks due to operational schedules, EPA required 9 test runs (3 runs for each of 3 cycles). All of the test data and results had to be submitted via the Electronic Reporting Tool (ERT). Test methods and run times are summarized in Tables 1 and 2.

Table 1: Furnace Stack - Test Matrix				
Parameter	**Test Method**	**# of Test Runs**	**Run Length**	**Units of Measure**
Aldehydes[1]	ASTM D6348-03	7	60 minutes	lb/hr and ppmvd @ 3% O_2
Ammonia	ASTM D6348-03	7	60 minutes	lb/hr and ppmvd @ 3% O_2
Organic HAP[2] (1,3-Butadiene[3])	EPA Method 18 (ASTM D-6348-03)	7	60 minutes	lb/hr and ppmvd @ 3% O_2
Polycyclic Aromatic Hydrocarbons (PAH)[4]	CARB Method 429	7	Minimum of 2 dry cubic meters (~2 hours)	lb/hr and µg/dscm @ 3% O_2
Total Hydrocarbons (THC)	EPA Method 25A	7	60 minutes	lb/hr and ppmvd @ 3% O_2
Methane Ethane Ethylene Propylene	ASTM D6348-03	7	60 minutes	lb/hr and ppmvd @ 3% O_2
Nitrogen Oxides (NOx)	EPA Method 7E	7	120 minutes, minimum	lb/hr and ppmvd @ 3% O_2
Filterable PM and Metal HAP[5]	EPA Method 5 EPA Method 29	7	Minimum of 4 dry cubic meters (~4 hours)	lb/hr and µg/dscm @ 3% O_2
Hydrogen Cyanide	ASTM D6348	7	60 minutes	lb/hr and µg/dscm @ 3% O_2
Carbon Monoxide	EPA Method 10	Concurrent with all test runs	4 hours	lb/hr and ppmvd @ 3% O_2
Flow Rate[6]	EPA Method 2 O_2-EPA Method 3A EPA Method 4 (CO_2-ASTM 6348-03)	Concurrent with all test runs (7)	4 hours (60 minutes)	

[1]Acrolein, Acetaldehyde, Formaldehyde, Propionaldehyde
[2]Benzene, Cumene, Ethylbenzene, Hexane, Styrene, Toluene, o-Xylene, m-Xylene, p-Xylene
[3]1,3-Butadiene measured by ASTM D6348-03
[4]The target analyte list is included as Table 1 of CARB Method 429
[5]Metal HAP includes compounds of Antimony, Arsenic, Beryllium, Cadmium, Chromium, Cobalt, Lead, Nickel, Manganese, Mercury, and Selenium and emissions of Phosphorus. Report concentrations and mass emissions rates of each pollutant separately.
[6]Measured in conjunction with M5/29

Table 2: Decoking Pot Stack - Test Matrix				
Parameter	**Test Method**	**# of Test Runs**	**Run Length**	**Units of Measure**
Aldehydes[1]	ASTM D6348-03	3 runs per cycle	60 minutes	lb/hr and ppmvd @ 3% O_2
Ammonia	ASTM D6348-03	7	60 minutes	lb/hr and ppmvd @ 3% O_2
Organic HAP[2] (1,3-Butadiene[3])	EPA Method 18 (ASTM D6348-03)	3 runs per cycle	60 minutes	lb/hr and ppmvd @ 3% O_2
Polycyclic Aromatic Hydrocarbons (PAH)[4]	CARB Method 429	3 runs per cycle	Minimum of 2 nominal actual cubic meters (~2 hours)	lb/hr and μg/dscm @ 3% O_2
Total Hydrocarbons (THC)	EPA Method 25A	3 runs per cycle	60 minutes	lb/hr and ppmvd @ 3% O_2
Methane Ethane Ethylene Propylene	ASTM D6348-03	3 runs per cycle	60 minutes	lb/hr and ppmvd @ 3% O_2
Nitrogen Oxides (NOx)	EPA Method 7E	3 runs per cycle	120 Minutes, Minimum	lb/hr and ppmvd @ 3% O_2
Filterable PM and Metal HAP[5]	EPA Method 5 EPA Method 29	3 runs per cycle	Minimum of 4 nominal actual cubic meters (~4 hours)	lb/hr and μg/dscm @ 3% O_2
Hydrogen Cyanide	ASTM D6348	3 runs per cycle	60 minutes	lb/hr and μg/dscm @ 3% O_2
Carbon Monoxide	EPA Method 10	Concurrent with all test runs	4 hours	lb/hr and ppmvd @ 3% O_2
Flow Rate[6]	EPA Method 2 O2-EPA Method 3A EPA Method 4 (CO2-ASTM 6348-03)	Concurrent with all test runs (7)	4 hours (60 minutes)	

[1]Acrolein, Acetaldehyde, Formaldehyde, Propionaldehyde
[2]Benzene, Cumene, Ethylbenzene, Hexane, Styrene, Toluene, o-Xylene, m-Xylene, p-Xylene
[3]1,3-Butadiene measured by ASTM D6348-03
[4]The target analyte list is included as Table 1 of CARB Method 429
[5]Metal HAP includes compounds of Antimony, Arsenic, Beryllium, Cadmium, Chromium, Cobalt, Lead, Nickel, Manganese, Mercury, and Selenium and emissions of Phosphorus. Report concentrations and mass emissions rates of each pollutant separately.
[6]Measured in conjunction with M5/29

EPA established an aggressive schedule for ICR completion as indicated in the following table:

Table 3: ICR Schedule	
ICR Element	**Due Date**
Submit Historical Stack Testing Results for Ethylene Cracking Furnaces, Decoking Operations, and Heat Exchanger Systems	July 29, 2016
Submit Heat Exchanger System Sampling and Analysis Data	September 30, 2016
Submit Ethylene Cracking Furnace Stack Testing Emissions Data	November 30, 2016
Submit Decoking Operations Stack Testing Emissions Data	November 30, 2016

EPA did not require affected facilities to submit test plans in connection with these test programs. Although test plans were not required, TRC developed test plans for ethylene production ICR test programs that conformed to the requirements of ASTM D7036 (Standard Practice of Competence for Air Emission Testing Bodies). Test plans are critically important for complex test programs that require large test teams, multiple and simultaneous test methods, specific unit operating conditions, the collection of defined operational data, and specific reporting requirements. Documenting test program requirements in advance of the test program allows the facility and test team to anticipate problems, develop contingency plans, and ensures that all ICR program requirements are satisfied.

Execution of the Ethylene Production Facility ICR

The ethylene production facility ICR test program presented safety, scheduling, staffing, method application challenges, logistical support, and reporting. This section highlights those challenges and TRC's approach to managing them.

- **Safety:** All emission testing work within a petrochemical facility requires heightened attention to safety. Safety management for the ethylene production facility ICR test program was especially challenging due to the heat at the sample location (120°F), the size of the test team (8-9 people), the crowding associated with multiple sample trains, and the 14-16-hour work days (in some cases, round-the-clock testing in two shifts). TRC assigned a dedicated to safety officer to the program, selected a test team with previous petrochemical experience, mandated hourly rest and hydration breaks in an air-conditioned area, and scheduled the test program over multiple mobilizations to satisfy both test program requirements and to provide a break for team members. In addition to the onsite safety issues, managing the front end safety processes associated with Transportation Worker Identification Credential (TWIC) requirements, background screening, employee drug testing, and site-specific requirements was challenging.
- **Scheduling and Staffing:** Each facility had to select one ethylene cracking furnace for emission testing during cracking and decoking operations and all identified heat exchange

systems. As directed in the ICR, three (3) runs had to be completed on the furnace stack and decoking pot stack simultaneously during three separate decoking cycles, for a total of eighteen (18) runs (9 runs per source) for all parameters listed in Table 1 (furnace stack) and Table 2 (decoke pot stack). To meet operational and personnel scheduling requirements, TRC completed the test programs in four separate mobilizations. The first mobilization lasted 2 days and required 3 people to complete testing of the heat exchange system. The second mobilization lasted 10 days and required 8 people to complete 7 furnace runs and 3 decoking runs. The third and fourth mobilizations each lasted 4 days and required 9 people to complete 3 decoking runs. Careful scheduling was required to coordinate operational requirements, assign testing team staff with petrochemical experience using complex and sophisticated test methods, and to ensure continuity of the test team for the four separate mobilizations.

- **Test Method Application:** The ICR identified 13 required test methods and provided alternates for 5 of the required test methods. Test method selection had to simultaneously satisfy source access, simultaneous testing requirements, and detection limits. After reviewing the list of compounds for which Method 320 was required, TRC opted for FTIR analysis by alternate ASTM Method D-6348. When Method 18 analysis was required and Method 320 and ASTM D-6348 were allowed alternates, TRC opted for the ASTM Method D6348 for only 1,3-butadiene. This approach allowed us to apply the robust quality assurance procedures in ASTM D-6348 where FTIR is the analytical method of choice and to apply conventional gas chromatography with flame ionization for hydrocarbon HAPs.
- **Logistical Support:** Each test program mobilization required considerable logistical support. Travel arrangements were required for test personnel coming from different TRC offices. Expendable supplies, particularly ice for the sampling trains and liquid nitrogen for the FTIR, had to be available onsite in sufficient quantities to support the test program. Additionally, couriers and offsite laboratory support were required to analytical requirements for filterable particulate matter, polyaromatic hydrocarbons, and metals.
- **Data Reduction and Reporting:** The ICR test program generated sampling and analytical data streams from a number for sources including handwritten physical measurements, continuous emission monitoring data, analytical data from the onsite FTIR and gas chromatograph, and analytical data from subcontract laboratories. Data reduction was labor-intensive requiring multiple quality control reviews to ensure accuracy of transcribed and merged electronic data. The final report for one test program spanned 3,254 pages. Importantly, the data had to be reported using EPA's Electronic Reporting Tool, a time-consuming process that is most efficiently performed when all data are available.

Although the ICR did not require a formal test plan, TRC believes that the successful completion of these test programs was accomplished by attentive test program planning. Detailed test planning, performed by an experienced project manager, ensured that the ICR test programs were completed safely by experienced personnel, without delays due to logistical support.

AIChE Paper Number: 165c

Predicting Long Term Evaporative Cooling Tower Water Usage: The Energy-Water Balance

William Celenza

Technical Professional Leader – Process

KBR, Inc.

Sarah Shank

Sr. Technical Professional – Process

KBR, Inc.

Prepared for Presentation at the 2018 Spring National Meeting
Orlando, Florida, April 25, 2018

Predicting Long Term Evaporative Cooling Tower Water Usage: The Energy-Water Balance

William Celenza
Technical Professional Leader – Process
KBR, Inc.

Sarah Shank
Sr. Technical Professional – Process
KBR, Inc.

Abstract:

Evaporative Cooling Towers are commonly used to dissipate heat from industrial process operations by cooling water through a combination of heat and mass transfer. An evaporative cooling tower can operate over a wide range of recirculating water rates, air velocity rates, and heat loads with variation in the approach of cold water to the ambient wet-bulb air temperature. Designed to manage summer worst case ambient conditions, the actual long term water and/or energy usage is overestimated, increasingly so for locations which experience extremes in seasonal weather. For instance, some evaporative cooling systems managing a constant industrial heat load can cool over 50% by non-evaporative air-cooled heat transfer; during reduced ambient air temperatures.

This paper presents a procedure for predicting long term water usage of a cooling system based on regional climactic data. The analysis focuses on industrial cooling systems consisting of wet towers that supply a common set of process loop heat exchangers, typically arranged in parallel; and will provide analysis for water or energy minimization. The estimated water usage on a typical monthly, seasonal or annual basis is presented to demonstrate the energy –water tradeoff for sites across the continental United States. The results estimate how much operational flexibility is available for a cooling system and offers an understanding of the interaction between water and energy requirements under varying climate conditions.

Introduction

Cooling towers are used for heat exchange, and when used to extract heat from operations in the oil refining and chemical process industries are often referred to as industrial process cooling towers. The design of cooling towers is based on a specified heat load, which in the ethylene industry is based on production and thus relatively constant throughout the year.

Water is the preferred heat-transfer medium because of availability, high heat capacity, and relatively low costs. Wet cooling towers dissipate the heat by recirculating water in direct contact with atmospheric air through convection and evaporation of water; measured as sensible and latent heat transfer, respectively. Both of these mechanisms operate at the air-water boundary. The driving force for heat and mass transfer is taken as difference in enthalpy between the water and air; although theoretically speaking the difference in water vapor pressure between the air and water phases is the actual driving force (1).

Figure 1 Typical Wet Cooling Tower System Arrangement

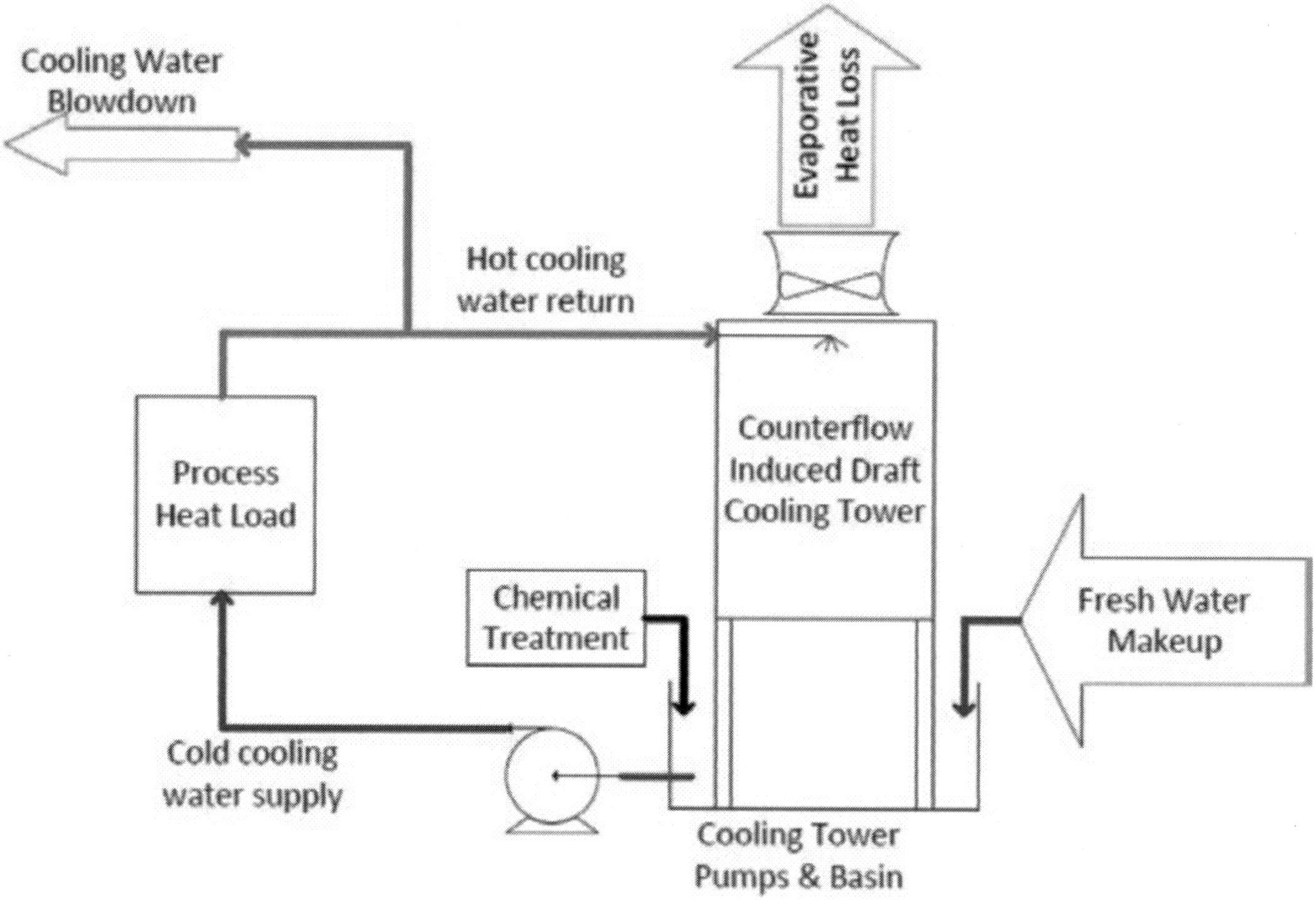

For maintenance reasons, heat transfer surfaces, such as heat exchanger tube walls, require a restriction be placed on cooling water velocity to prevent corrosion and fouling. Thus the combination of flow velocity and process heat load dictates the range, or fixed temperature reduction, the cooling tower's recirculating water must accommodate. A typical Wet Cooling Tower System Arrangement is depicted in Figure 1.

Open circuit counterflow-induced draft cooling towers, as shown in Figure 2, are predominant in the process industries; and only this type is considered further to simplify this analysis.

Figure 2 Typical Counterflow Induced Draft Cooling Tower Arrangement

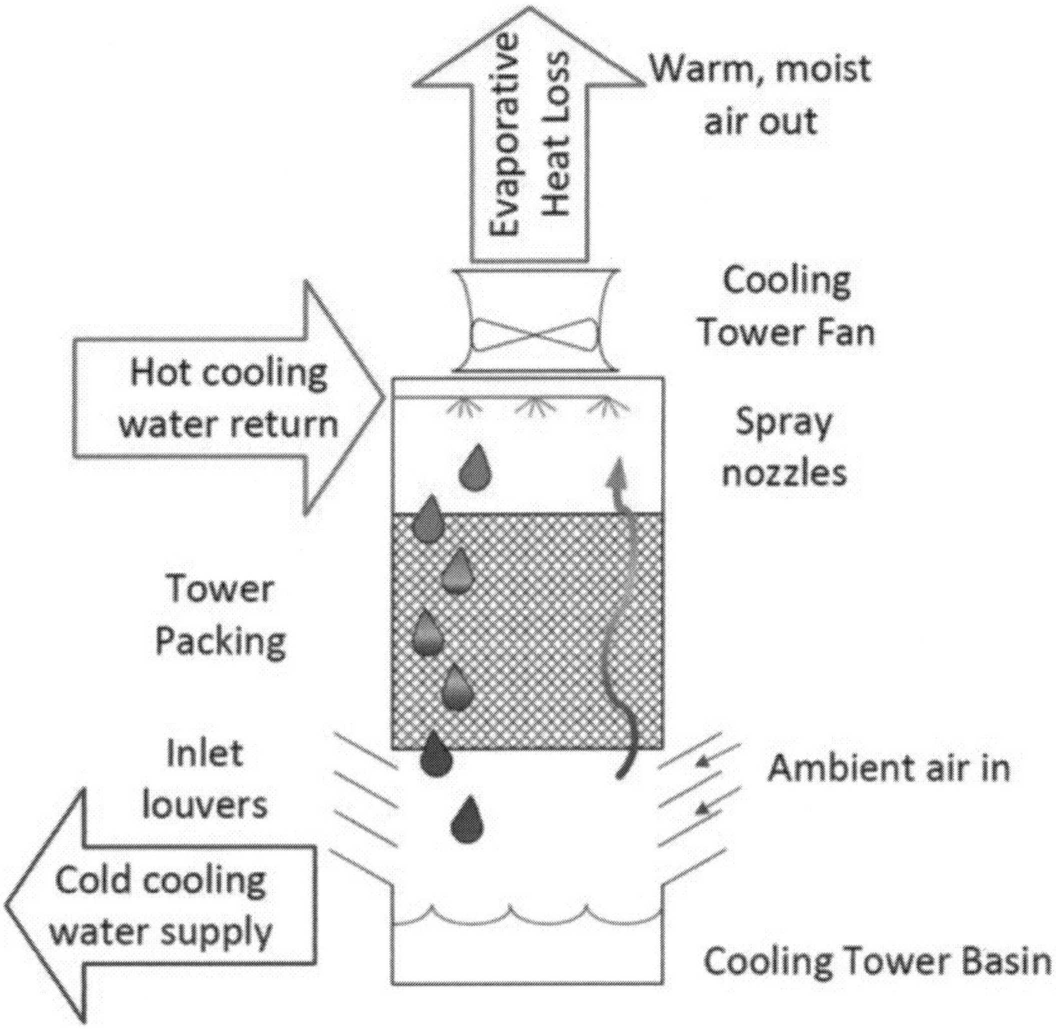

Cooling tower heat and mass transfer is based on evaporative cooling to the atmosphere. The maximum efficiency is limited by the local air temperature and ambient humidity, collectively measured as the wet-bulb temperature. The meteorological quantity called the wet-bulb temperature is measured with a normal thermometer that has a damp cloth covering the bulb which is fully ventilated. The wet-bulb temperature is always lower than the usually reported "dry-bulb" temperature; with how much lower dependent on the humidity. At 100% humidity the dry-bulb and wet-bulb temperatures are the same. Even though the amount of heat that needs to be extracted from the process is independent of the weather; the design and operation of wet cooling towers is dependent on ambient weather conditions. The design of wet cooling towers needs to ensure adequate cooling for continuous production under expected atmospheric conditions. However 99% of the time, cooling towers are not called upon to evaporate water at the design rate; creating an opportunity to conserve energy or water resources much of the year.

Predicting water usage in cooling systems can serve as a tool to provide for more efficient water resource management. Lakovic (2) reports that over 60% of the world's reliance on water is due to cooling purposes; and that by applying a re-circulation cooling system, such as a wet cooling tower, up to 98% savings in water consumption can be realized over once-through cooling (2,3,4). Though the design of wet cooling towers is a competitive field, where models and constants can be proprietary an approximate design is presented here using rules-of-thumb to provide for water evaporation rates and air blower power requirements. For this paper, the ethylene production facility wet cooling tower design outlined in Table 1 was used as a base case for a simplified model to predict water usage in wet cooling towers.

Table 1 Base case ethylene production facility wet cooling tower design

Parameter	**Value**	**Comment**
Ethylene Production Rate	1,000 MM lb/yr	Instantaneous
	890 MM lb/yr	Nameplate
Cooling Tower Water Recirculation Rate	105,000 gpm	Considered to be constant
Cooling Tower Design Temperature Range	15.8 F degrees	Design Hot Water Temp = 100 F, Design Cold Water Temp = 84.2
Design Outlet Air Temperature	92.1 F	Outlet Air temp = (Hot Water Temp + Cold Water Temp)/ 2 For more information see reference (5).
Design Liquid / Gas (Water to Air) ratio	1.1	L/G ratio = (Enthalpy Outlet Air - Enthalpy Inlet Air) / (Specific Heat * (Hot Water Temp – Cold Water Temp)). For more information see reference (5).
Ambient wet-bulb temperature increase to account for drift recirculation	1 F degree	Used as the cooling tower design wet-bulb temperature.

As tracking water consumption is just one piece of a water management program, this paper also presents how an energy management strategy can be applied to water conservation. The discussion first presents a water use analysis, then identifies key components of an energy strategy that can be applied to wet cooling tower water management.

Predicting Water Usage

Water is used in two ways in the ethylene production process: indirectly for heat transfer as cooling water or steam, and directly in the process. Most of the water is used for steam production and cooling towers, with up to 75% routinely used to replace water lost to evaporation in cooling towers (3). The remaining 25% is used to supply both process and other utilities, such as replacing cooling tower blowdown. Less than 10% of the supplied water sees a direct process use, and the supply water used as cooling tower makeup which is the sum of the evaporation, blowdown, and drift rates can account for 90% of water use in ethylene production. Note also that cooling tower blowdown can contribute over 75% of the total process wastewater, or dry weather flow.

Table 2 Summer and winter weather design conditions for evaporative cooling towers at nine US locations (6b, 6c)

	Ambient Air Temperature (F)				Cooling Water Cold Temperature			
Location	**Summer Wet Bulb, (a)**	**Summer Dry Bulb, (b)**	**Winter Wet Bulb, (a)**	**Winter Dry Bulb, (b)**	**Summer Supply F**	**Average Monthly Supply F**	**Winter Supply F**	**Max Season Variation %**
New Orleans, LA	81	89	31	34	84	74	55	35
Brownsville, TX	80	88	38	40	84	77	60	29
Houston, TX	79	88	29	31	84	75	55	35
Belleville, IL	78	88	8	9	84	68	43	49
Des Moines, IA	77	88	-4	-3	84	65	37	57
Phoenix, AZ	75	97	35	40	84	74	61	28
Pittsburgh, PA	73	82	6	7	84	69	45	47
Minot, ND	70	85	-18	-18	84	64	33	61
San Francisco, CA	63	74	38	40	84	78	70	17
Average	**75**	**87**	**18**	**20**	**84**	**71**	**51**	**40**

Notes:

a) For Summer Design Condition, the Wet Bulb Temp is 1.0% Occurrence Value. For Winter Design Conditions the Wet Bulb Temp is the Mean Coincident (Average) Value at the 99.0% Dry Bulb Temp Occurrence (6c).
b) For Summer Design Condition, the Dry Bulb Temp is the Mean Coincident (Average) Value at the 1.0% Wet Bulb Temp. For the Winter Design Condition, the Dry Bulb Temp is the 99.0% Occurrence (6c).

Summer and winter weather design conditions for evaporative cooling towers at nine US sites that are near to ethylene production facilities are included in Table 2. Design wet-bulb temperatures usually vary from 60 to 85 F in the US depending on the geographic area, and are usually equaled or exceeded only 2.5 percent to 5 percent of the total summer operating time (7). In the United States summer was standardized as June through September. However, the design values listed in Table 2 are annual values not seasonal values; with the 1% annual value usually about a degree cooler than the 2.5% seasonal value. The 99% winter design annual values are generally cooler than the 99% seasonal values that are based on the three coldest months of the year: December through February (6a). Recognizing that engineered solutions are seldom designed for the worst possible natural event; for petrochemical applications, a stringent 1% wet bulb exceedance level has been considered appropriate (8). Although not covered in this analysis, cooling tower performance during exceedance times is a concern for uninterrupted industry operation.

For development of seasonal or annual water usage patterns, mean monthly average wet bulb temperatures were collected for each site and used to model cooling tower operations. The mean monthly average wet-bulb temperature was derived from the mean monthly average dry-bulb temperature and the mean monthly average humidity ratio (6b). The cooling tower industry typically quotes the cooling in a cooling tower is on average 75% through an exchange of the latent heat of evaporation of water (5), with the balance through the sensible heat loss from the water. As further stated in reference 17, the percentage of heat removed by evaporative cooling in wet cooling towers varies between 60% and 90% during typical operating and ambient conditions. Analysis of the data in Figure 3 also indicates that on a yearly average basis across all the sites studied, 75% of the heat transfer is due to evaporative cooling.

Both forms of transferred heat together raise the dry-bulb and wet-bulb temperature of the air passed through the tower, which is then exhausted to the atmosphere (9). However, for the same heat load the ratio of latent to sensible heat transfer will vary over changing ambient air conditions. Figure 3 shows how the percentage of total heat transfer in a cooling tower due to water evaporation or latent heat can change during the months of the year. For most locations in Figure 3, the percentage varies from 60% to 90% over the course of the year; however, the total variation is from 40% to 105%. Note that under certain conditions, such as summer in Phoenix, both the water and air are cooled by evaporation so the evaporative cooling exceeds 100% of the cooling water effect measured only on the water.

The wet cooling tower size is a function of the water recirculation flow rate, the temperature change of the recirculating water, the wet bulb temperature, the approach to the wet bulb temperature of the returned cooled water, the air velocity through the tower, and heat transfer area provided by the tower dimensions with fill pack, if any. The hot water temperature minus cold water temperature is termed the cooling range and the difference between cold water and wet-bulb temperature is called approach. For a given cooling tower design, the quantity of water to be cooled and its inlet temperature is usually known. The outlet water temperature is specified, so the design range is set. With a design ambient wet-bulb temperature dictated by site location, only an approach would need to be chosen. In design practice the actual temperature of the cooled water is higher than the wet bulb temperature of the entering air usually by 10 to 15 degrees Fahrenheit, but seldom less than five degrees Fahrenheit.

Figure 3 Seasonal Variation of Percentage of Total Cooling Tower Heat Transfer Due to Latent Heat (Evaporative Cooling)

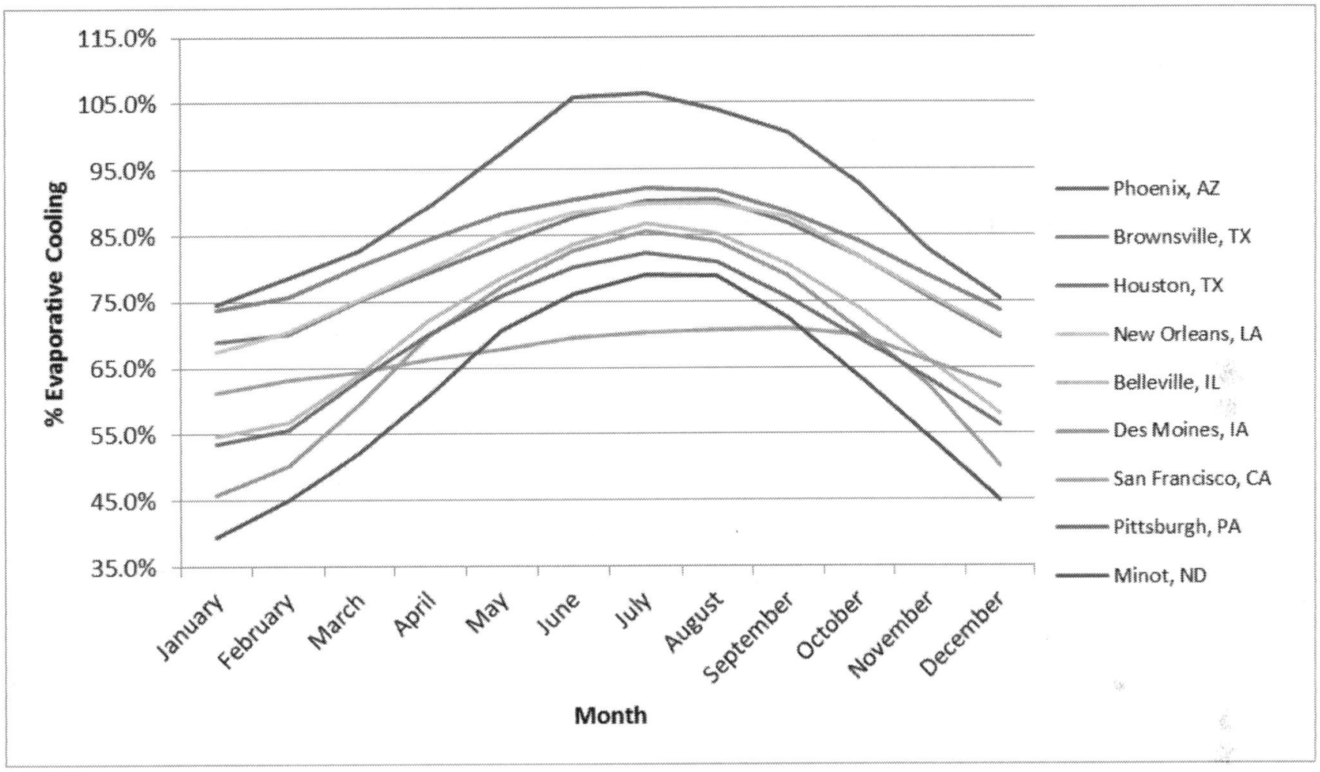

In turn for non-design ambient conditions, the size of the cooling tower, the recirculating water flow rate and the actual ambient wet-bulb temperature determine the inlet and outlet water temperatures, but not the difference between them. The amount of heat transferred to the atmosphere by the cooling tower is always equal to the heat load imposed on the tower, while the temperature at which the heat is transferred is determined by the thermal capability of the cooling tower and the entering wet-bulb air temperature. Increased tower performance can only be achieved by adding surface area or by boosting the air flow (10).

As stated earlier, cooling tower performance is related to ambient wet-bulb temperature. Wet bulb temperature is a site condition. At a constant ambient temperature, the lower the wet bulb, the drier the air, the more moisture the air can absorb and the more latent heat a given cooling tower can transfer. The highest wet-bulb temperatures occur during summer months, when air temperatures and humidity are the highest. During all other times, the cooling tower is likely oversized for the actual required heat transfer; and the cold water temperature will simply be less than the design supply temperature. Thus the cooling water temperature will vary by the season of the year, with all nine locations experiencing an average 40% difference between summer and winter design temperatures as shown in Table 2; even with modulating air flow to at least partially compensate for lower ambient temperatures. For this analysis the design cooling water supply temperature has been taken as 84 F as listed in Table 2, with the resulting monthly average supply temperature across all the sites being 71 F. The monthly average temperatures themselves can vary over 30 degrees F during the year, with an average site monthly variation of 20 degrees F during the year.

Note that in operation of cooling tower systems during subfreezing weather, a by-pass circulation method may be used to protect the tower from icing. For instance if the cooling water supply temperature should fall below 40F, a partial or full by-pass of the hot process cooling water may be implemented. In addition, induced draft fans may be reversed periodically in order to de-ice the air intakes. Thus the theoretical benefit of the ambient conditions towards process cooling may not be realized in practice.

Appreciating how seasonal design conditions influence the cooling tower operation is important to understand. Hot and cold water temperatures fall when the ambient wet-bulb temperature falls at constant heat load, range, and water flow rate. The cold water temperature from the cooling tower decreases in a linear relationship, reported as between 1.5 to nearly 2 degree wet-bulb temperature change resulting in a 1 degree cold water temperature change (5, 9, 11). This relationship is also a weaker function of the recirculating water to air ratio. For this analysis, a 1.7 ratio of the change in the wet-bulb temperature to the cold cooling water temperature is used and assumed constant over varying water to air ratios of 1 to 1.25. Thus when the wet-bulb temperature changes from 81F to 31F, or 50 degrees, from summer to winter for New Orleans as listed in Table 2 the predicted approach, the difference between the wet-bulb and cold water temperatures, would have changed only 50/1.7 or 29 degrees. The resulting cold water temperature would then be 84 – 29 = 55F at the 31F wet-bulb temperature.

However, the air's overall ability to absorb moisture reduces with temperature; the maximum absolute humidity level is halved for every 20 degree Fahrenheit temperature drop. Air at a higher wet-bulb temperature is capable of absorbing more heat according to the psychrometric properties of air. Therefore, at the lower wet-bulb temperatures a greater proportion of heat rejection is sensible; so the amount of water evaporated is reduced compared to the summer design condition (11). Figure 4 shows the predicted change in water lost to evaporation over the course of a year at the different locations. According to the ASHRAE (12) rule of thumb, the evaporation rate at typical summer design conditions is approximately 1% of the water flow rate for each 12.6F of water temperature range. Per the ASHRAE rule of thumb, the 15.8 F cooling water range listed in Table 1 for this analysis would require a 1.25% recirculating water evaporation rate. Although most sites in Figure 4 are within a 1.2 to 1.4% band during the summer, all are below 1.2% in the winter months and half are below 1% during winter. By analysis of Figure 4, the average monthly predicted evaporation rate for all sites over the year is 1.1%. This predicted rate is less than the ASHRAE predicted evaporation rate because the sensible component of the total heat transfer increases as the entering air bulb temperature decreases (4). The ratio of latent heat to sensible heat is important in analyzing water usage of a cooling tower. The ASHRAE rule of thumb appears to be more accurate for summer conditions, although even the Summer Design Conditions listed in Table 2 are predicted to average a 1.4% circulating water evaporation rate. For comparison, the Winter Design Conditions listed in Table 2 are predicted to average a 0.7% circulating water evaporation rate.

Figure 4 Seasonal Variation in Percentage of Recirculating Water Lost to Evaporation for Constant Cooling Load

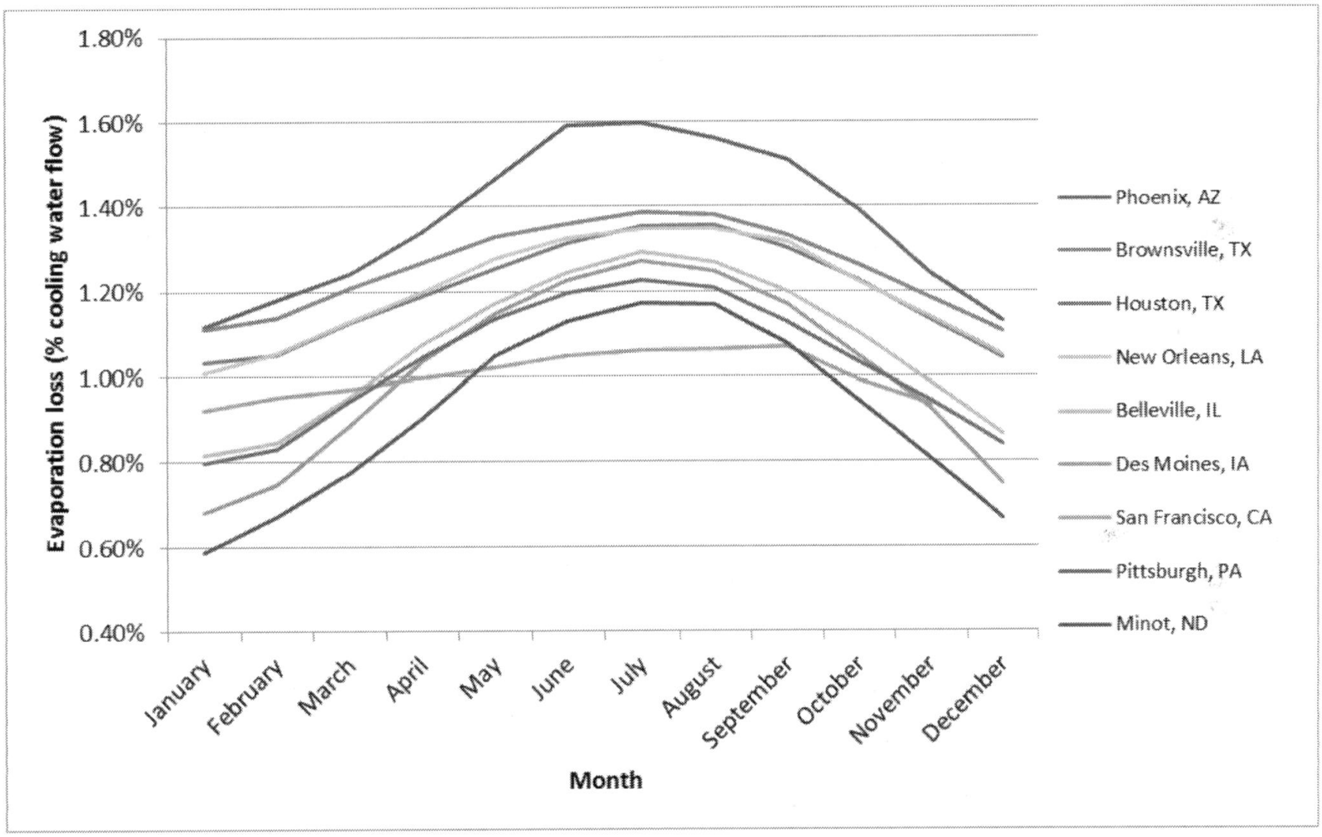

Based on the data presented in Table 3, average July summer conditions for a cooling tower in Pittsburgh, PA would be predicted to require 6% less water usage than at the Summer Design Conditions. For average January winter conditions, a cooling tower in New Orleans, LA would be predicted to require 30% less water usage than at the Summer Design Conditions. On an annual basis, across all these locations a cooling tower would be predicted to use on average 20% less water than under the Summer Design Conditions.

Table 3 Relative predicted evaporative cooling water loss vs. Summer Design Condition at nine US locations for the same cooling tower design

Location	Summer Design (% Evaporative Water Loss)	Relative Predicted Evaporative Water Loss (% difference from Summer Design)			
		Average July	Average January	Winter Design	Annual
New Orleans, LA	1.43	-6	-30	-40	-16
Brownsville, TX	1.42	-2	-22	-36	-11
Houston, TX	1.41	-4	-27	-41	-15
Belleville, IL	1.4	-8	-42	-57	-24
Des Moines, IA	1.4	-9	-51	-63	-28
Phoenix, AZ	1.6	0	-30	-41	-15
Pittsburgh, PA	1.31	-6	-39	-52	-21
Minot, ND	1.36	-14	-57	-73	-33
San Francisco, CA	1.13	-11	-23	-30	-16
Average	**1.4**	**-7**	**-35**	**-48**	**-20**

In addition, as the evaporation rate is lowered by site ambient weather conditions; related site water usage would also be lowered. For instance, the blowdown flowrate is a function of the cycles of concentration in the cooling tower; which is dependent on the quality of the cooling tower makeup water supply. The blowdown is related to the evaporation rate through the cycles of concentration. At cycles of 4 to 5, the incoming makeup water quality would be concentrated at 4 to 5 times supply concentration in the recirculating cooling water by control of the blowdown rate. When the evaporation rate reduces 10%, the blowdown rate would reduce 3% for a cycle of 4, and 2.5% for a cycle of 5. The utility water use and wastewater generation flows then would reduce in direct proportion that the blowdown represents these flows. The blowdown is usually less than 15% of total utility use and up to 75% of process or dry weather wastewater flows at an ethylene production facility. So in comparison, a 10% reduction in the evaporation rate could reduce the water supply demand up to 7.5% in addition to 0.3% from blowdown; however, the evaporation impact on the wastewater flows would be limited to only the reduction in the required blowdown rate, which could translate into up to 2% less generated wastewater.

Note that the values presented in Figure 4 and Table 3 assume that a cooling tower can effectively modulate the air blower flow rate to adjust cooling tower capacity to accommodate ambient conditions. The correlation of energy usage and water usage is discussed in the next section.

Energy Management

Since cooling is primarily by transfer of latent heat, cooling tower size is based on the total heat content or enthalpy of the entering air. At any one enthalpy condition, the wet-bulb temperature is constant. Thus cooling towers are selected and guaranteed to cool a specific volume of water from a hot water temperature to a cold water temperature while operating at a design wet-bulb temperature, or constant enthalpy. However, most cooling towers encounter substantial changes in ambient wet-bulb temperature during the year. A cooling tower design should provide for energy efficient operation during the majority of the year when atmospheric conditions do not approach design cases and the tower has excess capacity.

The major operating expense of a cooling tower is the fan power consumption (5), which can be balanced against water usage partly by design choices in the tower construction. A low air rate requires a large tower, while a high rate in a smaller tower requires greater fan power. Typical limits in air velocity are 300 to 700 fpm in counterflow towers (4). The Liquid/Gas (L/G) ratio of a cooling tower is the ratio between the water and the air mass flow rates. A selected L/G ratio for tower operation depends on the type of fill media, but ranges from 1 to 2.

Cooling rates of mechanical draft towers depend upon their fan diameter and speed of operation. To maintain energy efficient operation at lower ambient temperatures, reducing airflow would be an option to reduce capacity; with fan speed reduction usually desirable over closing dampers on air intakes which just reduce fan housing efficiency. The capacity reduction can be achieved with fan cycling, multi-speed motors, extra motors, or variable speed or frequency drives. Fan cycling, or turning off the motor of one or more of a series of fans, is an easy method but does not work well when close temperature control is required and can result in frequent motor starts (10).

Table 4 Cooling tower capacity, fan speed and energy (13)

Fan Speed	Tower Capacity	Energy Use
100%	100%	1^3 = 1 x1 x 1=100% hp
90%	90%	0.9^3 = 0.9 x 0.9 x 0.9 = 72.9% hp
80%	80%	0.8^3 = 0.8 x 0.8 x 0.8 = 51.2% hp

As shown in Table 4, Cooling Tower capacity varies directly with the fan speed, but energy consumption varies with the cube of the speed ratio. Thus reducing the fan speed to 80% would provide for a nearly 50% reduction in horsepower demand. For instance, variable frequency drives coupled with a standard fixed-pitch fan would provide the ability to modulate fan motor speed to provide capacity control along with energy savings (4). At a minimum, two-speed fans are typically recommended since the high and low speeds also allows for flexibility in controlling the return water temperature. For two-speed fans with single winding motors, the low speed is always half of full speed (10). However, for industrial facilities with constant heat loads, the change in ambient conditions usually does not require a compensation in the air flow less than down to 75% of the design rate. Thus two-speed fans would not be able to efficiently adjust to only ambient temperature driven variations. Even fan cycling would have limitations. Without fans operating, air flow through a tower will continue by the thermal draft at flow rates up to one fourth that of operating two-speed fans on low speed (14). Both would be sizeable reductions when only a maximum of 20 to 25% reduction in fan speed or around 50% reduction in fan power is typically required, as shown in Figures 5 and 6.

Figure 5 Seasonal Variation in Percentage of Required Fan Horsepower for Constant Heat Load (A)

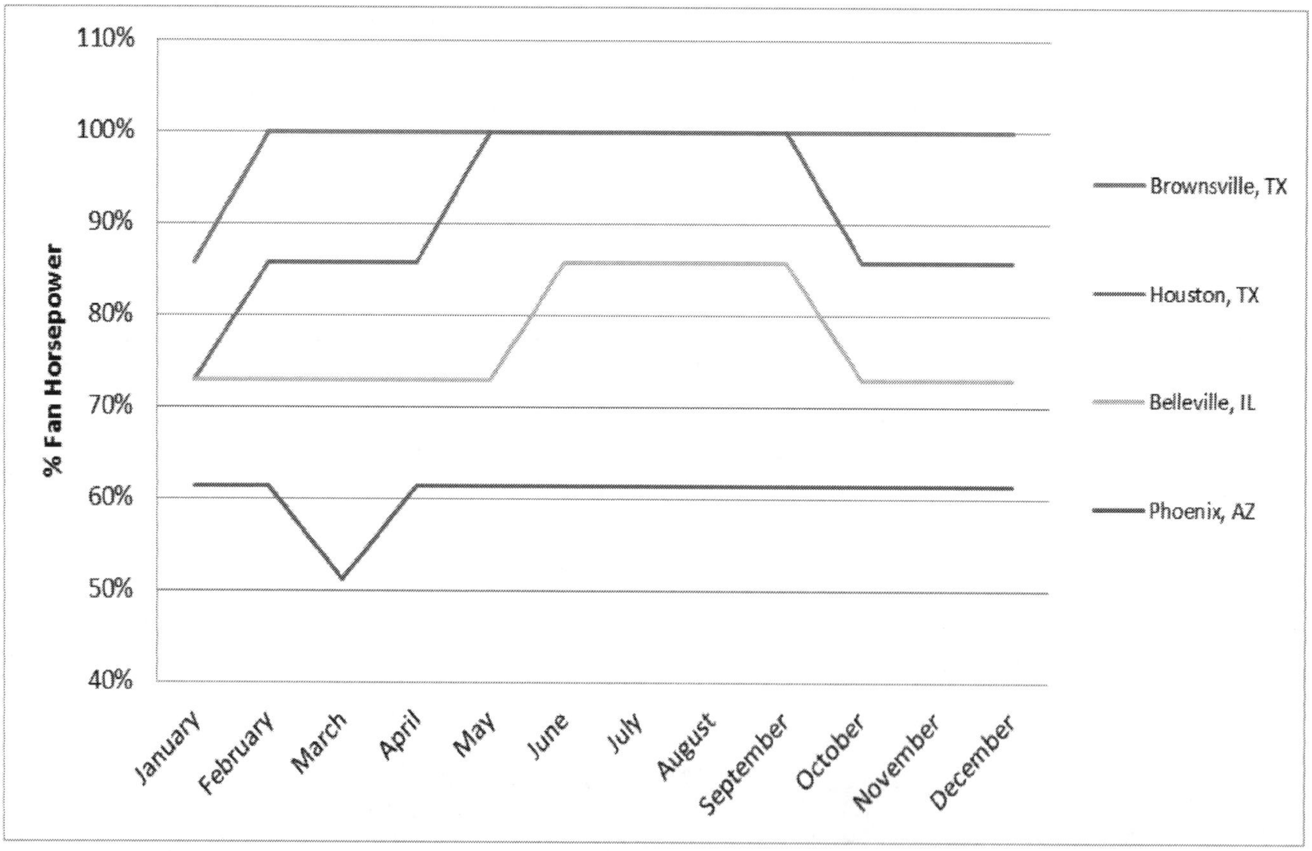

Figure 5 also summarizes the energy savings benefit of reducing tower capacity to match changing site ambient conditions. For these sites the energy demand is highest during summer conditions; with winter conditions being the season for lowest energy use, as well as the season for lowest water use. With the exception of the US Gulf Coast locations during summer, the monthly energy use usually averages more than 15% less than the equipment design capacity.

However, the sites graphed in Figure 6 either show constant energy use based on monthly average conditions or highest energy use during winter conditions.

Figure 6 Seasonal Variation in Percentage of Required Fan Horsepower for Constant Heat Load (B)

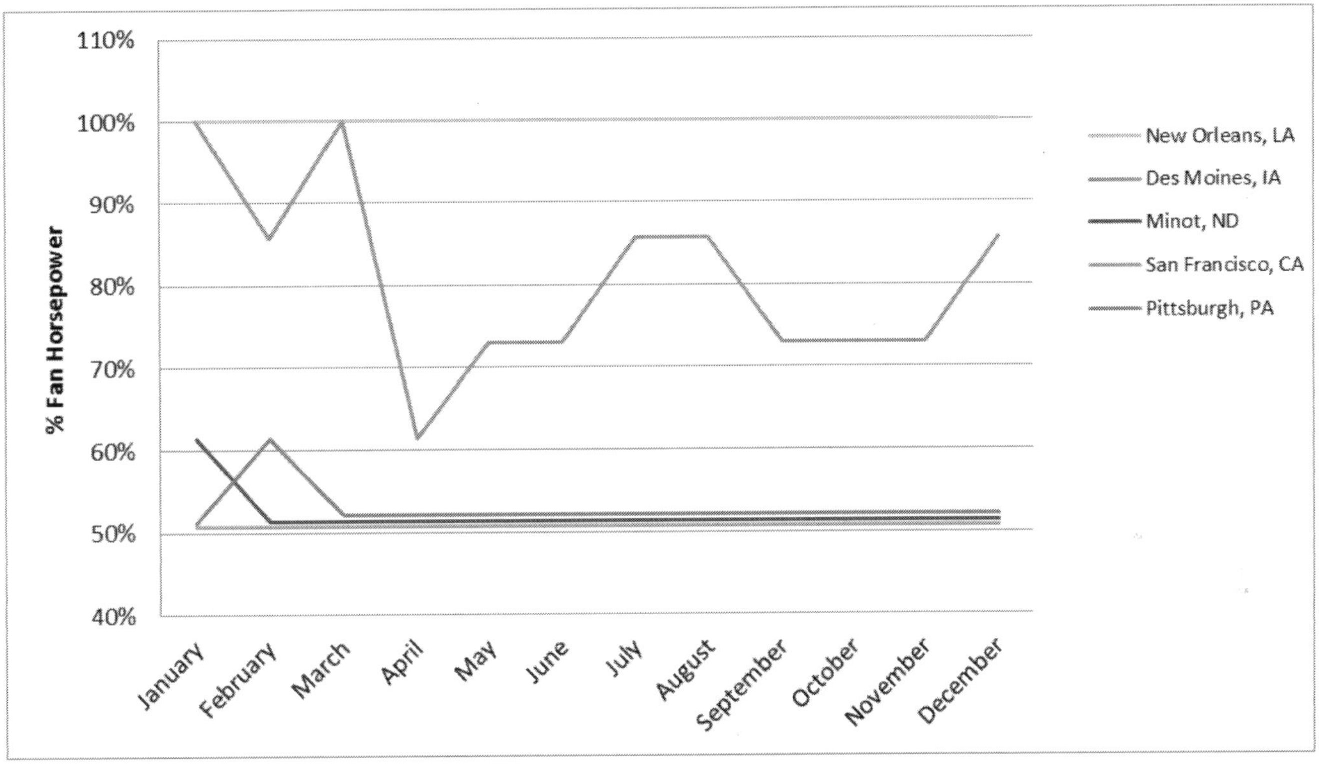

While New Orleans and San Francisco experience ambient weather conditions that trend to consistent energy use; sites such as Des Moines, Minot and Pittsburgh that experience cold and dry ambient weather trend a more complicated energy use graph. Although all the sites, except Minot, in Figures 5 and 6 experience maximum predicted fan energy use for the summer design condition; the average monthly predicted maximum energy use for Des Moines, Minot, and Pittsburgh might occur during the coldest winter months. This energy use pattern is attributed to the water absorption capacity of cold air decreasing quicker than the increasing ability of this cold air to remove heat by sensible heat transfer. More air flow would be required as the sensible heat transfer mechanism becomes dominant; even though less water is removed during these months. However, when the cooling water cold temperature falls into the low 40's F; many sites choose to save fan energy and reduce icing risk by allowing the cold water setpoint to rise. Most processes do not benefit from cooling water at these winter temperatures anyway (13). Therefore, Figure 6 likely overstates the predicted energy use for sites such as Des Moines, Minot and Pittsburgh during the winter months.

As shown in Table 5 for the same heat load, the predicted average fan energy required across the sites in July and January is similar; however, the percentage of water required to be evaporated in January averages 30% less than in July. Most of the sites require approximately the same fan energy in winter and summer, while two are predicted to require more in winter and two less.

Table 5 Comparison of predicted energy use vs. evaporative cooling water loss at nine US locations for the same cooling tower design

		Average July		Average January	
Location	Summer Design (% HP)	% Evaporation	% HP	% Evaporation	% HP
New Orleans, LA	100	1.3	100	1	100
Brownsville, TX	100	1.4	100	1.1	100
Houston, TX	100	1.4	100	1	85
Belleville, IL	100	1.3	100	0.8	75
Des Moines, IA	100	1.3	85	0.7	100
Phoenix, AZ	75	1.6	60	1.1	60
Pittsburgh, PA	60	1.2	50	0.8	50
Minot, ND	45	1.2	35	0.6	65
San Francisco, CA	35	1.1	30	0.9	30
Average	**80**	**1.3**	**75**	**0.9**	**75**

Climate

With cooling tower system performance dependent on local ambient weather conditions dealing with the weather is a daily business concern. However, a changing climate can take weather concerns to a different level especially when accompanied by growing regulatory, environmental, and supply pressures. A 2013 report (15a) highlighted that climate change is often perceived as a slow, gradual change in mean temperatures. However, a relatively small change in average temperatures results in large increases in the risk of extreme heat and substantial decreases in the risk of cold weather. The same report (15b) identified the Manufacturing and Industrials sector as the only major sector to expect both "increased operational costs" and "decreased production capacity" from a changing climate.

Climate forecasting can highlight long-term changes in such factors as temperature, humidity, or rainfall patterns to help quantify a changing risk profile. For assessing the sustainability of water supplies as part of a water management program; drought can raise the price of water and climate-related regulation can drive up the cost of energy (3).

Figure 7 Seasonal Variation of Percentage of Total Cooling Tower Heat Transfer Due to Latent Heat (Evaporative Cooling) for a Climate with a 2F Degree Rise in Ambient Wet-Bulb Temperature

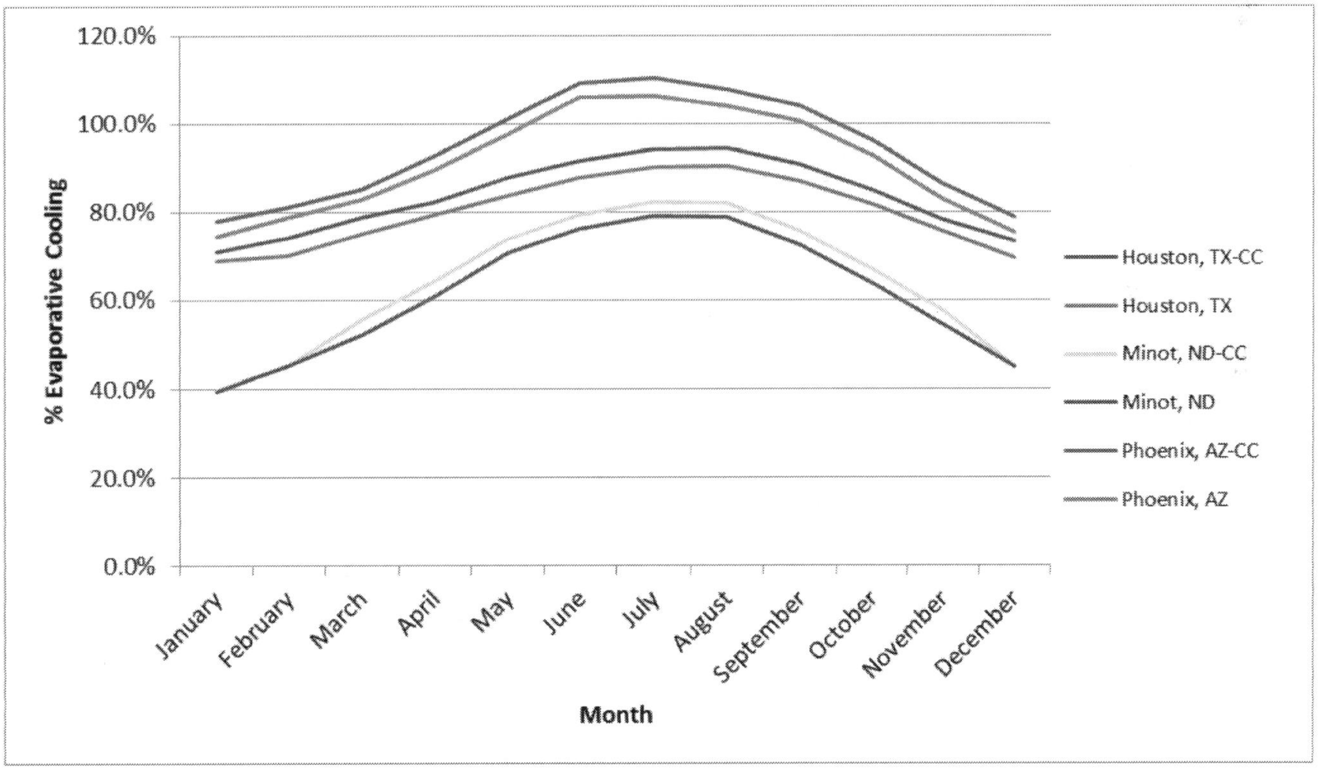

To project the impact of a warming, and for most regions moistening climate, over the next decades Figure 3 is modified as Figure 7 to show the projected increase in evaporative cooling required for a climate with a 2 F degree increase in ambient wet-bulb temperature. Though

more pronounced for the coldest part of the year, the predicted percentage increase in water use would average around 4% more than before the temperature change; as shown in Table 6. Assuming the same cooling tower equipment outlined in Table 1; the ability to increase the capacity of the cooling tower by this amount would rely entirely on more fan power use.

Table 6 Relative predicted evaporative cooling water loss with a 2F Degree Rise in Ambient Wet-Bulb Temperature at nine US locations for the same cooling tower design

Location	Summer Design (% Change)	Relative Predicted Evaporative Water Loss (% Change)			
		Average July	Average January	Winter Design	Annual
Des Moines, IA	4.1	3.7	9.3	0.3	4.4
Houston, TX	4.1	4.3	3.1	2.7	4.4
Brownsville, TX	4.1	4.2	3.6	6.6	4.6
New Orleans, LA	4.0	4.3	6.1	3.4	5.0
Belleville, IL	4.1	3.7	2.7	9.7	3.6
Minot, ND	3.4	4.0	0.7	14.5	4.0
Pittsburgh, PA	3.7	3.8	6.2	0.1	4.7
San Francisco, CA	3.9	4.5	5.2	5.9	4.8
Phoenix, AZ	3.8	3.7	4.6	2.4	3.7
Average	**3.9**	**4.0**	**4.6**	**5.1**	**4.3**

Concluding Thoughts

Based on a simplified model, the prediction of water evaporation loss in wet cooling towers as a function of ambient weather conditions has been discussed for constant heat load at ethylene production facilities. As shown in this paper, a cooling tower is designed to meet production cooling demands during extreme conditions and due to seasonal climate changes, the tower operates at design conditions infrequently leading to annual conservation of water and energy resources.

Historically, the primary factor considered in selecting cooling water systems was the availability of water and its associated cost. The analysis discussed here would enable the prediction of water and energy use of evaporative cooling tower systems to be available to assist in planning, siting, and permitting at the Front End Engineering Design level. As water and energy are being realized as essentially interdependent resources; considering monthly and annual use patterns provides an opportunity to approach the challenges of water resource management in a more integrated fashion.

References:

1. Gloyer, Walter (1978), Review of Cooling Tower Calculation, Cooling Tower Institute – TP78-06 (TP-194A), January.
2. Lakovic, L.S.; Mladen Stoiljkovic and Mirjana Lakovic (2004), Today's condition and the future of the cooling towers, Thermal Sciences 2004. Proceedings of the ASME - ZSIS International Thermal Science Seminar II, Bled, Slovenia, June 13-16, pages 701-704
3. Stone, John C. et al (1975), An Integrated Power Process Model for Ethylene Production, Water Resources Research Vol 11, No. 9 December.
4. ____ (2008), ASHRAE Handbook – HVAC Systems and Equipment, Chapter 39 – Cooling Towers.
5. Leeper, Stephen A. (1981), Wet Cooling Towers: Rule of Thumb Design and Simulation, US DOE EGG-GTH-5775 July
6. Engineering Weather Data Handbooks (2003) Air Force Combat Climatology Center (AFCCC) / National Climatic Data Center (NCDC). For each site:
    a. Section 2-2.2.1
    b. Figure 2-12. Sample Data Set Page 12
    c. Figure 2-1. Sample Data Set Page 1
7. ___ (1974) Development Document For Effluent Limitations Guidelines And New Source Performance Standards For The Major Organic Products Segment Of The Organic Chemicals Manufacturing Point Source Category, USEPA, April; Section III pg. 53.

8. Eftekharzadeh, Dr. Shahriar; Dr. Muin M. Baasiri; Paul Lindahl, Jr. (2003) Feasibility of Seawater Cooling Towers for Large-Scale Petrochemical Development, Paper TP03-17 Presented at the Cooling Technology Institute Annual Conference; San Antonio, Texas – February 10-13, 2003 p 11.

9. Jalal Engineering. Cooling Tower Basics and Common Misconceptions. oshuja@cyber.net.pk, pp. 1-6. Accessed online September 2016.

10. Cheremisinoff, Nicholas (2000), Handbook of Chemical Processing Equipment, Chapter 2 – Evaporative Cooling Equipment. Elsevier.

11. Schwedler, Mick (2014), Effect of Heat Rejection Load and Wet-Bulb on Cooling Tower Performance, ASHRAE Journal (January) pp. 16-22.

12. ASHRAE (2001), ASHRAE Handbook and Product Directory—Equipment, Chapter 21. American Society of Heating, Refrigerating and Air-Conditioning Engineers, Inc., USA

13. ___ (2013) Variable Flow Over Cooling Towers For Energy Savings, Publication TR-014 SPX Marley Cooling Technologies Inc., Overland Park, KS (March).

14. Midwest Cooling Towers Blog (2013), Cooling Tower Fundamentals: Cold Weather Operations and Components, January 22, http://midwesttowers.com/cooling-tower-fundamentals-cold-weather-operations-and-components/

15. Crawford, Meg and Stephen Seidel (July 2013), *Weathering the Storm: Building Business Resilience to Climate Change*, Center for Climate and Energy Solutions, c2es.org.
    a) Appendix B
    b) Figure 12

16. Engel, Hauke; Per-Anders Enkvist and Kimberly Henderson (July 2015), "How companies can adapt to climate change", McKinsey & Company.

17. Morrison, Frank (2014), Saving Energy with Cooling Towers, ASHRAE Journal (February) pp. 34-40.

AICHE PAPER NUMBER: 171d

OPTICAL GAS IMAGING: TAKING A DRIVE THROUGH IR CAMERA TECHNOLOGY AND HOW IT'S USED

Justin Lamb
Project Manager, GHG
Dexter ATC Field Services

Nick James
Operations Director
Dexter ATC Field Services

Prepared for Presentation at the 2018 Spring National Meeting
Orlando, Florida, April 25, 2018

OPTICAL GAS IMAGING: TAKING A DRIVE THROUGH IR CAMERA TECHNOLOGY AND HOW IT'S USED

Justin Lamb
Project Manager, GHG
Dexter ATC Field Services
Nick James
Operations Director
Dexter ATC Field Services

Abstract:

Have you heard about the latest Optical Gas Imaging (OGI) technology, but are still asking the question "What's the big deal?" Don't worry, you aren't alone! This presentation will drive you through federal and state OGI rules/regulations; exploring baseline and creative applications of this technology for industry use.

Along the way, we'll revisit the birth of infrared (IR) technology, how far we've come, and where OGI applications can take us in the future. We'll take a look at a few industry-leading brands of IR cameras; comparing operating capabilities and individual accessories/functions of each.

Real videos of recorded fugitive emissions will be presented to provide a visual representation of IR technology. Examples of leaks will be provided in multiple video layers to discuss the advantages and limitations of each filter applied. In addition, this discussion will address the concerns of OGI efficiency and/or effectiveness, as opposed to the traditional leak detection methods.

If you don't know where you're going, how can you get there? This presentation will map out the many different factors that can drive federal/state OGI monitoring requirements, as well as voluntary best practice programs. Topics such as safety considerations and product losses will be covered.

The road to OGI monitoring applications and programs should never be burdensome, but preparation is the key to success in this fast-moving industry. Join us as we go on this journey together because remember, protecting the environment is good business.

Optical Gas Imaging: Taking A Drive Through IR Camera Technology And How It's Used

Introduction

The goal of any industry, whether it be petro-chemical refining and production, storage and transfer, or exploration, is to optimize the overall process at the highest efficiency, with the lowest cost. Over the past few decades, world-leading engineers and scientists have developed new ways to produce and extract resources at rates previously not thought to be possible. As these advancements allow increased production rates and higher profits, the number of facilities designed to do so have grown exponentially. As most would agree, the ramped up production rates have caused public interest in the health of our environment. The U.S. Environmental Protection Agency (EPA) has responded to this interest by promulgating regulatory requirements to ensure these operations are limiting their effect on the atmosphere. Releases of volatile organic compounds (VOC) and other potentially harmful chemicals to workers, the public, and the atmosphere have been one of the leading factors that laid the foundation to Leak Detection and Repair (LDAR) requirements. As these requirements have been implemented, revised, and increased, the need for technology to provide us with the ability to do so has also grown.

Optical Gas Imaging (OGI) is a technology that has been designed help industry meet the demands of the EPA. The term OGI, coined by EPA in 2006, simply put is the act of making invisible gases, visible. This technology uses infrared (IR) waves to detect emissions or VOCs from leaking components on piping, vents on tanks, or combustion flares (to name a few) that are otherwise invisible to the naked eye. The IR camera is a non-contact device that detects infrared energy (heat) and converts into an electronic signal, which is then processed into a thermal image on a video monitor. To most, this cutting-edge technology has changed the way we see gases, almost as if we're viewing our surroundings in another dimension. OGI monitoring has traveled a long, not so narrow path from where its technology was discovered, to making it an industry-wide recognized way of performing and maintaining LDAR programs.

History Of The Camera

In order to understand how far this technology has come, it's important to know where it began. It in year 1800, Sir William Herschel of Germany, accidentally discovered IR light. Herschel, an astronomer by trade, wanted to know how much sunlight passed through different color filters on his telescope lens. He developed an experiment to record the temperature of light at each color on the visible light spectrum, and noted that as you move from the violet side of the spectrum all the way to the far red side, temperature recordings increased. He was curious to see what would happen if he went beyond the red spectrum, so he conducted an experiment he called "calorific rays", derived from the Latin word "heat". The results showed that beyond the red spectrum, there was radiation that displayed similar qualities (reflection, refraction, absorption, etc.) as that of the visible light spectrum. Herschel had officially discovered what today we call, infrared light or infrared radiation.

Fast forward 130 years, and we finally begin to see the application of this spectrum applied to an instrument, and used for a specific purpose. Following World War I, the first thermal imaging camera was produced by a Hungarian physicist for the British anti-aircraft defense. Also known as "night vision", the camera allowed the military to see thermal images that the rest of world didn't have access to. Following the 1930's, interest in the technology and possible advantages it could provide during war increased dramatically. In 1947 the United States and Texas Instruments developed the first ever infrared line-scanner, which was operated by a single element detector. It was soon realized that the primary limitation to the scanner was the time it took to produce a single image, which was roughly one hour at the time. As stays true for most inventions, once a limitation is evident, it's a figurative foot race to overcome and eliminate that road block. As new models were developed and the speed of image production increased, the use of thermal imaging cameras was applied to more practicable situations such as fighting fires or locating victims following a natural disaster. Even though applications for the technology were broadening, the military wanted sole possession. High prices and sensitive, possible dangerous applications were two primary reasons the military sought to keep it out of the public's hands. It wasn't until 1992 that the U.S. government officially declassified the technology, making it available for commercial sale and use. Shortly after the release to the public, Honeywell began licensing patents associated with microbolometers to several companies, which was originally developed in the late 1970's. While Honeywell focused on what was labeled vanadium oxide (VOx) microbolometer technology, Texas Instruments went after commercial production of infrared camera detectors. They were granted patents for detectors that were made of barium strontium titanate, or BST. BST was used to coat the thermal imager's sensor, which allowed for more absorption of light.

With the declassification of IR technology, following the Gulf War, demand for OGI cameras for commercial use increased dramatically. Commercialization forced production volume up, allowing for mass manufacturing, which in return minimized the cost per unit for both suppliers and consumers. As history tells it, this was a pivotal time in the life cycle of the camera's physical dimensions and practicality of use. Consistent with the evolution of most technology (personal computers, mobile phones, etc.), the desire to reduce the size and weight of the camera while maintaining core operating functionality saw a dramatic increase. Early equipment models were bulky, extremely heavy, and difficult to transport. The first industry-used cameras were so large in fact, that the only way to transport and ultimately use them was by stationary use. Some cameras were mounted to the top of vehicles (vans) and driven around facilities in order to perform OGI monitoring.

It's important to discuss what has been labeled as The Common Sense Initiative, an excerpt published by the U.S. EPA in April of 2006. As quoted by Carol M. Browner, an American lawyer, environmentalist, and businesswoman "We want to make good on what business and environmentalists have been telling us for two decades – that we must look at whole facilities, whole industries, and their overall impact on the environment. We must do a better job of cleaning up the environment and do it cheaper". Browner was referring to the burden of 25 plus years of environmental initiatives (regulations), the differences in cost required to comply for small businesses vs. large industry leaders, and the overall complexity and rigidness that came along with it. Such initiatives were founded on the intention of protecting our waterways, soils, and limiting air pollutants, but as history tells it, presented unintentional consequences. The goal of The Common Sense Initiative was to ensure a healthier environment, but without diminishing the health of the U.S. economy, which prior to this implementation, resulted in nearly 30 billion dollars spent for compliance by U.S. businesses. In order to comprehend the needs, limitations, and impact of this initiative, representatives from federal, state, environmental justice groups, labor force, industry, and community-based organizations teamed up to develop a plan that was all-inclusive and representative of nationwide operations. By developing a stream-line process to overall better the intentions of the agency, optical gas imaging popularity and discussions about its implementation were, for the first time, taken into serious consideration. Of the five major projects that the Common Sense Initiative (CSI) counsel undertook to resolve complexity and burdens, exploring alternative ways to meet regulatory compliance and standards gave OGI monitoring a solid footprint for future discussions. As we saw the momentum of these discussions increase by the CSI, it was now technology's turn to step up to the plate and make this alternative monitoring technique a practicable solution.

It wasn't until 2006 when FLIR Company (Forward Looking Infrared) introduced the benchmark OGI camera called the GasFindIR. This modern-looking, light-weight model was portable enough for users to carry in the field, while maintaining an all-inclusive battery system that, until then, wasn't feasible. It was the first model of its kind that was solely designed for the detection of VOCs inside refineries, chemical plants, gas processing facilities, transmission stations, storage facilities, and production wells. Following the release of the GasFindIR camera and the overwhelming feedback from users, the OGI manufacturing industry foot race was underway. This introduced opportunity for smaller start-ups to piggy-back off of existing technology, allowing for the development of similar instruments to be released to the market. Companies such as FLIR and OpGal (founded in 1983), who controlled the majority of the market, began to develop newer models that provided specific functions for all OGI needs. Cameras designed specifically for furnace and electrical inspections, refrigerant leak detection, continuous gas leak detection, and CO detection gave industry a wide variety of options to choose from, based on their specific needs. Recent years of advancement has given compatibility of cameras to smart phones, allowing real-time video of surveys to be transferred without the phone user to physically be there. Fast-forward to the present, and now we see various IR camera manufacturers and UAV drone companies working together to conduct surveys remotely. This development allows for operators to perform OGI monitoring in areas previously inaccessible by foot, which if you're familiar with the geographic locations of production well pads and pipelines in the U.S., is a major advantage. Not only did this transition increase productivity in scarce areas, it also provided an additional layer of safety for monitoring in areas known to be hazardous or immediately dangerous to life. As the old saying goes in the south "The proof is in the pudding". There has been years of research and development put into this technology to where we see it today, and if it wasn't a feasible, more efficient and safer practice, the evolution wouldn't have made it this far.

So what brings us to where we are now, and why does OGI monitoring tend to be in most major environmental discussions today? First and foremost, we've finally arrived at the "walking" stage in the life-cycle of this technology. As mentioned earlier, the sophistication of OGI technology has been steadily increasing over the past few decades. Numerous pilot studies such as the Common Sense Initiative, have proved that the use of infrared camera technology is capable of meeting the requirements of government agencies. Secondly, regulatory requirements are now finding themselves to be in parallel with the use of this monitoring technique. In the past, when OGI monitoring first made its debut in the Alternative Work Practice, it was merely an option to primary methods. The pressure that was put on EPA to release another method to maintain compliance was evident, and soon shown to be somewhat of a premature decision. While a new technique of doing things was underway, the requirements for reporting and record-keeping remained consistent with traditional practices. It was soon realized that though OGI monitoring was an acceptable practice, it only fulfilled a fraction of the various already-existing regulations. This meant that if operators wanted to use this surveying technique, they would still have to maintain their previous obligations of monitoring compliance, such as Method 21. Fairly enough, this discouraged a majority of companies to explore alternative methods, because of the simple fact that it didn't replace the system they were trying to escape from. Even though this presented a major roadblock in what was hoped to be a straight road, the use of OGI monitoring was still favorable in many situations. It wasn't until 2015, when EPA published the first ever, 100% OGI monitoring regulation, now commonly known as NSPS OOOOa. This requirement was aimed at reducing the amount of methane release into the atmosphere from the upstream and midstream oil & gas sectors nationwide. Probably the largest accomplishment in the eyes of manufacturers, who pushed their research for decades, to finally see a regulatory requirement for industry to use their product.

Rules And Regulations

It's no secret that the implementation of regulatory requirements comes at the expense of the businesses affected. It's reasonable to argue that by complying with these requirements, companies are losing profit. Costs associated with developing and maintaining environmental departments in the oil and gas industry, for example, have significantly affected the way these companies run their business. In the early days, there were simply no regulatory requirements to ensure extraction, transfer, and refining of crude oil was conducted in a manner that would limit contamination and/or pollution to the environment and surrounding public. Many could argue that the lack of environmental responsibility was due to limited research of chemical compounds handled, the public's elementary knowledge of the industry and how it worked, or by the bottlenecking and large donations for political gain. Whatever the reason(s), it's important to understand that most regulatory implications were birthed in a reactive manner. Reactive implications typically followed the release of an oil spill in our oceans, an explosion at a refinery, or long-term devastating contamination in our soils and rivers. Each time something like this happened, new regulations were put in place to limit the reoccurrence of similar events in the future. Consequently, every time a new regulation was born it added a layer of complexity for businesses to continue operations.

In regards to OGI monitoring regulations, we have seen a slower but just as complex life cycle. Industry has struggled to understand and comply with the multiple layers of regulatory sanctions imposed, but has played a pivotal role in this evolution. One simple way to break down how these requirements affect a business, is to break it down into agency levels. First and foremost, in regards to the U.S. EPA, we look at federal-wide regulations. These typically affect the lower 48 states in the same manner, no matter what. Now there are some exceptions for businesses that may have operations in multiple unique geographic locations. For example, the recent NSPS OOOOa federal regulation states that owners or operators are required to perform OGI surveys at compressor stations on a quarterly basis. The vast majority of businesses affected by this rule are capable of and required to do so every single calendar quarter. EPA does, however, provide an exemption from surveying compressor stations located in geographic regions with significantly lower average temperatures for a specific quarter. Reasons for this exemption may reflect inaccessible facilities with increased snow/ice on roads, and a drawback in actual compression operations, to name a few. After federal regulations, we see the next "level" down into state-wide regulatory requirements. States such as California (CARB-32), Colorado (Rule 7), and Pennsylvania (Exemption 38), have all led the way in developing their own set of requirements, in regards to OGI monitoring. In fact, all three of the states listed above were implemented before the federal NSPS OOOOa requirement, making it evident that a slow introduction was necessary before EPA felt they could gain national acceptance. Making our way down another level is where we can find tribal and local agencies implementing more specific rules. Typically, compliance requirements at this level are the result of public concerns and immediate need for some type of resolution. Although local (city, county, etc.) agencies don't have the level of authority that state and federal agencies do, we see organizations such as the Bureau of Land Management (BLM) retain a high level of respect. The Methane Rule, as named by BLM, mirrors that of NSPS OOOOa, but with a few different requirements for subjectivity. The compilation of multiple agency regulations has created pushback from industry, arguing that monitoring and reporting requirements are too complex, financially burdensome, and in many cases redundant. Businesses were finding themselves reporting the same information to multiple agencies, at multiple times of the year, with different record-keeping requirements. Even with the publication of NSPS OOOOa, industry is still required to follow the rules of whichever is most stringent, while maintaining compliance as stated by each agency level.

As industry has slowly adopted such regulatory requirements, an interesting and positive trend has come to light. While not all businesses and facilities have been subjected to maintaining OGI monitoring programs, a majority have adopted voluntary practices for proactive measures. Infrared technology has proved to be beneficial to businesses in ways not previously found. Today we see a majority of industry voluntarily investing in the technology to better their business, seeing positive results in both the short and long term aspects. Research has shown that implementing the use of OGI programs has consequently improved the financial health of businesses that opt into this practice. Conoco Phillips Company and FLIR teamed up to conduct a pilot study to determine the cost benefit analysis of implementing such a program, and found that the return on investment for OGI was extremely beneficial. This research illustrated that "The study identified 144 leaking components. Collectively, these leaks account 58.26 mmcf/y and $358,012.10 USD/year in lost product. The methane leak sources contribute 21,420.7 tonnes/year CO2e to GHG emissions. It is estimated that 92% of the 144 fugutive sources are economical to repair. Implementing all economical repairs would result in a net present savings of $2,002,602.72 USD". Another study, conducted by Jonah Energy and FLIR showed "The cumulative gas savings realized by the program has exceeded $5 million in the past 6 years, which has more than covered the overall program costs. This includes the Optical Gas Imaging equipment and associated operators, along with all repairs and maintenance, including labor and parts." Although these are only two examples of the benefits for voluntary implementation, the numbers speak for themselves.

Aside from limiting product loss and increasing revenue, the majority of businesses strive to do so in a safe manner. In recent years, we've seen worker safety pushed to the forefront of day-to-day operations, ensuring all employees return home safely to their families. Not only does implementing safer work practices benefit the health of those in immediate contact of hazardous processes, it also aids in ensuring the company is positively viewed in the public's eye. Utilizing OGI monitoring has allowed for surveying at safer distances, which isn't true for traditional M21 monitoring. It serves as an important tool during facility-wide turnarounds when pressuring up, allowing operators to quickly and safely detect possible releases from malfunctioning process equipment and loosely tightened connectors. Another example of how OGI monitoring has provided safety measures for both industry and surrounding communities is that of the infamous Aliso Canyon Blowout, near Los Angeles, California. During a routine flyover, scientists conducting OGI surveys on pipeline came across one of the largest known methane leaks known today. Also known as the Porter Ranch gas blowout, Southern California Gas Company's well was discovered leaking on October 23, 2015. During a 2018 earnings call, Sempra Energy (parent company) reported that costs associated to this catastrophe has reached nearly $1 Billion dollars. Lost revenue, costs associated to the remediation efforts, restitution, as well as penalties and fines all account for that figure. The leak was so large, it is estimated that more than 100,000 metric tons of methane were released into the atmosphere, requiring immediate evacuation from nearby communities in the San Fernando Valley.

OGI Monitoring vs Method 21

As industry either fought for or fought against the implementation of OGI monitoring, the technology's effectiveness for leak surveying was of high priority. The only feasible way to build an argument is to compare OGI monitoring to traditional and existing methods used in industry today. The technique used, for comparative purposes, is that of Method 21. Method 21 is a widely used term in the LDAR industry that describes the approved monitoring protocol on equipment (valves, connectors, pumps, PRVs, open-ended lines, etc.) that have the potential for emitting VOCs or HAPs (Hazardous Air Pollutants) into the atmosphere. One example of an instrument designed for Method 21 is that of the Thermal Toxic Analyzer 1000B, or TVA-1000B. Simply put, it's a machine that measures the concentration of a leak in parts per million (ppm) of any given leak (VOC/HAP) on a component. In general, facilities that are subjected to using this method, are required to perform inspections on all equipment that fall into this category, dependent on the stream composition of each. Since Method 21 is the primary option for LDAR compliance monitoring programs, it makes sense that all arguments be centered around it for OGI monitoring discussions. A simple yet effective way of comparing the two, was for EPA and industry to list the Pros and Cons of each. As previously suggested, OGI monitoring allows the user to visibly see the leak, the source of it, and the direction it comes from. Since "seeing is believing", this method allows for little to zero doubt that emissions exist, and provides real-time accuracy of the origination. While nobody is discrediting the accuracy of Method 21 monitoring, it's evident that it comes with its limitations. Compliance programs designed for Method 21 are required to have a complete inventory of all equipment subjected to monitoring, identified with a metal tag, and maintained in a robust database. After this phase is completed, technicians are required to monitor the components on a scheduled basis, depending on the regulation they fall into. The TVA-1000B for example, is designed to provide readings in ppm only, not always allowing the user to determine the exact point or origin of the leak. Furthermore, Method 21 is limited to locating fugitive emissions from only the tagged and documented components in the compliance database. Components that don't fall into the program still have the opportunity to emit harmful toxins, but there's no way to identify the geographic location of these if they aren't being regularly monitored. When performing Method 21, the user can sometimes pick up high readings near a tagged component that isn't actually leaking, which urges the question "Where is this coming from?" Utilizing OGI methods allows users to perform surveys on all components, not just the ones regulated in the program. This provides the capability to detect and resolve leaks that could be problematic for years, but otherwise wouldn't be found if only using traditional methods.

A major setback for using the camera to monitor is the reality of not being able to measure the concentration of the leak, as the TVA-1000B is designed to do. This presents a gap in the argument of what constitutes a leak, and what is an acceptable limit. In a controlled environment, the OGI camera is capable of detecting leaks at or around 1500 ppm. In personal experience using the camera for monitoring, most leaks aren't found unless they're at least 5000 ppm or higher in real-time weather conditions. Not only is the majority of Method 21 monitoring conducted in an uncontrolled environment, the benchmark leak limit nation-wide is set at 500 ppm. Obviously, the gap in detection limits between the two methods presents a hesitation for allowing either practice. Environmental conditions play a huge role in how effective and accurate an IR camera can be throughout a facility. The wind speed, wind direction, temperature, barometric pressure, and humidity all play a factor in how well the user can see emissions through the camera lens. In an effort to limit this concern, the EPA publishes the Alternative Work Practice (AWP) for the oil & gas sectors in 2006. In the AWP, if a business opts to use

OGI monitoring, they're required to perform a daily camera check before monitoring commences. Unlike the TVA-1000B, which requires morning and afternoon machine calibrations with methane gases, this camera check is designed to ensure the user is monitoring in a manner directly related to real-time weather conditions. For example, a calibration gas (concentration 99% methane) is set up in the field with a flow meter attached before camera surveying commences. The user enters in the current environmental data, then sets the flow meter to a specific range to represent real-time conditions. The user then stands back 15-30 ft from the calibration setup to ensure the methane plume is visible from a specific distance; doing so in the mode they intend on performing the survey. This task allows the user to determine their survey distance from components in the actual conditions they're monitoring in, ensuring they don't exceed the distance from what the camera is capable of "seeing". So unlike calibrating the TVA-1000B to specific gas ranges, performing the daily camera check is really just telling the user that they can monitor from a specific distance in specific conditions.

Just as the debate on whether or not OGI monitoring is an effective way to perform surveys, the conversation on how efficient this practice is followed. Before looking at ways to compare efficiency, it's important to distinguish the two types of efficiency discussed; monitoring/use efficiency and program efficiency. The monitoring efficiency of OGI practices in many ways, trumps that of Method 21 monitoring. Studies show that the average number of components monitored/day using Method 21 is anywhere from 200 to 400 components/day, with OGI monitoring being ~15,000 components/day. With Method 21 a user is required to physically walk up to each components to collect readings, while a technician using OGI can view each component from a distance of ~30 feet away (note: assuming leaks are medium in size). Another pro of using the camera vs. Method 21 has been shown during unit or facility-wide shutdowns and turnarounds. Something as intensive and high-risk as replacing major process equipment can present potentially catastrophic events if taken lightly. One important risk that OGI monitoring can help reduce is that of pressure-up tests, when the construction is completed and the unit is ready to be tested. During such pressure tests, performing OGI-monitoring surveys on newly added vessels and associated equipment can ensure everything is working as designed. The chance that not all equipment has been tightened down securely following construction is always there, especially with many contractors performing the work. Miscommunication between operations and contractors is common, so the ability to utilize OGI technology to ensure proper maintenance is conducted can be a huge benefit during these times.

For purposes of this paper, program efficiency is referred to the techniques, training and overall continuation of maintaining an OGI program. Facilities required to maintain an LDAR program via Method 21, know that monitoring technicians are put through official Method 21 training. This training educates them on the proper technique for surveying each component, the speed of how to do so, and the time required to spend on each. As the majority of technicians performing these inspections are entry-level candidates, usually not required to have any sort of technical degree, the expectations for process knowledge and chemicals within facilities are limited. Although basic chemical awareness is required, technicians aren't expected to understand the advanced engineering methods and chemical mixtures of each process lines in which they're monitoring. LDAR software now has the capability to direct technicians to each component, provides them with a time lapse of monitoring requirements, and automatically logs leaks if they exceed the defined emission limit. With OGI monitoring, the training expectations are much more intensive than that of Method 21. Since not all cameras are created equal, it's important for the technicians to possess knowledge of stream speciation and how the overall process is engineered. All chemicals have what is called a Minimum Detected Leak Rate (MDLR) in accordance with the camera. In order to perform OGI surveys in a satisfactory manner, the technicians should know exactly what's in the streams they're monitoring. Likewise, it's also very important to understand the chemicals you're working with because there isn't one camera that can detect everything. Some instruments are made specifically for electrical and thermal inspections, some are designed for Carbon Monoxide, CO2 or Ammonia only. The need for initial training and education is higher and more intensive for OGI monitoring, which could pose higher administrative costs.

Recordkeeping And Reporting

There's an old saying in the LDAR industry that suggests "You're only as good as your documentation". By this we mean, it doesn't matter what you do with your compliance program unless you can prove it. Over the years, software databases geared to house large LDAR programs utilizing Method 21 have evolved in a manner which takes a majority of the risk away from businesses. The databases work directly with the field software used, and compile monitoring data in an easy-to-read format for review. Although somewhat sophisticated programs, traditional LDAR databases are limited to traditional monitoring methods. With the introduction of OGI monitoring techniques, we're now presented with the task of managing our data. Such record-keeping requirements can include: monitoring dates, equipment used, calibration records, list of components monitored, leaks detected, dates of repair, GPS coordinates of leaks, pictures of leaks, videos of leaks, videos of repaired leaks, videos of entire surveys, deviation from monitoring plans, etc. Until recently, there weren't any LDAR databases or software capable of storing these records. Most businesses had to create large files to retain hours of video footage, hand drawn maps, and custom leak documentation. Not only was this burdensome to hold large files of data, but it was unorganized and hard to report if requested. With the introduction of new OGI requirements and the overall sense of its long-term stay, new databases have emerged that have the capability of managing this data. Field tablets with GPS capability linked to GoogleEarth can now map out survey routes, with time stamps and leak footage. The capability of compiling this data into the expected reporting format is now available with the click of a button, which has been a main point of concern for businesses. The argument that it's impossible to streamline OGI monitoring data to meet reporting expectations no longer exists, and we're still only in the beginning stages.

Choosing Your Equipment

Before making the decision to purchase your first IR camera, there's homework to be done first. As mentioned earlier, not at cameras are created equally. As the IR manufacturing industry began to lift off in the early 90's, it was merely a race to the consumer. The goal was to provide the smallest, lightest camera, capable of monitoring in the infrared mode before the next guy. Now that we've come to understand the ability of the technology and made it a reality, we see a rise in expectations of using it. Two of the industry's leading manufacturers, as mentioned previously, are the FLIR and OpGal companies. In 2006 when FLIR introduced the GasFindIR model, it was designed specifically for the detection of VOCs in the oil & gas industry. As luck would have it, not all industries are created equal either. Different chemical compositions, processes, and engineering are seen throughout the world, facility to facility. Implementing OGI usage from a crude oil refinery to an Ethylene plant introduces the possibility of needing specific camera models, dependent on gases to be monitored. While the FLIR GF-320 and Opgal EyeCGas models are the most sophisticated and popular, they aren't a one-stop shop for meeting your goals. It's extremely important to know the exact chemicals you wish to detect, and then steer your shopping in that specific direction. I encourage you to visit the manufacturer specifications of each model, studying the limitations and advantages of each. Not only will this ensure you have the right equipment to fulfill the desired goal, but it may save on costs associated with each. If the determination is to survey for a single chemical composition or a specific task, you may find a particular model cheaper that's more affordable while still accomplishing the task. As we continue to make stride towards this "new age" technology and ways to implement into every day operations, there's little doubt that our workplace can be more efficient, effective, safer, and profitable.

References:

1. United States Environmental Protection Agency (October, 2007), *Leak Detection and Repair: A Best Practices Guide*, Office of Compliance, EPA-305-D-07-001
2. Watton, R., Smith, C., Jones, G.R., (1974, Dec. 9-11). *The pyroelectric vidicon: Pyroelctric materials for operation at cathode potential in a "Hard" tube.* Retrieved from https://ieeexplore.ieee.org/document/1477887/.
3. Know Made Patent & Technology Intelligence. (2015, July). Honey Microbolometer Patent Portfolio Analysis [Web Log Post]. Retrieved April 17, 2018, from http://www.knowmade.com/downloads/honeywell-microbolometer-patent-portfolio-analysis/
4. National Service Center for Environmental Publications. (April, 1996), *The Common Sense Initiative: A New Generation of Environmental Protection.* Retrieved from https://nepis.epa.gov
5. Grigoryants, O. (2018, February 28). *Here's the bill so far for the Aliso Canyon gas leak: Nearly $1 billion – and counting.* Retrieved from https://www.dailynews .com
6. Hazburn, C. (2018, February 19). *The History of Thermal Imaging Cameras.* Retrieved from https://www.ecamsecure.com/blog/history-thermal-imaging-cameras/
7. U.S. EPA, Code of Federal Regulation, Title 40, Part 60, Appendix A, Method 21 "Determination of Volatile Organic Compound Leaks". 2014 Ed.
8. McKenna, J., Gromicko, N. *The History of Infrared Thermography.* Retrieved April 1, 2018, from https://www.nachi.org/history-ir.htm
9. Palmer Wahl Instrumentation Group. (May 2017). *The Evolution of Thermal Imaging Cameras.* Retrieved from http://instrumentation.com/PDFS/ Evolution ThermalImagingCameras.pdf
10. Trefiak, T. (2015, December 14). *LDAR Case Study Comparison of Conventional Method 21 vs Alternative Work Practice (Optical Gas Imaging).* Retrieved from https://www.epa.gov/sites/production/files/2016-04/documents/20trefiak.pdf
11. O'Neill, G.R., Natural Gas STAR Methane Challenge Program. (2017, October 26). *Optical Gas Imaging History, Innovations and Industry Results.* Retrieved from https://www.epa.gov/sites/production/files/201711/documents/17.oneil_2017aiw.pd

514115 Implementing a Benzene Fenceline Passive Sampling Program

Abstract:

EPA has implemented a rule that requires refineries to conduct continuous passive monitoring of benzene at multiple fenceline locations using a new passive sorbent based Method (EPA Method 325) in order to control fugitive emissions that migrate past a facility border. Although, the rule is currently only applicable to petroleum refineries, this approach to monitoring may be extended to other facilities in the future. In this presentation we will touch on passive sampling theory, provide a regulatory breakdown. We will also discuss the current rule strategy or requirements (EPA Method 325a) for number and placement of samplers using specific site examples. We will discuss pre regulation information gathering or pilot study strategies as well as some discussion of corrective action if a â€œHOTâ€ spot or High benzene levels are found.

511581 Control Valve Packing Follower Failure in Cold Service

Abstract:

â€œCONTROL VALVE PACKING FOLLOWER FAILURE IN COLD SERVICEâ€

By

Richard Kratzer

BASF Corp

P.O. Box 2506

Port Arthur, TX 77643

Phone : 409-960-5239

Email : Richard.Kratzer@BASF.COM

Prepared for

AIChE 2018 Spring EPC Conference

Abstract

The authors will detail the incident of a control valve packing follower failure in cold service. While attempting a repair, the follower failed on a process service @ 465 psig / -150 degrees Fahrenheit which was eminent based on an operations / maintenance action. The failure potential of failing without any interaction was highly possible due to the broken follower.

Some of the topics that will be covered by the authors are:

1. The root cause analysis findings from the incident.
2.
3. The immediate corrective actions taken as a result of the findings.
4.
5. The long term corrective action items and the path forward to implement them.

512604 CSB Case Study - Summary of Williams Plant Incident

Abstract:

On June 13, 2013, the Williams Geismar Olefins Plant experienced a catastrophic accident: a reboiler for the propylene fractionator overpressured and ruptured, resulting in two employee fatalities. The U.S. Chemical Safety and Hazard Investigation Board (CSB) investigated the incident and published an investigation report, which issued recommendations to Williams and presented Key Lessons for the petrochemical industry. In this paper and presentation, the CSB presents its investigation findings. Topics discussed to prevent a similar incident include using the hierarchy of controls, establishing a strong process safety culture, and developing a robust process safety management program.

513257 Synopsis of Risk Assessment and Analysis Techniques for Brittle Fracture and Current Vulnerabilities

Abstract:

Olefins production facilities have significant risk for failure due to brittle fracture. Various techniques are available for risk assessment and analysis of brittle fracture. Different analysis techniques are covered in codes API 579 Part 3 and 9, ASME Sec VIII Div 1 and ASME B31.1. Owner-users and contractors have devised various techniques for risk assessment and MI programs that address brittle fracture concerns. This paper seeks to summarize the assessment methods, focusing on the risk of carbon steel materials, such as SA-212B.

In addition to discussing the approaches used for evaluating the risk for brittle fracture, this paper seeks to highlight current shortfalls in assessment methodologies and techniques and how they affect implementation of MI programs for brittle fracture. Examples of where these obstacles have arisen in implementing an MI program focused on brittle fracture will also be discussed.

AIChE Paper Number: 203b

AVOIDING RISKS OF NEW BRITTLE FRACTURE SUSCEPTIBILTY IN CARBON STEEL PIPE, FITTINGS AND FLANGES THAT MEET ASME CODE

Barry Messer,
Director, Metallurgical Engineering
Fluor Corp., Calgary, Alberta, Canada

Shahab Soltaninia
Senior Metallurgical Engineer
Fluor Corp., Calgary, Alberta, Canada

Dishoo Randhawa
Metallurgical Engineer
Fluor Corp., Calgary, Alberta, Canada

Prepared for Presentation at the 2018 Spring National Meeting
Orlando, FL, April 22-26, 2018

AVOIDING RISKS OF NEW BRITTLE FRACTURE SUSCEPTIBILTY IN CARBON STEEL PIPE, FITTINGS AND FLANGES THAT MEET ASME CODE

Barry Messer,
Director, Metallurgical Engineering
Fluor Corp., Calgary, Alberta, Canada
Shahab Soltaninia
Senior Metallurgical Engineer
Fluor Corp., Calgary, Alberta, Canada
Dishoo Randhawa
Metallurgical Engineer
Fluor Corp., Calgary, Alberta, Canada

Abstract

Severe low toughness properties of some new fabrication carbon steel piping components have created a new brittle fracture risk in the last few years. The issue is becoming widespread globally, affecting up to 30 percent of materials tested, with a smaller percentage exhibiting cast iron like properties. This phenomenon contradicts the ASME Code which assumes, based on historical material properties, that these materials behave in a ductile manner, where instead they behave in a brittle fashion at temperatures between -29°C to +15°C. Ethylene facilities that operate at the low end of this temperature range are at an increased risk for this "break before leak" brittle phenomenon, with documented plant failure explosions and fatalities.

ASTM and ASME do not require testing to mitigate risks to brittle fracture at these temperatures and wall thicknesses using ASME impact testing exemption curves that are based on fracture mechanics theory and assumptions.

Laboratory test results of parent carbon steel pipe, fitting, and flanges in ASME B31.3 refinery and gas plant construction and facility start-ups reveal unexpected low toughness of 3J to 7J at -10°C to -29°C. Other end users and laboratories are independently testing and globally reporting similar findings.

Analysis of these failures shows no obvious correlation with location of manufacture or country of origin. Some manufacturers are not experiencing this issue. This uncertainty needs to be addressed due to increase of real risk to reliability and safety. Hydrostatic test, cold start-ups, cold operation and low temperature upset in-service failures are surfacing with low toughness and brittle transgranular cleavage cracks.

Additionally, a small percentage of ASTM A350 LF2 CL1 forged flanges certified to -46°C, experienced parent metal hydrostatic test failure and examination found brittle failure with 3J to 7J at -46°C. These unacceptable results conflict with the acceptable values of certified material test reports. Correlation to trace elements and forging technique was found. ASTM impact testing methodology was found insufficient to identify sub-standard material and the authors addressed by developing a modified sampling and Charpy location methodology for forged flanges.

An advanced new technology non-destructive ultrasonic technique was also developed to accurately test and evaluate components subject to brittle fracture and has demonstrated greater than 95% correlation with laboratory test results.

This paper briefly discusses historical literature, metallurgical investigations, findings, and factors that contribute to susceptibility to brittle fracture. Guidance to avoid low toughness and mitigate risks of in-service brittle fracture to ensure that components are suitable for service is provided.

Introduction

Brittle fracture of carbon steel parent material at low temperatures and the effects of chemistry and heat treatments were studied in depth in the early 1950's [5, 6, and 7]. Carbon steel manufacturing has changed in the last decade. Most billet and ingots are made using the electric arc furnace, with nearly 100 percent recycled charge material, whereas in the past, the blast- furnace was used with mostly mined iron ore charge. Additionally the fabricators of pipe, fittings and flanges are optimizing their process factors such as chemistry and trace element alterations, grain size, melting practice, forming parameters, and heat treatments to reduce the cost.

Over the last 20 years industry has experienced a degradation of Charpy impact properties. Historically, these carbon steel components have inherent good toughness down to -29°C. This assumption has led ASME B31.3 to exempt impact testing for carbon steel materials per Fig. 323.2.2A and ASME VIII Div I Fig. UCS-66. However there have been a few low toughness Charpy in-service failures [11]. For example A105N forgings, where the low toughness was attributed to incomplete normalizing heat treatment process after forming. This was sometimes associated with the lower cost intentional lean Mn:C ratio, and resulting upward shift of the brittle to ductile transition temperature. These exemption curves were introduced after the 1987 ASME study and subsequent reports [8], where the fracture mechanical approach was accepted by these Codes. Until then carbon steel pipe, fittings and flanges were acceptable down to -29°C regardless of thickness and is still the case for ASTM A105 [4] forgings. Since then, ASME Codes have made minor modifications and allow carbon steel without impact testing down to -48°C using stress reduction ratios. These exemption rules have been based on the historic properties of the materials and fracture mechanical crack propagation theory.

Parent metal brittle failures were extremely rare and primarily related to equipment hydrotest and pressure relief auto-refrigeration. The most common low toughness issues were related to the weld metal deposit and Heat Affected Zone (HAZ) and not the parent material. ASME VIII Div I UCS-66(a) note states: "The use of provisions in UCS-66 which waive the requirements for impact testing does not provide assurance that all test results for these materials would satisfy the impact energy requirements of UG-84 if tested". ASTM A707 "Forged Carbon and Alloy Steel Flanges for Low-Temperature Service" note "C" of table 2 and supplementary requirement S4 advise that the specifying minimum Charpy absorbed energy value at temperature is intended to mitigate crack initiation and specifying the minimum Charpy shear value is to mitigate crack propagation.

However in the last 20 years there has been a dramatic global increase in use of recycled steels with many areas of the world now converted to 100% recycled steels. The product of scrap feed to the Electric Arc Furnace (EAF) introduces several trace elements to carbon steel, which are difficult to remove easily and can strongly affect base material low temperature toughness properties and susceptibility to service failure. The trace elements content of these recycled steel varies considerably from low to high levels. These raw ingots are being marketed globally and the issue is no longer isolated to specific manufacturers. Until the code is revised to accommodate new baseline, it is the responsibility of the Engineer, Owner and Regulator to address this issue.

Discussion

In 2013 the authors observed a significant increase of pipe, fittings, and flanges parent metal brittle failures during hydrotest and ambient temperature facility start-up at various facilities worldwide. The vast majority of these materials met the specified Code requirements, but when additionally, laboratory Charpy tested, they exhibited brittle failure with transgranular cleavage with below 12J (typically 3J to 7J) at test temperatures of -10°C to -29°C, shear less than 20% (typically 0% to 10%), and lateral expansion below 0.30 mm. These components were all expected to pass Charpy tests at -29°C.

Based on the preliminary screening data it was initially believed that the anomaly was only associated with a few manufacturing facilities. However, the investigation expanded globally as it became apparent that this is not an isolated case. In 2014 authors started involving several certified laboratories and several hundred piping component manufacturers from various global regions. A large quantity of Charpy test results surprisingly had laboratory failures below 12J (Table1, Steel grades of concern) of thickness range from 3.9 mm to 23 mm.

Table 1- Charpy percentage test failure at -29°C

	Sample quantity	Failure % Note 1	Less than 12J	Less than 7J
Pipe	122	30%	19%	10%
Fitting	82	18%	15%	8%
Flanges	50	38%	21%	9%

Note 1: Not meeting ASME B31.3 Table 323.3.5
Note 2: Initial year 2014 test results at -29°C from 30 pipe manufacturers, 20 fitting manufacturers and 15 flange manufacturers, from several heat numbers with nominal wall 3.9 mm to 23 mm.

Affected component manufacturers were found to be scattered globally and the raw ingots came from various global sources. Test results are showing the brittle to ductile transition curve appears to be shifting up from historic values by about 20oC to a higher temperature and there may be issues up to 0°C. Some engineering companies, Owners and Regulators are now addressing this issue by overlaying addendum to the ASME Code and ASTM material standards. ASME is now reviewing the issue.

The steel grades of concern are:

a) Pipe: ASTM/ASME A106 Gr.B, A53 Gr.B, API 5L Gr. B and similar
b) Fittings: ASTM/ASME A234 Gr. WPB and similar

c) Flanges: ASTM/ASME A105 and A105N and similar

Acceptable steels with Impact Testing at -46°C:

a) Pipe: ASTM/ASME A333 Gr.6
b) Fittings: ASTM/ASME A420 WPL6
c) Flanges: ASTM/ASME A350 LF2 CL1 (some failures experienced, additional test recommendations discussed in this paper).

In 1998 Borealis Beringen (Belgium) experienced a transgranular brittle fracture in an A105N 24 NPS Class 600 weld neck flange, when the operating temperature dropped to -16°C due to auto-refrigeration and resulted in a catastrophic failure. The 2003 Brussel report summarizes the investigation and findings and suggests that A105N be limited to 0°C unless additional testing is carried out [11].

With the introduction of the European directive, 97/23/EC, which requires impact testing for all carbon steel representative sample at minimum operating temperature to achieve 27J Europe no longer has this problem to the same extent [10].

Components made from these steels with poor low temperature toughness are directionally going to other markets such as North America where there are no code requirements for Charpy impact testing. As more end users and regulators specify minimum Charpy supplemental requirements, there is an increased probability of suspect material being directed to those who do not take similar precautions.

Data Analysis

All testing was done by several global experienced, qualified, and certified laboratories. Table 2 shows the first test results that indicated there was a low Charpy toughness parent material concern.

The high percentage of laboratory failure was unexpected and triggered further investigation including, laboratory calibration verification, third party laboratory audit and double blind test replication at other laboratories. Investigation proceeded to evaluate pipe, fittings, and flanges from various manufacturers from local stockists, and then expanded to global stockists.

There are limited published data on Charpy test results for these materials as they are mostly exempt from testing by the Code. A baseline was established with Charpy tests on components made over 10 to 30 years ago, the low toughness at -29°C was not evident. The finding clearly showed that 10% of laboratory tested randomly selected piping components as well as several recent field failures had absorbed energy values of 3J - 7J at -29°C. It became evident that a new brittle-fracture risk has emerged.

Through discussions with other EPCs, facility owners and testing facilities, who are experiencing recent, similar issues, it was realized that this phenomenon is becoming wide spread and not addressed by the Code. Investigation included exploration of root cause and mitigation [12].

Table 2 - Pipe, fitting, flanges parent material failed on Charpy test

Specimen	Test Temp ^{o}C	Impact Value J	Shear %
Pipe A106 Gr. B - 168.3 mm OD x 10.97 mm Charpy Specimen Size 10x10 mm			
1	21	62.4	40
2	21	124	55
3	21	100	50
1	-21	11.7	6
2	-21	10.8	11
3	-21	9.5	11
1	-29	6.2	6
2	-29	6	6
3	-29	4.6	6
Tee A234 WCB - 8"x 8" XS 12.7 mm Charpy Specimen Size 10x10 mm			
1	21	98.7	50
2	21	101	44
3	21	144	60
1	-21	5.7	6
2	-21	8.1	6
3	-21	4.1	6
1	-29	3.5	0
2	-29	<3.4	0
3	-29	<3.4	0
Flange A105N - 8" 600# XS Charpy Specimen Size 10x10 mm			
1	21	108	70
2	21	77.6	60
3	21	88.4	62
1	-21	7.3	6
2	-21	7.6	6
3	-21	7.1	0
1	-29	5.7	0
2	-29	6.0	0
3	-29	6.2	0

Field brittle failures were also observed for pipe and fittings, down to 3.9 mm (0.154 inch) wall thickness, which is on the flat portion of Curve B of Fig. 323.2.2A ASME B31.3 and should not be subject to brittle crack propagation [13, 14].

Each analysis was broken out by component type, pipe, fitting, and flange and included multiple manufacturers, heat numbers, and wall thickness. Several elements, mechanical properties and their interactions were statistically evaluated for the effect on Charpy results at -29°C.

It was also observed, additionally, that some ASTM A350 LF2 CL1 forged flanges with impact test certification above 20J at -46°C exhibited the same 3J to 7J at -46°C failures [9]. One owner has recently published similar laboratory findings [15].

The primary factors affecting toughness are summarized as follows:

Grain Size: Historically there has been a linear relationship between grain size and Charpy. All materials for testing were fine grain with approximate grain sizes of 6.5-10. Surprisingly, a poor correlation to Charpy was found for pipe, fittings, and flanges (Fig. 1).

All of the in-service cracking failed with transgranular cleavage cracks (Fig. 2) with under 12 J and less than 20% shear at temperature up to 10°C. For all testing done, no intergranular cracks were found.

NACE TM-0177 "Laboratory testing for metals for resistance to sulfide stress cracking and stress corrosion cracking in H2S" test was performed on the additional samples that previously failed 3J - 7J Charpy test, and these sample all passed this NACE test.

Chemistry: The following elements were tested to evaluate the effect on Charpy: carbon, manganese, phosphorous, sulfur, silicon, nickel, chromium, molybdenum, copper, vanadium, titanium, aluminum, boron, calcium, tantalum, and nitrogen. The only statistically valid correlation found was Mn:C ratio (Fig. 3, 4, 5). The Mn:C ratio effect on Charpy was studied and well characterized and reported in the early 1950s [5,6,7].

Manganese is added to improve carbon steel strength and toughness; additionally, the Mn:C ratio has a strong effect on transformation temperature (Ac_3). Increasing Mn promotes formation of austenite. It lowers the transformation temperature which promotes formation of fine ferrite grains. Lowering C lowers the transition temperature and increases the Charpy upper shelf energy (Fig. 6).

Some carbon steel ingot manufacturers have reduced the amount of manganese added to the steel to reduce cost and improve other mechanical properties such as strength and formability. Other elements are also being added such as vanadium, niobium, titanium and boron. High manganese to carbon ratio is known to contribute to high toughness and high shear at low temperatures when other manufacturing parameters are optimized. For example, ASME A105N allows a Mn range of 0.6-1.05% which permits a minimum Mn:C ratio of 1.7. Various end users have received piping component with Mn:C ratio less than 3 with very poor toughness properties.

Normalizing and tempering heat treatments: All items were tested as received. Low toughness pipe and fittings were additionally normalized at 900°C with negligible improvement. Low toughness flanges were renormalized with some minor improvement by approximately 5J. Subsequent tempering had no improvement on any of the flanges tested.

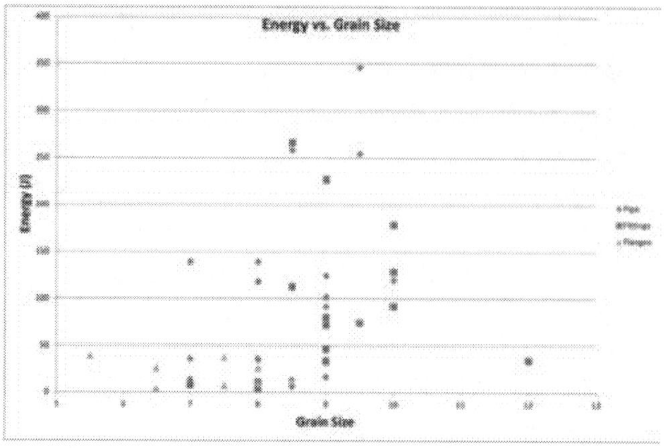

Fig. 1- Impact energy vs grain size at -29°

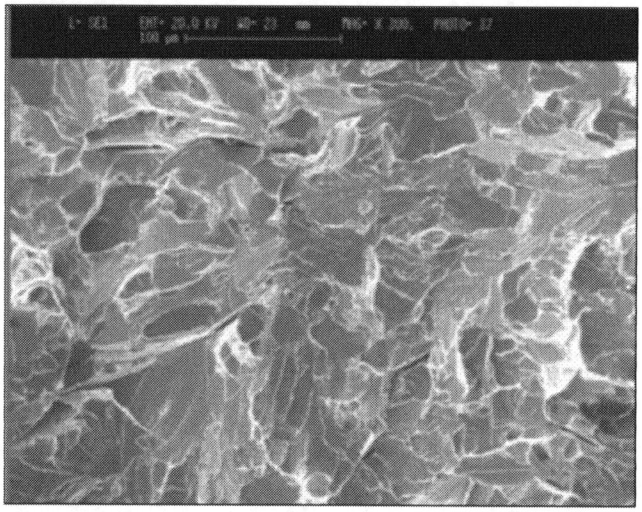

Fig. 2- A105N 600 Cl 10 NPS In-service cleavage fracture at 10°C

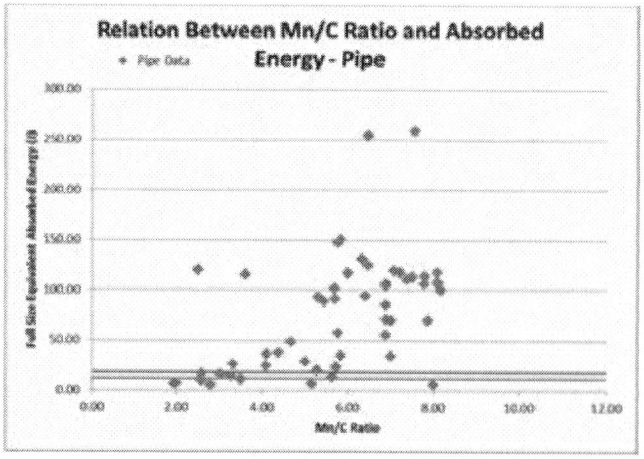

Fig. 3 – Pipe impact energy vs Mn:C at -29°C

(red: ASME B31.3 18J min., green: authors 12 J min)

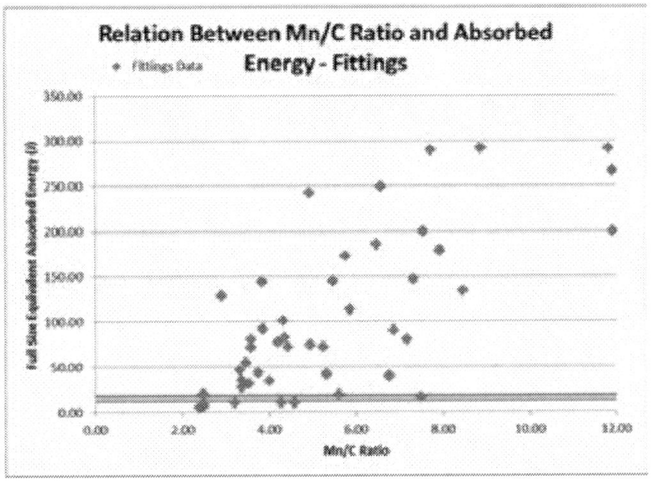

Fig. 4 – Fitting impact energy vs Mn:C at -29°C

(red: ASME B31.3 18J min., green: authors 12 J min)

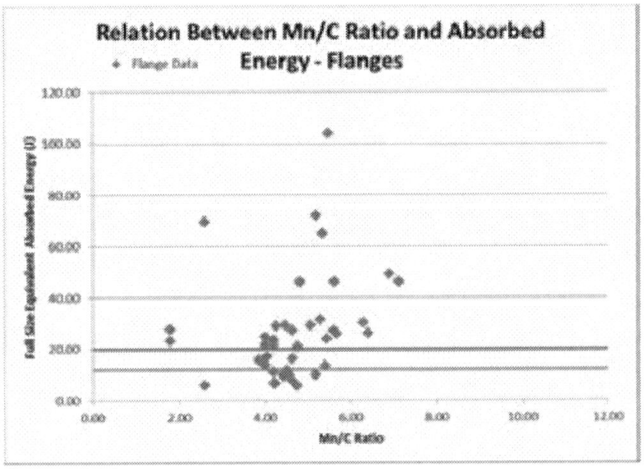

Fig. 5 – Flange impact energy vs Mn:C at -29°C

(red: ASME B31.3 20J min., green: authors 12 J min)

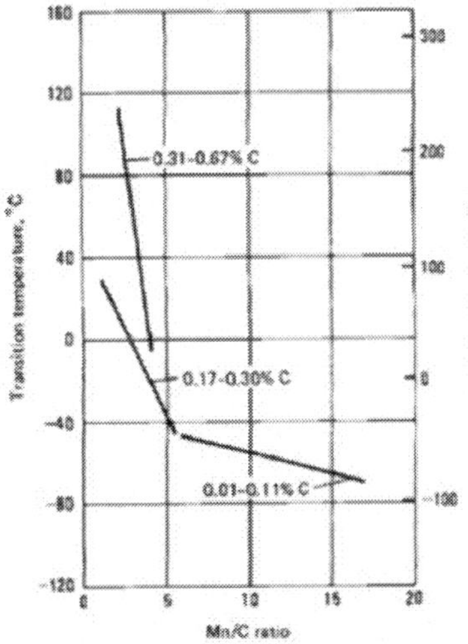

Fig. 6- Effect of carbon and manganese on notch toughness [20]

Lateral expansion and shear: As expected, there was a direct correlation between Charpy to lateral expansion and no correlation to yield and tensile strength. Similarly, there was no correlation between Charpy and Brinell hardness testing.

Structure: The microstructures of both failed and passed Charpy tests were typically equi-axed ferrite and pearlite. Visual and 50X, 300X and 500X micro comparisons of the good and bad Charpy samples were made and there were no significant noticeable differences on micro-structures. This was surprising as we were expecting to see abnormalities in grain size and ferrite:pearlite distribution.

Angles of 45 degrees to rolling or forging direction: Susceptibility to brittle fractures was identified by Charpy coupon orientation (Fig. 7 and Table 3), these findings was unexpected, as it was expected the longitudinal specimen would have slightly higher absorbed impact energy versus the transverse specimens and that the impact specimen at 45° angle would be the average of longitudinal and transverse test results. Further evaluation using a newly developed advanced Phased Array microscopy technique using multiple angles collected signal attenuation of the 1.0.0 plane versus a function of Nb, Ti, V, B and Mn:C ratio (Fig. 8) [18].

Table 3 – Charpy test result of cracked A350 LF2 Cl1 at different rolling or forging orientations

Orientation to rolling direction	-46°C	-29°C	0°C
L-T (Longitudinal)	10J	14J	32J
T-L (Transverse)	7J	12J	28J
D-D (45 degree)	3J	3J	8J

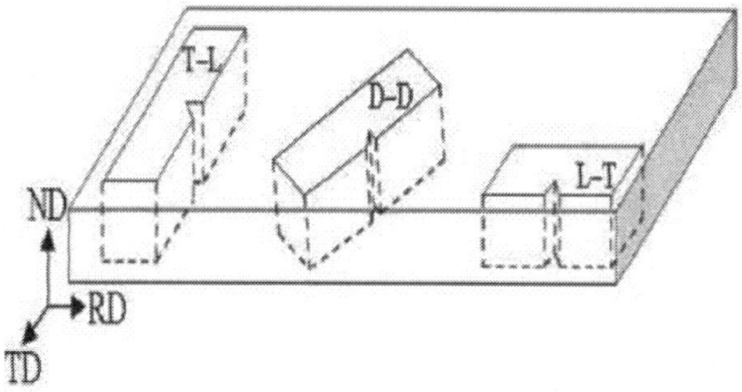

Fig. 7 – Charpy orientation to rolling or forging direction

The actual angle can vary slightly as the predominant amorphous crystal structure orientation is not exactly at 45° to the rolling or forging direction and depends on the forming process.

This technique revealed that the crystal orientation of the 1.0.0 plane plays an important role in the crack initiation and propagation. This brittle plane at 45° behaves like a fragile near crack-like zone in the angular transition area of the component such as hub to the body for flanges, the branch connection at a tee, or the longitudinal plane of spiral-wound pipe. The ASME Code Charpy exemption curve is based on fracture mechanics with the assumption that material is mostly homogenous [8]. This data helps explain why this assumption is not always reasonable as the effect of 1.0.0 plane is dependent on the chemistry and forming factors. It also helps explain why failures at thinner wall and higher temperatures than what the ASME Codes Charpy exemption curves allow, are sometimes encountered.

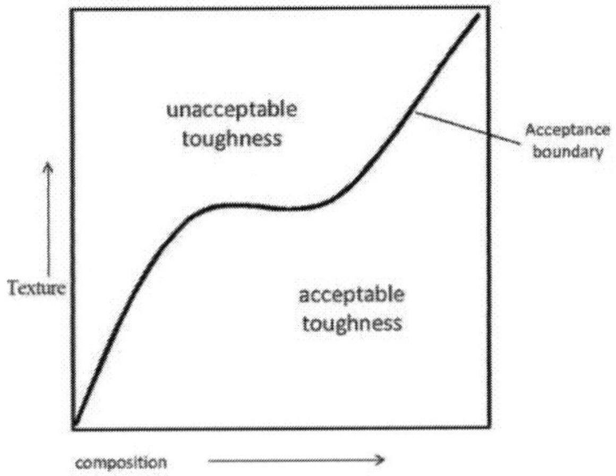

Fig. 8 – 12J toughness at -29°C as function of texture (from phased array) and chemical composition [18]

Charpy test coupon for A105 and A350 flanges: Literature shows the absorbed impact energy values for forged flanges varies depending on the location and orientation of Charpy test coupon specimen [16, 17]. ASTM A105 has no test requirements and specimens for Charpy test of A105 have been taken based on ASTM A350 and ASTM A370 [19]. However our finding suggests that for forged flanges, a more reliable Charpy test coupon location and orientation that meets the standard is recommended in order to obtain a test result that better represents fit for purpose engineering. The coupon is orientated with the notch centered at the inner hub edge as described in Fig. 9.

Field low toughness failure examples: Fig. 10, 11 and 12 show typical flange and fitting parent metal hydrotest failures with ambient temperature of 5°C-15°C. The cracks originate on the outside diameter. For flanges many of the cracks originate in the crotch area between the hub and the body. These field brittle failures were analyzed in the laboratory and exhibited transgranular cleavage with full size Charpy equivalent absorbed energy values of 3J - 7J, below 20% shear, and below 0.30 mm lateral expansion.

Some of the field failures involved wall thicknesses down to 3.9 mm where brittle crack propagation is not expected per Fig. 323.2.2A ASME B31.3.

For failures in thick materials (i.e. flanges with thick bodies) the constraint is not reflected in the Charpy coupon specimen size, 10 mm x 10 mm and less for sub-size coupons, and partially explains how field brittle failure could occur at warmer temperature than Charpy test temperatures.

Mechanisms Associated with Low Toughness

Three predominant mechanisms affecting toughness have been identified for the pipe, fittings, and flanges.

a) Mn:C ratio

Chemistry plays a significant role with heating operations a close second follower. At one facility the entire manufacturing process for making A106 Gr. B and A333 Gr.6 were identical except for the incoming ingots.

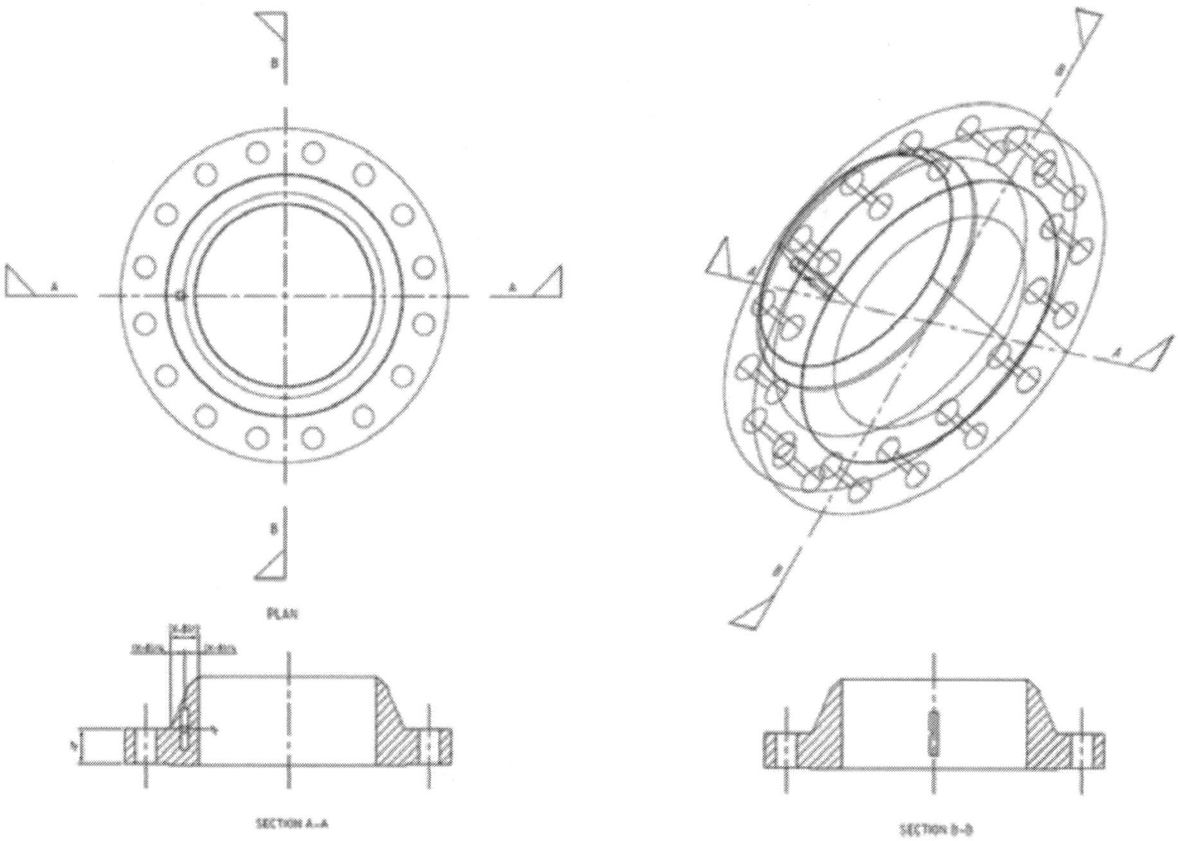

Fig. 9 – Recommended location and orientation of V-notch specimen for forged flanges with notch normal to crotch

A Mn:C ratio of 3.0 gave 3J at -29°C and a Mn:C ratio of 7.0 resulted in 40 J at -46°C. Even with good chemistry, the heating and cooling rates have to be suitable. Mintz's decade long investigations into the normalized carbon steel was able to mathematically reduce the key parameters affecting toughness including pearlite:ferrite ratio, Manganese and Carbon content, and grain size [16, 17].

b) Pinning effect during normalizing heat treatment process

This pinning effect collars the austenitic grain, and helps prevent grain growth during the cooling transformation to ferrite and promotes good toughness. The higher Mn:C ratio lowers the Ac_3 temperature, and allows for more effective normalization at lower temperatures.

Fig. 10 – Failed CS A234-WPB at 15°C hydrotest

Fig. 11 – Failed A350 LF2 CL1 at 0°C hydrotest

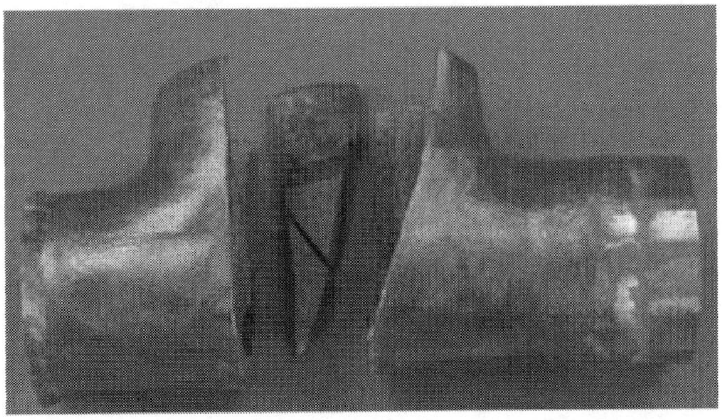

Fig. 12 – Failed A234 WPB Brittle failure at 10°C

The aluminum nitride (AlN) locks the re-crystalized grain refinement by precipitating and effectively pinning grains. If the normalizing temperature is too high (i.e. to accommodate a low Mn:C ratio) then more AlN goes into solution and the pinning effect of AlN precipitates is reduced. The cooling rate needs to be optimized to achieve fine-grain distribution and the optimal proportion of the various ferrite structures. For complex geometries such as flanges the location and position of the Charpy test coupon needs to be at the least optimum grain structure to prevent false positive values in comparison to the rest of component shape. The authors suggest that "V" notch coupons be placed near the crotch area in the longitudinal axis of flange hub in addition to A350 specified location, typically tangential to the axis in the flange body. The low toughness in A105N and A350 LF2 has similar mechanisms, but is not reported for A105N as Charpy testing is not mandatory per ASTM A105.

c) Ac_3 temperature effect on grain size

The typical normalizing temperature is 900-920°C. The Ac_3 temperature at higher Mn:C ratios above 5:1 is typically 820°C or lower (Fig. 13). Adding 30-50C° gives a normalizing temperature of 850-870°C. Normalizing at too high temperature has a negative effect. The high normalizing temperature increases the prior austenite grain size and puts more Al/N into solution preventing the pinning effect to collar grain size growth thus leading to a lower toughness.

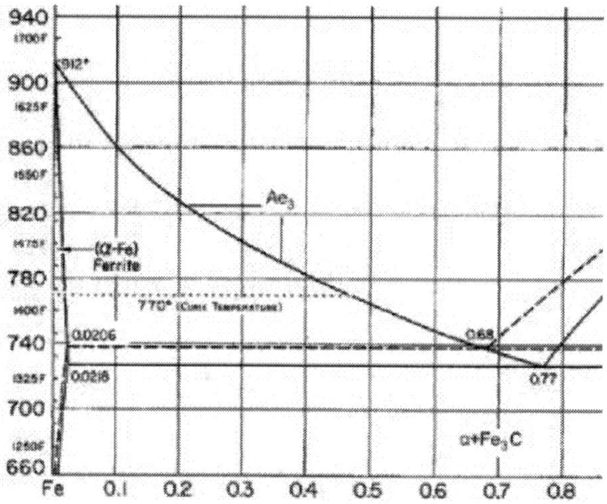

Fig. 13- Ac_3 temperature vs Carbon wt% - Fe-C Phase Diagram

d) Recycled steel trace element pinning inhibition effect

This anti-pinning effect is complex and not fully understood. Recycled steels contain variable trace element such as Aluminum (Al), Antimony (Sb), Arsenic (As), Bismuth (Bi), Boron (B) (Fig. 14), Chromium (Cr), Copper (Cu), Hydrogen (H), Lead (Pb), Nickel (Ni), Nitrogen (N), Tin (Sn), Titanium (Ti), Vanadium (V), and Zinc (Zn). Additionally some trace elements are added to the steel to improve tensile and yield mechanical properties.

The authors noticed significant differences between virgin and recycled steels with respect to ppm level of these elements. There are reports describing similar low toughness issues for A350 LF2 forgings [9, 11, 15].

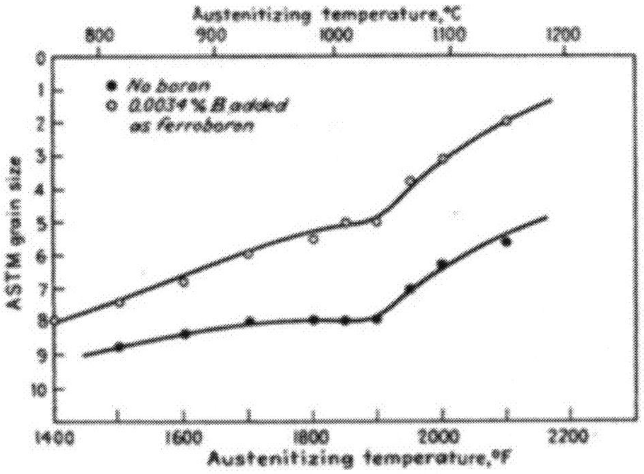

Fig. 14- Effect of boron on the austenite grain coarsening of two induction-furnace heats steel containing 0.26% C and 1.65% Mn. Each was deoxidized with 2 lb per ton of aluminum. [21]

Recently the authors have investigated the low toughness of A350 LF2 Cl1 flanges that failed during hydrotest. This material had been normalized at 900°C for 3 hours and then tempered at 650°C for additional 3 hours, and had been certified to 26J at -46°C. Laboratory testing confirmed the absorbed energy of 4 - 6J, 5% shear and 0.02 mm lateral expansion and ASTM grain size of 7.5. This 300 Class flange had good forging flow lines (Fig. 15).

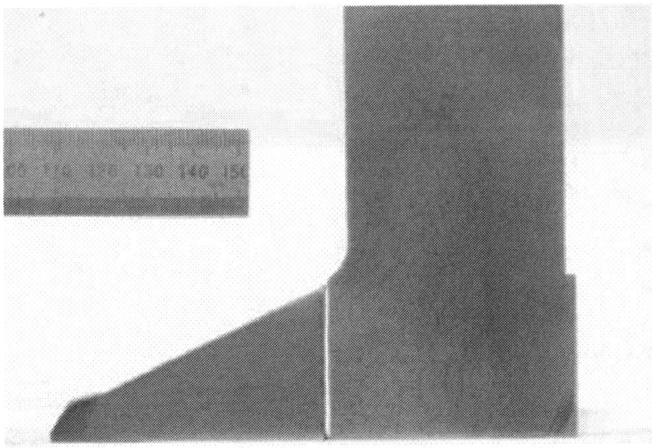

Fig. 15 – 10NPS 300 Class Sch 40 A350 LF2 Flange - 6J at -46°C tangential body and longitudinal hub

Historically poor results like this were attributed to incorrect normalizing heat treatment process or incorrect sample location to properly reflect the parts dimension or the work reduction. In this case there were also other A350 LF2 forging heats in the furnace with similar chemistries, sizes and work reduction but with the very high absorbed energy of over 60J.

The low toughness in this case was attributed to the pinning neutralization effect of some of the trace impurities of the recycled steel ingots. This strong effect was able to overwhelm both the Mn:C ratio effect and pinning effect discussed in the previous paragraphs, even with the desired high Mn:C ratio between 5 to 8.

Depending on the normalizing temperature some of these trace elements compete with Silicon and Aluminum in the oxidization process and after consuming the Oxygen then compete with Aluminum for the Nitrogen to form nitrides. The fine grain precipitated AlN pins the austenite grain above the Ac_3 and prevents austenite grain growth during the normalizing air cool transformation to ferrite. Competition by the other trace elements for Nitrogen inhibits the effectiveness of the pinning effect, as they interfere with AlN effectiveness, and promote undesirable coarser precipitates which disrupt the ferrite texture as austenite cools and promotes low toughness. The level of Nitrogen in recycled steel is typically double that of virgin steel (40 to 120 ppm versus below 40 ppm). Any free nitrogen is known to compromise good toughness and it is essential to tie up all the Nitrogen with element such as Aluminum.

Certain global ingot manufacturers control their chemistry to accommodate the recycled steel and need for good toughness. Chemistry restriction may include:

a) Mn:C ratio minimum 5:1 for carbon steel and Mn:C ratio minimum 6:1 for low temp carbon steel.

b) Aluminium minimum 0.020 wt% c) Titanium maximum 0.0040 wt% d) Niobium maximum 0.0040 wt% e) Phosphorous maximum 0.015 wt% f) Tin+Zinc+Lead maximum 0.0010 wt% g) Boron maximum 0.0005 wt% h) Vanadium maximum 0.030 wt%.

Suggested Mitigations

For ethylene facilities there are several locations where chilling process as well as overhead vaporization is occur down to -23°C where non impacted carbon steel is acceptable to code down to -29°C, for these areas additional mitigations to reduce this real low toughness cracking risk above code minimum is imperative.

Several ethylene operating plant incidents with explosions and fatalities have been attributed to carbon steel brittle failures at operating temperatures between -5°C and -29°C see reference for a few of these cases. [22]

For example a brittle fracture that caused the 1989 Quantum Chemical Morris ILL explosion where -12°C overhead carried liquid C2's from deethanizer reflux drum towards the reactors with non-impact tested regular carbon steel. In these cases low temp carbon steel would have been a better choice. For ethylene plants the piping transitions from the cold service to warm service and vice versa need to be evaluated carefully to mitigate the risk.

Options:

There are three basic options that can help mitigate the risk of low toughness brittle fracture. Without these precautions these carbon steel components are likely to be more susceptible to brittle fracture if operated at temperatures below 10°C.

General Requirements:

For each option, the following general requirements apply; reevaluate raising MDMT to better reflect real conditions. Historically equipment nameplate and piping line class Minimum Design Metal Temperature (MDMT) specified based on the lowest temperature that Code allows the material without additional impact testing. With the new susceptibility this may no longer be appropriate. All A105 forging shall be normalized. For Charpy testing of all A105N and A350 LF2 flanges, the location position of the Charpy test coupon is critical. The authors recommend that the "V" notch be positioned close to the crotch area. For A350 LF2 forgings, a better sampling criteria and location and orientation of the test coupon needs to be standardized to address the inconsistency of Charpy test results for this certified material at -46°C. The authors recommend that for standardization Fig. 9 showing the detailed location and "V" notch orientation be adopted. Similarly for A105/N material where no testing is required down to -29°C service, as one option Charpy testing should be mandated by representative sample reflecting the final geometry to ensure the integrity. Also, the normalizing heat treatments and grain size distribution needs to be optimized. For forgings ensure a minimum forging reduction ratio of 3.5 to 1.

Option 1:

a) General requirements and

b) Impose Charpy test supplemental requirements at -29°C for carbon steels with MDMT 10°C and lower to achieve code minimum absorbed energy requirement.

Option 2:

a) General requirements and

b) Minimum Mn:C ratio of 6, as well as ASTM grain size of 9 or finer.

c) In order to achieve effective normalization and pinning for all forgings and some pipe and fittings depending on the manufacturer's process, the following chemical restriction should be considered.

Aluminium minimum 0.020wt%
Titanium maximum 0.0040 wt%
Niobium maximum 0.0040 wt%
Phosphorous maximum 0.015 wt%
Tin+Zinc+Lead maximum 0.0010 wt%
Boron maximum 0.0005 wt%
Vanadium maximum 0.030 wt%

Option 3:

a) General requirements and

b) Specify low temperature carbon steel certified to -46°C.

Conclusion

Reduced toughness properties of recycled carbon steels increase brittle fracture risk during hydrotest, start-up and depressurization auto-refrigeration. The new era of recycled carbon steels with the trend of increasing trace elements and proliferation of recycled steel ingots and billets globally has resulted in a baseline shift which is starting to show up more frequently. The toughness properties characterizations of 50 years ago may no longer be fully applicable as susceptibility is seen in both laboratory testing and during construction and startup of facilities. Some of fracture mechanical assumptions do not consider factors such as 45 degree susceptibility and combined trace elements effects with hot working and heat treatment. Charpy impact test exemptions need to be relooked as design Codes and material standards currently do not fully address these new risks. Until this situation is better understood and addressed by the Codes; Engineers, Owners and Regulators need to take additional precautions to ensure the material is resistant to brittle fracture. This can involve process operational modifications, specifying materials with certified toughness properties, controlling parent material chemistry, grain size, and post-forming heat treatments.

Owners need to impose requirements that are more conservative than the code to ensure reliability and safety. Take the leading role in establishing a better practice for Charpy impact test exemptions to ensure reliability and safety.

References

1. ASME: Section VIII Division 1: Rules for construction of Pressure Vessel. American Society of Mechanical Engineering. 2015 Edition.
2. ASME: B31.3: Process Piping. American Society of Mechanical Engineering. 2016 Edition.
3. ASTM: A350/A350M - 15: Standard Specification for Carbon and Low-Alloy Steel Forgings, Requiring Notch Toughness Testing for Piping Components. ASTM International. 2015.
4. ASTM: A105/A105M - 14: Standard Specification for Carbon Steel Forgings for Piping Applications. ASTM International. 2014.
5. Tensile & Impact Properties of Iron and some Iron Alloys of High Purity, W. P. Reese, B. E. Hopkins and H. R. Tipler, J. Iron Steel Inst., October 1951.
6. Behavior of Metals at Low Temperatures, R. M. Brick, J.R. Low, C.H. Lorig, ASM 34th National Metal Congress, October 1952.
7. The Low Temperature Properties of Relatively High Purity Iron-Carbon Alloys, R. L. Smith, R. V. Fostini, R. M. Brick, National Research Council, Washington D. C. August 1952.
8. New Toughness Rules in Section VIII, Division 1, A. Selz, BPV Task Group on Toughness, Mechanical Engineering, April 1988.
9. Safeguarding Mechanical Properties Carbon Steel - ASTM: Flanges – Fitting – Pipes – Valves, John Houben, MTI, EuroTAC March 2014
10. European Directive 97/23/EC of the European Parliament and of the Council, May 1997
11. Recommendation for Pipe Flanges Made in Forged Steel Complying with ASTM A105, Chemical Risk Directorate, Brussel, January 2003
12. New Low Temperature Impact Toughness Issues with carbon Steel Piping Components, B. Messer, T. Hamre, IIW/CWC Conference Vancouver, September 2014
13. Carbon Steel Low Toughness Issues, B. Messer, ASME VIII Committee, November 2015
14. Update Carbon Steel Low Toughness CS Pipe, Flanges and Fittings, B. Messer, ASME B31.3 Committee, April 2016
15. Determination of the Cause of Low Temperature Charpy Toughness Values in ASTM

A350 LF2 Flanges, A. B. Nissan, K. C. Baker, Proceedings of the 28th ASM. October 2015.
16. Structure/Property Relationship for Normalized Pipe Flanges (Grade LF2) Made to ASTM S350 Specification A350/A350M, B. Mintz, Journal of Testing and Evaluation, November 1988.
17. Structure/Property Relationships for Normalized Flanges for Low Temperature Service, B. Mintz, Ironmaking & Steel making Vol.26 No. 1, 1999.
18. Nondestructive Determination of toughness of Metal, Plastic and Composite materials, Patent Pending, Fluor Corp. Ref. 100325.0602PRO July 2015.
19. ASTM: A370 - 17: Standard Test Methods and Definitions for Mechanical Testing of Steel Products1. ASTM International. 2017.
20. ASM Handbook, Volume 1: Properties and Selection: Irons, Steels, and High-Performance Alloys.
21. Boron, Calcium, Columbium, Zirconium in Iron and Steel, R. A Grange, John Wiley & Sons, 1957
22. Design Guidelines For Safety In Piping Networks, K. Kolmetz, Engineering Practice, January 2016

512433 Chrome VI in Cracking Furnaces

Abstract:

Chrome VI is a well-known and toxic material that is present in many industries and OSHA has set standards for the safe handling and exposure to it. Recent findings at NOVA Chemicalsâ€™ have revealed the presence of chrome VI at various locations within the cracking furnaces. This paper discusses areas of the furnaces where the chrome VI was found, the methods used to detect it, safe handling of contaminated materials, suggested PPE to protect workers from exposure and ways to neutralize the chrome VI to safer products. The paper will also describe how chrome VI was found and the observed concentration levels. Some potential sources of alkali metal and alkaline earth metals that are believed to have resulted in chrome VI formation within the furnaces are also identified

513103 Quench Systems for Gas and Liquids Crackers - a Primer

Abstract:

The quench system is a unique part of an ethylene plant. The oil and water quench towers are some of the largest in the plant and represent a significant fraction of the total investment cost. They also can be the source of some of the largest operating problems in the plant. Olefin plant engineers, operators and managers must have an understanding of how the quench system is designed.

The oil and water quench towers and their related systems are critical for the safe, efficient, and reliable operation of the ethylene plant. This paper presents fundamental principles for the design of these systems such as performance objectives, configuration choices, and design procedures. This paper is focused on providing basic knowledge to people relatively new to the ethylene business.

514239 Gas Cracking Quench Systems Tutorial

Abstract:

This tutorial will review the design fundamentals for quench systems for gas cracking ethylene plants. It will include discussion of the overall system design basis and considerations, and specific procedures and particulars regarding the design of major equipment such as the Quench Water Tower, the Quench Water Separator and the Quench Water Circulation loop.

514238 Quench Water Quality Challenges in Ethane Cracker

Abstract:

Quench Water system is a core loop in the processing of cracked gas prior separation train in the ethylene plant, the variety and nature of the contaminants founded in the circuit depends on the feedstock, the process condition at furnaces and the control of quench water tower temperature, all of this variables should be taken into account in order to operate it in a reliable way and maintain the operational continuity in good shape. This paper share the experience of Braskem Idesa 1050 kty ethane cracker quench water system problems and some solutions that improve the quality of the quench water, and maintain fouling and plugging controlled.

513122 EPC Sasol Quench Tutorial Fundamentals Presentation

Abstract:

EPC Sasol Quench Tutorial Fundamentals Presentation

Abstract

An Ethylene Pyrolysis Cracking Quench System is designed to cool the hot furnace effluent gases by direct counter-current contact with the circulating quench water prior to entering the charge gas compression system. As a result of the direct contact water cooling, most of the dilution steam and some heavier hydrocarbons are condensed and exit at the bottom of the quench tower along with the heated quench water to the Quench Settler for an oil/water phase separation. The hot quench water is circulated through a series of exchangers before being recirculated back to the Quench Tower to be used as the cooling medium for the hot cracked gas. The cooled charge gas vapors exit the overhead of the Quench Tower and flow to the charge gas compressor system.

The presentation will cover the fundamental basics of a Quench System by providing a general flow scheme of Sasolâ€™s Lake Charles Chemical Plant Quench System, a timeline / evolution of changes from Sasolâ€™s plant inception from the mid 1960â€™s to present day configuration, good troubleshooting methods, and general problems that can occur. These problems include emulsion issues, typical turnaround drivers and solids / tar handling removal options which come with inherit risks and anticipative rewards.

512250 Primary Fractionator Fundamentals & Flexibility to Feedstocks

Abstract:

The Primary Fractionator is a typical component of the quench system in liquid crackers. Besides quenching pyrolysis gases from the furnaces, it provides fractionation of raw pyrolysis gasoline (RPG) and ensures heat integration between the furnace effluent and dilution steam generation. Over the years, many liquid crackers have developed feed flexibility by processing a mix of liquid and natural gas liquids (NGL) feedstocks. In the last 5 to 7 years, the availability and use of ethane feedstock has pushed the feed flexibility to challenging levels for reliable operation of the Primary Fractionator due to the changes in quench oil make and composition. Known to be a fouling service the Primary Fractionator could become a constraint to achieving desired plant run length. This paper discussed the governing factors that affect operation of Primary Fractionators. The impacts of feedstock, tower internals and metallurgy are also covered.

AUTHOR INDEX

AUTHOR INDEX